Aquaculture Handbook

Aquaculture Handbook

Edited by **Roger Creed**

R CALLISTO REFERENCE

New York

Published by Callisto Reference,
106 Park Avenue, Suite 200,
New York, NY 10016, USA
www.callistoreference.com

Aquaculture Handbook
Edited by Roger Creed

International Standard Book Number: 978-1-63239-081-3 (Hardback)

Printed in the United States of America.

Contents

Preface   IX

Part 1   Nutrition   1

Chapter 1   **Unsupplemented *Artemia* Diet Results in Reduced Growth and Jaw Dysmorphogenesis in Zebrafish**   3
Michael P. Craig, Mitul B. Desai, Kate E. Olukalns, Scott E. Afton, Joseph A. Caruso and Jay R. Hove

Chapter 2   **Development of Biopolymers as Binders for Feed for Farmed Aquatic Organisms**   11
Marina Paolucci, Adele Fabbrocini, Maria Grazia Volpe, Ettore Varricchio and Elena Coccia

Chapter 3   **Replacement of Fishmeal with Plant Protein Ingredients in Diets to Atlantic Salmon (*Salmo salar*) – Effects on Weight Gain and Accretion**   43
Marit Espe, Adel El-Mowafi and Kari Ruohonen

Chapter 4   **Nutritional Value and Uses of Microalgae in Aquaculture**   59
A. Catarina Guedes and F. Xavier Malcata

Part 2   Genetics   79

Chapter 5   **Determination of Fish Origin by Using 16S rDNA Fingerprinting of Microbial Communities by PCR-DGGE: An Application on Fish from Different Tropical Origins**   81
Didier Montet, Doan Duy Le Nguyen and Amenan Clementine Kouakou

Chapter 6   **Validation of Endogenous Reference Genes for
qPCR Quantification of Muscle Transcripts in Atlantic
Cod Subjected to Different Photoperiod Regimes**   **97**
Kazue Nagasawa, Carlo Lazado and Jorge M. O. Fernandes

Chapter 7   **Mitochondrial DNA Variation
as a Tool for Systematic Status
Clarification of Commercial Species –
The Case of Two High Commercial
Flexopecten Forms in the Aegean Sea**   **109**
Anastasia Imsiridou, Nikoleta Karaiskou,
Elena Aggelidou, Vassilios Katsares and Sofia Galinou-Mitsoudi

Chapter 8   **Genomics and Genome
Sequencing: Benefits for Finfish Aquaculture**   **127**
Nicole L. Quinn, Alejandro P. Gutierrez,
Ben F. Koop and William S. Davidson

Part 3   **Post-Harvest**   **155**

Chapter 9   **Novel Approach for Controlling Lipid
Oxidation and Melanosis in Aquacultured
Fish and Crustaceans: Application of Edible
Mushroom (*Flammulina velutipes*) Extract *In Vivo***   **157**
Angel Balisi Encarnacion, Huynh Nguyen Duy Bao
Reiko Nagasaka and Toshiaki Ohshima

Part 4   **Culture Techniques and Management**   **185**

Chapter 10   **Improving Larval Culture and
Rearing Techniques on Common
Snook (*Centropomus undecimalis*)**   **187**
Carlos Yanes Roca and Kevan L. Main

Chapter 11   **Measurements Population Growth and
Fecundity of Daphnia Magna to Different
Levels of Nutrients Under Stress Conditions**   **217**
Lucía E. Ocampo Q., Mónica Botero A. and Luis Fernando Restrepo

Chapter 12   **Advances in Domestication and Culture
Techniques for Crayfish *Procambarus acanthophorus***   **245**
Martha P. Hernández-Vergara and Carlos I. Pérez-Rostro

Chapter 13   **An Updating of Withebait**
              Farming (*Galaxias maculatus*) in Chile   **269**
              Alfonso Mardones and Patricio De los Ríos-Escalante

Chapter 14   ***In Vitro* Culture of Freshwater**
              Pearl Mussel from Glochidia to Adult   **279**
              Satit Kovitvadhi and Uthaiwan Kovitvadhi

Chapter 15   **Potency of Barnacle in Aquaculture Industry**   **295**
              Daniel A. López, Boris A. López,
              Christopher K. Pham and Eduardo J. Isidro

Chapter 16   **Culture of Harpacticoid Copepods: Understanding**
              the Reproduction and Effect of Environmental Factors   **317**
              Kassim Zaleha and Ibrahim Busra

Chapter 17   **New Developments in Biotechnology and**
              IPR in Aquaculture – Are They Sustainable?   **335**
              Anne Ingeborg Myhr, G. Kristin Rosendal and Ingrid Olesen

Chapter 18   **Omics Methodologies:**
              New Tools in Aquaculture Studies   **361**
              María-José Prieto-Álamo, Inmaculada Osuna-Jiménez,
              Nieves Abril, José Alhama, Carmen Pueyo and Juan López-Barea

              Permissions

              List of Contributors

Preface

Aquaculture is the practice of raising fish, aquatic plants and other aquatic species for commercial purposes. This book offers an insight into a diversity of topics related to aquaculture. It discusses fish nutrition, application of genetics in agriculture, present techniques for limiting lipid oxidation and melanosis in aquacultural products, and culture techniques and management. The text of this book has been written by leading experts from their respective areas. It will be a beneficial resource for students and professionals in aquaculture and biotechnology.

This book has been the outcome of endless efforts put in by authors and researchers on various issues and topics within the field. The book is a comprehensive collection of significant researches that are addressed in a variety of chapters. It will surely enhance the knowledge of the field among readers across the globe.

It is indeed an immense pleasure to thank our researchers and authors for their efforts to submit their piece of writing before the deadlines. Finally in the end, I would like to thank my family and colleagues who have been a great source of inspiration and support.

Editor

Part 1

Nutrition

Unsupplemented *Artemia* Diet Results in Reduced Growth and Jaw Dysmorphogenesis in Zebrafish

Michael P. Craig, Mitul B. Desai, Kate E. Olukalns,
Scott E. Afton, Joseph A. Caruso and Jay R. Hove
University of Cincinnati
USA

1. Introduction

The number of laboratories using zebrafish as an experimental animal model has risen tremendously over the past two decades (Craig et al., 2006). As a result, the number of zebrafish facilities around the world has dramatically increased to meet the elevated demand for proper animal care and maintenance. In order to meet this demand, aquaculture facilities must employ husbandry protocols designed to produce a constant supply of healthy, viable eggs. Surprisingly, many husbandry strategies, particularly feeding protocols, are frequently passed down from members of one lab to another in a colloquial fashion without rigorous experimental validation. An ideal diet should consist of a minimal variety of foodstuffs designed to be nutritionally complete, simple to prepare, non-fouling, and cost-effective. Previous studies aimed at streamlining adult zebrafish feeding strategies in large aquaculture facilities have emphasized cost-effective, single-food models, but such diets lead to diminished survivorship and reproductive capacity (Goolish et al., 1999; Meinelt et al., 1999; Barnard & Bagatto, 2002), suggesting that these diets are lacking in some key nutritional component(s). Restricting adult fish diets to single foodstuffs, while desirable from a time and cost perspective, may not provide the trace mineral balance needed for adequate hormone and enzyme production, proper skeletal formation, and other biochemical or physiological needs. Nonetheless, given the intense breeding schedules many facilities are forced to adopt to meet research needs, a single component, high protein diet for enhanced egg production is frequently adopted.

Processed flake food diets are generally rich in proteins and essential trace elements required for normal physiological function. Despite their critical role in modulating somatic growth and basal metabolic rate in homeotherms (Eales & Brown, 1993), recent studies aimed at identifying alternative diets for improving the growth and reproductive capacity of zebrafish have paid little attention to minimum dietary iodine requirements. This may be due, at least in part, to the fact that zebrafish maintain a large reserve of T4 in the follicular colloid effectively allowing the animal to accommodate transient iodine deprivation (Brown et al., 2004). Additionally, iodized salts in system water and iodine-fortified processed foodstuffs are commonly used to avoid iodine-deficiency pathologies such as hyperthyroidism and goiter. Nevertheless, environmental conditions such as poor water quality or malnutrition can lead to marked proliferation of thyroid tissue in some fishes (Hoover, 1984) indicating that care must be taken to maintain a minimum iodine level in artificial environments including zebrafish aquaculture facilities. Dietary supplements have been shown to prevent thyroid hyperplasia in the saltwater cyprinodont *Fundulus* (Pickford, 1954) and goiters in the Chinook salmon (Woodall & La Roche, 1964), but no study to date has effectively assessed the minimum iodine level for maintaining normal development and thyroid health in zebrafish.

While experimenting with varied diets in our facility, we observed that conditions of reduced iodine result in less robust somatic growth and an elevated incidence of jaw dysmorphogenesis. This prompted us to evaluate the developmental consequences of rearing zebrafish on a single live food or a mixed food diet with varying levels of iodine. Herein we demonstrate that a strict single-food *Artemia* diet, although high in the protein necessary for robust egg production, may also result in a morphological alteration of the zebrafish jaw architecture formed in response to swelling and hypervascularization of adjacent tissues.

2. Experimental methods

2.1 Animal housing and maintenance

Adult WIK wild-type zebrafish were raised, maintained, and spawned in the Zebrafish Model Organism Laboratory, located at the University of Cincinnati (UC) Genome Research Institute according to previously published protocols (Craig et al., 2006; Kimmel et al., 1995; Westerfield, 2000). Zebrafish fry were reared in a multi-rack system from Aquatic Habitats (Apopka, FL) with automated control of water chemistry (YSI Model 5200 control system, Yellow Springs, OH) utilizing a 14-h light/10-h dark cycle. Water chemistry was maintained at the following conditions to ensure optimal fish health: temperature: 28.5 ± 1 ºC; pH: 7.5 ± 0.2; ammonia: 0 ± 0.25 mg/l; nitrite: 0 ± 0.20 mg/l; nitrate: 5 ± 5 mg/l; total hardness: 102 ±17 mg/l; alkalinity 50 ± 17 mg/l. Conductivity was maintained at 750 ± 50 µS. Adult breeding stock fish were fed a standard diet of 3 day old HUFA-enriched (Super SELCO, Aquatic Ecosystems) *Artemia* in the morning plus a balanced flake food (Aquatox, Aquatic Ecosystems) in the afternoon.

2.2 Dietary paradigm

Fry were fed a mixed diet of *Paramecium*, rotifers, and Zeigler's larval diet (Zeigler Bros, Inc.) until they were capable of eating *Artemia* (approximately 1 month of age) and to ensure proper early development. Beginning at 1 month of age, fish were placed on either a live study diet of three *Artemia* feedings or a mixed control diet of two Aquatox flake and one *Artemia* feedings per day.

Dietary *Artemia* intake was measured by feeding *Artemia* to overnight fasted fish (17.2 mg per tank, 15 minutes feeding time to approach satiety). Fish were euthanized in concentrated 250 mg/l MS-222, eviscerated and the alimentary tract was removed. The gut was everted, and the total number of ingested shrimp counted under a stereomicroscope. Aquatox flake food uptake was similarly measured by feeding fish to satiety (4.7 mg per tank, given 15 minutes feeding time) and then vacuuming up the excess food with a siphon, drying and weighing the uneaten excess.

2.3 Iodine analyses

Quantification of iodine in the foodstuffs was performed using ICP-MS as described previously for other heavy metals (Figueroa et al., 2008). Briefly, *Artemia* samples were cultured for 72 h, harvested and freeze-dried overnight in a lyophilizer (FreeZone 4.5 Liter Benchtop System, Labconco Corporation, Kansas City, MI). Three replicates of dried *Artemia* and system water samples (~30 mg) were placed in septum sealed glass tubes, treated with HNO_3 and microwave digested: power, 130 W; ramp, 1:00 min; hold, 1:00 min; temp, 75 °C. Subsequently, 0.1 mL of 30 % H_2O_2 was added and the samples were processed through the second digestion stage: power, 130 W; ramp, 1:00 min; hold, 1:00 min; temp, 110 °C. Samples were diluted with DDW to 10 mL and analyzed for total iodine content in continuous flow mode using an Agilent 7500ce ICP-MS (Agilent Technologies, Tokyo, Japan) equipped with a quadrupole mass analyzer.

2.4 Histology

Standard hematoxylin and eosin staining was performed at Oregon State University on both phenotypic and control juvenile zebrafish in order to ascertain the nature of the jaw dysmorphogenesis. Sagittal jaw sections were cut 4 micron thick, fixed in paraformaldehyde and stained with Alizarin Red according to previously published protocols (Cailliet et al., 1986). Adult fish were cleared and stained with Alizarin Red dye in order to visualize bone and pigment structure. Fish were fixed for 3 days in 10% phosphate buffered formalin. Fish were first rinsed in deionized water, bleached in 3% hydrogen peroxide / 2% potassium hydroxide to remove pigment, rinsed a second time in DI water, trypsin digested for eight days (3375 U/ml in 30% Sodium Borate) until a majority of the spine was visible, soaked in 0.5% KOH, soaked in Alizarin Red for several hours and destained in 2% KOH overnight. Stained fish were then transferred stepwise into increasing glycerol concentrations (90% final) for imaging and long-term storage.

2.5 Morphometrics

The effects of diet on normal fish growth were determined with standard growth metrics. Body mass measurements were made on anesthetized (125 mg/l MS-222) fish. Water accumulated on the body surface can result in significant overestimation of body mass and so it was removed by gently blotting the animal with absorbent paper (Kimwipe). Individual fish were then weighed three times on a 5-place analytical balance (Mettler-Toledo, Inc., Columbus, Ohio) and the average calculated. Total lengths (tip of the nose to the tip of the tail fin) were obtained from still images using a calibrated measurement tool in Rincon 7.1 software (Optronics Inc., Goleta, California) or using a manual (dial) caliper. Craniofacial development was similarly assessed from still images of Alizarin Red bone stained adult fish by measuring the lengths and widths of the Meckel's and ceratohyal complexes in fish exposed to the different dietary regimens.

2.6 Statistics

All statistics were performed in SigmaStat XI (Systat Software Inc., San Jose, California). Means were analyzed using Student two-tailed unpaired t-tests or nonparametric rank sum tests as needed. Probability (P) values of <0.05 were considered significant. Incidence levels were compared using a Yates-corrected Chi-square test.

3. Results and discussion

3.1 Dietary paradigm results in a 6-fold difference between *Artemia*-only and mixed food diets

Fish fed *Artemia* three times daily consumed an average of 1.2 mg of *Artemia* (51.6 mg wet weight) per feeding for a total of 3.6mg dry mass. Fish reared on the mixed food diet consumed a total of 4.8mg of dry mass per day (1.2mg *Artemia* twice and 2.37mg of Aquatox flakes once). Dietary and environmental iodine levels measured by ICP-MS (Table 1) indicated that the *Artemia* only diet contained just under one-sixth of the total iodine present in the mixed diet (11.48ng and 72.6ng, respectively).

Source	Concentration (ppm)
Crystal Sea Marine Mix[a]	0.12
Instant Ocean Sea Salt[a]	0.26
Seawater[b]	0.49-0.58
System Water (Instant Ocean + salts)	0.73 ± 0.26
SuperSelco Enriched *Artemia*[b]	3.19 ± 1.04
Aquatox Flakes[b]	27.38 ± 1.56

Table 1. Iodide concentrations of study-related salts, feed and supplements. Mean values are listed ± 1 s.d. [a](Bidwell and Spotte, 1985), [b]ICP-MS quantified.

The reduction in somatic growth may be the result of diminished endocrine output, possibly caused by an iodine or other nutrient deficiency (Ikeda et al., 1973). The smaller body size was not likely due to underfeeding since the fish were fed three satiating meals of *Artemia* daily. Further, there was no indication of impaired feeding capacity or precipitous wasting condition in grossly phenotypic fish, despite our expectation that the displacement of the lower jaw bones would negatively affect feeding ability. The *Artemia* fed fish were, in fact, visibly smaller throughout the study.

3.2 *Artemia*-only diet results in smaller, less sexually dimorphic individuals compared with mixed diets

The body mass and length of fish raised on the *Artemia* diet was significantly reduced compared with fish reared on a mixed diet (Table 2, P<0.0001) despite having made morphometric measurements well after the fish should have normally reached their terminal length (~150 dpf, unpublished data). Sexual dimorphism was clearly evident in the mixed diet population by 4 months of age, but the *Artemia* fed fish were only minimally dimorphic at 10 months of age. Interestingly, a diet of processed flake food alone also

	Artemia Only	Mixed Diet	P-value
Mass (mg)	139 ± 64 (78)	278.80 (15)	0.079
Length (mm)	25.4 ± 1.9 (78)	60.4 ± 2.1 (15)	<0.001
Meckle's Length (mm)	1.23 ± 0.29 (4)	1.52 ± 0.75 (2)	0.499
Ceratohyal Length (mm)	1.43 ± 0.53 (4)	1.79 ± 0.84 (2)	0.541
Iodine Content (mg/day)	0.11	0.28	n/a

Table 2. Body morphometrics of 160-180 day old fish on study diets. Mean values listed ± 1 s.d.

resulted in reduced terminal lengths when compared to fish fed *Artemia* only or mixed food diets (data not shown).

3.3 Thyroid enlargement leads to jaw bone displacement in fish fed low-iodine diets

The protruded mandibular jaw phenotype (Fig. 1) was observed within 1 week of initiating the single food *Artemia* diet, with the incidence level increasing logarithmically to 6.3% at 180dpf and reaching a plateau at 10.5% at 10 months of age (n=828 total fish). The incidence level in the population receiving the mixed diet was significantly lower throughout, reaching 3.4% at 10 months (n = 507 total fish, Yates-corrected Chi-square stat: 21.4, α = 0.05, 95% CI: 4.3 – 10.1%). A gradual increase in the extent of mandibular protrusion and degree of localized vascularization was observed throughout the 10 month test period with 78% of the phenotypic fish exhibiting large vascularized growths that protruded an average of 2.1% of their total body length (Fig. 1A). Histologic analyses revealed clear glandular involvement (Fig. 1B) with a distinct rotation of the basihyal jawbone from a sagittal to a more transverse orientation (Fig. 1C). The jaw dysmorphogenesis appeared to be the result of bone displacement by the enlarged thryroid adjacent to the basihyal none rather than a deficiency in osteogenesis, although the normalized Meckel's cartilage and ceratohyal bone lengths were marginally reduced compared to fish fed the mixed diet (Table 2, 19% and 20% reductions for Meckel's and ceratohyal lengths).

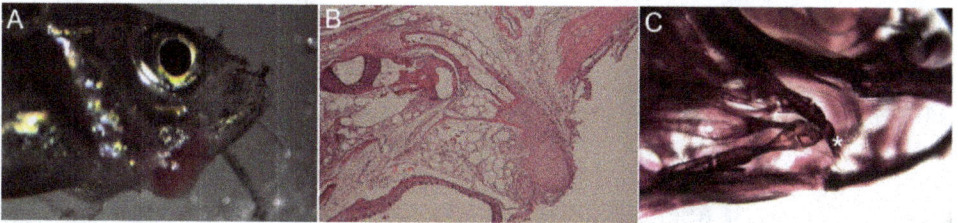

Fig. 1. Phenotypic characterization of low iodine diet-reared fish. Gross phenotype (A), histological section stained with hematoxilin and eosin (B), and Alizarin Red bone stain (C). Ventral rotation of the basihyal bone is indicated by an *.

The presence of an enlarged, hypervascularized tissue mass in proximity to the thyroid gland prompted us to speculate that the overall phenotype was due to an iodine deficiency. Because the teleost thyroid is an unencapsulated colloid-filled follicle (Gundernatsch, 1911), it is often difficult to classify the exact nature of piscine thyroid phenotypes. Nevertheless, several studies in other fish species have shown lower jaw deformities similar to those seen in the *Artemia* fed zebrafish in this study (Gaylord & Marsh, 1912; Marine & Lenhart, 1911; Marine, 1914). The enlarged hypervascularized masses protruding beneath the jaw of New York hatchery salmon were attributed to thyroid carcinomas (Gaylord & Marsh, 1912). Brook trout in a Pennsylvania hatchery also appeared to have simple goiters (Marine, 1914) externally similar in appearance to the phenotype observed in our study. Histochemical analysis of fish reared on our low-iodine diet yielded no indication of cell over proliferation, nuclear atypia, or extra cellular layers typical of carcinomas, but instead showed evidence of pushing borders, organ involvement and hypervascularization typical of simple goiter (Gaylord & Marsh, 1912). Histological analysis also clearly indicated a ventrally-oriented bone displacement associated with the observed lower jaw protrusion, suggesting that the glandular tissue was impinging upon the jawbones and altering their normal orientation. Although our studies terminated after six months, chronic iodine deprivation would have eventually led to 100% incidence of thyroid malformations, however the gradual hyper-vascularization appeared to exhibit an endpoint at 10.5% of the study fish. Other contributing factors such as environmental stresses and individual fish health may also have modulated the observed phenotypes.

4. Conclusion

Freshwater fish generally require at least 1 – 4 mg total I kg^{-1} (Watanabe et al., 1997) and a dietary minimum of 2.8 mg I kg^{-1} (Lovell, 1979), but plasma iodide levels in freshwater fish vary greatly indicating a wide spectrum in the efficiency of iodine uptake and its use. Given the low 0.1 – 10 µg/l iodine content of freshwater (Eales, 1979) and a metabolic need not met by many natural or artificial environments (Schlumberger, 1954), these fish generally rely more heavily on dietary sources of iodine than do marine fish. Freshwater fish obtain iodine from food by transport through the gut, environmental uptake across the gills, and a very small portion through hormone recycling. Our data suggests that the minimum iodine level required to prevent diminished growth and other apparent iodine-related phenotypes in zebrafish is somewhat lower than for other freshwater fish, falling between the iodine content in our *Artemia* and mixed food diets (being in the range of 0.11 to 0.28 mg I kg^{-1} per day above the negligible 0.7 ppm environmental exposure). It remains unclear whether iodine supplementation in excess of the mixed diet level (e.g. two *Aquatox* flake feedings per day or *Artemia* supplemented with Kent's iodine) would correct the observed gland enlargement and bone displacement, but previous studies have shown that trout goiters can be attenuated with a potassium iodide supplement in as little as 10 days of exposure time (Marine & Lenhart, 1911; Marine, 1914).

We support the notion that the minimal costs associated with adding iodine-rich flake food to the diet of these fish both is well worth the effort as it results in improved fish health and reduces the time needed to reach sexual maturity.

5. Acknowledgment

This work was supported by grants from the NIH (#1R01RR023190-01) and AHA (#0555236B) to J.R.H. All experiments were conducted according to guidelines reviewed by the University of Cincinnati Institutional Animal Care and Use Committee under protocol 07-07-30-01.

6. References

Barnard, E. & Bagatto, B. (2002). Effects of maternal diet on reproductive strategies and embryo development in the zebrafish, Danio rerio. *Integrative and Comparative Biology* 42:1191

Bidwell, J. P. & Spotte, S. (1985). *Artificial Seawaters: Formulas and Methods.* Jones and Bartlett, Boston, Massachusetts.

Brown, S.B.; Adams, B.A.; Cyr, D.G.; & Eales, J.G. (2004). Contamination effects on the teleost fish thyroid. *Environmental Toxicology and Chemistry* 23:1680-1701.

Cailliet, G. M.; Love, M.S.; & Ebeling, A.W. (1986). *Fishes. a field and laboratory manual on their structure, identification, and natural history.* Wadsworth Publishing Co, Belmont, California.

Craig, M. P.; Gilday, S.D.; & Hove, J.R. (2006). Dose-dependent effects of chemical immobilization on the heart rate of embryonic zebrafish. *Lab Animal (NY)* 35:41-7.

Eales, J. G. (1979). *Thyroid function in cyclostomes and fishes.* Academic Press, London.

Eales, J.G.; & Brown, S.B. (1993). Measurement and regulation of thyroidal status in teleost fish. *Rev. Fish Biol Fish* 3:299-347.

Figueroa, J. A.; Wrobel, K.; Afton, S.; Caruso, J.A.; & Corona Felix Gutierrez, J. (2008). Effect of some heavy metals and soil humic substances on the phytochelatin production in wild plants from silver mine areas of Guanajuato, Mexico. *Chemosphere* 70:2084-91.

Gaylord, H. R.; & Marsh, M.C. (1912). Carcinoma of the thyroid in salmonid fishes. *Bureau of Fisheries* 32:363-524.

Goolish, E. M.; Okutake, K; & Lesure, S. (1999). Growth and survivorship of larval zebrafish Danio rerio on processed diets. *North American Journal of Aquaculture* 61:189-198.

Gundernatsch, J. F. (1911). The thyroid gland of teleosts. Journal of Morphology 21:709-782.

Hoover, K.L. (1984). Hyperplastic thyroid lesions in fish. *National Cancer Institute Monograph* 65:275-289.

Ikeda, Y., H. Ozaki, & H. Yasuda. 1973. Effects of potassium iodide on growth of body and scale in goldfish. *Journal of Tokyo University of Fisheries* 59:333-342.

Kimmel, C. B.; Ballard, W.W.; Kimmel, S.R.; Ullmann, B.; & Schilling, T.F. (1995). Stages of embryonic development of the zebrafish. *Developmental Dynamics* 203:253-310.

Lovell, R. T. (1979). Formulating diets for aquaculture species. *Feedstuffs* 51:29-32.

Marine, D. (1914). Further observations and experiments on goitre (so-called thyroid carcinoma) in brook trout (Salvelinus fontinalis): III. Its prevention and cure. *Journal of Experimental Medicine* 19:70-88.

Marine, D.; & Lenhart, C.H. (1911). Further observations and experiments on the so-called thyroid carcinoma of the brook trout (Salvelinus fontinalis) and its relations to endemic goitre. *Journal of Experimental Medicine* 13:455-475.

Meinelt, T.; Schulz, C.; Wirth, M.; Kurzinger, H.; & Steinberg, C. (1999). Dietary fatty acid composition influences the fertilization rate of zebrafish (Danio rerio Hamilton-Buchanan). *Journal of Applied Ichthyology* 15:19-23.

Pickford, G.E. (1954). The response of hypophysectomized male killifish to prolonged treatment with small doses of thyrotropin. *Endocrinology* 55:274-287

Schlumberger, H. G. (1954). Spontaneous hyperplasia and neoplasia in the thyroid of animals. *Brookhaven Symposia in Biology* 7:169-191.

Watanabe, T.; Kiron, V.; & Satoh, S. (1997). Trace minerals in fish nutrition. *Aquaculture* 151:185-207.

Westerfield, M. (2000). *The zebrafish book: a guide for the laboratory use of zebrafish.* University of Oregon Press, Eugene, OR.

Woodall, A.N.; & La Roche, G. (1964). The nutrition of salmonoid fishes. XI. Iodide requirements of Chinook salmon. *Journal of Nutrition* 82:475-482.

Development of Biopolymers as Binders for Feed for Farmed Aquatic Organisms

Marina Paolucci[1], Adele Fabbrocini[2],
Maria Grazia Volpe[3], Ettore Varricchio[1] and Elena Coccia[1]
*[1]Department of Biological, Geological and
Environmental Sciences, University of Sannio
[2]C.N.R. National Research Council, I.S.MAR
[3]C.N.R. National Research Council, I.S.A.
Italy*

1. Introduction

Diets for aquatic animals are numerous. They differ from species to species and may change to meet varied nutritional requests during the life cycle, and may be designed for larvae, juveniles, adults and breeders. In this review we will focus on a particular aspect of aquaculture feed represented by binders. Binders can be liquids or solids with the capacity of forming bridges, coatings or films that make strong inter-particle bonding. Binders are used to improve feed manufacture and to stabilize diets in water. Differently from feed for livestock, feed for aquaculture requires an adequate level of processing to guarantee good stability in water, long enough for animals to consume it. For this reason the role of binder is crucial in determining variable levels of firmness adequate to specific feeding behaviour. Although the problem of feed stability is far more crucial with crustaceans than with fish, some fish are benthic and small pellets that sink rapidly to the bottom where they can be located and recognized by the chemoreceptors of the fish are highly sought. Usually commercial feed for fish is stable after extrusion and binders are not requested to improve water stability. In some recent experiments binders are included in practical diets for fish to generate firmer feces when emitted into water to reduce pollution (Brinker, 2007). Among crustaceans crayfish are slow feeders with a characteristic tendency, that they share with prawns and shrimps, to manipulate food using mouth appendages before ingestion (Holdich, 2002). Thus, in aquatic animal feed preparation, to stabilize feed pellets and to ensure minimum nutrient leaching and disintegration appear to be crucial. Feed stability is considered a crucial requirement also in the echinoculture. Indeed, sea urchin are grazers and need time to eat the offered feed, so that it must remain intact for several days, in order to limit the loss of nutrients and to make rearing structure management easier (Caltagirone et al., 1992; Mortensen et al., 2004; Pearce et al., 2004). In addition, prepared diets frequently lead to poor gonad quality in terms of texture, firmness, colour and taste (Pearce et al., 2002a), that means low marketability of the product. For these reasons research focused on the selection of appropriate binders to ensure consistence to the experimental feed must take into account their effects not only on feed stability but also on gonad yield and sensorial quality. Since a binder may not be optimal for all species, and even for the same species the feeding behaviour may change during the life cycle numerous studies have been conducted to evaluate the type of binder, the optimal level of inclusion and its effect on growth and digestibility in different species of aquatic animals.

In this review, feed will be examined from the point of view of binder selection, method of preparation, assessment of water stability and nutrient leaching and with reference to the binder effect on growth and utilization by the animal. Purposefully, this review does not deal with diet composition for aquatic animals. Research on nutritional requirements of aquatic animals is a highly competitive field in constant and rapid evolution and it is outside the scope of the present article. For readers interested in this topic we suggest referring to the numerous reviews and books available.

2. Feed binders

Food is the material which, after ingestion by the animal, is capable of being digested absorbed and utilized. However, not all components of food are digestible. The main components of food are as follows:

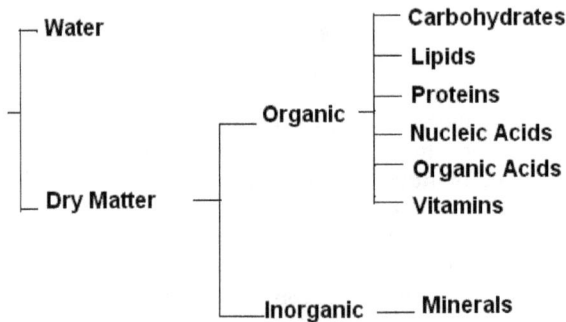

```
            ┌─ Water                          ┌─ Carbohydrates
            │                                 ├─ Lipids
            │                                 ├─ Proteins
            │                       ┌─ Organic ┤
─┤                                  │         ├─ Nucleic Acids
            │                       │         ├─ Organic Acids
            └─ Dry Matter ─┤        │         └─ Vitamins
                           │
                           └─ Inorganic ── Minerals
```

Nutrients are components of food which can be utilized by the animal either as energy sources or for metabolic processes. The energy source components are the proteins, lipids and carbohydrates, while the vitamins and minerals are need to be present in small quantities in the feed for metabolism and life maintenance. There are various factors influencing the choice of feed for farmed fish. Of these, the economic factor is usually given precedence. Feeds and feeding can represent about half the operating costs in farming. Other important factors in manufacturing aquatic animal feeds are the stability of the feed in water and its acceptability. Inclusion of a binder is a necessity to ensure water stable feed. Many substances have been selected for their binding properties. Numerous natural binders have been employed to manufacture hard pellets with the purpose of increasing water stability with a concomitant decrease in nutrient loss. Some of these binders, such as wheat gluten or starches, have a nutritive value for the animals, while others are inert raw materials without nutritional value (Sinha et al., 2011). More than 50 organic and inorganic binders have been employed in feed industry (Kalian & Morey, 2009). Among natural binders examined and employed by researchers, biopolymers are by far the most studied. The most common biopolymers derive mainly from plants and animals of marine and agricultural origin. Examples are cellulose, starch, pectin, chitin and proteins such as casein, whey, collagen and soy proteins. In this review we will focus on the most employed for aquatic feed.

According to De Silva and Anderson (1995) binders can be grouped in three different classes:

1. binders of protein origin

2. carbohydrate source binders

3. binders with no nutritional value

From a nutritional perspective a protein binder may be more appropriate than a carbohydrate binder for a carnivorous species, while the contrary holds true for a herbivorous species. Carbohydrates are often included in crustacean artificial diets for their protein-sparing effect (Shiau & Peng, 1992; Rosas et al., 2000). According to the theory of Minimal Total Discomfort (Forbes, 2001), the ingestion of carbohydrates of vegetal origin brings about a protein sparing effect in wild animals and therefore enhances growth without wasting proteins. In fact, the intake of food has physiological consequences via physical (e.g. distension) and chemical (e.g. glucose) stimulation of receptors in the viscera and, in the longer term, by changes in signals from adipose tissue (e.g. leptin), integrated by the CNS (Paolucci, 2010). These consequences are associated with the sensory properties of the food such that repeated exposure to a food generates a conditioned acceptance or rejection reflex with the physiological consequences of eating as the unconditioned stimulus (US) and the sensory characteristics of the food as the conditioned stimulus (CS). Such learnt preferences and aversions occur throughout the animal kingdom, from nematodes to human beings, with much of the research being carried out with insects, laboratory animals and farm animals. Preferences for and aversions to particular foods are manifested in non-random choices between two or more foods on offer but also influence the quantity eaten when only one food is available. These considerations have been developed into a theory of Minimal Total Discomfort (Forbes, 2001) which proposes that an animal experiments with the amount eaten per day, and its selection between different foods, until the total of the signals generated from excesses or deficiencies of food components is minimised. Changes in food composition and/or nutrient requirements can therefore be matched by appropriate changes in intake and selection.

2.1 Carbohydrates

Polysaccharides are natural biopolymers formed by high molecular weight carbohydrates (Aspinall, 1982). They are biodegradable, biocompatible and non-toxic polymers showing peculiar physical-chemical properties and environmentally sustainable features. The specific macromolecular structures, characterized by the presence of several polar functional groups, allow polysaccharides to retain significant amounts of water or biological fluids, thus providing the formation of hydrogels, i.e. three-dimensional, reliable networks, able to become water resistant throughout chemical or physical phenomena, such as gel formation, retrogradation process, pH-changing and cross linking processes (Coviello et al., 2007). The molecular diversity of carbohydrates allows a large array of functions of great significance. In the food industry, polysaccharides play important roles as phytocolloid as well as emulsifying agents (Farris et al., 2009). Non-digestible carbohydrates are insoluble, like cellulose, or soluble, like pectin, gum, β-glucan, mucilage, algal polysaccharides (Atkins, 1985; Clark & Ross-Murphy, 1987). Technologically, they are needed for altering the texture and consistency of foods (Drochner et al., 2004) and show properties that make them suitable for use in increasing food shelf-life (Volpe et al., 2010). Carbohydrates represent the most abundant biological molecules, covering a large array of fundamental roles in living things: from the reserve and transport of energy, (starch and glycogen), to the development of structural components (cellulose in plants, chitin in animals), to the linking between intercellular walls (hemicellulose) (Kennedy & White, 1983). The high molecular weight carbohydrates derived, are known as polysaccharides. They may be viewed as condensation polymers in which carbohydrates have been glycosidically joined, with the elimination of molecules of water, according to the empirical equation

$$n\ C_6H_{12}O_6 \rightarrow (C_6H_{10}O_5)n + (n-1)H_2O$$

The different macromolecular structures and chemical compositions of polysaccharides are responsible for the large array of their physical and biochemical applications. A wide range of polysaccharides, such as agar, alginate, chitin and pectin are able to hydrate in cold and hot water, thus giving rise to both viscous solutions or dispersions and gels. The great interest with these polysaccharides in aquatic animal feed is strictly related to their gelling properties.

2.1.1 Agar

Agar is a polymer extracts from agarophyte, a seaweed, typically a red alga (Usov, 1998). It consists of at least two separate polymers that could be fractionated into agarose andagaropectin. Agarose is high in molecular weight and low in sulphate. Agaropectin is low in molecular weight and high in sulphate. Agarose is the gelling fraction of agar. Agarose is a linear polymer structure consisting of alternating D-galactose and 3,6 anhydro-L-galactose as shown in Fig. 1.

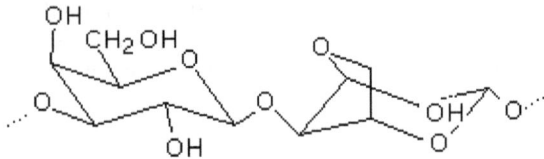

Fig. 1. Molecular structure of agarose

The diverse forms of agarose determine the physicochemical characteristics of agar such as gelling and melting temperature and reactivity. Agar is a mixture of agarose and agaropectin in variable proportions depending on the original raw material and the manufacturing process employed. Agar gelation occurs only by its agarose content and is produced exclusively by hydrogen bonds. Agar does not need any other substance to gel, and therefore it has an enormous potential in applications such as foodstuff ingredient, for biotechnology uses, for cell and tissue cultures or as a support for electrophoresis or chromatography. Agarose produces "physical gels" which means that these aqueous gels are formed only by the polymer molecules united solely by hydrogen bonds. Due to this unique gelling property, such gels hold in the interior network a great amount of water which can move freely through the macroreticulum. Each molecule maintains its structure in complete independence so that the process is not a polymerization but a simple electrostatic attraction. Agar is actually a very complex polysaccharide and varies considerably depending on the source. Back in 1991, eleven different agarose structures were identified in different agar bearing weeds depending on gender, species environmental conditions and time of the year. Moreover, agar may be modified by substitution of sulphate, pyruvate, uronate or methoxyl groups. Modern alkali treatment methods tend to increase the level of anhydrous bridging in the molecule with subsequent improvement of gel strength.

2.1.2 Alginate

Sodium Alginate is the sodium salt of alginic acid, a complex mixture of oligo-polymers, polymannuronic acid (MM), polyguluronic acid (GG) and a mixed polymer (MG) where sequences like GGM and MMG co-exist (Smidsrod et al., 1966). The mannuronic acid forms β (1-4) linkages, so that M-block segments show linear and flexible conformation. The guluronic acid instead gives rise to α (1-4) linkages, introducing a steric hindrance around the carboxyl groups; for this reason the G-block segments provide a folded and rigid structural conformation, responsible for a pronounced stiffness of the polymer. M and G blocks of the alginic acid salts are reported in Fig. 2.

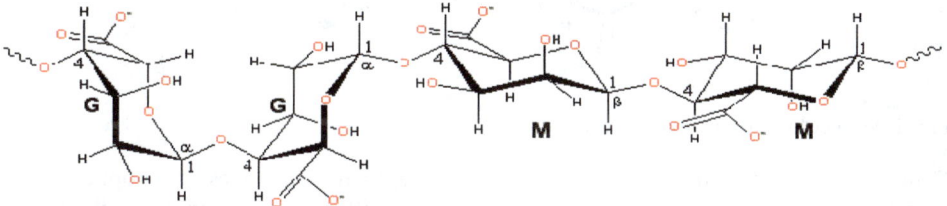

Fig. 2. Molecular structure of alginic acid

In a water solution and in the presence of divalent cations, such as calcium ions, the peculiar buckled backbone of G segments gives rise to water insoluble gel due to the strong interactions between the divalent cations and the COO- groups of the base residual of guluronic acid. Cations can be trapped in a stable, continuous and thermo-irreversible three-dimensional network, whose conformation is typical of an "egg box" (Grant et al., 1973; Grasdalen, et al., 1981) as shown in Fig. 3.

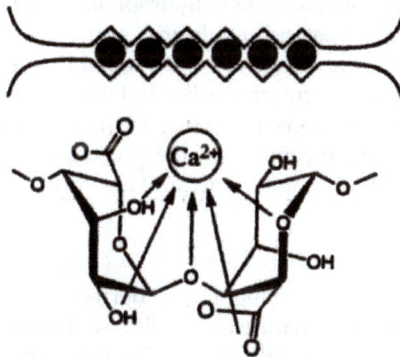

Fig. 3. Alginic acid "egg box" structure.

As a consequence, calcium alginate gel does not dissolve in water. Nevertheless, a re-hydration process occurs, allowing the gel to swell and to modulate the releasing of entrapped substances.

2.1.3 Pectin

Pectin is a polysaccharide extracted from plant cell walls and is used in many industrial applications as a thickener or gelling agent (Stephen, 1983). Pectin is a polymer of α-galacturonic acid with a variable number of methyl ester groups (Fig. 4).

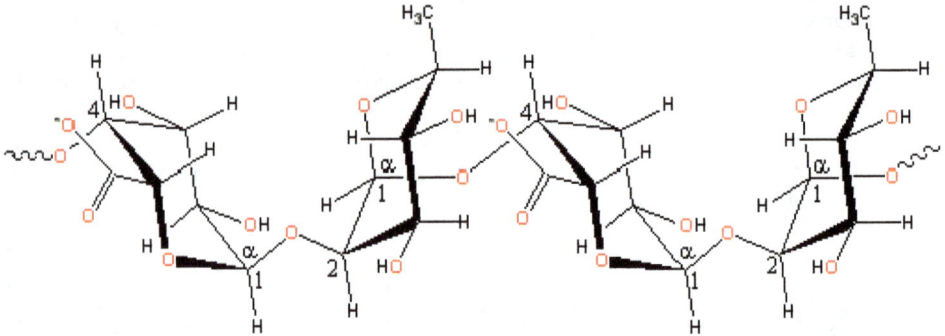

Fig. 4. Molecular structure of pectin

Generally, pectin does not possess an exact structure, therefore, it does not adopt a straight conformation in solution, but is extended and curved with a large amount of flexibility. The methylation of the carboxylic acid groups forms methyl esters which are much more hydrophobic and consequently have a different effect on the structuring of the surrounding water. The properties of pectin depend on the degree of esterification, which is normally about 70%. Gel strength increases with increasing Ca^{2+} concentration but it reduces with temperature and acidity increase (pH < 3). In the absence of added cations, low methoxy-pectin (~35% esterified) gels by the formation of cooperative "zipped" associations at low temperatures (~10°C) forming transparent gels. This hydrogen-bonded association is likely to be similar to that of alginate. High methoxyl-pectin (> 43% esterified, usually ~67%) gels by the formation of hydrogen-bonding and hydrophobic interactions in the presence of acids (pH ~3.0, to reduce electrostatic repulsions) and sugars (for example, about 62% sucrose by weight, to reduce polymer-water interactions). While the gel formed by low methoxyl-pectin results as being thermoreversible and not stable in water, the gel formed by high methoxyl-pectin is thermoirreversible mostly in the presence of high content of sugar (sucrose) and at low pH. In these conditions, the process of re-hydration of pectin gel is partially avoided by the firm hydrophobic structural network.

2.1.4 Chitosan

Chitosan is a cationic carbohydrate biopolymer derived from chitin, the second most abundant polysaccharide present in nature after cellulose. The main sources of chitin are the shell wastes of shrimps, lobsters and crabs (Johnson & Peniston, 1982). Upon the removal of most of the acetyl groups of chitin by treatment with strong alkali, chitosan yields. So chitosan may be considered as a family of linear binary copolymers of (1→4)-linked 2-acetamido-2-deoxy-β-D-glucopyranose (GlcNAc) and 2-amino-2-deoxy- β-D-glucopyranose (GlcN) (Fig.5).

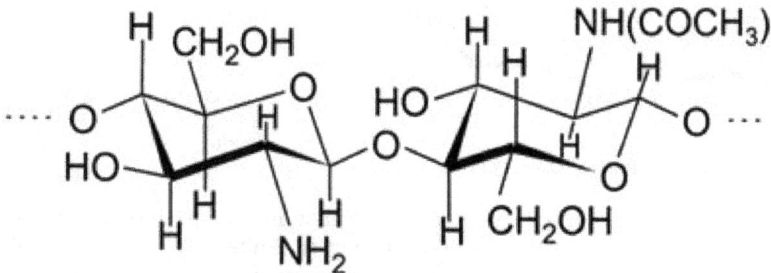

Fig. 5. Molecular structure of chitosan

Chitosan has a very diversified range of established and potential applications related to its polycationic properties, which are unique among polysaccharides and natural polymer in general (No & Meyers, 1995). It is well known that chitosan may complex with certain metal ions and this property is used for removal of traces of heavy metals or radio isotope in waste water. Another well documented application of chitosan is as a cholesterol-lowering agent but the much more controversial use of it, is as a weight reducing agent. It is also used as a potential vehicle for orally administered controlled-release drugs. So this material is proposed for preparing micro spheres and microcapsules (El-hefian et al., 2011). These are only few of the applications of chitosan. Although chitin is insoluble in most solvents, the properties of chitosan are related to its polyelectrolyte polymeric carbohydrate character. It is insoluble in water while it readily dissolves in dilute solutions of most organic acids such as acetic, citric, tartaric acids. Chitosan gel does not undergo a substantial swelling process when placed in water; it is instead stable because the polymer itself results as being insoluble in water (Cerver et al., 2004).

2.1.5 Carrageenan

Carrageenan is the hydrocolloid obtained from some red seaweeds by extraction with water or aqueous alkali (Murano et al., 1992; Kalia, 2005). Carrageenan consists chiefly of potassium, sodium, calcium, magnesium, and ammonium sulfate esters of galactose and 3,6-anhydrogalactose copolymers. These hexoses are alternately linked α-1,3 and β-1,4 in the polymer(Fig. 6). The threemain copolymers in the hydrocolloid are designated kappa-carrageenan (k-carrageenan), iota-carrageenan (ι-carrageenan) and lambda-carrageenan (λ-carrageenan). K-carrageenan is mostly the alternating polymer of D-galactose-4-sulfate and 3,6-anhydro-D-galactose. I-carrageenan is similar, except that the 3,6-anhydrogalactose is sulfated at carbon 2. Between k-carrageenan and ι-carrageenan there is a continuum of intermediate compositions differing in the degree of sulfation at carbon 2. In λ-carrageenan, the alternating monomeric units are mostly D-galactose-2-sulfate (1,3-linked) and D-galactose-2,6-disulfate (1,4-linked). The ester sulfate content for carrageenan ranges from 18 to 40%. In addition, carrageenan contains inorganic salts that originate from the seaweed and from the process of recovery of the extract. Carrageenan is recovered by either alcohol precipitation, or drum drying, or freezing. Primary differences which influence the

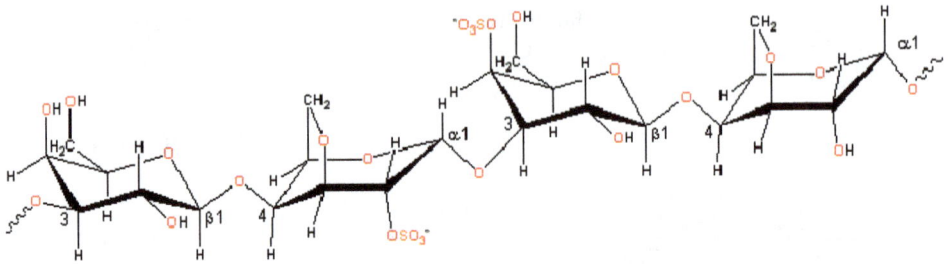

Fig. 6. Molecular structure of carrageenan

properties of k-, ι- and λ-carrageenan are the number and the position of the ester sulfate groups on the repeating galactose units (Kennedy & White, 1983). Higher levels of ester sulfate lower the solubility temperature of the carrageenan and produce lower strength gel, or contribute to gel inhibition.

The comparative properties of the three types of carrageenan are reported in Table 1.

Kappa carrageenan	Iota carrageenan	Lambda carrageenan
Soluble in hot water.	Dilute solutions exhibit thixotropic characteristics.	Free flowing, non-gelling pseudo-plastic solutions in water.
The addition of potassium ions induces the formation of a durable, brittle gel; it also increases the gelling and melting temperatures.	Soluble in hot water; sodium iota carrageenan is soluble in cold and hot water.	Partially soluble in cold water, fully soluble in hot water.
Strong, rigid gel, some syneresis, forms helix with K+ ions. Ca++ causes helices to aggregate and the gel to contract and become brittle.	The addition of calcium ions will induce the formation of a durable, elastic gel, and increase gelling and melting temperatures.	No gel, random distribution of polymer chains.
		Range from low to high viscosity.
Slightly opaque gel. Becomes clear with sugar.	Elastic gels, forms helix with Ca++. Limited aggregation contributes to elasticity, no syneresis.	Addition of cations has little effect on viscosity.
Approximately 25% ester sulfate and 34% 3,6-AG.	Clear gel.	Compatible with water miscible solvents.
Compatible with water miscible solvents.	Freeze/thaw stable.	Insoluble in most organic solvents.
Insoluble in most organic solvents.	Insoluble in most organic solvents.	Stable over a wide range of temperatures, including freeze/thaw cycles.
Typical use levels — 0.02 to 2.0%.	Approximately 32% ester sulfate and 30% 3,6-AG.	Soluble in 5% salt solution, hot or cold.
	Typical use levels — 0.2 to 2.0%.	Approximately 35% ester sulfate and little or no 3,6-AG.
		Typical use level — 0.1 to 1.0%.

Table 1. Comparative properties of types of carrageenan

2.1.6 Carboxymethylcellulose

Carboxymethylcellulose (CMC) is a derivative of cellulose formed by its reaction with alkali and chloroacetic acid. The CMC structure is based on the β-(14)-D-glucopyranose polymer of cellulose (Fig. 7).

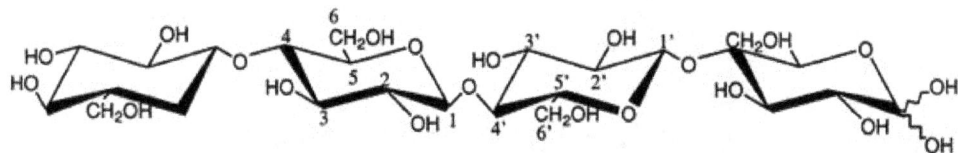

Fig. 7. Molecular structure of carboxymethylcellulose.

CMC molecules are shorter than native cellulose with uneven derivatization giving areas of high and low substitution. Substitutions are mostly 2-O- and 6-O-linked, followed in order of importance by 2,6-di-O- then 3-O-, 3,6-di-O-, 2,3-di-O- lastly 2,3,6-tri-O-linked. It appears that the substitution process is slightly cooperative (within residues) rather than a random process, which gives slightly higher than expected unsubstituted and trisubstituted areas. CMC molecules are highly extended (rod-like) at low concentration, while at higher concentration the molecules overlap and coil up and then entangle, to become a thermoreversible gel. Both increasing ionic strength and reducing pH decrease the viscosity as they cause the polymer to become more coiled. CMC dissolves rapidly in cold water and is mainly used for controlling viscosity without gelling. As its viscosity drops during heating, it may be used to improve the volume yield during baking by encouraging gas bubble formation. Its control of viscosity allows its use as thickener, and phase or emulsion stabilizer (for example, with milk casein), and suspending agent. CMC can be also used for its water-holding capacity as this is high even at low viscosity particularly when used with Ca2+ salt (Arisz et al., 1995).

2.1.7 Guar gum

Guar has been cultivated in India since centuries. India accounts for 80% of the total guar produced in the world. Historically, guar has been an important source of nutrition for humans as well as animals. Guar gum (galactomannan) is a high molecular weight carbohydrate derived from the natural seed of guar plant (*Cyampopistetragonolobus*). Structurally, guar gum is a polysaccharide consisting of a mannose backbone with a galactose side chain. The galactose is randomly placed on the mannose backbone with an average ratio of 1:2 galactose to mannose. The polymeric structure of guar gum containing numerous hydroxyl groups, has long since been exploited by industry (Goldstein et al., 1973). In Fig. 8 the guar gum chemical structure is reported.

The most important property of guar gum is its ability to hydrate rapidly in cold or hot water to attain uniform and very high viscosity at relatively low concentration. Another advantage associated with guar gum is that it provides full viscosity even in cold water. Although guar gum is one of the most cost-effective stabilizers and emulsifiers, its applications in food industry are numerous, due to its capacity to enhance texture, mouth feel, and to control crystal formation due to superior water-binding properties (Castillo-Garcia et al., 2005). It is inert in nature and it is resistant to oil, greases, and solvents. It also has excellent synergy with several other hydrocolloids, particularly xanthan gum (Ahmed et al., 2005). It has a very high viscosity even when very little is used.

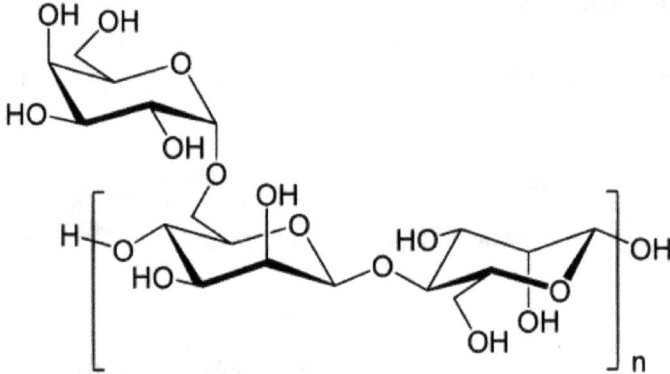

Fig. 8. Molecular structure of guar gum

2.2 Proteins

Among proteins, the mostly widely used as feed binders are urea formaldehyde, wheat gluten and gelatin. Gelatin-based binders are a good alternative to gluten and urea formaldehyde. Gelatin and gluten have the advantage of being fully digestible and contain proteins (De Muylder et al., 2008), however, gluten employment may have some drawbacks due to its antigenicity (Shewry, 2009).

2.2.1 Gelatin

Gelatin is a mixture of water-soluble proteins derived from collagen by hydrolysis. The protein fraction consists almost entirely of amino acids. These amino acids are joined by amide linkage to form a linear polymer ranging from 15,000 to 250,000 Mw. Gelatin is particularly attractive for forming hydrogel packaging because it is relatively inexpensive and biodegradable, and its structure enables multiple combinations of molecular interactions (Farris et al., 2009). There are two types of gelatins which are characterized by their mode of manufacture. Type A gelatin (pH 3.8–6.0; isoelectric point 6–8) is derived by acidic hydrolysis of pork skin and contributes to increasing the plasticity and elasticity to the blend. Type B gelatin (pH 5.0–7.4; isoelectric point 4.7–5.3) is derived by basic hydrolysis of bones and animal skin and contributes to giving high gel strength to the blend. Various purity grades of gelatin are sold commercially in the form of translucent sheets, granules, or powders. Gelatin is practically odorless and tasteless. It is insoluble in acetone, chloroform, ethanol (95%), ether, and methanol. It is soluble in glycerin, acids, and alkalis, although strong acids or alkalis cause it to precipitate. It swells and softens in water, gradually absorbing 5 to 10 times its own weight in water. It solubilizes in hot water. Upon cooling to 35–40 °C, it forms a jelly or gel. At temperatures ≥40 °C, the system exists as a sol. A gel of higher viscosity is formed in alkaline media as compared with acid media (Robinson et al., 1975). Since it is a protein, gelatin exhibits chemical properties characteristic of proteins (e.g., gelatin is hydrolyzed by most of the proteolytic systems to yield its amino components). Gelatin reacts with acids and bases, aldehydes and aldehydic sugars, anionic and cationic polymers, electrolytes, metal ions, plasticizers, preservatives, and surfactants.

3. Feed manufacture

Feed used in small-scale experiments is usually prepared by simple operations. Ingredients are ground and mixed by hand or by extrusion at room temperature (cold-extrusion). Grinding of feed ingredients is a crucial operation since pellets made with finely ground ingredients will be more durable and hard (Lim & Cuzon, 1994). Cold extrusion has the advantage of preserving the nutritional properties of the diet in comparison to extrusion at high temperature. Although extrusion technology is widely employed in the production of aquaculture feeds (Chevanan et al., 2009) the high temperature and pressure may negatively affect the nutritional value of ingredients (Cuzon et al., 1994; Watanabe, 2002). On the other hand, heat-extrusion of macro algae based prepared feed makes carbohydrates more digestible compared with fresh algae ones, leading to better gonad yields in reared sea urchins (Akiyama et al., 2001).

Binders usually employed at a concentration to reach a final percentage ranging from 1 to 10% are dissolved in either warm or hot water. As the temperature decreases and the solution turns into a gel, diet ingredients are added (Ruscoe et al., 2005; Volpe et al., 2008; Simon, 2009). Diversely, binders are added in their powdery form to the diet ingredients, mixed and hot water is poured on the mixture, then stirred thoroughly to obtain a dough (Orire et al., 2010). Some binders such as zein are first dissolved in ethanol due to their water insolubility (Genodepa et al., 2004; Johnston & Johntson, 2007). The outcome is a microbound diet. Pellets are obtained by passing the dough in a mincer to form noodles that are cut in pieces of few mm, or the dough is spread on a tray and cut into pieces. Alternatively, the mixture is put in a marumerizer and spheroid fed particles are produced (Liu et al., 2008). Steam pellets can be obtained using a laboratory pellet mill (Palma et al., 2008). Pellets are either stored at low temperature (from 4°C to -20°C) until use or dried in order to decrease the moist content and then stored (Ruscoe et al., 2005). The former can be defined as "moist pellets" and the latter as "dry pellets". A third type of pellet can be defined as "gelatinous pellets" and are obtained by mixing the dietary ingredients with oil (Johnston & Johnstson, 2007; Liu et al., 2008). Pellets can also be coated with binders, leading to what is then termed a microcoated diet (Bautista et al, 1989).

4. Assessment of feed stability in water

Water stability of feed is of paramount importance in the manufacture of aquaculture diets. Water stability is greatly influenced by the properties of binders, although the ingredients themselves have a direct influence on the characteristics of the binders (Dominy & Lim, 1991). Durazo-Beltran and Viana (2001) report difficulty in bonding the constituents of fish silage to generate pellets for abalone farming, probably because of the high content of hydrolyzed protein. Although the water stability of aquatic feed is a major concern of the aquaculture industry, there is no standard method to determine feed water stability. It is usually estimated by the method of dry matter weight loss, according to which a certain amount of feed, usually in the form of pellets, is placed in a water containing beaker and allowed to stay for a variable length of time with occasional shaking. Pellets are then decanted and dried. Variability can be found in literature about the amount of water in which pellets are placed, temperature, time intervals, methods in pellet drying after incubation in water (Ruscoe et al., 2005; Johnston & Johnston, 2007; Palma et al., 2008; Orire et al., 2010). A slight modification of the dry matter weight loss is suggested by the

American Feed Industry Association (AFIA 1999 in Liu et al., 2008). According to the AFIA water stability is expressed as a percentage of pellets retained on a wire mesh sieve after immersion in a shaking water bath for a period on time after which the pellets are retrieved and dried. Obaldo et al. (2002) developed three methods for measuring the water stability of shrimp pelleted feeds: a static method, a horizontal method and a vertical shaking method. Several factors were taken into consideration (water and pellet agitation, direction of agitation, water temperature, salinity, leaching container, filtration medium) among which temperature and water salinity influenced the rates of dry matter retention the most. Higher water temperatures and lower levels of salinity produced more leaching and lower dry matter retention. In general, the three methods appeared to provide an accurate and precise means of determining the water stability of shrimp feeds. The employment of water shaking is particularly interesting inasmuch as it may mimic the actual indoor and outdoor culture conditions and can be usefully adjusted to variable conditions.

Water stability is expressed as a percentage of immersed diet weight/initial sample weight (Liu et al., 2008; Orire et al., 2010), or as percent loss of dry matter (%LDM) calculated as percent difference in sample weight (minus the initial diet moisture) after reweighing (Fagbenro & Jauncey, 1995; Johnston & Johnston, 2007), or as percentage of dry matter remaining (%DMR) calculated with the formula %DMR = Wo X (1-M) – Wt/Wo X (1-M) X 100 where Wo = pellet weight as-fed, Wt = weight after immersion and drying, and M = moisture content of diet as a proportion (Ruscoe et al., 2005).

Caltagirone et al. (1992) suggest a subjective parameter, to be taken into account together with the %LDM, in the evaluation of binder efficiency, that is the consistency of food, defined according to the following scale: 1) inconsistency (feed completely disintegrates on removal from the water; 2) weak consistency (feed partially disintegrates on removal from the water; 3) good consistency (feed does not disintegrate on removal from the water but disaggregates after the application of a weak pressure; 4) very good consistency (feed does not disintegrate on removal from the water and resists weak pressure). According to these evaluation parameters Caltagirone et al. (1992) evaluated six binding agents after 48 hours of water immersion: agar-agar, cellulose, CMC, guar gum, gelatine and sodium alginate. Cellulose and CMC were unsuitable as binders even at strong concentrations; sodium alginate retained a good consistency but showed a high loss of weight, while guar gum, gelatine and agar-agar appeared to be good binders in terms of pellet stability. Pearce et al. (2002b) tested prepared feed with different binders (gelatin, guar gum, sodium alginate, and corn starch), evaluating their effectiveness according to the above described methodology up to 216 hours of immersion in sea water. Guar gum and starch-based pellets became inconsistent after 24 hours of immersion; sodium alginate retained a good consistence for up to 48 hours, while gelatine remained firm and intact even after 216 hours in sea water. Again with the same evaluation procedure Mortensen et al. (2004) tested the stability of a pellet in which gelatine from fish-skin was strengthened with the enzyme transglutaminase. Feed retained a good consistency for up to 7 days in sea water and the loss in dry matter, even increasing with time, reached 17.6%, with respect to 60.1% of the salmon feed used as control.

In our laboratory, we adopted a rather different methodology, according to which pellet water stability was analyzed by monitoring the diameter of the released particles in water over progressive time intervals, employing a Low Angle Laser Light Scattering Technique. (Volpe et al., 2008; Coccia et al., 2010; Fabbrocini et al., 2011). The diameter of particles released by pellets are continuously monitored over time, thus providing a time-course indication about the water stability as a function of the released particle diameter inasmuch as pellets that disaggregate into small particles are less stable in water than pellets that disaggregate into particles of a larger diameter. According to this evaluation methodology the binder capacity of two algal polysaccharides (alginate and agar) and one polysaccharide from fruit (pectin) revealed that pectin showed better water stability than alginate and agar (Volpe et al., 2008). When a coating was added to the microbound pellets water stability was improved. Indeed, microbound pellets disaggregated into particles of a derived diameter almost twice as large as pellets manufactured without coating (Coccia et al., 2010). Palma et al. (2008) report that pellets without binder, but with a microcoat of lignosol behaved less well in water than pellets made with lignosol as a binder, indicating that the presence of an internal binder is necessary to provide firmness to the feed. Using the Low Angle Laser Light Scattering Technique 3% and 6% agar-based round pellets specifically designed for sea urchin feeding, proved to have good water stability with a significant increase in the diameter of the released particles recorded only after 5 days of immersion in sea water for 3% agar- and 6 days for 6% agar-based pellets (Fabbrocini et al., 2011).

Due to the ample variability in feed ingredients, percentage of binders included and manufacture technology, it is impossible to come to the conclusion that a certain binder is better than another with respect to its water stability performances. Moreover, experimental outcomes are reported in literature as relative, that is a binder behaving better than another under specific conditions, making impossible an objective evaluation of binder performances. Nonetheless, in table 2, literature data on feed water stability are reported.

WATER STABILITY TEST	MAX TIME INTERVAL CONSIDERED	% OF BINDER EMPLOYED	OUTCOME[1]	REFERENCE
Dry matter weight loss	10 min	3% CMC, corn starch, guar gum, wheat gluten	CMC, guar gum, wheat gluten > corn starch	Fagbenro and Jauncey, 1995
Dry matter weight loss	2 h	2% agar, lignin	Lignin>agar	Palma et al., 2008
Dry matter weight loss	12 h	5% carrageenan 3% zein, agar 2% alginate, gelatin	Alginate,gelatin> carrageenan, zein, agar	Johnston and Johnston, 2007
Dry matter weight loss	12 h	10% Wheat flour 2% Whole cassava meal, dried molasses, agar, Langobin°	Wheat flour>whole cassava meal> dried molasses>agar> langobin	Seixas Filho et al., 1997a
Dry matter weight loss	24 h	0.5-2.5% Agar, Alginate, Carrageenan	All effective at 0.5%	Durazo-Beltran and Viana, 2001
Dry matter weight loss	48h	Agar-agar, carboxyl-methyl-cellulose (CMC), guar gum, gelatin, sodium alginate, cellulose	Agar-agar, gelatin, guar gum > sodium alginate > CMC, cellulose	Caltagirone et al., 1992

WATER STABILITY TEST	MAX TIME INTERVAL CONSIDERED	% OF BINDER EMPLOYED	OUTCOME[1]	REFERENCE
Dry matter weight loss	168h	Fish-skin gelatin+0.13% transglutaminase (TG)	Gelatin + TG > commercial salmon feed	Mortensen et al., 2004
Dry matter weight loss	216h	Guar gum, gelatin, sodium alginate, corn starch	Gelatin > sodium alginate > guar gum, corn starch	Pearce et al., 2002b
Dry matter retention	30 min	7% starch 2% carrageenan, CMS, alginate, gelatin	Carrageenan> alginate> gelatin> CMS	Liu et al., 2008
Dry matter remaining after immersion in water	180 min	3 and 5% carrageenan, CMC, agar, gelatin	Carrageenan , CMC> agar, gelatin	Ruscoe et al., 2005
Variation in weight after soaking in water	24 hours	10% Gelatin 30% alginate	Alginate>gelatin	Valverde et al., 2008
Variation in weight after soaking in water	24 hours	2% alginate plus 1% sodium hexametaphosphate or D-gluconic acis as sequestrants 9 and 13% agar 6 and 15% gelatin 20% agar/gelatin mixture	Agar/gelatin > agar> alginate>gelatin	Knauer et al., 1993
Long Angle Laser Light Scattering Technique	24 hours	2.5% Agar, alginate, pectin	Pectin>agar, alginate	Volpe et al., 2008
Long Angle Laser Light Scattering Technique	24 hours	1% chitosan, pectin, agar, alginate	Chitosan, pectin> agar, alginate	Coccia et al., 2010
Long Angle Laser Light Scattering Technique	96h 144h	3% agar, 6% agar	No differences agar 6% > agar 3%	Fabbrocini et al., 2011

[1] The reported outcome refers to the maximum time interval tested . > means that the binder(s) on the left perform better that the binder(s) on the right. * a commercial binder.

Table 2. Effect of binders on feed water stability.

5. Assessment of nutrient leaching

Binder type may affect feed stability in water, and by leaching of attractant molecules may determine attractiveness of microbound diets. Partridge and Southgate (1999) report that leaching increased with decreasing binder concentration. O'Mahoney et al. (2011) in a study aimed at investigating the characteristics of new substances to be employed as binders for abalone farming, found no dry matter leaching after up to 4 days of water immersion of feed containing glucomannan-xanthan gum. In spite of the importance of assessing nutrient leaching there is still no standard method to determine it. Some authors report nutrient leaching and dry matter loss after water immersion as synonyms. Other authors use instead different procedures to determine nutrient leaching such as radiolabelled substances included into the feed, internal markers, detection of specific nutrients, protein or lipid. Genodepa et al. (2007) included ^{14}C-labelled rotifers in the feed and tested the effect of zein, agar, alginate and carrageenan on nutrient leaching. They report that leaching occurred primarily within the first 30 minutes of immersion and longer immersion periods (up to 240 min) only resulted in a very modest increase in leaching rates. The leaching rate of zein-bound diets was significantly lower than that of agar-, alginate- and carrageenan-bound diets. On the contrary, Fagbenro and Jauncey (1995) report that the leaching of total protein content was very low, after 10 min immersion in water of pellets made with 3% CMC, corn starch, guar gum or wheat gluten. Accordingly, Simon (2009) found no significant difference in chromic oxide (as internal standard) and nutrient concentrations after 1 hour immersion in water of pellets containing gelatine (8%), alginate (7%), or agar (8%). Kovalenko et al. (2002) evaluated dietary water soluble vitamins riboflavin and thiamine levels before and after water immersion of feed containing 5.38% of alginate and report a substantially high loss of both vitamins after 90 min of immersion, with the greatest percent loss within 1 to 15 min interval. Falayi et al. (2006) evaluated total protein and lipid content in pelleted feed samples containing 10% wheat grain starch and cassava tuber starch as both binding agents and carbohydrate source. They found that after 1 hour of water immersion protein retention dropped to about 80% while lipid retention was about 97-95%.

6. Binder effect on growth

The relationship between the availability and composition of food and the growth rate of a given aquatic species is of a crucial importance for the optimization of the rearing conditions. In order to understand this relationship the main physiological processes of the organism (ingestion, assimilation, respiration, growth and reproduction) should be taken into account and evaluated by means of integrated models (Van der Meer, 2006). Theories of dynamic energy budgets (DEB) received great impetus in the last decades, against the previous prevalent static approach. Among the various models proposed, the Kooijman κ-rule DEB theory (Kooijman, 2001) is one of the most encompassing. As reported in Figure 9, this model assumes that a part of ingested food is assimilated, then it enters the reserves and therefore a part is spent for maintenance and growth (prioritary flux), while the rest is utilized for maturation and reproduction.

Being all DEB models relatively simple, they did not take into account species-specific physiological aspects of the single considered species, which therefore need to be considered each time. On the basis of these considerations and given that binders often account for a consistent part of a prepared pellet, the effect of these substances on the ingestion, digestion

Ingested food ⟶ Faeces
Assimilated food

Reserves

Somatic maintenance
Growth

Maturation
Reproduction

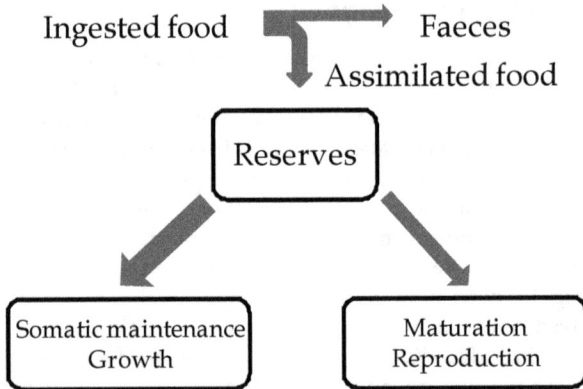

Fig. 9. Schematic representation (modified from van der Meer, 2006) of the Koijman Dynamic Energy Budget model.

and assimilation processes of each reared species must be carefully taken into account in order to maximize the experimental feed efficiency.

6.1 Fish

A limiting factor in larvicolture of many fish and crustaceans is the availability and nutritional value of live food. Larvae of many species have been raised using *Artemia salina* naupli and rotifers. However, their nutritional quality may change according to the source and the time of harvest of cysts. Formulated feed is by far more attractive due to the possibility to adjust its composition and size to the nutritional needs of the species. Most commercial larval diets are microbound diets, where binders are used to stabilize feed pellets and ensure minimum leaching and disintegration (Hashim and Saat, 1992). Carrageenan containing pellets are suitable for the ayu, *Plecoglossus altivellis* (Teshima et al., 1982), but not for postlarval tongue sole (*Cynoglossus semilaevis*) (Liu et al., 2008). Both alginate and zein appear to be suitable for farming larvae of sea bass (*Dicentrarchus labrax*) (Person Le Ruyet et al., 1993) but not barramundi (*Lates calcalifer*) (Partridge & Southgate, 1999), although Lee et al. (1996) previously reported that *Lates calcifer* larvae were able to digest both alginate and gelatine bound diets. Sodiumcarboxymethilcellulose, sodium alginate and gelatin included in feed for postlarval tongue sole (*Cynoglossus semilaevis*) brought about a growth similar to controls (Liu et al., 2008). Recently, a microbound diet containing soy lecithin, wheat gluten, and alginate has been patented for the culture of larval fish and crustaceans (D'Abramo, 2003).

6.2 Crustaceans

Dietary binders have been tested for growth performance in several species, with often conflicting results. Wheat flour, whole cassava meal and dry molasses were effective binders and caused the highest weight gain in crayfish *Macrobrachium rosenbergii* post-larvae, in comparison to agar (Seixas Filho et al., 1997a), although agar containing pellets had the best texture (Seixas Filho et al., 1997b). Unfortunately, no comparison with a control group fed natural diet was provided in this study. In contrast, Kovalenko et al. (2002) report that growth of larvae fed alginate containing feed, was only 90% of that achieved for larvae fed natural diet (newly hatched nauplii of *Artemia*). Both lignosol and agar brought about a significant increase in weight in juvenile shrimp *Palaemonetes varians* and *Palaemon elegans* (Palma et al., 2008). Alginate, agar and pectin caused good growth performance in both *Cherax albidus* juveniles and adults with respect to control animals fed natural diet, with pectin giving the best results in both juveniles and adults (Volpe et al., 2008; Coccia et al., 2010).

6.3 Molluscs

Alginate and gelatin have been tested on cephalopod growth. The presence of alginate as food binder in *Octopus maya* negatively affected growth rate and survival even when added at a very low percentage. Indeed, *Octopus maya* fed alginate-bound crab showed a survival rate of only 15% with respect to the control group fed with crabs (Rosas et al., 2008). Both *Octopus maya* and *Octopus vulgaris* fed diets containing alginate and gelatin as binders grew significantly less than the controls fed natural diets (Rosas et al., 2008; Valverde et al., 2008; Garcia et al., 2011).

Konjac glucomannan and xanthan gum were used together in different configurations (1:1; 1:3) in feed for abalone (*Haliotis discus hannai*). Although no differences in growth have been recorded between treatments, growth parameters were significantly higher in animals fed a natural diet (O'Mahoney et al., 2011).

6.4 Sea urchins

The rapid expansion of the culture of sea urchins has increased the demand for formulated feed, specifically developed taking into account the echinoids needs. It has been widely demonstrated as prepared diets increase both test growth and gonad yield with respect to natural food (Pearce et al., 2004; Otero-Villanueva et al., 2004; Schlosser et al., 2005; Azad et al., 2011) and also feed initially formulated for other species than echinoids proved to successfully support urchins' growth and gonad maturation (Fabbrocini & D'Adamo, 2010). On the other hand, these diets frequently lead to poor gonad quality in terms of texture, firmness, colour and taste (Pearce et al., 2002b; Schlosser et al., 2005), that means low marketability of the product. For these reasons research focused on the formulation of suitable feed for sea urchin must take into account the effects on both gonad yield and sensory quality of both the components of the experimental diets and the binders used to give consistence to the pellets. Numerous binders have been successfully employed in the formulation of feed for different sea urchin species. In many echinoid species gelatin based pellets lead to a greater urchin growth (Fernandez & Pergent, 1998; Daggett et al., 2005) and a higher gonad yield (Pearce et al., 2004; Barker et al., 2006; Phillips et al., 2009) with respect to urchin fed on fresh macroalgae; also a pellet based on gelatin strengthened with the enzyme transglutaminase (γ-glutamyltransferase) gave better results in terms of gonad index with respect to *Laminaria hyperboreus* diet in *Strongylocentrotus droebachiensis* (Mortensen et al., 2004). Similarly, a feed containing a mixed binder gelatin-sodium alginate supported juveniles growth of *Strongylocentrotus droebachiensis* better than fresh *Laminaria* (Kennedy et al., 2007). Also agar (Barker et al., 1998), sodium alginate (Akiyama et al., 2001; Pearce et al., 2002b), guar gum and corn starch (Pearce et al., 2002b) based pellets gave better results in terms of gonad indices with respect to a natural diet. Despite the fact that much of the research on sea urchin rearing in confined conditions has focused on the formulation of prepared feed, few published studies have examined the effects of the binder nature and concentration on urchins' gonad growth and sensory quality.

Akiyama et al (1997) tested the effectiveness as a binder of 30% sodium alginate and of 15% Curdlan™, a polysaccharide produced by the bacteria *Alcoligenes faecalis* var. *myxogenes* and commonly used in human foods. They found that gonad growth in *Pseudocentrotus depressus* was higher with alginate based feed, with respect to Curdlan™ based pellet. Pearce et al. (2002b) tested different experimental feed which differed only for the binder type (gelatin, guar gum, sodium alginate, and corn starch) and concentration (3% and 5%), in order to evaluate the binder effect on gonad growth of *Strongylocentrotus droebachiensis*. Even if gonad growth was faster with gelatin and alginate based pellets, at the end of the trial (12 weeks) gonad indices (GI) were similar for the various binders at the higher concentration (5%), while it was slightly lower with 3% guar gum and corn starch. Fabbrocini et al. (2011) tested the effectiveness of agar based pellets for gonad growth and gamete production in *Paracentrotus lividus*. Pellets made of commercial feed Classic© (hendrix) and 3% agar gave the same results of the Classic© alone, therefore agar did not hamper gonad growth; in addition, urchins fed on *Ulva*-agar pellets in a 4 weeks rearing trial progressed in the reproductive cycle as those fed on commercial feed, showing comparable gonad indices levels. Regarding gonad sensory quality, a prepared diet with a high content of gelatin gave good results in *Evechinus chloroticus* (Phillips et al., 2009). On the contrary, in *Strongylocentrotus droebachiensis* a gelatin-transglutaminase based diet gave poor results in terms of gonad taste (Mortensen et al., 2004), while the best results in terms of gonad colour were obtained with a starch-based pellet by Pearce et al. (2002b). All these considered, when choosing the binder it is crucial to take into account the urchin species to be reared and the objective to be reached: urchin growth, gonad yield and gamete production, or gonad sensory quality.

In Table 3 the effect of feed binders on growth are reported.

7. Binder effect on feed digestibility

Several reports indicate a relationship between binders and nutrient digestibility. Different techniques are used for evaluating feed digestibility (Khan et al., 2003) among which the total collection technique, although time-consuming and stressful for the animal, is the most reliable method. The feed is fed in known quantities to the animal and accurate records of feed intake and fecal output are kept. Thus, it is foreseeable that digestibility experiments in aquatic animals are prone to errors related to the difficulty of measuring the correct amount of feed ingested and the methods employed to collect feces (i.e., settlement, filtration) which result in leaching losses (Irvin & Tabrett, 2005). It has been shown that the settlement collection method results in an overestimation of protein digestibility by up to 6% when feces are collected once every 6 h rather than hourly post-feeding in shrimp (Smith & Tabrett, 2004). Alternatively, *in vitro* techniques may be used to provide a quick and low cost method to predict nutrient digestibility. The digestibility of several carbohydrate sources for juveniles of the spiny lobster *Jasus edwardsii* has been estimated by measuring their rates of hydrolysis *in vitro* using enzyme homogenates and postprandial haemolymph glucose concentrations following ingestion of semi-purified diets (Simon, 2009). Many different factors influence the efficiency of digestion. The impact of binder type on feed digestibility

BINDER	% Binder employed	SPECIES	GROWTH	Reference
CARBOHYDRATES				
Agar	2	*Palemonetes varians* (juveniles)	Positive	Palma et al., 2008
	2	*Palemon elegans* (juveniles)	Positive	
	2.5	*Cherax albidus* (adults)	Positive	Volpe et al., 2008
	2	*Cherax albidus* (juveniles)	Positive	Coccia et al., 2010
	3	*Paracentrotus lividus* (adults)	Same as control	Fabbrocini et al., 2011
Alginate	2.5	*Cherax albidus* (adults)	Positive	Volpe et al., 2008
	5.38	*Macrobrachium rosembergi* (larvae)	Negative	Kovalenko et al., 2002
	10	*Octopus vulgaris*	Negative	Garcia et al., 2011
	1	*Octopus Maya*	Negative	Rosas et al., 2008
	30	*Octopus vulgaris*	Negative	Valverde et al., 2008
	2	*Cynoglossus semilaevis* (postlarvae)	Same as control	Liu et al., 2008
	2	*Cherax albidus* (juveniles	Positive	Coccia et al., 2010
	30	*Pseudocentrotus depressus* (juveniles)	Negative vs control Positive vs curdlan®	Akiyama et al., 1997
	30	*Pseudocentrotus depressus* (juveniles)	Positive	Akiyama et al., 2001
	3; 5	*Strongylocentrotus droebachiensis* (adults)	Positive	Pearce et al., 2002b
Carboxymethyl cellulose	2	*Cynoglossus semilaevis*	Same as control	Liu et al., 2008
Carrageenan	2	*Cynoglossus semilaevis*	Negative	
Chitosan	1	Cherax albidus (juveniles)	Positive	Coccia et al., 2010
Konjac glucomannan-xanthan gum	0.1; 7,5	*Haliotis discus hannai*	Negative	O'Mahoney et al., 2011

BINDER	% Binder employed	SPECIES	GROWTH	Reference
Guar gum	3; 5	*Strongylocentrotus droebachiensis* (adults)	Positive	Pearce et al., 2002b
Lignosol	2	*Palemonetes varians* (juveniles)	Positive	Palma et al., 2008
	2	*Palemon elegans* (juveniles)	Positive	
Corn starch	3; 5	*Strongylocentrotus droebachiensis* (adults)	Positive	Pearce et al., 2002b
Pectin	2.5	Cherax albidus (adults)	Positive	Volpe et al., 2008
	1	Cherax albidus (juveniles)	Positive	Coccia et al., 2010
PROTEINS				
gelatin	10	*Octopus vulgaris*	Negative	Garcia et al., 2011
	2	*Octopus Maya*	Same as control	Rosas et al., 2008
	10	*Octopus vulgaris*	Negative	Valverde et al., 2008
	2	*Cynoglossus semilaevis* (postlarvae)	Same as control	Liu et al., 2008
	3; 5	*Strongylocentrotus droebachiensis* (adults)	Positive	Pearce et al., 2002b
	5	*Strongylocentrotus droebachiensis* (juveniles)	Positive	Pearce et al., 2002b
	5	*Strongylocentrotus droebachiensis* (juveniles)	Positive	Daggett et al., 2005
MIXED BINDERS				
Gelatin+ transglutaminase	? 0.13%	*Strongylocentrotus droebachiensis* (juveniles)	Positive	Mortensen et al., 2004
Gelatin+ alginate	5 2	*Strongylocentrotus droebachiensis* (juveniles)	Positive	Kennedy et al., 2007

Table 3. Effect of feed binders on growth

is another of a series of factors that must be taken into consideration when designing a formulated feed type that is intended to be successful. High concentrations of binder may in fact cause a reduction in diet digestibility. Measuring the apparent digestibility of binders is necessary in order to better understand their bioavailability when incorporated in formulated diets and their potential influence on the digestibility of other major macro-ingredients such as protein. Unfortunately, the effect of binders on diet digestibility has been largely overlooked. Simon (2009) found that gelatin, but not agar or alginate (7-8% inclusion level), positively affected digestibility in the juveniles of the lobster *Jasus edwardsii*. Similarly, *Octopus maya* fed alginate containing pellets lost weight and died, suggesting that this binder negatively affects nutrient absorption, while gelatin promoted absorption of nutrients (Rosas et al., 2008). It seems that binders have detrimental effects on nutrient digestibility in fish, as they accelerate gastrointestinal transit time (Storebakken, 1985). The inclusion of guar gum to tilapia feed has nutrient digestibility coefficients lower than diets containing CMC, corn starch and wheat gluten (Fagbenro & Jauncey, 1995). This is in line with previous results showing that fish feed containing highly effective binders had negative effects on macronutrient digestibility probably due to physical effects, such as changes in viscoelastic properties (Storebakken & Austreng, 1987; Storebakken et al., 1998). Recently, the effect of two natural binders (fava and kidney beans) on protein and lipid digestibility in the Atlantic salmon *Salmo salar* showed no significant differences between the dietary treatments (Pratoomyot et al., 2011) and the addition of guar gum to the diet of rainbow trout positively affected digestibility when employed with plant meal (Brinker & Reiter, 2011). The pattern of food residence in the gut was found to vary little with diet or urchin size (38-42h) in *Psammechinus miliaris* (Otero-Villanueva et al., 2004). Lawrence et al. (1989) found that in *Paracentrotus lividus* gut passage-time and apparent dry matter digestibility were similar with agar-based feed containing either soybean or fish meals. Similar results were reported by Klinger et al. (1994) for *Litechinus variegatus*. No differences in absorption efficiency were recorded for *Evechinus chloroticus* fed on extruded or agar-based pellets (Barker et al., 1998). In *Pseudocentrotus depressus* feed efficiency was higher for alginate- based pellets than for kelp (Akiyama et al., 2001), while regarding the effect of different binders, alginate-based pellets showed a lower feed efficiency than Curdlan™-based ones (Akiyama et al., 1997).

8. Binder effect on digestive enzymes

It is commonly agreed that species exhibit a particular suite of digestive enzymes that reflect their different life history (Figueiredo & Anderson, 2009). Consequently, from the digestive enzyme profile it is possible to predict the ability of the species to use different nutrients. Since the digestive enzyme activity is high for those substrates that are more common in the diet (Moss et al., 2001) there is a common belief that the activity of digestive enzymes can be boosted by providing certain nutrients into the diet for an adequate amount of time.

It is well known that crustaceans adapt their digestive enzymatic profile and activity to the diet composition (Muhlia-Almazan et al., 2003; Johnston & Freeman, 2005; Pavasovic et al., 2007a), as a consequence of an enzymatic battery capable of hydrolyzing the wide variety of substrates that they encounter in their natural diet (Linton et al., 2009). Modifications in digestive enzyme activity in different penaeid shrimp larvae have been related to the quantity or quality of some component of the diet (Le-Vay et al., 1993; Rodriguez et al., 1994; Le Moullac et al., 1997; Lemos & Rodriguez, 1998; Brito et al., 2000). Protein composition of the diet increased protease activity in *Marsopenaeus japonicas* juveniles and *Litopenaeus vannamei* larvae (Van Wormhoudt et al., 1986; Le Moullac & Van Wormhoudt, 1994; Le Moullac et al., 1997). Some authors suggest that the increase in amylase activity found in

Penaeus monodon (Fang & Lee, 1992), and *Litopenaeus vannamei* (Brito et al., 2000), may be related to the low level of carbohydrate in the diet (Le Vay et al., 1993; Rodriguez et al., 1994; Kumlu & Jones, 1995; Lemos & Rodriguez, 1998). Protease, amylase and cellulase activities have been found to be positively correlated with dietary protein levels (Pavasovic et al., 2007b), and negatively correlated with high cellulose levels (Pavasovic et al., 2006). In contrast, in the white shrimp *Penaeus vannamei* dietary protein level did not influence total protease activity, although modulation of trypsin- and chymotrypsin-like activities has been documented (Muhlia-Almazan et al., 2003). Moreover, no significant difference was observed in protease activity of *Litopenaeus vannamei* (Rivas-Vega et al., 2006) and *Scylla serrata* subjected to different dietary treatments (Pavasovic et al., 2004). In *Cherax quadricarinatus* the digestive enzyme secretion is modified by the diet composition, such as the presence of animal- or plant-derived ingredients (Lopez-Lopez et al., 2005; Pavasovic et al., 2007a). In the mollusk *Babylonia areolate* a positive correlation between dietary lipid level and lipase activity has been reported (Zhou et al., 2007). Among teleosts, proteolytic activity has been found unresponsive to the protein content of the diet in *Pseudoplatystoma corruscans* (Lundstedt et al., 2004) and in early weaned sea bass (*Dicentrarchus labrax*) larvae (Cahu & Zambonino-Infante, 1994), but not in *Colossoma macropomum*, where digestive protease increased along with dietary protein content (De Almeida et al., 2006). On the other hand, alkaline protease activity was higher in *Dentex dentex* fed a diet with less protein and more carbohydrates (Perez-Jimenez et al., 2009). Amylase activity was higher in *Pseudoplatystoma corruscans* fed diets containing 13-25% of starch and similarly, it could be adjusted through the starch level of the diet in *Colossoma macropomum* (De Almeida et al., 2006) and *Dentex dentex* (Perez-Jimenez et al., 2009). A positive correlation has been observed between lipase and dietary lipid levels in *Colossoma macropomum* (De Almeida et al., 2006) and *Lateolabrax japonicus* (Luo et al., 2010). In any case the emerging picture in teleosts is one of wide complexity and variability due to the different of habitats and life histories of such vertebrates.

Virtually no studies have been carried out with the main purpose of investigating the effect of feed binders on digestive enzyme activity. The effect of different binders on digestive enzymes profile has been studied by us in *Cherax albidus* (Volpe et al., submitted). In an attempt to understand if the presence of binders influenced the digestive enzymatic profile we undertook a screening of carbohydratases, lipases and proteases along the digestive tract of juvenile *Cherax albidus* under both natural diet (Coccia et al., 2011) and different experimental diets based on pellets containing different binders but the same proximate composition. We found that digestive enzyme activities did not statistically show significant differences in the digestive tract except for amylase that was significantly higher in the intestine compared to the gastric juice and hepatopancreas of animals fed pectin containing pellets for 12 weeks.

9. Conclusion

Good water stable pellets are a major concern for the aquaculture industry. Binder agents have been widely employed to improve the stability of aquatic animal feed. Binders are believed to prevent or at least reduce major nutrient loss and feed softness which is inevitable when pellets stay in water for long periods of time. Preventing or reducing such an inconvenient would have beneficial consequences on the aquaculture industry. Binder prices may significantly affect the cost of the diet, therefore the utilization of natural substances, biodegradable and renewable may be advantageous from an environmental and economical point of view. Although there is not such a thing as the best binding agent, the research in this field is actively ongoing, since different species and even the same species during different phases of the life cycle have different feeding behaviours and nutritional needs. However, despite progress in recent years, knowledge of microbound diets and their impact on digestive physiology and growth performances of reared aquatic species is still not sufficient in light of the industry's needs and thus further research is required.

10. References

Ahmed, J.; Ramaswamy, H.S. ; Ngadi, M.O.(2005). Rheological characteristics of Arabic gum in combination with guar and xanthan gum using response surface methodology: Effect of temperature and concentration. *International Journal of Food Properties*, Vol. 8, No. 2, (April 2005), pp. 179-192, ISSN 1094-2912

Akiyama, T.; Unuma, T.; Yamamoto, T.; Furuita, H. & Konishi, K. (1997). An evaluation of amino acid sources and binders in semipurified diet for red sea urchin *Pseudocentrotus depressus*. *Fisheries Science* , Vol. 63, No. 6, pp. 881-886, (December 1997), ISSN 1444-2906

Akiyama, T.; Unuma T. & Yamamoto, T. (2001). Optimum protein level in a purified diet for young red sea urchin *Pseudocentrotus depressus*. *Fisheries Science*, Vol. 67, No. 2, (April 2001), pp. 361-363, ISSN 1444-2906

Aspinall, G.O. (1982) Chemical characterization and structure determination of polysaccharides. In: *The Polysaccharides*, Aspinall G.O. (ed.), Vol. 1, pp. 35-131, Academic Press, Inc., New York

Arisz, P.W.; Kauw, H.J.J. & Boon, J.J. (1995) Substituent distribution along cellulose backbone in O-methylcelluloses using GC and FAB-MS for monomer and oligomer analysis. *Carbohydrate Research*, Vol. 271, No. 1, (July 1995), pp. 1-14, ISSN 0008-6215

Atkins, E. D. T. (Ed.). (1985). *Polysaccharides: Topics in structure and morphology*. Wiley-VCH Verlag GmbH, ISBN 9783527264100, Germany

Azad, A.K.; Pearce, C.M. & McKinley R.S. (2011). Effects of diet and temperature on ingestion, absorption, assimilation, gonad yield, and gonad quality of the purple sea urchin (*Strongylocentrotus purpuratus*). *Aquaculture*, Vol. 317, No. 1-4, (July 2011), ISSN 0044-8486

Barker, M.F.; Keogh, J.A.; Lawrence, J.M. & Lawrence, A.L. (1998). Feeding rate, absorption efficiencies, growth, and enhancement of gonad production in the New Zealand sea urchin *Evechinus chloroticus* Valenciennes (Echinoidea: Echinometridae) fed prepared and natural diets. *Journal of Shellfish Research*, Vol. 17, No. 5, (December 1998), pp. 1583-1590, ISSN 0730-8000

Barker, M.F.; Bremer, P.; Silcock, P.; Delahunty, C. & Sewell, M. (2006). Enhancement of yield and quality of *Evechinus chloroticus* roe through controlled diet. *Journal of Shellfish Research*, Vol. 25, No. 2, (August 2006), p. 711, ISSN 0730-8000

Bautista M.N.; Millamena, O.M. & Kanazawa, A. (1989). Use of kappa-carrageenan microbound diet (C-MBD) as feed for *Penaeus monodon* larvae. *Marine Biology*, Vol. 103, pp. 169-173, ISSN 0025-3162

Brinker, A. (2007). Guar gum in rainbow trout (*Oncorhynchus mykiss*) feed: the influence of quality and dose on stabilization of faecal solids. *Aquaculture*, Vol. 267, No.1-4, (July 2007), pp. 315-327, ISSN 0044-8486

Brinker, A. & Reiter, R. (2011). Fish meal replacement by plant protein substitution and guar gum addition in trout feed, Part I: Effects on feed utilization and fish quality. *Aquaculture*, Vol. 310, No. 3-4, (January 2011), pp.350-360, ISSN 0044-8486

Brito R.; Chimal, M.E.; Gaxiola, G. & Rosas, C. (2000). Growth, metabolic rate, and digestive enzyme activity in the white shrimp *Litopenaeus setiferus* early postlarvae fed different diets. *Journal of Experimental Marine Biology and Ecology*, Vol. 255, No. 1, (December 2000), pp. 21-36. ISSN 0022-0981

Cahu, C.L. & Zambonino-Infante, J.L. (1994). Early weaning of sea bass (*Dicentrarchus labrax*) larvae with a compound diet: effect on digestive enzymes. *Comparative Biochemistry and Physiology part A*, Vol. 109, No. 2 (October 1994), pp. 213–222, ISSN 1095-6433

Caltagirone, A.; Francour, P. & Fernandez C. (1992). Formulation of an artificial diet for the rearing of the urchin *Paracentrotus lividus*: I. Comparison of different binding agents. In: *Echinoderm Research* L. Scalera-Liaci, C. Canicattì (Eds), pp. 115-119, ISBN 90-5410-049-4, Balkema, Rotterdam

Castillo-García, E.; López-Carballo, G. & López-Castellano, A. (2005). Guar gum characteristics and applications. *Ciencia y Tecnologia Pharmaceutica*, Vol. 15, No. 1, (March 2005), pp. 3-10, ISSN 1575-3409

Cerver, M.F.; Heinämäki, J.; Räsänen, M.; Maunu, S.L.; Karjalainen, M.; Acosta, O.M., Colart, A. I. & Yliruusi, J. (2004). Solid-state characterization of chitosan derived from lobster chitin. *Carbohydrate Polymers*, Vol. 58, No. 4-7, (December 2004), pp. 401-408, ISSN 0144-8617

Chevanan, N.; Muthukumarappan, K. & Rosentrater, K. A. (2009). Extrusion Studies of Aquaculture Feed using Distillers Dried Grains with Solubles and Whey. *Food and Bioprocess Technology*, Vol. 2, No. 2, (March 2009), pp. 177–185, ISSN 1935-5130

Clark, A. H. & Ross-Murphy, S. B. (1987). Structural and mechanical properties of biopolymers gels. In: *Advances in Polymer Science*, Vol. 83 (October 1987), pp. 57–192, ISSN 0065-3195

Coccia, E.; Santagata, G.; Malinconico, M.; Volpe, M.G.; Di Stasio, M. & Paolucci, M. (2010) *Cherax albidus* juveniles fed polysaccharide-based pellets: rheological behavior and effect on growth. *Freshwater crayfish*, Vol. 17, pp.13-18, ISBN 978-951-27-1322-6

Coccia, E.; Varricchio, E. & Paolucci, M. (2011). Digestive enzymes in the crayfish *Cherax albidus*: polymorphism and partial characterization. *International Journal of Zoology*, Vol. 2011, (March 2011), pp.1-9, ISSN 1687-8477

Coviello, T.; Alhaique, F.; Dorigo, A.; Matricardi, P. & Grassi, M. (2007). Two galactomannans and scleroglucan as matrices for drug delivery: preparation and release studies. *European Journal of Pharmaceutics and Biopharmaceutics*, Vol. 66, No. 2, (May 2007), pp. 200-209, ISSN 0939-6411

Cuzon, G.; Guillaume, J. & Cahu, C. (1994). Composition, preparation and utilization of feeds for Crustacea. *Aquaculture*, Vol. 124, No.1-4, (June 1994), pp. 253–267, ISSN 0044-8486

D'Abramo, L.R. (2003). Microparticulate microbound diet for the culture of larval fish and crustaceans. US 6645536 (2003)

Daggett, T.L.; Pearce, C.M.; Tingley, M.; Robinson, S.M.C. & Chopin, T. (2005). Effect of prepared and macroalgal diets and seed stock source on somatic growth of juvenile green sea urchins (*Strongylocentrotus droebachiensis*). *Aquaculture*, Vol. 244, No. 1-4, (February 2005), pp. 263-281, ISSN 0044-8486

De Almeida, L.C.; Lundstedt, L.M. & Moraes, G. (2006). Digestive enzyme responses of tambaqui (*Colossoma macropomum*) fed on different levels of protein and lipid. *Aquaculture Nutrition*, Vol. 12, No. 6, (December 2006), pp. 443-450, ISSN 1353-5773

De Muylder, E.; Hage, H. & van der Velden, G. (2008). Binders: Gelatin as alternative for urea formaldehyde and wheat gluten in the production of water stable shrimp feeds. *Aquafeed International magazine*, Volume 11, No. 2, (April 2008), ISSN 1464-0058

De Silva, S.S. & T.A. Anderson. (1995). *Fish Nutrition in Aquaculture*. Chapman & Hall, ISBN 0412 55030X, London

Dominy, W.G. & Lim, C. (1991). Performance of binder in pelleted shrimp diets. *Proceedings of the aquaculture feed processing and nutrition workshop*, pp. 149-157, Thailand and Indonesia, September 1991

Drochner, H.; Kerler, A. & Zazharias, B. (2004). Pectin in pig nutrition, a comparative review. *Journal of Animal Physiology and Animal Nutrition*, Vol. 88, No. 11-12, (April 2004), pp. 367-380, ISSN 0931-2439

Durazo-Beltran, E. & Viana, M.T. (2001). Effect of the concentration of agar, alginate and carrageenan on the stability, toughness and nutrient leaching in artificial diets for abalone. *Ciencias marinas*, vol. 27, No. 1, pp. 1-19, ISSN 0185-3880

El-hefian, E.A.; Nasef, M.M. & Yahaya, A.H. (2011). Chitosan physical forms: A short review. *Australian Journal of Basic and Applied Sciences* Vol.5, No. 5, (May 2011), pp. 670-677, ISSN 19918178677

Fabbrocini, A. & D'Adamo, R. (2010). Gamete maturation and gonad growth in fed and starved sea urchin *Paracentrotus lividus* (Lamarck, 1816). *Journal of Shellfish Research*, Vol. 29, No. 4, (December 2010), pp. 1051-1059, ISSN 0730-8000

Fabbrocini, A.; Volpe, M.G.; Di Stasio, M.; D'Adamo, R.; Maurizio, D.; Coccia, E. & Paolucci, M. (2011). Agar-based pellet as feed for sea urchins (*Paracentrotus lividus*): rheological behaviour, digestive enzymes and gonad growth. *Aquaculture Research*, (in press, DOI: 10.1111/j.1365-2109.2011.02831.x), ISSN 1365-2109

Fagbenro, O. & Jauncey, K. (1995). Water stability, nutrient leaching and nutritional properties of moist fermented fish silage diets. *Aquaculture Engineering*, Vol. 14, No. 4, pp. 143–153, ISSN 0144-8609

Falayi, B.A.; Sadiku, S.O.; Eyo, A.A. & Okaeme, A.N. (2006). Water stability and floatation of fish diets bound with different agents and *Saccharomyces cerivisiae*. *Nigerian Journal of Fisheries*, Vol 2, No. 3, pp. 181-196, ISSN 1597-443.

Fang, L.S. & Lee, B.N. (1992). Ontogenic change of digestive enzymes in *Penaeus monodon*. *Comparative Biochemistry and Physiology Part B: Comparative Biochemistry*, Vol. 103, No. 4, (December 1992), pp. 1033-1037, ISSN 1096-4959

Farris, S.; Schaich, K. M.; Liu, L.; Piergiovanni, L. & Yam, K. L. (2009). Development of polyion-complex hydrogels as an alternative approach for the production of bio-based polymers for food packaging applications: a review. *Trends in Food Science and Technology*, Vol. 20, No. 2, (February 2009), pp. 316-332, ISSN 0924-2244

Fernandez, C. & Pergent, G. (1998). Effect of different formulated diets and rearing conditions on growth parameters in the sea urchin *Paracentrotus lividus*. *Journal of Shellfish Research*, Vol. 17, No. 5, (December 1998), pp. 1571-1581, ISSN 0730-8000

Figueiredo, M.S.R.B., & Anderson, A.J. (2009). Digestive enzyme spectra in crustacean decapods (Paleomonidae, Portunidae and Penaeidae) feeding in the natural habitat. *Aquaculture Research*, Vol. 40, No. 3, (February 2009), pp. 282–291, ISSN 1365-2109

Forbes, J.M. (2001). Consequences of feeding for future feeding. *Comparative Biochemistry and Physiology Part A: Molecular & Integrative Physiology*, Vol. 128, No. 3, (March 2001), pp. 463-70, ISSN 1095-6433

Garcia, S.; Domingues, P.; Navarro, J.C.; Hachero, I.; Garrido, D. & Rosas, C. (2011). Growth, partial energy balance, mantle and digestive gland lipid composition of *Octopus vulgaris* (Cuvier, 1797) fed with two artificial diets. *Aquaculture Nutrition*, Vol. 17, No. 2 (April 2011), pp. e174-e 187, ISSN 1353-5773

Genodepa, J.; Zeng, C. & Southgate, P. C. (2004). Preliminary assessment of a microbound diet as an *Artemia* replacement for mud crab, *Scylla serrata*, megalopa. *Aquaculture*, Vol. 236, No. 1-4 (June 2004), pp. 497-509, ISSN 0044-8486

Genodepa, J.; Zeng, C. & Southgate, P. C. (2007). Influence of binder type on leaching rate and ingestion of microbound diets by mud crab, *Scylla serrata* (Forsskal), larvae. *Aquaculture Research*, Vol. 38, No. 14, (October 2007), pp. 1486-1494, ISSN 1365-2109

Goldstein, A.M.; Alter, E.N. & Seaman, J.K. (1973). Guar gum, *In: Industrial Gums, Polysaccharides and their Derivatives* Whistler R.L. Ed., pp. 303–321, Academic Press, New York

Grant, N. H.; Clark, D. E. & Rosanoff, E. I. (1973). *Biochemical and Biophysical Research Communications*, Vol. 51, pp. 100-106, ISSN 0006-291X

Grasdalen, H.; Larsen, B. & Snidsrod, O. (1981). 13C-N.M.R. studies of monomeric composition and sequence in alginate *Carbohydrate Research*, Vol. 89, No. 2, (January 1981), pp. 179-191, ISSN 0008-6215

Hashim, R. & Saat, M.A.M. (1992). The utilization of seaweed meals as binding agents in pelleted feeds for snakehead (*Channa striatus*) fry and their effects on growth. *Aquaculture*, Vol. 108, No. 3-4, (December 1992), pp. 299-308, ISSN 0044-8486

Holdich, D.M. (2002). *Biology of freshwater crayfish*. Blackwell Science, ISBN 978-0-632-05431-2, United Kingdom.

Irvin, S.J. & Tabrett, S.J. (2005). A novel method of collecting fecal samples from spiny lobsters. *Aquaculture*, Vol. 243, No. 1-4, (January 2005), pp. 269-272, ISSN 0044-8486

Johnson, E.L. & Peniston, Q.P. (1982). Utilization of Shellfish Waste for Chitin, Chitosan Production, In: *Chemistry and Bio-chemistry of Marine Food Products*, Martin, R.E.; Flick, G.J.; Hebard C.E. & Ward D.R. eds, pp. 415-428, AVI Publishing Co, ISBN 0870554085, 9780870554087, Westport, CT, USA

Johnston, D.J. & Freeman, J., (2005). Dietary preference and digestive enzyme activities as indicators of trophic resource utilization by six species of crab. *The Biological Bulletin*, Vol. 208, No. 1, (February 2005), pp. 36–46, ISSN 0006-3185

Johnston, M. D. & Johnston, D.L. (2007). Stability of Formulated Diets and Feeding Response of Stage I Western Spiny Lobster, *Panulirus cygnus*, Phyllosomata. *Journal of the World Aquaculture Society*, Vol. 38, No.2, (June 2007), pp. 262–271, ISSN 1749-7345

Kalia, A.N. (2005). Textbook of Industrial Pharmacognosy. New Delhi CBS Publishers and Distributors, ISBN 8123912099, India

Kalian, N. & Morey, R.V. (2009) Factors affecting strength and durability of densified biomass products. *Biomass and Bioenergy*, Vol. 33, No. 3, (March 2009), pp. 337-359, ISSN 0961-9534

Kennedy, J.F. & White, C.A. (1983). Bioactive Carbohydrates, In: *Chemistry, Biochemistry and Biology*, Kennedy, J.F. & White, C.A. Eds., pp. 163, Ellis Horwood Publishers, Chichester, United Kingdom

Kennedy, E.J.; Robinson, S.M.C.; Parsons, G.J. &. Castell, J.D. (2007). Effects of dietary minerals and pigment on somatic growth of juvenile green sea urchins, *Strongylocentrotus droebachiensis. Journal of the World Aquaculture Society*, Vol. 38, No. 1, (March 2007), pp. 36-48, ISSN 1749-7345

Khan, M.A.; Nisa, M.U. & Sarwar, M., (2003). Techniques Measuring Digestibility for the Nutritional Evaluation of Feeds. A Review in *International Journal of Agriculture and Biology*, Vol. 5, No. 1, pp. 91-94, ISSN 1560-8530.

Klinger, T.S.; Lawrence, J.M. & Lawrence A.L. (1994). Digestive characteristics of the sea-urchin *Lytechinus variegatus* (Lamarck) (Echinodermata: Echinoidea) fed prepared feeds. *Journal of the World Aquaculture Society*, Vol. 25, No. 4, (December 1994), pp. 489-496, ISSN 1749-7345

Knauer, J.; Britz, P.J. & Hecht, T. (1993). The effect of seven binding agents on 24-hour water stability of an artificial weaning diet for the South African abalone, *Haliotis midae* (Haliotidae, Gastropoda). *Aquaculture*, Vol. 115, No. 3-4, (September 1993), pp. 327-334, ISSN 0044-8486

Koojiman, S.A.L. (2001) Quantitative aspects of metabolic organization: A discussion of concepts. *Philosophical Transactions of the Royal Society B: Biological Sciences*, Vol. 356, No. 1407, (March 2001), pp. 331-349, ISSN 0962-8436

Kovalenko, E. E.; D'Abramo, L. R.; Ohs, C. L. & Buddington, R. K. (2002). A successful microbound diet for the larval culture of freshwater prawn *Macrobrachium rosenbergii. Aquaculture*, Vol. 210, No. 1-4 (July 2002), pp. 385-395. ISSN 0044-8486

Kumlu, M. & Jones, D.A. (1995). The effect of live and artificial diets on growth, survival, and trypsin activity in larvae of Penaeus indicus. *Journal of World Aquaculture Society*, Vol.26, No. 4, (December 1995), pp 406-415, ISSN 0893-8849

Lawrence, J.M.; Regis, M.B.; Delmas, P.; Gras, G. & Klinger T.S. (1989). The effect of quantity and quality of food on feeding and digestion in *Paracentrotus lividus* (Lamarck) (Echinodermata: Echinoidea). *Marine Behaviour and Physiology*, Vol. 15, No. 2, (August 1989), pp. 137-144, ISSN 0091-181X

Lee, P.S.; Southgate, P.C. & Fielder, D.S. (1996). Assessment of two microbound artificial diets for weaning Asian sea bass (*Lates calcarifer*, Bloch). *Asian Fisheries Science*, Vol. 9, No.1, pp. 115-120

Lemos, D. & Rodriguez, A. (1998). Nutritional effects on body composition, energy content and trypsin activity of *Penaeus japonicus* during early postlarval development. *Aquaculture*, Vol. 160, No. 1-5 (January 1998), pp. 103-116, ISSN 0044-8486

Le Moullac, G.; Klein, G.B.; Sellos, D. & Van Wormhoudt, A. (1997). Adaptation of trypsin, chymotrypsin anda-amylase to casein level and protein source in *Penaeus vannamei* (Crustacea, Decapoda). *Journal of Experimental Marine Biology and Ecology*, Vol. 208, No. 1-2 ,(January 1997), pp. 107-125, ISSN 0022-0981

Le Moullac, G. & Van Wormhoudt, A. (1994). Adaptation of digestive enzymes to dietary protein,carbohydrate and fibre levels and influence of protein and carbohydrate quality in *Penaeus vannamei* larvae(Crustacea, Decapoda), *Aquatic Living Resources*, Vol. 7, No. 3 (July 1994), pp. 203-210, ISSN 0990-7440

Le-Vay, L.; Rodriguez, A.; Kamarudin, M.S. & Jones, D.A. (1993). Influence of live and artificial diets on tissue composition and trypsin activity in *Penaeus japonicus* larvae. *Aquaculture*, Vol. 118, No. 3-4 (December 1993), pp. 287–297, ISSN 0044-8486

Lim C. & Cuzon, G. (1994). Water stability of shrimp pellet: a review. *Asian Fisheries Science*, Vol 7, No. 2-3, (October 1994), pp. 115-126, ISSN 0116-6514

Linton, S.M.; Allardyce, B.J.; Hagen, W.; Wencke, P. & Saborowski, R. (2009). Food utilization and digestive ability of aquatic and semi-terrestrial crayfishes, *Cherax destructor* and *Engaeus sericatus* (Astacidae, Parastacidae). *Jounal of Comparative Physiology, part B*, Vol. 179, No. 4, (April 2009), pp. 493-507, ISBN 0174-1578

Liu, F.; Ai, Q.; Mai, K.; Tan, B.; Ma, H.; Xu, W.; Zhang, W. & LiuFu, Z. (2008). Effects of Dietary Binders on Survival and Growth Performance of Postlarval Tongue Sole, *Cynoglossus semilaevis* (Günther). *Journal of the World Aquaculture Society*, Vol. 39, No. 4, (August 2008), pp. 441–571, ISSN 1749-7345

Lopez-Lopez, S.; Nolasco, H.; Villarreal-Colmenares, H. & Civera-Cerecedo, R. (2005). Digestive enzyme response to supplemental ingredients in practical diets for juvenile freshwater crayfish *Cherax quadricarinatus*. *Aquaculture nutrition*, Vol. 11, No. 2, (April 2005), pp. 79–85, ISSN 1353-5773

Lundstedt, L.M.; Melo, J.F. & Moraes, G. (2004). Digestive enzymes and metabolic profile of *Pseudoplatystoma corruscans* (Teleostei: Siluriformes) in response to diet composition. *Comparative Biochemistry and Physiology, Part B Biochemistry and Molecular Biology*, Vol. 137, No. 3 (March 2004), pp. 331-339, ISSN 1096-4959

Luo, G.; Xu, J.; Teng, Y.; Ding, C. & Yan, B. (2010). Effects of dietary lipid levels on the growth, digestive enzyme, feed utilization and fatty acid composition of Japanese sea bass (*Lateolabrax japonicus* L.) reared in freshwater. *Aquaculture research*, Vol. 41, No. 2 (January 2010), pp. 210-219, ISSN 1365-2109

Mortensen, A.; Siikavuopio, S.I. & Raa, J. (2004). Use of transglutaminase to produce a stabile sea urchin feed. In: *Sea Urchins-Fisheries and Ecology* Lawrence, J.D. Ed., pp. 203-213. DEStech Publications Inc., ISBN 13 978-1-932078-32-9, ,Landcaster, PA, USA.

Moss, S.M.; Divakaran, S. & Kim, B.K. (2001). Stimulating effects of the pond water on digestive enzyme activity in the Pacific white shrimp, *Litopenaeus vannamei* (Bonne). *Aquaculture Research*, Vol. 32, No. 2, (February 2001), pp. 125–131, ISSN 1365-2109

Muhlia-Almazan, A.; Garcìa-Carreño, F.L.; Sanchez-Paz, J.A.; Yepiz-Plascencia, G. & Peregrino-Uriarte, A.B. (2003). Effects of dietary protein on the activity and mRNA level of trypsin in the midgut gland of the white shrimp *Penaeus vannamei*. *Comparative Biochemistry and Physiology, part B*, Vol. 135, No. 2, (June 2003), pp. 373–383, ISSN 1096-4959

Murano, E.; Toffanin, R.; Zanetti, F.; . Knutsen, S.H.; Paoletti, S. & Rizzo,R. (1992). Chemical and macromolecular characterisation of agar polymers from Gracilariadura (C. Agardh) J. Agardh (Gracilariaceae, Rhodophyta). *Carbohydrate Polymers,*Vol. 18, No. 3, (October 1992), pp. 171-178, ISSN 0144-8617

No, H.K. & Meyers, S.P. (1995). Preparation and Characterization of Chitin and Chitosan-A Review. *Journal of Aquatic Food Product Technology*, Vol. 4, No. 2, (June 1995), pp. 27-52, ISSN 1049-8850

Obaldo, L. G.; Divakaran, S. & Tacon, A. G. (2002). Method for determining the physical stability of shrimp feeds in water. *Aquaculture Research*, Vol. 33, No. 5, (April 2002), pp. 369-377, ISSN 1365-2109

O' Mahoney, M.; Mouzakitis, G.; Doyle, J. & Burnell, G. (2011). A novel konjac glucomannan–xanthan gum binder for aquaculture feeds: the effect of binder configuration on formulated feed stability, feed palatability and growth performance of the Japanese abalone, *Haliotis discus hannai*. *Aquaculture Nutrition*, Vol. 17, No. 4, (August 2011), pp. 395-407, ISSN 1353-5773

Orire, A.M.; Sadiku, S.O.E. & Tiamiyu, L.O. (2010). Evaluation of Yam Starch (Discorea rotundata) as Aquatic Feed Binder. *Pakistan Journal of Nutrition*, Vol. 9, No. 7, (July 2010), pp. 668-671, ISSN 1680-5194

Otero-Villanueva, M.; Kelly, M.S. & Burnell, G. (2004). How diet influences energy partitioning in the regular echinoid *Psammechinus miliaris*; constructing an energy budget. *Journal of Experimental Marine Biology and Ecology*, Vol. 304, No. 2, (June 2004), pp. 159-181, ISSN 0022-0981

Palma, J.; Bureau, P. & Andrade, J. P. (2008). Effects of binder type and binder addition on the growth of juvenile *Palaemonetes varians* and *Palaemon elegans* (Crustacea: Palaemonidae). *Aquaculture International*, Vol. 16, No.5, (October 2008), pp. 427-436, ISSN 0967-6120

Paolucci, M. Ed. (2010). *Leptin in non mammalian vertebrates*. Transworld Research Network. ISBN 978-81-7895-436-3, Trivandrum-695 023, Kerala, India.

Partridge, G.J. & Southgate, P.C. (1999). The effect of binder composition on ingestion and assimilation of microbound diets (MBD) by barramundi *Lates calcarifer* Bloch larvae. *Aquaculture Research*, Vol. 30, No. 11-12, (November 1999), pp. 879-886, ISSN 1365-2095

Pavasovic, M.; Richardson, N.A.; Anderson, A.J.; Mann, D. & Mather, P.B. (2004). Effect of pH, temperature and diet on digestive enzyme profiles in the mud crab, *Scylla serrata*. *Aquaculture*, Vol. 242, No. 1-4 , (December 2004), pp. 641-654, ISSN 0044-8486

Pavasovic, A.; Richardson, N.A.; Mather, P.B. & Anderson, A.J. (2006). Influence of insoluble dietary cellulose on digestive enzyme activity, feed digestibility and survival in the red claw crayfish, *Cherax quadricarinatus* (von Martens). *Aquaculture Research*, Vol. 37, No. 1,, (January 2006), pp. 25-32, ISSN 1365-2109

Pavasovic, A.; Anderson, A.J.; Mather, P.B. & Richardson, N.A. (2007a). Effect of a variety of animal, plant, and single cell-based feed ingredients on diet digestibility and digestive enzyme activity in redclaw crayfish, *Cherax quadricarinatus* (Von Marten 1868). *Aquaculture*, Vol. 272, No. 1-4, (November 2007), pp. 564-572, ISSN 0044-8486

Pavasovic, A.; Anderson, A.J.; Mather, P.B. & Richardson, N.A. (2007b). Influence of dietary protein on digestive enzyme activity, growth and tail muscle composition in redclaw crayfish, *Cherax quadricarinatus* (von Martens). *Aquaculture Research*, Vol. 38, No. 6, (April 2007), pp. 644-652, ISSN 1365-2109

Pearce, C.M.; Daggett, T.L. & Robinson, S.M.C. (2002a). Effect of protein source ratio and protein concentration in prepared diets on gonad yield and quality of the green sea urchin, *Strongylocentrotus droebachiensis*. *Aquaculture*, Vol. 214, No. 1-4, (November 2002), pp. 307-332, ISSN 0044-8486

Pearce, C.M.; Daggett, T.L. & Robinson, S.M.C. (2002b). Effect of binder type and concentration of prepared feed stability and gonad yield and quality of the green sea urchin, *Strongylocentrotus droebachiensis*. *Aquaculture*, Vol. 205, No. 3-4, (March 2002), pp. 301-323, ISSN 0044-8486

Pearce, C.M.; Daggett, T.L. & Robinson S.M.C. (2004). Effect of urchin size and diet on gonad yield and quality in the green sea urchin (*Strongylocentrotus droebachiensis*). *Aquaculture*, Vol. 233, No. 1-4, (April 2004), pp. 337-367, ISSN 0044-8486

Perez-Jimenez, A.; Cardenete, G.; Morales, A. E.; Garcia-Alcazar, A.; Abellan, E. & Hidalgo, C. (2009). Digestive enzymatic profile of Dentex dentex and response to different dietary formulations. *Comparative Biochemistry and Physiology - Part A: Molecular & Integrative Physiology*, Vol. 154, No. 1 (September 2009), pp 157-164, ISSN 1095-6433

Person Le Ruyet, J.P.; Alexander, J.C.; Thebaud, L. & Mugnier, C. (1993). Marine fish larvae feeding: formulated diets or live prey? *Journal of World Aquaculture Society*, Vol. 24, No.2, (June 1993), pp. 211– 224, ISSN 1749-7345

Phillips, K.; Bremer, P.; Silkock, P.; Hamid, N.; Delahunty, C.; Barker, M. & Kissick, J. (2009). Effect of gender, diet and storage time on the physical properties and the sensory quality of sea urchin (*Evechinus chloroticus*) gonads. *Aquaculture*, Vol. 288, No. 3-4, (March 2009), pp. 208-215, ISSN 0044-8486

Pratoomyot, J.; Bendiksen, E.Å; Campbell, P. J.; Jauncey, K. J.; Bell, J.G. & Tocher, D.R. (2011). Effects of different blends of protein sources as alternatives to dietary fishmeal on growth performance and body lipid composition of Atlantic salmon (*Salmo salar L.*). *Aquaculture*, Vol. 316, No. 1-4, (June 2011), pp. 44-52, ISSN 0044-8486

Rivas-Vega, M.E.; Goytortúa-Bores, E.; Ezquerra-Brauer, J.M.; Salazar-García, M.G.; Cruz-Suárez, L.E.; Nolasco, H. & Civera-Cerecedo, R. (2006). Nutritional value of cowpea (*Vigna unguiculata L.* Walp) meals as ingredients in diets for Pacific white shrimp (*Litopenaeus vannamei* Boone). *Food Chemistry*, Vol. 97, No. 1, (July 2006), pp. 41–49, ISSN 0308-8146

Robinson, J.A.J.; Kellaway, I.W. & Marriott, C. (1975). Effects of Blending on the Rheological Properties of Gelatin Gels. *Journal of Pharmacy and Pharmacology*, Vol. 27, No 11, (November 1975), pp. 188–824, ISSN 0022-3573

Rodrıguez, A.; Le Vay, L.; Mourente, G. & Jones, D.A. (1994). Biochemical composition and digestive enzyme ´activity in larvae and postlarvae of *Penaeus japonicus* during herbivorous and carnivorous feeding, *Marine Biology*, Vol. 118, No. 1, (March 1994), pp. 45–51, ISSN 0025-3162

Rosas, C.; Cuzon, G.; Gaxiola, G.; Arena, L.; Lemaire, P.; Soyez, C. & Van Wormhoudt , A. (2000). Influence of dietary carbohydrate on the metabolism of juvenile Litopenaeusstylirostris. *Journal of Experimental Marine Biology and Ecology* , Vol. 249, No. 2, (June 2000), pp. 181–198, ISSN 0022-0981

Rosas, C.; Tut, J.; Baeza, J.; Sanchez, A.; Sosa, V.; Pascual, C.; Arena, A.; Domingues, P. & Cuzon, G. (2008). Effect of type of binder on growth, digestibility, and energetic balance of *Octopus maya*. Aquaculture, Vol. 275, No. 1-4, (March 2008) pp. 291-297, ISSN 0044-8486

Ruscoe, I.M; Jones, C.M.; Jones, P.L. & Caley, P. (2005) The effects of various binders and moisture content on pellet stability of research diets for freshwater crayfish . *Aquaculture Nutrition*, Vol 11, No.2, (April 2005), pp. 87-93, ISSN 1353-5773

Schlosser, S.C.; Lupatsch, I.; Lawrence, J.M.; Lawrence, A.L. & Shpigel, M. (2005). Protein and energy digestibility and gonad development of the European sea urchin *Paracentrotus lividus* (Lamarck) fed algal and prepared diets during spring and fall. *Aquaculture Research*, Vol. 36, No. 10, (July 2005), pp. 972-982, ISSN 1365-2109

Seixas Filho J.T.; Rostagno, H.S.; Queiroz, A.C.; Euclydes, R.F. & Barabarino, P. Jr. (1997a). Evaluation of the performance of post-larvae of the freshwater prawn *Macrobrachium rosenbergii* fed with balanced diets containing different binders. *Revista Brasiliana de Zootecnia*. Vol. 26, No. 4, pp. 638-644, ISSN 1806-9290

Seixas Filho J.T.; Rostagno, H.S.; Euclydes, R.F.; Queiroz, A.C. & Barabarino, P. Jr. (1997b). Effect of binders on the hydrosolubility of balanced diets for freshwater prawns (*Macrobrachium rosenbergii*, de Man) in the post-larval stage. *Revista Brasiliana de Zootecnia*. Vol. 26, No. 4, pp. 638-644, ISSN 1806-9290

Shewry, P.R. (2009). Wheat. *Journal of Experimental Botany*, Vol. 60, No. 6, (April 2009) pp. 1537-53, ISSN 0022-0957

Shiau, S.Y. & Peng, C.Y. (1992). Utilization of different carbohydrates at different dietary protein levels in grass prawn, *Penaeus monodon*, reared in seawater. *Aquaculture*, Vol. 101, No. 3-4, (February 1992), pp. 241-250, ISSN 0044-8486

Simon, C.J. (2009). The effect of carbohydrate source, inclusion level of gelatinised starch, feed binder and fishmeal particle size on the apparent digestibility of formulated diets for spiny lobster juveniles, *Jasus edwardii*. *Aquaculture*, Vol. 296, No. 3-4 , (November 2009), pp. 329-336, ISSN 0044-8486

Sinha, A.K.; Kumar, V.; Makkar, H.P.S.; De Boeck, G. & Becker, K. (2011). Non-starch polysaccharides and their role in fish nutrition - A review *Food Chemistry*, Vol. 127, No. 4, (August 2011) pp. 1409-1426, ISSN 03088146

Smidsrod, O.; Haug, A. & Larsen, B. (1966). The influence of pH on the rate of hydrolysis of acidic polysaccharides. *Acta chemical Scandinavica*, Vol. 20, No. 4, pp. 1026-1034, ISSN 0302-4377

Smith, D.M. & Tabrett, S.J. (2004). Accurate measurement of in vivo digestibility of shrimp feeds. *Aquaculture*, Vol. 232, No.1-4, (April 2004), pp. 563-580, ISSN 0044-8486

Stephen, A.M. (1983). Other plant polysaccharides, In: The polysaccharides. Aspinall, G.O., pp. 97-193, Academic Press, New York

Storebakken, T. (1985). Binders in fish feeds I. Effect of alginate and guar gum on growth, digestibility, feed intake and passage through the gastrointestinal tract of rainbow trout. *Aquaculture*, Vol. 47, No. 1, (July 1985), pp. 11-26, ISSN 0044-8486

Storebakken, T. & Austreng, E. (1987). Binders in fish feeds II. Effect of different alginates on the digestibility of macronutrients in rainbow trout. *Aquaculture*, Vol. 60, No. 2, (February 1987), pp. 121-131, ISSN 0044-8486

Storebakken, T.; Shearer, K.D. & Roem, A.J. (1998). Availability of protein, phosphorus another elements in fish meal, soy-protein concentrate and phytase-treated soy protein-concentrate-based diets to Atlantic salmon, *Salmo salar*. *Aquaculture*, Vol. 161 No. 1-4, (February 1998), pp. 365-379, ISSN 0044-8486

Teshima. S.; Kanazawa, A. & Sakamoto, M. (1982). Microparticulate diets for the larvae of aquatic animals. Mini review and data file of Fisheries Research, Kagoshima University 2, pp. 67-86

Usov, A I. (1998). Structure analysis of red seaweed galactan of agar and carrageenan groups. *Food Hydrocolloids*, Vol. 12, No 3, (July 1998), pp. 301-308, ISSN0268-005X

Valverde, J. C.; Hernández M. D.; Aguado-Gimenez, F. & Garciá Garciá, B. (2008). Growth, efficiency and condition of common octopus (*Octopus vulgaris*) fed on two formulated moist diets, Aquaculture, Vol. 275, No. 1-4, (March 2008), pp. 266-273, ISSN 0044-8486

Van der Meer, J. (2006). Metabolic theories in ecology. *Trends in Ecology and Evolution*. Vol. 21, No. 3, (November 2006), pp. 136-140, ISSN 0169-5347

Van Wormhoudt, A.; Cruz, E.; Guillaume, J. & Favrel, P. (1986). Action de l'inhibiteur trypsique de soja sur lacroissance et l'activite des enzymes digestives chez ´ *Penaeus japonicus* (Crustacea, Decapoda): role ^eventuel des hormones gastro-intestinales. *Oceanis*, Vol. 12, pp. 305–319, ISSN 0182-0745

Volpe, M.G.; Monetta, M.; Di Stasio, M. & Paolucci, M. (2008). Rheological behavior of polysaccharide based pellets for crayfish feeding tested on growth in the crayfish *Cherax albidus*. *Aquaculture*, Vol. 274, No. 2-4, (February 2008), pp. 339–346, ISSN 0044-8486

Volpe, M.G.; Malinconico, M.; Varricchio, E. & Paolucci, M. (2010). Polysaccharides as biopolymers for food shelf-life extension: Recent patents. *Recent Patents on Food, Nutrition & Agriculture*, Vol. 2, No. 2, (June 2010), pp. 129-139, ISSN 1876-1429

Watanabe, T. (2002). Strategies for further development of aquatic feeds. *Fisheries Science*, Vol. 68, No. 2, (May 2002), pp. 242-252, ISSN 1444-2906

Zhou, Q-C.; Zhou, J-B.; Chi, S-Y.; Yang, Q-H; Liu, C-W. (2007). Effect of dietary lipid level on growth performance, feed utilization and digestive enzyme of juvenile ivory shell, *Babylonia areolate*. *Aquaculture*, Vol. 272, No. 1-4 (November 2007), pp. 535-540, ISSN 0044-8486

3

Replacement of Fishmeal with Plant Protein Ingredients in Diets to Atlantic Salmon (*Salmo salar*) – Effects on Weight Gain and Accretion

Marit Espe[1], Adel El-Mowafi[2] and Kari Ruohonen[2]
[1]National Institute of Nutrition and Seafood Research (NIFES), Bergen Norway
[2]Ewos Innovation AS, N-4335 Dirdal
Norway

1. Introduction

Farming of fish and especially farming of carnivorous fish depends on high protein diets. Atlantic salmon (*Salmo salar*) generally require 55-60% of dietary protein in juvenile stages thereafter the protein requirement declines (NRC, 1993). Historically the protein source has been fishmeal, produced from wild caught fish, while the lipid source has been the fish oil from the same source. Production of farmed Atlantic salmon in Norway has increased steadily during the last decade (from 440061 tons in year 2000 to 927876 metric tons in year 2010) while the wild fish catch has been stable during the same period of time (statistics from Directorate of Fisheries, Norway; www.fiskeridir.no). This also holds for the worldwide wild caught fish and aquaculture production (FAO; www.fao.org/fishery; Tacon, 1995; FAO, 2006; Tacon & Metian, 2008). Therefore if a continued increase of farmed fish is to occur, both alternative protein ingredients as well as lipid sources need to be assessed as feed ingredients. This is to be done by supporting both the growth and at the same time not compromising the health of the farmed fish. In future aquaculture these novel ingredients have to constitute the main protein and lipid sources in the fish diets. The main such sources are likely to be of plant origin but some animal- by-products and microbiologically produced proteins might be allowed by national and/or international legislation as well. Such animal-by-products may arrive from processing of poultry and swine (Sugiura et al., 1998; Yanik et al., 2003; Rahnema et al., 2005) and microbiological produced protein sources as single cell proteins (Storebakken et al., 2004; Ozyurt & Deveci, 2004; Berge et al., 2005). Plant proteins are the most likely candidates because of their abundance and relatively low cost. However, upon substituting the fishmeal with plant protein ingredients reduced growth performance generally occurs in Atlantic salmon (Olli et al., 1995; Refstie et al., 1998; 2000; Storebakken et al., 1998; Carter & Hauler, 2000; Krogdahl et al., 2003; Opstvedt et al., 2003) as well as in other fish species of commercial interest such as Atlantic cod, *Gadus morhua* (von der Decken & Lied, 1993; Hansen et al., 2007), sea bream, *Sparus aurata* (Robaina et al., 1995; Gòmez-Requeni et al., 2004), rainbow trout, *Oncorhynchus mykiss* (Pongmaneerat & Watanabe, 1992; Gomes et al., 1995; Kaushik et al., 1995) and turbot, *Psetta maxima* (Regost et al., 1999; Fournier et al., 2004). Increased feed conversion ratio (FCR) and reduced protein utilization are of major concern when fishmeal is replaced with plant ingredients (Robaina et al., 1995; Refstie et al., 1998; 2000; Opstvedt et al., 2003). The reduced performance might be due to an imbalanced amino acid profile or simply due to lower voluntary feed intake in fish fed the diets based on plant ingredients.

Protein synthesis requires that all amino acids needed for the protein synthesis are present in the cell. If one amino acid is absent or present in lower concentrations than required for the synthesis to occur in any particular tissue, the protein cannot be synthesised and the constituent amino acids will be metabolised and used as fuel due to the simple fact that free amino acids are never stored in tissues (Geiger, 1947). The main amino acid depot is the muscle (Houlihan et al., 1995), but these amino acids are not available for protein metabolism until after the proteins are hydrolysed to free amino acids. Generally the free amino acids constitute less than 0.1% of the protein bound amino acids in animals (Njaa, 1990; Millward, 1988; 1999; Espe, 2008). Although the proteins are constantly turned over fish needs a continuous supply of indispensable amino acids (IAAs) as only about 50% of the protein bound amino acids are reutilised for protein synthesis (Houlihan et al., 1995). Recently, it was found that in addition to the IAAs, fish also seem to require the dispensable amino acids (DAA, Abbouti et al., 2009). Further, the delivery of adequate amounts of amino acids with a balanced profile is essential for maximal protein accretion and growth in fish (El-Mowafi et al., 2010). Animal derived protein ingredients have a balanced amino acid profile independent of being a beef muscle or a fish muscle and provide the indispensable amino acid profile required by the consumer or animal. For the formulation of diets this balanced amino acid profile is adopted by using the ideal protein concept in which the amount of each indispensable amino acid is presented relative to the amount of lysine (Wang & Fuller, 1987; 1989; Rollin et al., 2003). Further, any reduction in the apparent digestibility or presence of possible anti nutrients might add to the reduced absorption of nutrients in the fish fed the plant based diets and might add to the reduced performance. Both the reduced feed intake and the reduced absorption in fish fed such diets might change the availability of the balanced amino acid profile in tissue compartments. Any imbalances of amino acids in tissues may have a negative impact on the metabolism and significantly affect both the health and the protein and/or lipid deposition in the farmed fish and animals (Wu, 2009).

In fish as in domestic animals, the feed intake determines the weight gain. Traditionally feed intake has been of less concern when the diet fishmeal inclusion was high as fish had satisfactory voluntary feed intake to maximise the growth potential. But upon replacing the fishmeal with plant protein ingredients, the acceptability of the feed may reduce resulting in a reduced growth performance (Kaushik et al., 1995; Fournier et al., 2004; Glencross et al., 2004; Dias et al., 2005). Thus one of the greatest challenges in the formulation of high plant protein diets for farmed Atlantic salmon has been to secure that the fish accept the feed offered to them equally well as the fishmeal based diets. Therefore upon formulating the diets one needs to seek for ingredients that may help in maintaining the feed intake. Both fish protein hydrolysate (FPC) and stick water (the water soluble protein fraction obtained during fishmeal production) contain several non protein nitrogen compounds (NPN) that the plant ingredients contains in very different concentrations or are devoid of. Examples of NPN are taurine, betaine, trimethylamine oxide, spermine, spermidine among others (Espe & Lied, 1999; Liaset & Espe, 2008; historical data NIFES). As addition of both FPC (Espe et al., 1992a) and stick water (Laksesvela, 1958; 1960) improved the voluntary feed intake in poultry and FPC improved feed intake in rats (Espe et al., 1989; 1992b; Liaset et al., 2000)

Replacement of Fishmeal with Plant Protein Ingredients in Diets to Atlantic Salmon (Salmo salar) –
Effects on Weight Gain and Accretion

45

and Atlantic salmon (Espe et al., 1999; Refstie et al., 2004; Hevrøy et al., 2005) both components seem to have the potential to help in maintaining the intake of fishmeal replacement diets for Atlantic salmon. Hydrolysed squid protein is known to contain quite high concentration of NPN compounds (~6%) of which about 1/6th is taurine (historical data NIFES) making it another possible candidate with the benefit of increasing dietary taurine concentration of the diets. Additionally squid contains quite high concentration of betaine (trimethylglycine, Treberg & Driedze, 2007). Betaine is consumed in the remethylation of homocysteine to regenerate methionine in the choline-betaine pathway and is crucial for the methylation capacity in animals (Finkelstein, 1990; Mato et al., 2002).

Therefore when we first formulated diets devoid of fishmeal (Espe et al., 2006) or containing low fishmeal (Espe et al., 2007) either stick water (5%) or hydrolysed squid (3%) were used with the aim to assess the effect on voluntary feed intake and growth. Additionally 5% FPC was included. Fish fed the total replacement diet consumed slightly less feed as compared to the fishmeal fed Atlantic salmon and those fish fed the diet containing hydrolysed squid consumed more feed than did the fish offered the diets containing the stick water. Fish fed diets not added any fishmeal grew less than the fish fed the reference diets in which most of the protein arrived from fishmeal. The reduced growth observed was due to a general reduced lipid gain (Espe et al., 2006). Addition of 5% fishmeal to the diets improved the lipid accretion to similar levels as the fish meal fed control (Espe et al., 2007). Therefore one probably should not formulate diets without any fishmeal for Atlantic salmon, but inclusion of low fishmeal (5%) together with the marine ingredients as squid or stick water and FPC seems to work well. It should be noted that Fournier et al. (2003) also added FPC in diets fed to turbot without any fishmeal and reported similar voluntary feed intake as compared to fish fed the fishmeal control, but in those experiments the growth was reduced. Experiments with different species fed total or high fishmeal replacement diets are summarised in **Table 1**, and generally total or high replacement of marine protein with plant proteins reduced the performance. As Atlantic salmon fed FPC and squid meal performed better than the fish fed the stick water (Espe et al., 2006; 2007), the test diets used by us in trials where Atlantic salmon were fed high fishmeal replacement diets were added FPC and hydrolyzed squid meal or krill meal and fishmeal was never less than 5%.

It should be noted that replacement of fish oil also depend on blends of plant oils to make sure the requirement of polyunsaturated fatty acids is met (Torstensen et al. 2004; 2005; 2008). However, in all our studies with Atlantic salmon, the lipid source was fish oil thus any discussion of alternative lipid sources in fish feed is not included in the current chapter. The effect of plant oil substitution for fish oil was recently reviewed by Leaver et al. (2009).

2. The effect of fish meal replacement on growth and accretion

Six experiments in which high fishmeal replacement diets were fed to Atlantic salmon were compared for the wet weight gain, as well as protein and lipid gain. The diets tested contained balanced amino acid profiles (**Figure 1**), but were based on different plant protein ingredients (**Table 2**) and were supplemented with low concentration of crystalline amino acid to balance the amino acid profiles not to jeopardize growth (Espe & Njaa 1991; Espe et al., 1992a; Espe & Lied 1994; Berge et al., 1994; Cowey 1995; Dabrowski & Guderley 2002; Rollin et al., 2003). Daily voluntary feed intake was monitored carefully to make sure that

Species	Fishmeal protein (% of total)	Marine protein (% of total)	Voluntary feed intake	Growth	PER	PPV	FCR	References
Atlantic salmon (*Salmons salar*)	0	7	Less or equal	Reduced	Equal	Equal	Equal	**Espe et al., 2006**
	7	17	Equal	Equal	Equal	Equal	Equal	Espe et al., 2007
	11-14	17-18	Equal	Equal	Equal	-	Equal	Kousoulaki et al., 2009
	20	28	-	Equal	Equal	Equal	Equal	Torstensen et al., 2008
Sea bass (*Dicentrarchus labrax*)	0	17	Reduced	Reduced	Reduced	Reduced	Increased	**Dias et al., 2005**
	7	7	Equal	Equal	Reduced	-	Equal	Kaushik et al.,2004
Seabream (*Sparus aurata*)	27	27	Reduced	Reduced	Reduced	-	Increased	Gomez-Requeni et al., 2004
	0	0	Reduced	Reduced	Reduced	-	Increased	**Gomez-Requeni et al, 2004**
Atlantic cod (*Gadus morhua*)	24	24	-	Reduced	Reduced	Reduced	Increased	Hansen et al. 2007
Turbot (*Psetta maxima*)	12	17	Reduced	Reduced	Equal	-	-	Fournier et al., 2004
	0	6	Reduced	Reduced	Equal	-	-	**Fournier et al, 2004**
	0	6	Equal	Reduced	Reduced	Reduced	Increased	**Fournier et al, 2003**
Rainbow trout (*Oncorhynchus mykiss*)	26	26	Reduced	Reduced	-	-	Increased	Glencross et al., 2004
	0	8	Equal	Reduced	Equal	Reduced	Equal	**Fournier et al, 2003**
	0	0	Reduced	Reduced	Equal	-	Equal	**Gomes et al, 1995**
	34	34	Equal	Equal	Equal	-	Equal	Gomes et al., 1995
	10	34	Reduced	Reduced	Equal	Reduced	Increased	Aksnes et al., 2006

Table 1. The effect on voluntary feed intake, growth performance, protein accretion and feed utilisation in fish fed high plant protein diets compared to fishmeal control fed fish in some studies testing total or high fishmeal replacement diets. -No information given, Total fishmeal replacement trials are marked with bold letters

Replacement of Fishmeal with Plant Protein Ingredients in Diets to Atlantic Salmon (Salmo salar) –
Effects on Weight Gain and Accretion

47

IAA (g/16g N):

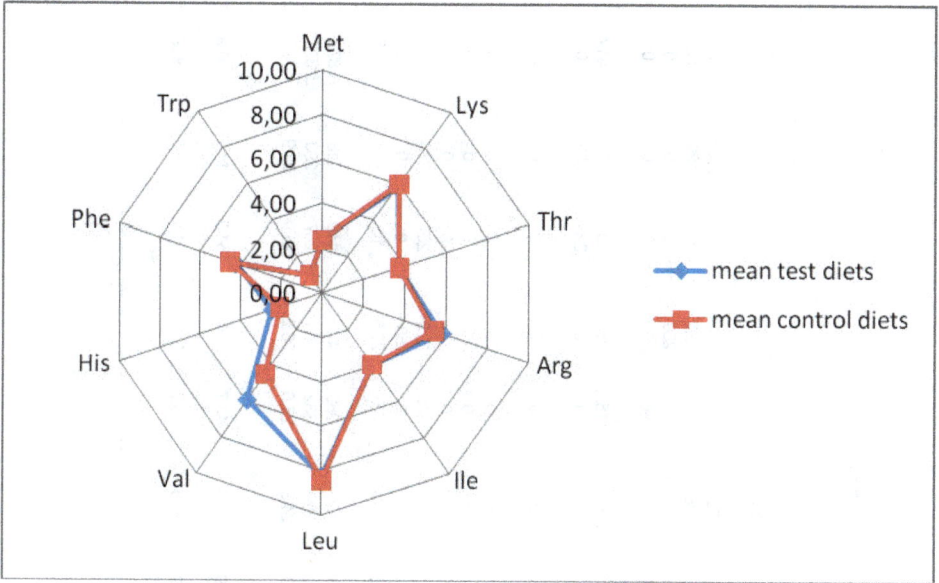

DAA (g/16gN):

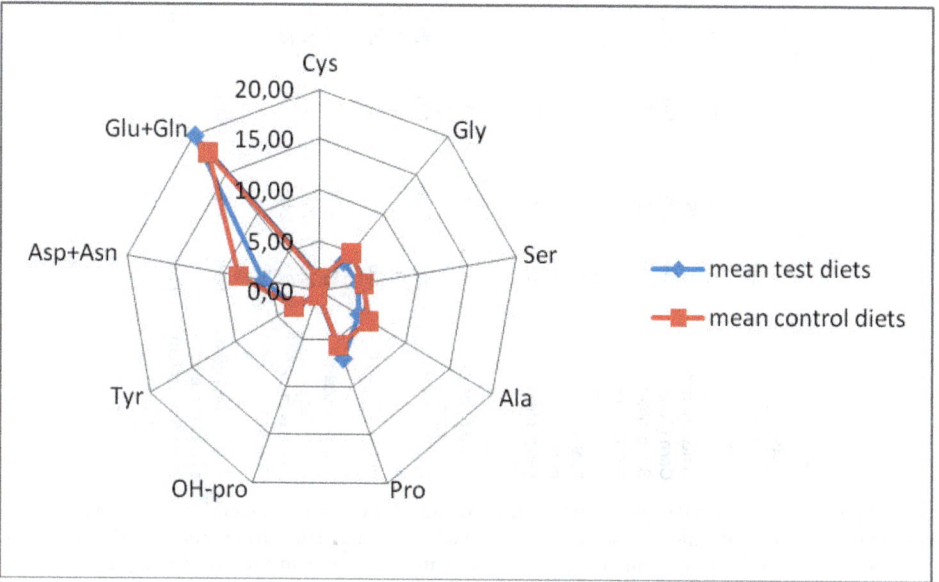

Fig. 1. The diets were balanced in both indispensable (IAAs) and dispensable (DAAs) amino acids towards the controls used in each trial and equalled the requirement of amino acids. Here is shown the mean values over all trials (g amino acid/16gN).

Diets	Trial I C	Trial I T	Trial II C	Trial II T	Trial III C	Trial III T	Trial IV C	Trial IV T	Trial V C	Trial V T	Trial VI C	Trial VI T
Fishmeal (Norse LT-94)	490	50	400	50	260	50	250	50	354	87.5	280	50
FPC (C.P.S.P.)	0	50	0	50	0	50	0	50	0	112.5	0	59
Wheat gluten	0	151	0	135	80	238	70	75.3	40	120	50	150
Corn gluten	100	222	170	190	140	0	150	250	0	0	0	0
Soy protein concentrate	55	0	35	0	123	11	96	0	0	0	0	0
Wheat grain	106.4	112.4	156.4	190.4	106.4	225.8	115.4	146.1	0	0	144.4	0
Pea/ soy protein mix	0	0	0	0	0	0	0	0	323	382	0	0
Pea protein conc. 50	0	0	0	0	0	0	0	0	0	0	120	130
Tapioca	0	0	0	0	0	0	0	0	0	0	0	155.9
Squid hydrolysates	0	30	0	30	0	30	0	30	0	0	0	30
Krill meal	0	0	0	0	0	0	0	0	0	0	0	50
Fish oil	240	258	230	253	272	280	310	325	267	273	312	293.9
AA mix	0	126.6	0	71	0	71.6	0	50	0	0	0	4.4
micronutrients*	8.6	30.5	8.6	30.6	19.5	43.6	8.6	23.6	16	25	13.6	42.1
Crude protein	477	504	424	446	433	430	389	400	440	418	373	360
Crude fat	248	224	241	239	274	289	332	331	287	290	373	367
Gross energy (MJ/kg)	23.6	24.3	23.7	23.8	24.0	24.3	25.6	26.2	24.0	23.9	24.7	25.8
% fishmeal protein of diet protein	69.8	6.7	64.1	7.6	40.9	7.9	43.7	8.0	54.8	14.1	50.9	9.4
% marine protein of total diet protein	69.8	17.0	64.0	18.5	40.9	20.0	43.7	21.5	54.8	21.7	50.9	18.6

*crystalline amino acid to balance the dietary amino acids and vitamin + mineral mixtures to fulfil the requirement (for more details see Espe et al., 2007; 2008). Percentage of marine protein in the diets was calculated assuming fishmeal, krill meal and FPC contained 68% protein and squid 60% protein

Table 2. The compositions (g/kg) in the diets used in the different trials (I-VI). For each trial the control (C) contained more fishmeal, while the test (T) diets are based on different blends of plant proteins containing lower fishmeal inclusion.

Replacement of Fishmeal with Plant Protein Ingredients in Diets to Atlantic Salmon (Salmo salar) –
Effects on Weight Gain and Accretion

49

any differences in feed intake should have any impacts on growth or accretion. Some of the general growth data of these trials have been published previously assessing the requirement of amino acids in Atlantic salmon (Espe et al., 2007; 2008) but here only fish fed the control and the test diets with similar amino acid profiles within each experiment are included. Additionally some unpublished experiments are included. The results were sorted after high protein low fishmeal (HPLF) and medium protein medium fishmeal (MPMF) replacement. Each response was modelled statistically with a mixed-effects model using the level of fishmeal (HPLF or MPMF) as a fixed effect and trial as a random effect.

To be able to compare the growth and accretion data obtained in the six experiments included, growth data was calculated g/kg average BW/day as both the duration of the experiments and the initial body weight of the fish varied. Likewise the protein and lipid gains were calculated as gain/kg average BW/day. The mean wet weight gain, the protein and lipid gains obtained in the six trials are listed in **Table 3**. The replacement ratios of both the fishmeal protein and the marine protein ingredients with plant proteins are summarised in **Table 2**. The test diets contained from 6.7 to 14.1% of dietary protein as fishmeal, while the control diets (MPMF) contained from 40.9 to 69.8% of the dietary protein as fishmeal. As the test diets were added FPC and squid or krill meal the marine protein content in test diets varied from 17.0 to 21.5 % of total protein while the medium fishmeal replacement diets contained from 40.9 to 69.8 % marine protein of the total dietary protein.

Weight gain varied from 6-12g/kg average BW/day in the trials included (**Table 3**), being highest in trials III and V. The protein gain varied from 0.6 to 1.8g/kg Average BW/day being highest in trials III and V (**Table 3**). The lipid gain varied from 0.8 to 3.2g/kg average BW/day, being highest in trials I, III and V (**Table 3**). There was a high probability (>99%) that lipid gain was reduced when high plant protein diets were fed to Atlantic salmon (**Table 4**). The 95% credible interval (CI) of this difference ranges from about -3 to -16%, i.e. it is clearly negative. Similarly, there was a high probability (>99%) that protein gain was reduced when fishmeal was replaced with plant ingredients (95% CI of the effects ranges from about -3 to -16%, **Table 4**). The estimated mean wet weight gain in the MPMF was 8.1g/kg BW/day and there was a high probability (>99%) that wet weight gain decreased with the HPLF fed groups (95% CI of the effect ranges from about -2 to -13%, **Table 4**). Although the effects on protein and lipid gain as well as the wet weight gain are clearly negative they are all less than 10% (i.e. effect of size median; **Table 4**). Any reduced performance of about 10% is however not devastating and show that probably the balancing of the dietary amino acids and addition of marine ingredients improved the performance of the fish fed the diets with HPLF.

Generally wet weight gain decrease when high plant protein diets are fed to fish (Gomes et al., 1995; Kaushik et al., 1995; 2004; Mambrini et al., 1999; Dias et al., 2005; Torstensen et al., 2008). This may of course be due to the fact that the diets do not contain a balanced amino acid profile, but even when the amino acid profile is balanced the performance declines as the plant ingredients increase (Fournier et al., 2003, 2004). However, when juvenile Senegalese sole were fed diets with very low fishmeal inclusion and balanced in all dietary amino acids and supplemented with taurine, the wet weight gain was equal to fish fed the diet containing high fishmeal inclusion (Silva et al., 2009). Also taurine supplementation in high fishmeal replacement diets fed to juvenile Atlantic salmon was found to reduce the whole body lipid to protein ratio (Espe et al., 2011). The higher the accretion of protein the

Diets	Trial I		Trial II		Trial III		Trial IV		Trial V		Trial VI	
	C	T	C	T	C	T	C	T	C	T	C	T
Initial BW	333±4	327±8	633±31	619±50	481±25	495±5	1123±9	1128±55	371±11	383±6	1441±25	1402±30
End BW	763±37	735±15	1108±23	986±48	1235±94	1125±64	1891±25	1928±77	1065±93	965±91	2435±88	2497±72
Weight gain	8.7±0.4	8.3±0.2	6.7±0.7	5.3±0.4	10.3±0.3	9.5±0.7	6.0±0.3	6.0±0.8	12.3±0.9	10.8±1.0	5.8±0.3	6.3±0.4
ADC-N	89±1	92±2	88±1	93±1	93±2	93±1	86±5	92±1	88±1	89±1	92±1	94±1
Protein gain	1.13±0.1	1.18±0.1	1.14±0.1	0.84±0.1	1.72±0.1	1.43±0.1	0.77±0.1	0.75±0.1	1.85±0.1	1.73±0.1	0.62±0.1	0.74±0.1
Lipid gain	3.20±0.2	2.81±0.2	1.23±0.1	1.02±0.1	2.40±0.1	1.90±0.2	1.30±0.1	1.24±0.3	1.71±0.2	1.53±0.1	0.75±0.1	0.84±0.1

Arbitrary BW (ABW) is the mean wt during the trials (end BW+initial BW/2). In trials I to IV is mean of 3 tanks per treatment while trials V and VI is the mean of 4 tanks per treatment. Values are the tank means±standard deviation

Table 3. The mean initial and end body wt (BW, g), mean wet wt gain (g/kg ABW/day), mean apparent digestibility of nitrogen and mean protein and lipid gain (g/kg ABW/day) in the six trials included in the statistics to test the effects of high or medium fishmeal replacement diets fed to Atlantic salmon.

Replacement of Fishmeal with Plant Protein Ingredients in Diets to Atlantic Salmon (Salmo salar) –
Effects on Weight Gain and Accretion

51

	MPMF	Diff with HPLF	SE diff	Effect size median	Effect size 95% CI low	Effect size 95% CI high	Probability HPLF<MPMF
	g/Kg ABW/day	g/Kg ABW/day	g/Kg ABW/day	% of MPMF	% of MPMF	% of MPMF	
Wet wt gain	8.07	-0.59	0.23	-7.24	-13.14	-1.87	0.995
Protein gain	1.19	-0.11	0.04	-9.47	-16.10	-2.77	0.998
Lipid gain	1.64	-0.15	0.06	-9.47	-16.48	-3.01	0.997

MPMF medium protein, medium fishmeal
HPLF high protein, low fishmeal

Table 4. Statistical estimates of the mean wet weight gain, protein and lipid gain of the MPMF feeds together with the estimated differences of the HPLF feeds (g/Kg ABW/day). Median effect size and its 95% credible intervals are given as percentage of the MPMF estimate. The estimated probability that HPLF is smaller than MPMF is also given in the last column.

more cost effective is the feed. Also less nitrogen will be spilled to the ocean when accretion is high (Kaushik et al., 2004). Therefore also from an environmental aspect, the novel fishmeal replacement diets have to result in protein accretion close to what is the case in using the traditional fishmeal based diets. Upon feeding rainbow trout diets with high plant proteins the protein accretion declines (Fournier et al., 2003; Aksnes et al., 2006). This generally also is the case when any marine species is fed increasing amounts of plant proteins. Atlantic cod reduce protein accretion significantly when plant protein exceeds 50% of the dietary protein (Hansen et al., 2007) while the European seabass did not reduce protein accretion until plant protein inclusion exceeded 80% of the dietary protein (Kaushik et al., 2004). Also seabream shows reduction in protein gain when fishmeal was replaced with plant proteins (Robaina et al., 1995; Gomez-Requeni et al., 2004) and this was described to be due to an imbalanced amino acid profile. However, in turbot fed diets containing less than 20% fishmeal even though the amino acid profile in the diets was similar to the fishmeal based control, the protein accretion was less than the control fed turbot (Fournier et al., 2004).

All animals, man included, have a tremendous ability to adapt to a low protein intake (Millward, 1999). This adaption also seems to be the case in Atlantic salmon as protein gain was not affected even though the feed intake declined and lipid gain reduced when the fish was fed diets without any fishmeal (Espe et al., 2006).

The reduced lipid deposition also was reported by Dias et al. (2005) when European seabass was fed plant protein based diets as compared to fish fed a fishmeal diet. The protein source in the current diets was based on variable blends of plant proteins (Table 2), but the amino acid profiles were pretty similar within all trials (Figure 1) and well above the anticipated amino acid requirement. Any reduced availability of dietary amino acids cannot explain the reduced performance as the digestibility in all trials was high (Table 3). Therefore the reduced performance when Atlantic salmon are fed plant protein based diets cannot be explained by any reduced availability of amino acids. However, the metabolic cost following high replacement diets might be increased due to the fact that NPN metabolites that are present in fishmeal either are not delivered or has to be synthesised by the fish.

Previously we found less fat in whole fish, muscle and liver of Atlantic salmon fed hydrolysed proteins as compared to fish fed diets containing the intact control protein (calculated from Espe et al., 1992c). As the fishmeal replacement diets compared in the current chapter contained 5-11.25% FPC and the control diet did not contain any FPC, the lesser fat deposition might have been due to the inclusion of the FPC in the test diets. Thus probably the reduced lipid deposition in total fishmeal replacement diets is due to metabolic changes or higher metabolic costs. How the fish sense the energy and choose either protein or fat as fuels needs to be focussed to understand how to formulate diets for Atlantic salmon that burn the lipid and deposit as much as possible of the dietary protein. Interestingly, we recently found that FPC affected viscera mass without affecting the dress out weight in Atlantic salmon fed diets containing medium fishmeal inclusion of which was replaced with FPC (Espe et al., In prep) supporting an altered deposition pattern in Atlantic salmon when FPC is added to the diets. Further, when both the protein and the lipid source were of plant origin more adipose tissue were present in Atlantic salmon (Torstensen et al., 2011) implying that the deposition pattern also should be addressed when the effects of high fishmeal replacement diets are assessed.

3. Future aspects

Even though Atlantic salmon grew and utilise diets with total replacement of the fish meal with plant ingredients or very high replacement levels recently well, the compartmentalisation and lipid deposition may change even though the voluntary feed intakes are similar. This might be due to altered concentration of NPN in the replacement diets. In the current studies the NPN was higher in the replacement diets in trials III and VI (**Figure 2**), while the wet weight gain was best in trials III and V, thus NPN cannot alone explain the reduced performance. Atlantic salmon fed diets low in methionine had reduced concentration of taurine in both muscle and liver (Espe et al., 2008). The reduced methionine intake, and the lower liver taurine, resulted in higher accumulation of lipid in the liver (Espe et al., 2010) while lysine limitations increased the lipid concentration in the whole fish (Espe et al., 2007). Also in the rainbow trout the methionine and taurine was found to affect the lipid deposition, but methionine rather than taurine was responsible (Gaylord et al., 2007). In juvenile Atlantic salmon we reported that taurine supplementation to plant protein based diets reduced the whole body lipid deposition without affecting protein deposition (Espe et al., 2011). Thus the altered lipid deposition pattern following limitations in amino acids most probably is affected through interactions between metabolites when present in excess or in limited concentration in the diets. In the studies described on Atlantic salmon all diets were balanced in amino acids, but as amino acid requirements might change when fishmeal is replaced, the future research needs to focus metabolic consequences of feeding the Atlantic salmon total or almost total fish meal replacement diets and determine whether the NPN compounds are conditionally indispensable. Also the aspects of any increased requirement of indispensable amino acids and or dispensable amino acids should be addressed as synthesis of metabolites not present in the diets may increase the requirement of the precursor amino acids.

Replacement of Fishmeal with Plant Protein Ingredients in Diets to Atlantic Salmon (Salmo salar) –
Effects on Weight Gain and Accretion

53

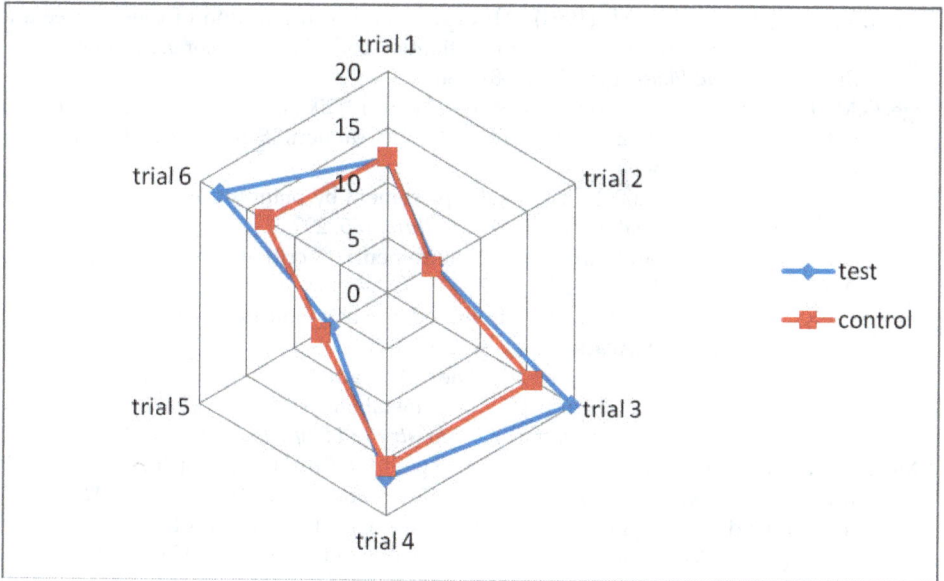

Fig. 2. The non amino acid nitrogen (NPN) differed between the diets in trial III and VI, otherwise the NPN was pretty similar. Here is shown the NPN in the relative high fishmeal diets (control) vs the NPN in the replacement diets.

As we previously reported that adding all protein as hydrolysed proteins, reduced lipid deposition in whole fish, fillet and liver (calculated from Espe et al., 1992c), the optimisation of FPC inclusion in producing a leaner fish might be a strategy if a leaner Atlantic salmon should be a goal in fish farming. Also in the rat the visceral lipid decreased upon offering them a diet in which the dietary protein source was FPC as compared to both casein and soy protein (Liaset et al., 2009). The results from the last years in replacing fishmeal by plant protein ingredients in diets to Atlantic salmon show clearly that a future sustainable aquaculture is possible. Meantime carefully use of the marine ingredients the fishmeal included to stimulate the voluntary feed intake and thus the growth, a normal healthy metabolism and the general health of the fish should be intensified to secure increased production of farmed fish without being too dependent on the limiting marine ingredients available.

4. References

Abbouti, T., Mambrini, M., Larondelle, Y. & Rollin, X. (2009). The effect of dispensable amino acids on nitrogen and amino acid losses in Atlantic salmon (*Salmo salar*) fry fed a protein-free diet. *Aquaculture* 289, 327-333

Aksnes, A., Hope, B. & Albrektsen, S. (2006). Size-fractionated fish hydrolysate as feed ingredient for rainbow trout (*Oncorhynchus mykiss*) fed high plant protein diets. II: Flesh quality, absorption, retention and fillet levels of taurine and anserine. *Aquaculture*, 261, 318-326

Berge, G.E., Lied, E. &, Espe, M. (1994). Absorption and incorporation of dietary free and protein bound (U[14]C)-lysine in Atlantic cod (*Gadus morhua*). *Comparative Biochemistry and Physiology* 109A, 681-688

Berge, G.M., Baeverfjord, G., Skrede, A. & Storebakken, T. (2005). Bacterial protein grown on natural gas as protein source for Atlantic salmon, *Salmo salar*, in saltwater. *Aquaculture* 244, 233-240

Carter, C.G. & Hauler, R.C. (2000). Fish meal replacement by plant meals in extruded feeds for Atlantic salmon, *Salmo salar* L. *Aquaculture*, 185, 299-311.

Cowey. C.B. (1995). Protein and amino acid requirements: A critique of methods. *Journal of Applied Ichthyology* 11, 199-204

Dabrowski, K, & Guderley, H.Y. (2002). Intermediary metabolism. In: *Fish nutrition* (Ed: Halver J & Hardy R), Academic press, pp 309-365

Dias, J., Alvarez, M.J., Arzel, J., Corraze, G., Diez, A., Bautista, J.M. & Kaushik, S.J. (2005). Dietary protein source affects lipid metabolism in the European seabass (*Dicentrarchus labrax*). *Comparative Biochemistry and Physiology* 142, 19-31.

El-Mowafi, A., Ruohonen, K., Hevrøy, E.M. & Espe, M. (2010). Impact of three amino acid profiles on growth performance in Atlantic salmon. *Aquaculture Research* 41, 373-384

Espe, M. (2008). Understanding factors affecting flesh quality in farmed fish. In: *Improving farmed fish for the consumer* (Ed: Ø Lie), Woodhead Publishing, Cambridge UK pp 241-269

Espe M & Njaa LR (1991) Growth and chemical composition of Atlantic salmon (*Salmo salar*) given a fish meal diet or a corresponding free amino acid diet. *Fiskeridirektoratets Skrifter Serie Ernaering* 4, 103-110

Espe, M. & Lied, E. (1994). Do Atlantic salmon (*Salmo salar*) utilize mixtures of free amino acids to the same extent as intact protein sources for synthetic purposes? *Comparative Biochemistry and Physiology* 107A, 249-254.

Espe, M. & Lied, E. (1999). Fish silage prepared from different cooked and uncooked raw materials. Chemical changes during storage at different temperatures. *Journal of Food Science and Agriculture* 79, 327-332.

Espe, M., Raa, J. & Njaa, L.R. (1989). Nutritional value of stored fish silage as a protein source for young rats. *Journal of Food Science and Agriculture* 49, 259-270

Espe, M., Haaland, H. & Njaa, L.R. (1992a). Substitution of fish silage protein and a free amino acid mixture for fish meal protein in a chicken diet. *Journal of Food science and Agriculture* 58, 315-320

Espe, M., Haaland, H. & Njaa, L.R. (1992b). Growth of young rats on diets based on fish silage with different degrees of hydrolysis. *Food Chemistry* 44, 195-200.

Espe, M., Haaland, H. & Njaa, L.R. (1992c). Autolysed fish silage as a feed ingredient for Atlantic salmon (*Salmo salar*). *Comparative Biochemistry and Physiology* 103A, 369-372

Espe, M., Sveier, H., Høgøy, I. & Lied, E. (1999). Nutrient absorption and growth of Atlantic salmon (*Salmo salar* L.) fed fish protein concentrate. *Aquaculture* 174, 119-137

Espe, M., Lemme, A., Petri, A. & El-Mowafi, A. (2006). Can Atlantic salmon (*Salmo salar*) grow on diets devoid of fish meal? *Aquaculture* 255, 255-262

Espe, M., Lemme, A., Petri, A. & El-Mowafi, A. (2007). Assessment of lysine requirement for maximal protein accretion in Atlantic salmon using plant protein diets. *Aquaculture* 263, 168-178

Espe, M., Hevrøy, E.M., Liaset, B., Lemme, A. & El-Mowafi, A. (2008). Methionine intake affect hepatic sulphur metabolism in Atlantic salmon, *Salmo salar*. *Aquaculture* 274, 132-141.

Espe, M., Rathore, R.M., Due, Z-Y., Liaset, B. & El-Mowafi, A. (2010). Methionine limitation results in increased hepatic FAS activity, higher liver 18:1 to 18:0 fatty acid ratio and hepatic TAG accumulation in Atlantic salmon, *Salmo salar*. *Amino Acids* 39, 449-460

Espe, M., Ruohonen, K. & El.Mowafi, A. (2011). Effect of taurine supplementation on the metabolism and body lipid-to-protein ratio in juvenile Atlantic salmon (*Salmo salar*). *Aquaculture Research* In press Doi: 10.1111/j.1365-2109.2011.02837.x

FAO. (2006). The State of World Fisheries and Aquaculture. FAO, Rome, 3 pp.

Finkelstein, J.D. (1990). Methionine metabolism in mammals. *Journal of Nutritional Biochemistry* 1, 228-237.

Fournier, V., Gouillou-Coustans, M.F., Metailler, R., Vachot, C., Moriceau, J., LeDelliou, H., Huelvan, C., Desbrugeres, E. & Kaushik, S.J. (2003). Excess arginine affects urea excretion but does not improve N utilisation in rainbow trout *Oncorhynchus mykiss* and turbot *Psetta maxima*. *Aquaculture* 217, 559-576

Fournier, V., Huelvan, C. & Desbruyeres, E. (2004). Incorporation of a mixture of plant feedstuffs as substitute for fish meal in diets of juvenile turbot (*Psetta maxima*). *Aquaculture*, 236, 451–465.

Gaylord, T.G., Barrows, F.T., Teague, A.M., Johansen, K.A., Overturf, K.E. & Shepherd, B. (2007). Supplementation of taurine and methionine to all-plant protein diets for rainbow trout (*Oncorhynchus mykiss*). *Aquaculture* 269, 514-524

Glencross, B., Ewans, D., Hawkins, W. & Jones, B. (2004). Evaluation of dietary inclusion of yellow lupin (*Lupinus luteus*) kernel meal on the growth, feed utilisation and tissue histology in rainbow trout (*Oncorhynchus mykiss*). *Aquaculture* 235, 411-422

Geiger, E. (1947). Experiments with delayed supplementation of incomplete amino acid mixtures. *Journal of Nutrition* 34, 97-111

Gomes, E.F., Remam P., Gouveia, A. & Teles, A.O. (1995). Replacement of Fish-Meal by Plant-Proteins in Diets for Rainbow-Trout (*Oncorhynchus-mykiss*) - Effect of the Quality of the Fish-Meal Based Control Diets on Digestibility and Nutrient Balances. *Water Science and Technology* 31, 205-211.

Gomez-Requeni, P., Mingarro, M., Calduch-Giner, J.A., Medale, F., Martin, S., Houlihan, D.F., Kaushik, S. & Perez-Sanchez, J. (2004). Protein growth performance, amino acid utilisation and somatotropic axis responsiveness to fish meal replacement by plant protein sources in gilthead sea bream (*Sparus aurata*). *Aquaculture* 232, 493-510.

Hansen, A.C., Karlsen, O., Rosenlund, G., Rimbach, M. & Hemre, G-I. (2007). Dietary plant protein utilization in Atlantic cod, *Gadus morhua* L. *Aquaculture Nutrition* 13, 200-215.

Hevrøy, E.M., Espe, M., Waagbø, R., Sandnes, K., Ruud, M. & Hemre, G-I. (2005). Nutrient utilization n Atlantic salmon (*Salmo salar* L.) fed increased levels of fish protein hydrolysate during a period of fast growth. *Aquaculture Nutrition* 11, 301-313.

Houlihan, D.F., Carter, C.G. & McCarthy, I.D. (1995). Protein synthesis in fish. In: *Biochemistry and Molecular Biology in Fishes* (Eds: Hochachka, PW & Mommsen ID), Elsevier Science, Amsterdam, pp 191-219.

Kousoulaki, K., Albrektsen, S., Langmyhr, E., Olsen, H.J., Campbell, P. & Aksnes, A. (2009). The water soluble fraction in fish meal (stickwater) stimulates growth in Atlantic salmon (*Salmo salar*, L) given high plant protein diets. *Aquaculture* 289, 74-83.

Kaushik, S.J., Cravedi, J.P., Lalles, J.P., Sumpter, J., Fauconneau, B. & Laroche, M. (1995). Partial or total replacement of fish meal by soybean protein on growth, protein utilization, potential estrogenic or antigenic effects, cholesterolemia and flesh quality in rainbow trout (*Oncorhynchus mykiss*). *Aquaculture* 133, 257 – 274.

Kaushik, S.J., Covès, D., Dutto, G. & Blanc, D. (2004). Almost total replacement of fish meal by plant protein sources in the diet of a marine teleost, the European sea bass (*Dicentrarchus labrax*). *Aquaculture* 230, 391 – 404.

Krogdahl, A., Bakke-McKellup, A.M. & Baeverfjord, G. (2003). Effects of graded levels of standard soybean meal on intestinal structure, mucosal enzyme activities, and pancreatic response in Atlantic salmon (*Salmo salar* L.). *Aquaculture Nutrition* 9, 361-371.

Laksesvela, B. (1958). Protein value and amino-acid balance of condensed herring soluble and spontaneously heated herring meal. Chick experiments. *Journal of Agricultural Science* 51, 164-176

Laksesvela, B. (1960). The potency of balancing interactions between dietary proteins. Chick experiments on the significance for prior judgment and use of herring proteins. *Journal of Agricultural Science* 55, 215-223

Leaver, M.J., Bautista, J.M., Bjornsson, B.T., Jonsson, E., Krey, G., Tocher, D.R. & Torstensen, B.E. (2009). Towards fish lipid nutrigenomics: Current state and prospects for finfish aquaculture. *Reviews in Fisheries Science* 16, 73-94

Liaset, B. & Espe, M. (2008). Nutritional composition of two main peptide fractions obtained in the enzymatic hydrolysis of fish raw material. *Process Biochemistry* 43, 42-48

Liaset, B., Lied, E. & Espe, M. (2000). Enzymatic hydrolysis of by-products from the fish-filleting industry; chemical characterisation and nutritional evaluation. *Journal of Food Science and Agriculture* 80, 581-589.

Liaset, B., Madsen, L., Qin, H., Criales, G., Mellgren, G., Marschall, H.U., Hallenborg, P., Espe, M., Frøyland, L. & Kristiansen, K. (2009). Fish protein hydrolysate elevates plasma bile acids and reduces visceral adipose tissue mass in rats. *Biochimica et Biophysica Acta – Molecular and Cell Biology of Lipids* 1791, 254-262

Mambrini, M., Roem, A.J., Cravèdi, J.P., Lallès, J.P. & Kaushik, S.J. (1999). Effects of replacing fish meal with soy protein concentrate and of DL-methionine supplementation in high-energy extruded diets on the growth and nutrient utilization of rainbow trout, *Oncorhynchus mykiss*. *Journal of Animal Science* 77, 2990-2999.

Mato, J.M., Corrales, F.J., Lu, S.C. & Avila, M.A. (2002). S-adenosylmethionine: a control switch that regulates liver function. *FASEB Journal* 16, 15-26.

Millward, D.J. (1988). Metabolic demand for amino acids and the human dietary requirement: Millward and Rivers (1988) revisited. *Journal of Nutrition* 128, 2563S-2576S

Millward, D.J. (1999). The nutritional value of plant based diets in relation to human amino acid and protein requirements. *Proceedings of the Nutritional Society* 58, 249-260.

Njaa, L.R. (1990). Amino acid contents of fillet protein from 13 species of fish. *Fiskeridirektoratets Skrifter Serie Ernaering*, III, 43-45

National Research Council (NRC) (1993). Nutritional Requirements of Fish, National Academy Press, Washington DC, USA 114 pp.

Olli, J.J., Krogdahl, Å. & Våbenø, A. (1995). Dehulled solvent-extracted soybean meal as a protein source in diets for Atlantic salmon, *Salmo salar*, L. *Aquaculture Research* 26, 167-174.

Opstvedt, J., Aksnes, A., Hope, B. & Pike, I.H. (2003). Efficiency of feed utilization in Atlantic salmon (*Salmo salar* L.) fed diets with increasing substitution of fish meal with vegetable proteins. *Aquaculture* 221, 365-379.

Ozyurt, M. & Deveci, U.D. (2004). Conversion of agricultural and industrial wastes for single cell production and pollution potential reduction: A review. *Fresenius Environmental Bulletin* 13, 693-699

Pongmaneerat, J. & Watanabe, T. (1992). Utilization of soybean-meal as protein-source in diets for rainbow trout. *Nippon Suisan Gakkaishi* 58, 1983-1990

Rahnema, S., Borton, R. & Shaw, E. (2005). Determination of the effects of fish vs plant vs meat protein based diets on the growth and health of rainbow trout. *Journal of Applied Animal Research* 27, 77-80

Refstie, S., Storebakken, T. & Roem, A.J. (1998). Feed consumption and conversion in Atlantic salmon (*Salmo salar*) fed diets with fish meal, extracted soybean meal or soybean meal with reduced content of oligosaccharides, trypsin inhibitors, lectins and soya antigens. *Aquaculture* 162, 301-312.

Refstie, S., Olli, J.J.& Standal, H. (2004). Feed intake, growth and protein utilisation by post-smolt Atlantic salmon (*Salmo salar*) in response to graded levels of fish protein hydrolysate in the diet. *Aquaculture* 239, 331 – 349.

Refstie, S., Korsoen, O.J., Storebakken, T., Baeverfjord, G., Lein, I. & Roem, A.J. (2000). Differing nutritional responses to dietary soybean meal in rainbow trout (*Oncorhynchus mykiss*) and Atlantic salmon (*Salmo salar*). *Aquaculture* 190, 49-63.

Regost, C., Arzel, J. & Kaushik, S.J. (1999). Partial or total replacement of fish meal by corn gluten meal in diet for turbot (*Psetta maxima*). *Aquaculture* 180, 99-117.

Robaina, L., Izquierdo, M.S., Moyano, F.J., Socorro, J., Vergara, J.M., Montero, D. & Fernandezpalacios, H. (1995). Soybean and lupin seed meals as protein-sources in diets for gilthead seabream (*Sparus-aurata*) - Nutritional and histological implications. *Aquaculture* 130, 219-233.

Rollin, X., Mambrini, M., Abboudi, T., Larondelle, Y. & Kaushik, S.J. (2003). The optimum dietary indispensable amino acid pattern for growing Atlantic salmon (*Salmo salar* L.) fry. *British Journal of Nutrition* 90, 865-876

Rowling, M.J., McMullen, M.H., Chipman, D.C. & Schaliniske, K.L. (2002). Hepatic glycine N-methyltransferase is up-regulated by excess dietary methionine in rats. *Journal of Nutrition* 132, 2545-2550.

Sugiura, S.H., Dong, F.M., Rathbone, C.K. & Hardy, R.W. (1998). Apparent protein digestibility and mineral availabilities in various feed ingredients for salmonid feeds. *Aquaculture* 159, 177-202

Silva, J.M.G., Espe, M., Conceção, L.E.C., Dias, J. & Valente, L.M.P. (2009). Senegalese sole juvenile (*Solea senegalensis* Kaup, 1858) grow equally well on diets devoid of fish meal provided the dietary amino acids are balanced. *Aquaculture* 296, 309-317.

Storebakken, T., Shearer, K.D., Refstie, S., Lagocki, S. & McCool, J. (1998). Interactions between salinity, dietary carbohydrate source and carbohydrate concentration on the digestibility of macronutrients and energy in rainbow trout (*Oncorhynchus mykiss*). *Aquaculture* 163, 3-4.

Storebakken, T., Baeverfjord, G., Skrede, A., Olli, J.J. & Berge, G.M. (2004). Bacterial protein grown on natural gas in diets for Atlantic salmon, *Salmo salar*, in freshwater. *Aquaculture*, 241, 413-425

Tacon, A.G.J. (1995). Feed ingredients for carnivorous fish species: Alternatives to fishmeal and other fishery resources. In: *Sustainable Fish Farming* (Reinertsen H. & Haaland H. eds.) pp. 89-114.)

Tacon, A.G.J. & Metian, M. (2008). Global overview on the use of fish meal and fish oil in industrially compounded aquafeeds: Trends and future prospects. *Aquaculture* 285, 146-158.

Torstensen, B.E., Frøyland, L., Ørnsrud, R. & Lie, Ø. (2004). Tailoring of a cardioprotective fillet fatty acid composition of Atlantic salmon (*Salmo salar*) fed vegetable oils. *Food Chemistry* 87, 567-580.

Torstensen, B.E., Bell, J.G., Rosenlund, G., Henderson, R.J., Graff, I.E., Tocher, D.R., Lie, Ø. & Sargent, J.R. (2005). Tailoring of Atlantic salmon (*Salmo salar* L.) flesh lipid composition and sensory quality by replacing fish oil with a vegetable oil blend. *Journal of Agricultural Food Chemistry* 53, 10166-10178

Torstensen, B.E., Espe, M., Stubhaug, I., Waagbø, R., Hemre, G-I., Fontanillas, R., Nordgarden, U., Hevrøy, E.M., Olsvik, P. & Berntssen, B.H.G. (2008). Combined maximum replacement of fish meal and fish oil with plant meal and vegetable oil blends in diets for Atlantic salmon (*Salmo salar* L.) growing from 0.3 to 4 kilo. *Aquaculture* 285, 193-200.

Torstensen, B.E., Espe, M.., Stubhaug, I. & Lie, Ø. (2011). Combined dietary plant proteins and vegetable oil blends increase adiposity and plasma lipids in Atlantic salmon (*Salmo salar* L.). *British Journal of Nutrition* 106, 633-647. doi: 10.1017/ S0007114511000729

Treberg, J.R. & Driedze, W.R. (2007). The accumulation and synthesis of betaine in winter skate (*Leucoraja ocellata*). *Comparative Biochemistry and Physiology* 147, 475-483

Von derDecken, A. & Lied, E. (1993). Metabolic effects on growth and muscle of soybean protein feeding in cod (*Gadus morhua*). *British Journal of Nutrition* 69, 689-697.

Wang, T.C. & Fuller, M.F. (1987). An optimal dietary amino acid pattern for growing pigs. *Animal Production* 44, 486.

Wang, T.C. & Fuller, M.F. (1989). The optimum dietary amino acid pattern for growing pigs. 1. Experiment by amino acid deletion. *British Journal of Nutrition* 62, 77-89.

Wu G. (2009) Amino acids: metabolism, functions, and nutrition. *Amino Acids* 37, 1-17

Yanik, T., Dabrowski, K. & Bai, S.C. (2003). Replacing fish meal in rainbow trout (*Oncorhynchus mykiss*) diets. *Israeli Journal of Aquaculture* 55, 179-186

4

Nutritional Value and Uses of Microalgae in Aquaculture

A. Catarina Guedes[1] and F. Xavier Malcata[1,2,3*]

[1]CIMAR/CIIMAR – Centro Interdisciplinar de Investigação Marinha e Ambiental, Porto
[2] ISMAI – Instituto Superior da Maia, Castelo da Maia, Avioso S. Pedro
[3]Instituto de Tecnologia Química e Biológica, Universidade Nova de Lisboa, Avenida da República, Oeiras Portugal

1. Introduction

Microalgae (i.e. single-celled algae or phytoplankton) represent the largest, yet one of the most poorly understood groups of microorganisms on Earth. As happens with plants relative to terrestrial animals, microalgae represent the natural nutritional base and primary source of bulk nutrients in the aquatic food chain.

Microalgae play indeed a crucial nutritional role with regard to marine animals in the open sea, and consequently in aquaculture. Most marine invertebrates depend on microalgae for their whole life cycle, so commercial and experimental mollusc or fish hatcheries have included a microalga production system in parallel to their animal production itself. Microalgae are utilized as live feed for all growth stages of bivalve molluscs (e.g. oysters, scallops, clams and mussels), for larval/early juvenile stages of abalone, crustaceans and some fish species, and for zooplankton used in aquaculture food webs at large. It should be emphasized that the productivity of any hatchery is directly related to the quantity and quality of the food source used therein.

On the other hand, the concept of aquaculture as a set of engineered systems in terms of wastewater treatment and recycling has received an impetus over the past few years. They are designed to meet specific treatment and wastewater specifications, and may simultaneously solve environmental and sanitary problems along with economic feasibility [1,2]. A renewed interest has also been experienced by high rate microalgal ponds for treatment of wastewater – where photosynthetic microalgae supply oxygen to heterotrophic bacteria, and where wastewater-borne nutrients are converted into biomass protein [2,3]. Therefore, microalga culturing is likely to play an increasingly important role in aquatic food production modules, specifically to produce (or be used as) feed for fish, convert CO_2 to O_2 and remediate water quality.

---

* Corresponding Author

2. General attributes of microalgal species in aquaculture

Unlike air-breathing animals, those living in aquatic media and used for large scale human consumption as food are seldom herbivorous at the adult stage; most farmed animals are indeed carnivorous from their post-larval stage on, or omnivorous at best. The associated food web is accordingly longer, so only filtering molluscs and a few other animals truly depend on plankton throughout their lifetime. However, microalgae are required for larva nutrition during a brief period – either for direct consumption in the case of molluscs and penaeid shrimps, or indirectly for the live prey fed to small-larval fish. In these cases, the post-larvae specimens are hatched, bred and raised by specialized establishments (hatcheries) – which are particularly complex to operate because they involve intensive production of microalgae and, in the case of small-larval fish, production of such small live prey as rotifers. Aquacultured animals for which rearing does not exhibit these constraints are seldom found; this is the case of salmonids, whose eggs have sufficient reserves to hatch big larvae capable of feeding directly on dry particles [4].

Over the last decades, several hundred microalga species have been tested as feed, but probably less than twenty have experienced a widespread application in aquaculture. In fact, microalgal species vary significantly in their nutritional value – which is also dependent on culturing conditions [5,6]. To provide a better balanced nutrition package and more effectively improve animal growth, a carefully selected mixture of microalgae should be fed to fish, directly or indirectly (through enrichment of zooplankton) – as this leads to better results than a diet composed of a single microalga [7,8].

Microalga production for use as feed is divided into intensive monoculture – for larval stages of bivalves, shrimp and certain fish species, and extensive culture – for growth of bivalves, carp and shrimp. Favored genera for the former include *Chaetoceros*, *Thalassiosira*, *Tetraselmis*, *Isochrysis*, *Nannochloropsis*, *Pavlova and Skeletonema* [6,9,10]. These organisms are fed directly or indirectly to the cultured larval organism; indirect means of providing them are usually through artemia, rotifers and *Daphnia* – which are, in turn, fed to the target larval organisms. It is widely accepted that microalgae are actively taken up by shrimp larvae, and play an important role in nutrition at that life stage; however, it is uncertain whether juveniles and adults do actively feed on microalgae as well. Some reports suggest that microalgae are found in their gut because shrimp accidentally ingest them together with debris [11,12].

The nutritional value of a microalgal diet is critically related to its ability to supply essential macro- and micronutrients to the target animal consumer. As emphasized above, a mixed microalgal diet – as routinely used in the hatchery and nursery phases of oyster cultivation [13], is likely to outperform monoalgal diets [14]. However, the nutritional requirements of bivalves are poorly defined; feeding experiments with microalgae of partially defined compositions have shown that carbohydrate and polyunsaturated fatty acid (PUFA) levels are major factors for growth of oysters [5]. Supply of additional dietary carbohydrate was found to increase oyster growth rate, provided that adequate protein and essential fatty acids were concomitantly supplied [9]. Supplementation of juvenile mussel diets with protein microcapsules led to a positive growth response, and indicated that a protein content below 40 %(w/w) significantly contrains mussel growth rates [15].

Diatoms and haptophytes (prymnesiophytes) are nutritious microalgae that are frequently used as feed for oysters [13]. The prymnesiophytes *Isochrysis* sp. and *P. lutheri* are rich sources of docosahexaenoic acid (DHA, 22:6n-3) – comprising 8-10% total fatty acids [16], while diatoms are a rich source of eicosapentaenoic acid (EPA, 20:5n-3) [17]. Mixed microalgal diets of prymnesiophytes and diatoms are common in bivalve hatcheries, and considered as highly nutritious in terms of requirements for essential PUFAs [18].

Microalgae should, in general, possess a number of key attributes to be useful for aquacultured species: they should be of an appropriate size and shape for ingestion and ready digestion (i.e. they should have a digestible cell wall to make nutrients easily available); they should undergo fast growth rates, and be amenable to mass culture; they should be stable to fluctuations in temperature, light and nutrient profile, as often occur in hatchery systems; and they should exhibit appropriate nutritional qualities, including absence of toxins (that might otherwise accumulate through the food chain). A major challenge faced by algologists is thus to reduce production costs, while maintaining reliability of microalgal feed.

Microalgae provide food for zooplankton, but they can also help stabilize (and even improve) the quality of the culture medium. For numerous freshwater and seawater animal species, introduction of phytoplankton to rearing ponds (the so-called green-water technique) produces much better results in terms of survival, growth and transformation index than the classical clear-water technique [19-21]. The rationale behind this observation is not entirely known, yet it may include water quality improvement by oxygen production and pH stabilization, and action of some excreted biochemical compounds, along with induction of behavioral processes such as initial prey catching and regulation of bacterial population [4,22], probiotic effects [23], and stimulation of immunity [24].

3. Nutritional features of microalgae

Microalgal species can vary significantly in nutritional value, as a function of the prevailing culture conditions. Only a reduced number of species have been used, primarily for historical reasons and ease of cultivation – rather than supported by scientific evidence of any superior performance as nutritional or therapeutical supplements. Hence, formulations more carefully selected of microalgal origin may offer the opportunity for development of improved nutritional packages aimed at larval animals.

Several factors contribute to the nutritional value of a microalga – including its size and shape, and digestibility as related to cell wall structure and composition (as mentioned above), as well as biochemical composition (e.g. accumulation compounds, enzymes and toxins) and specific requirements of the target animal. For this reason, several studies have attempted to correlate the nutritional value of microalgae to their chemical profile. However, results from feeding experiments are often difficult to interpret because of the confounding effects of other formulation additives. An examination of literature data – including those pertaining to microalga-based, compounded diet emulsions, have meanwhile allowed a few general conclusions to be reached [25].

As primary producers in the aquatic food chain, microalgae provide many phytonutrients, including in particular PUFAs – e.g. EPA, arachidonic acid (AA) and DHA, which are known to be essential for various marine animals [25], as well as for growth and metamorphosis of many larvae [8,26]. However, the ratios of DHA, EPA and AA may actually be more important than their absolute levels [24,27]. Most microalgal species exhibit moderate to high percents of EPA (7 to 34%); and prymnesiophytes (e.g. *Pavlova* spp. and *Isochrysis* sp.) and cryptomonads are relatively rich in DHA (0.2 to 11%), whereas eustigmatophytes (e.g. *Nannochloropsis* spp.) and diatoms have the highest percentages of AA (up to 4%). Chlorophytes (*Dunaliella* spp. and *Chlorella* spp.) are deficient in both C20 and C22 PUFAs, although some species have small amounts of EPA (up to 3.2%); because of such a PUFA deficiency, chlorophytes are in general ascribed a poor nutritional value, so they are not suitable for use as single species-diet [6]. Prasinophyte species contain significant proportions of C20 (*Tetraselmis* spp.) or C22 (*Micromonas* spp.), but rarely of both. Therefore, the fatty acid contents of microalgae exhibit systematic differences according to taxonomic group – although there are examples of significant differences between microalgae, even within the same class.

The contents of antioxidants are also not uniform among microalgae; e.g. the concentrations of vitamins and carotenoids convey significant variations among species. Note that any mixed-algal diet should provide adequate concentrations of vitamins and carotenoids to be effective in aquaculture; unfortunately, the nutritional requirements of larval or juvenile animals that feed directly on microalgae are still poorly understood at present. In fact, artificial diets often lack natural pigments that allow such organisms as salmon or trout acquire their characteristic red color (muscle), which, in nature, is a result of eating microalgae containing red pigments; without such a color, a lower market value results. One way to alleviate this shortcoming is by adding astaxanthin to fish feed, with a consequently growing market for microalga-based sources, e.g. *Haematococcus pluvialis* [24,28].

On the other hand, the amino acid composition of microalgal proteins is rather similar between species [29], and relatively unaffected by their intrinsic growth phase and extrinsic light conditions [30,31]. Furthermore, the content in essential amino acids of microalgae is similar to that of oyster larvae. Overall, this indicates that protein quality is unlikely a factor that contributes to differences in nutritional value among microalgae. Finally, sterols [32], minerals [33] and pigments of microalgae also contribute to their nutritional performance in aquaculture.

Several studies have indicated that, in the late-logarithmic growth phase, microalgae contain typically 30-40 %(w/w) protein, 10-20 %(w/w) lipids and 5-15 %(w/w) carbohydrates [6,34]. When cultured through the stationary phase, the proximate composition of microalgae may significantly change; e.g. nitrate limitation leads carbohydrate levels to double at the expense of protein [31,35]. Hence, a strong correlation exists between composition of microalgae and their measurable nutritional value – even though diets containing high levels of carbohydrates have been reported to produce the best growth of juvenile oysters [9] and larval scallops [36], as long as PUFAs are also present to adequate proportions. Conversely, high dietary protein provides maximum growth for juvenile mussels [15] and oysters [18].

Another relevant issue is that marine environments are typically filled with bacteria and viruses that can attack fish and shellfish, and thus potentially devastate aquaculture farms. Bacteria and viruses can also attack single-celled microalgae, so these microorganisms have developed biochemical mechanisms for self-defense; such mechanisms involve secretion of compounds that inhibit bacterial growth or viral attachment. For instance, compounds synthesized by *Scenedesmus costatum*, and partially purified from its organic extract exhibited activity against aquacultured bacteria because of their fatty acids longer than 10 carbon atoms in chain length – which apparently induce lysis of bacterial protoplasts.

The ability of fatty acids at large to interfere with bacterial growth and survival has been known for some time, and recent structure-function relationship studies have proven that said ability depends on both their chain length and degree of unsaturation. Cholesterol and other compounds can antagonize antimicrobial features [37], so both composition and concentration of free lipids should be taken into account [38]. The activity of extracts of *Phaeodactylum tricornutum* against *Vibrio* spp. was attributed to EPA – a compound synthesized *de novo* by diatoms [39]; this PUFA is found chiefly as a polar lipid species in structural cell components (e.g. membranes) and is toxic to grazers [40], as well as a precursor of aldehydes with deleterious effects upon such consumers as copepods [41]. Similarly, unsaturated and saturated long chain fatty acids isolated from *S. costatum* [42] and organic extracts from *Euglena viridis* [43] display activity against that bacterial genus.

4. Microalgal biomass production systems

Commercial culture of microalgae targeted at their metabolites has been taking place for over 40 years, and the main microalgal species grown are *Chlorella* and *Spirulina* for healthy foods, *Dunaliella salina* for β-carotene, *H. pluvialis* for astaxanthin and several species for aquaculture [44].

There are several reactor configurations that met with success in mass cultivation of microalgae – chosen according to such factors as physiology of the microalga, cost of layout land, intensity of labour, cost of energy, availability of water, cost of nutrients, suitability of climate (if the culture is implemented outdoors) and specification of final product(s). Large-scale culture systems should be compared on the basis of such indicators as efficiency of light utilization, controllability of temperature, hydrodynamic stress allowable, ability to maintain unimicroalgal and/or axenic cultures and feasibility of scale-up.

A major decision to be made is whether to use closed photobioreactors (PBRs) or open ponds to cultivate a given microalga. The latter may entertain a large area, and are relatively cheap to build and easy to operate – but contamination is hard to control, stable environmental conditions (particularly temperature) are difficult to maintain, and the attainable cell density is relatively low because of mutual shading effects. On the other hand, extensive areas of land will be needed for commercial exploitation, besides substantial costs of harvesting afterwards [45,46]. The final choice of system is always a compromise between these parameters, aimed at achieving an economically acceptable outcome [44].

A common feature of most microalgal species produced commercially (i.e. *Chlorella*, *Spirulina* and *Dunaliella*) is that they grow in highly selective environments – which means that they can be grown in open air cultures and still remain relatively free of contamination by other microalgae and protozoa [47-49]. Species of microalgae that do not possess this selective advantage must be grown in closed systems; this includes most marine algae grown as aquaculture feeds (e.g. *Skeletonema*, *Chaetoceros*, *Thalassiosira*, *Tetraselmis* and *Isochrysis*), as well as the dinoflagellate *Crypthecodinium cohnii* [44].

Typical systems used indoors for microalgal mass culture include carboys (10 to 20 L), polythene bags (100 to 500 L) and tubs (1000 to 5000 L); these are usually operated batch- or continuouswise [44]. For larger volumes, outdoor tanks or ponds are preferred, which are operated semicontinuously; depending on their scale, hatcheries may produce between several hundreds to tens of thousands of liters of microalgal biomass per day. However, the culture systems employed at present are still fairly unsophisticated: e.g. *D. salina* is cultured in large (up to ca. 250 ha) shallow open-air ponds with no artificial mixing; and *Chlorella* and *Spirulina* are also grown outdoors, in either paddle-wheel mixed ponds or circular ponds (up to 1 ha each) with a rotating mixing arm. The production of microalgae for aquaculture occurs generally on a much lower scale. Other commercial large-scale systems include tanks used in aquaculture, the cascade system developed in the Czech Republic [50] and heterotrophic fermenter devices used for culture of *Chlorella* in Japan and Taiwan [47,51], and for culture of *C. cohnii* in USA [52,53].

The choice of which configuration is preferable depends obviously on the objective function; e.g. wastewater treatment might preclude open systems, owing to the unacceptably high costs arising from the large volumes to be processed and the low added value of the resulting products [54]. There has been a major effort directed at examining alternatives for the production of fresh microalgae, and also at more cost-efficient production systems.

4.1 Open cultivation systems

Microalgae cultivation in open ponds has been in current used since the 1950s [44]; these systems have been categorized as natural waters (lakes, lagoons and ponds), and artificial ponds or containers – with raceway ponds being the most frequently used artificial system [55]. The four major types of open-air systems currently in use (i.e. shallow big ponds, tanks, circular ponds and raceway ponds) have all advantages and disadvantages. This type of system usually consists of either circular ponds with a rotating arm to mix the culture, or long channels in a single or multiple loop configuration stirred by paddle wheels [56] – although simpler configurations have also met with sucess [54]. Raceway ponds are usually built in concrete, but compacted earth-lined ponds with (white) plastic have also been proposed. In a continuous production cycle, broth and nutrients required by microalgal growth are introduced in front of the paddlewheel, and circulated through the loop to the harvest extraction point; said paddlewheel undergoes a continuous motion to prevent sedimentation. The CO_2 requirement is usually satisfied using the open atmosphere as source – yet submerged aerators may be installed to enhance CO_2 supply, and thus absorption yield [57].

Compared to closed photobioreactors, open ponds represent a less expensive investment for large-scale production of microalgal biomass. On the other hand, open pond production does not necessarily compete with agricultural crops for land, since it can be implemented in areas with marginal crop production potential [58]. Open ponds also have low energy input requirements [59], and regular maintenance and cleaning are easier [60].

Open ponds and raceways were the first large-scale designs implemented, and are still the most widely applied in industrial processing. The main constraints related to their operation are the difficulty to control contamination and to keep the culture environment steady, and the cost associated with harvesting. Furthermore, the open character of the system makes it possible for naturally occurring microalgae or their predators to infiltrate, and thus compete with microalgae intended for cultivation. Therefore, a monoculture can only be maintained under extreme conditions of pH, salinity or temperature that guarantee dominance by the desired strain (e.g. *D. salina* dominance requires highly salted media, whereas *Spirulina platensis* demands high pH values). Unfortunately, high pH, temperature and salt concentration are not compatible with most microalgal species of interest.

Regarding biomass productivity, however, open pond systems are less efficient than closed photobioreactors [61]. This can be attributed to such parameters as evaporation losses, temperature variation, CO_2 deficiency, inefficient mixing and light limitation. Although evaporation losses make a net contribution to cooling, they may also cause significant changes in the ionic composition of the medium – with detrimental effects upon microalga growth [62]. Although this type of reactor is extensively used in industrial microalgal production – e.g. to produce *Spirulina* and *Dunaliella* spp. up to worldwide totals of 5000 and 1200 ton/yr, respectively [24], open systems have apparently reached their upper limit – with little room for further technological improvement.

4.2 Closed cultivation systems

Despite the success of open systems, future advances in microalgal mass culture will require improved closed systems, as the most interesting microalgal species cannot grow in highly selective environments [44]. Hence, photobioreactor technology is on the rise, which is designed to overcome the major constraints associated with open pond production systems [63]; recall that both pollution and contamination risks preclude use of open ponds to prepare high-value products for eventual use as active ingredients in aquaculture feed formulation [60].

Closed systems include tubular, flat plate and fermenter types, among other possibilities. The former two are specifically designed for efficient recovery of sunlight, whereas the latter may require artificial illumination. Owing to the higher cell mass productivities attained, harvesting costs can be significantly reduced. Closed photobioreactors also provide reproducible cultivation conditions, good heat transfer, better biomass yield, higher product quality and opportunity for flexible technical design [44,60]. Note, however, that the costs of closed systems are substantially higher than their open pond counterparts [54]. A variety of closed photobioreactors have been tested (or at least proposed) for industrial microalgal biomass production [64,65], but engineering and economic analyses of such reactors still lag behind the open ponds [66-70].

A typical photobioreactor is essentially a four-phase system, consisting of solid microalgal cells, a liquid growth medium, a gaseous phase and a light radiation field [71]. Its productivity is limited by various design features – but, most importantly, the reactor is to be operated under favorable illumination conditions, with optimized surface-to-volume ratio and light/dark cycle, coupled with adequate mass transfer features [72].

The current consensus is that commercial (photoautotrophic) production of metabolites with interest for aquaculture by microalgae should resort to outdoor enclosed photobioreactors [56,62,65,73,74]. Tredici [65] reviewed the development of those type of reactors over the last decade; while many types of experimental PBRs have been considered, built and tested, very few have actually succeeded on a commercial level. Commercial application of photobioreactor technology remains indeed restricted to the production of two Chlorophyte microalgae: *Chlorella* and *Haematococcus* [62,75].

Scale up of photobioreactors from bench to commercial scale is not trivial – since it needs changes in illumination, gas transfer and temperature to be taken into account, all of which are severely affected by turbulence in the reactor, and consequently require a tight control. Therefore, scale up appears to be much more of an engineering problem than a biological one; and general recommendations as to possible maximum scales have accordingly been produced [75,76].

5. Alternatives to fresh microalgae

Marine microalgae have been the traditional food component in finfish and shellfish aquaculture, e.g. for larval and juvenile animals [77]; they are indeed essential in hatchery and nursery of bivalves, shrimp and some finfish cultures. Microalgae are also used to produce zooplankton – typically rotifers, which are in turn fed to freshly hatched carnivorous fish [78]. As aquaculture industry expands [79] – and since microalgal biomass cultivation on-site may represent up to 30% of the operating costs [13], there is a demand for marine microalgae that cannot be met by the conventional methods used in hatcheries – thus forcing one to resort to substitutes with mediocre results that bring about several problems [44,80].

Despite the obvious advantages of alive microalgae in aquaculture, the current trend is to avoid using them because of their high cost and difficulty in producing, concentrating and storing them [8,81]. Alternatives that are potentially more cost-effective have been investigated – including nonliving food, viz. microalga pastes, dried microalgae, microencapsulates, cryopreservation, flocculation, bacteria or yeasts; they have been tested *in vitro* and in actual hatcheries, but met with variable degrees of success [82,83]. For instance, in Japan, where *Nannochloropsis oculata* is the most important cultured feed for the rotifer *Brachionus plicatilis*, concentrated suspensions and frozen biomass of this microalga are commercially available [84]; and partial replacement of alive microalgae by microencapsulated and yeast-based diets is indeed a routine practice in hatcheries for penaeid shrimp [24,85]. However, most these approaches have proven unsuitable as major dietary components, because of their lower nutritional value than mixtures of alive microalgae.

Several criteria should be addressed in attempts to find substitutes for alive microalgae as diet in aquaculture. From a nutrition standpoint, alive microalgae possess higher nutritional value and better digestibility than most substitutes; note that the nutritional quality depends critically on such biochemical constituents as PUFAs, vitamins, sterols and carbohydrates [86].

Useful bacteria can provide only a part of the metabolic requirements in aquaculture – by supplying a few organic molecules and vitamins. Under conditions close to those found in rearing facilities, the bacterial input should not represent more than 15% of the microalgal contribution for mollusk larvae and juveniles of many species [87,88]. Yeasts were as well investigated as an alternative food source – but poor results were observed [83,89]. Therefore, these two alternatives are not suitable to fully replace alive microalgae.

An alternative diet with an apparently better potential is microalgal pastes or concentrates [90-92]; these are prepared by centrifugation (up to 1:500 concentration) or flocculation (up to 1:100 concentration). Concentrates prepared from distinct microalgae vary in their suitability – with diatoms being the most promising; and they have a shelf life of between 2 and 8 weeks, when stored below 4°C. Commercially, microalgal concentrates can be prepared under two different scenarios: (a) by hatcheries on-site, which prepare concentrates as back-up or as a means to store overproduction; or (b) by remote production, centralized at a large facility – with a greater economy of scale, with the resulting concentrates dispatched to hatcheries upon request.

The advantage of such concentrates is that they can be used "off-the-shelf", thus contributing favorably to the cost-efficiency in hatcheries. On the other hand, the lower nutritional value of most dried microalgae compared to alive feed, and the limited availability of commercial dried products appear as main shortcomings. Globally speaking, concentrates have low levels (or even absence) of ω3-PUFA, and lead to a difficult digestion by bivalve larvae [93]. The genus *Tetraselmis* seems to be a good candidate for microalgal paste, but its nutritional quality deteriorates quite rapidly [94]; experiments have indicated that such substitutes should be used as supplement only when rations of live microalgae are insufficient. Furthermore, spray-dried microalgae and microalga paste may be useful to replace up to 50% of alive microalgae. Coutteau and Sorgeloos [13] reported that artificial or non-living diets are rarely applied in routine processing of bivalves, and are mostly considered as a backup food source only. Centrifuged concentrates of *P. lutheri*, in combination with *Chaetoceros calcitrans* or *S. costatum*, lead to 85-90% of the growth when a mixed diet of alive microalgae for oyster *Saccostrea glomerata* larvae is used [92].

Centrifugation has been successfully applied to prepare concentrates, but it has some limitations – i.e. the process involves exposing cells to high gravitational and shear forces that damage the cell structure. On the other hand, processing of large culture volumes is time-consuming and requires costly equipment, i.e. a specialized continuous centrifuge. Research on post-harvest preservation is required to extend shelf-life beyond 4 to 8 weeks, and also to prepare concentrates from flagellate species (e.g. *Isochrysis* sp. and *P. lutheri*).

Alternative processes have meanwhile been developed that are potentially less damaging to cells – including foam fractionation [95], flocculation [96,97] and filtration [98]. Sandbank [99] fed microalgae, grown in waste-water and flocculated with aluminium sulfate, to common carp (*Cyprinus carpia*); a diet containing 25% of microalgal meal led to a growth comparable to that by the control diet, with no harmful effects detected upon long term health of the fish. Millamena et al. [96] successfully fed *Penaeus monodon* larvae with dried, flocculated *C. calcitrans* and *Tetraselmis chuii* cells. However, a common disadvantage encountered was that the harvested cells are difficult to disaggregate back to single cells, which is a requirement to feed them to filter-feeding species such as bivalves [100].

A novel technique was developed for flocculation of marine microalgae that appears useful in aquaculture: it entails adjustment of pH of the culture using NaOH, followed by addition of a non-ionic polymer, Magnafloc LT-25; the ensuing flocculate is then harvested and neutralized, thus leading to a final concentration of between 200- and 800-fold. This process was successfully applied to harvest cells of *C. calcitrans* and *C. muelleri, Thalassiosira pseudonana, Attheya septentrionalis, Nitzschia closterium, Skeletonema* sp.*, Tetraselmis suecica* and

Rhodomonas salina, with efficiencies above 80%; it proved rapid, simple and inexpensive, and relatively independent of processed volume (unlike concentration by centrifugation). The harvested material was readily disaggregated to single cell suspensions by dilution in seawater, coupled with mild agitation. Microscopic examination proved that the final cells are indistinguishable from the nonflocculated ones; and assay for chlorophyll of the concentrates prepared from cultures of up to 130 L showed marginal degradation by 2 weeks of storage [100].

Cryopreservation has been thoroughly adopted by culture collections to preserve strains, but may also find an application in aquaculture [80]. Viable cryostorage of biological specimens has followed various protocols of cooling/thawing rates and cryoprotectant addition, which have been developed and tuned more or less empirically [101]. Recall that temperatures used for cryostorage are well below freezing – down to even -196 °C in liquid helium, when biological specimens are to be stored without limit [102]. While cryostorage is generally thought to be innocuous to the cell, the events occurring upon freezing or thawing can lead to severe damage, or even cell death. Moreover, cryoprotectants that enhance the cell viability at cryogenic temperatures are usually toxic at physiological temperatures [103] – an obstacle that is overcome by reducing the exposure time or the temperature of incubation prior to cryopreservation [104]. Knowledge of cryoprotectant tolerance levels for microalgae is still limited [105], as well as for early larval stages and for zooplankton that are cultivated and rely on the availability of microalgae for growth. In general, cryopreservation possesses a high potential for culture collections, and may also offer a solution for reliable supply of microalgae in aquaculture. For instance, marine microalgae used in aquaculture were successfully cryopreserved under 4, -20 and -80 °C using common cryoprotectants (i.e. methanol, dimethylsulfoxide, propylene glycol and polyvinylpyrrolidone), with promising results at least for *Chlorella minutissima*, *Chlorella stigmatophora*, *Isochrysis galbana* and *Dunaliella tertiolecta* [80].

Several products based on thraustochytrids (i.e. microorganisms with a taxonomy related to certain microalgal classes), from the genus *Schizochytrium*, have been marketed through Aquafauna Biomarine and Sanders Brine Shrimp. These products have high concentrations of DHA [106], and have accordingly been applied as alternatives to commercial oil enrichment of zooplankton fed to larvae. As direct feeds, most such products have a lower nutritional value than mixtures of microalgae, yet some performed well as components of a mixed diet with alive microalgae [83,107].

In general, substitutes of alive microalgae should present an appropriate physical behavior – and this constitutes a significant challenge; in particular, they should not aggregate or easily break apart. Drying microalgae can cause, due to oxidation, a loss of PUFAs [108], which are essential components for larval growth [87]; the poor performance reported for dried microalgae was associated chiefly with the difficulty to keep cells in suspension without disintegrating them, so as to avoid said oxidation [13]. Moreover, when cell walls are broken, a high fraction of water-soluble components cannot be ingested by the organism, and may consequently interfere with the water quality of the aquaculture [109]. Therefore, pathogenic bacterial proliferation may occur, and cause costly production losses. Similar difficulties arise when using microalga paste, because the preparation procedures (i.e. centrifugation, flocculation or filtration) and/or preservation techniques (i.e. additives or freezing) must ensure that cell wall integrity is essentially preserved.

Products other than alive microalgae must obviously be free of bacterial contamination and devoid of toxicity. Consequently, the use of alive bacteria as a food source in hatcheries seems somehow inappropriate, since physical and chemical treatments are often used to limit bacterial contamination that would otherwise be responsible for drastic larval mortality [110]. Oyster larvae fed with alive microalga diets underwent improved growth via addition of some bacterial isolates [111,112], but this advantage may obviously not be possible in a treated microalgal product. However, in alive microalgae, the natural bacterial flora was proven to enhance the health of molluscs. Langdon and Bolton [88] showed that antibiotic suppression of the bacterial flora in artificial feed of juvenile oysters reduced their growth.

In conclusion, mitigated or unsuccessful results when using nonliving microalgae have turned alive microalgae into the first choice in aquaculture feeding. Only partial replacement thereof has been possible in studies encompassing preserved non-living algae [113], microencapsulated diets [88] or spray-dried algae [114]; but no whole replacement can be recommended, despite intensive research efforts in that direction [107]. Consequently, novel solutions to totally replace microalgae in aquaculture diets cannot at present be widely adopted [4,24,81].

6. Use of microalgae to enrich zooplankton

Microalgae have an important role in aquaculture, also as a means to enrich zooplankton for feeding fish and other larvae afterwards. In addition to providing proteins (that contain essential amino acids) and energy, they carry such other key nutrients as vitamins, essential PUFAs, pigments and sterols – which are transferred up through the food chain. For instance, PUFA-rich microalgae, such as *Pavlova* sp. and *Isochrysis* sp., have been successfully fed to zooplankton to enrich them in DHA [115]. However, when the level of enrichment attained is not sufficient, commercial oil-emulsions are often used. Recently, such products as dried preparations of *Schizochytrium* sp. (which contain 5-15% of their DW as DHA) have been utilized, which produce levels of DHA enrichment in zooplankton comparable to use of commercial oils [116] – and also produce DHA to EPA ratios of 1-2, which are considered favorable for fish larval nutrition [117].

Brown, Skabo & Wilkinson [118] described that rotifers fed with microalgae (e.g. *Isochrysis* sp. and *N. oculata*) become rapidly enriched with ascorbic acid (AsA), whereas rotifers fed on baker's yeast (which itself is deficient in AsA) contained only residual amounts of AsA.; after 16 h of starving, rotifers lost ca. 10% of their AsA, while retaining ca. 50% of the total AsA ingested. Similarly, the concentration of AsA in *Artemia* sp. may be increased by feeding with microalgae [119]. However, little information is available on the transfer of other vitamins from microalgae to fish larvae.

Rønnestad, Helland & Lie [120] demonstrated that microalgal pigments transferred to zooplankton may add to their nutritional value; recall that the dominant pigments in the copepod *Temora* sp. are lutein and astaxanthin, whereas in *Artemia* it is canthaxanthin. When these microalgae were fed to copepods and then to halibut larvae, adequate amounts of vitamin A were found, but not when halibut was fed on *Artemia*; this was attributed to the ability of larvae to convert lutein and/or astaxanthin, but not canthaxanthin to vitamin A. They accordingly recommended that *Artemia* should routinely be enriched with astaxanthin and lutein (the latter pigment is common in "green" microalgae, e.g. *Tetraselmis* sp.) to improve their nutritional value.

A common procedure during culture of both larval fish and prawns is to add microalgae (i.e. "green water") to intensive culture systems, together with the zooplankton prey [121]. The most popular microalga species used for this purpose are *N. oculata* and *T. suecica*. Addition of microalgae to larval tanks can also improve the production of larvae, but their exact mechanism of action remains unclear. Light attenuation (i.e. shading effects) may have a beneficial effect on larvae; however, maintenance of nutritional quality of the zooplankton, excretion of vitamins or other growth-promoting substances by the microalgae, and probiotic effects of the microalgae have also been hypothesized. Maintenance of NH_3- and O_2-balances has also been proposed, but this assumption failed to be supported by experimental evidence [121]. More research is still needed on the application of other microalgae – especially those species rich in DHA, to green water systems. Green water may also be applied to extensive outdoor production facilities, by fertilizing ponds in attempts to stimulate microalgal growth, and consequently zooplankton production.

7. Avenues for future research on microalgae

The high production costs of microalgae remain a constraint to many hatcheries. Despite efforts developed over the latest decades toward cost-effective artificial diets to replace microalgae, on-site microalgal production still remains a critical element for operation of most marine hatcheries. Improvements in alternative diets will surely continue, but production costs of microalgae will also likely decrease – so it is not expected that microalgae will be replaced in full, at least on the medium run. A wide selection of microalgal species is already available to support aquaculture activities. However, specific applications in industrial subsectors demand novel species with improved nutritional quality or growth characteristics, which are compatible with attempts to improve hatchery efficiency and yield.

Appart from improvements in cost-efficiency of on-site microalgal production, an alternative is centralizing microalga production in dedicated mass-culture facilities, using heterotrophic methods or nonconventional photobioreactors. These technologies may be coupled with post-harvest processing (e.g. spray-drying) or concentration (e.g. centrifugation or flocculation) to develop off-the-shelf microalgal biomass for ready distribution to hatcheries.

On the other hand, antifouling activity of extracts from some microalgae has been observed in microalga culture tanks, which are better (and less toxic) than common biocides. Those natural compounds could therefore be considered as good substitutes of commercial biocides in antifouling paints. Furthermore, as paint coatings remain the predominant preventative technique of marine biofouling, coatings adapted to the needs of aquaculture apparatuses – containing an active product from microalgae and able to inhibit the major microorganisms causing trouble in cultivation, are a potentially good solution to fight fouling. The exact substances that exhibit antifouling activities in microalgae are not yet known, so this type of study is warranted – to purify and identify the active compounds involved.

The need to reduce water consumption in aquaculture has long been recognized, so a great deal of effort has been directed toward development of recirculating systems. Unfortunately, current research and development encompassing aquaculture water re-use is largely devoted to bacteria-based systems – and the possibility of using microalga-based water re-use has been essentially neglected. The bacterial component in a water re-use system dedicates itself in full to excessive nutrient removal; conversely, a microalga-based water re-use system produces microalgae that can be used to produce a second crop, such as bivalve seed or *Artemia* – which may thus be sold to generate extra income. The main difficulty faced in development of microalga-based water re-use systems is the inability to maintain the desired microalgal species in an open system. A breakthrough in marine diatom production technology may allow one to focus on development of water re-use systems where the 'effluent' becomes itself a valuable resource: an integrated shrimp/microalga/oyster production system reduces water consumption, and turns effluent 'waste' into a profitable item – while taking advantage of the antibacterial properties of the marine diatom to control diseases, and thus reduce susceptibility of the shrimp to viral infections.

8. Acknowledgements

A postdoctoral fellowship (ref. SFRH/BPD/72777/2010), supervised by author F.X.M., was granted to author A.C.G., under the auspices of ESF (III Quadro Comunitário de Apoio) and the Portuguese State.

9. References

[1] G. Oron, L. R. Wildschut, D. Porath, "Wastewater recycling by duckweed for protein production and effluent renovation," *Water Science and Technology*, vol. 17, pp. 803-818, 1985.

[2] O. Hammouda, A. Gaber, N. Abdel-Raouf, "Microalgae and wastewater treatment," *Ecotoxicology and Environmental Safety*, vol. 31, pp. 205-210, 1995.

[3] W. J. Oswald, "Algal Pond Systems–Habitats," in *Proceedings of a Seminar held at Murdoch University*, Western Australia, 1991, pp. 61-69.

[4] A. Muller-Feuga, "The role of microalgae in aquaculture: situation and trends," *Journal of Applied Phycology*, vol. 12, pp. 527-534, 2000.

[5] C. T. Enright, G. F. Newkirk, J. S. Craigie, J. D. Castell, "Evaluation of phytoplankton as diets for juvenile *Ostrea edulis* L.," *Journal of Experimental Marine Biology and Ecology*, vol. 96, pp. 1-13, 1986.

[6] M. R. Brown, S. W. Jeffrey, J. K. Volkman, G. A. Dunstan, "Nutritional properties of microalgae for mariculture," *Aquaculture*, vol. 151, pp. 315-331, 1997.

[7] K. Yamaguchi, "Recent advances in microalgal bioscience in Japan, with special reference to utilization of biomass and metabolites: a review," *Journal of Applied Phycology*, vol. 8, pp. 487–502, 1997.

[8] W. Becker, "Microalgae for aquaculture. The nutritional value of microalgae for aquaculture." In Richmond, A. (ed.), *Handbook of Microalgal Culture*. Blackwell, Oxford, pp. 380–391, 2004.

[9] C. T. Enright, G. F. Newkirk, J. S. Craigie, J. D. Castell, "Growth of juvenile *Ostrea edulis* L. fed *Chaetoceros gracilis* Schütt of varied chemical composition," *Journal of Experimental Marine Biology and Ecology*, vol. 96, pp. 15-26, 1986.

[10] P. A. Thompson, M. Guo, P. J. Harrison, "The influence of irradiance on the biochemical composition of three phytoplankton species and their nutritional value for larvae of the Pacific oyster (*Crassostrea gigas*)," *Marine Biology*, vol. 117, pp. 259-268, 1993.

[11] L. R. Marínez-Córdova, E. Peña-Messina, "Biotic communities and feeding habits of *Litopenaeus vannamei* (Boone 1931) and *Litopenaeus stylirostris* (Stimpson 1974) in monoculture and polyculture semi-intensive ponds," *Aquaculture Research*, vol. 36, pp. 1075-1084, 2005.

[12] M. Kent, C. L. Browdy, J. W. Leffler, "Consumption and digestion of suspended microbes by juvenile Pacific white shrimp *Litopenaeus vannamei*," *Aquaculture*, doi: 10.1016/j.aquaculture.2011.06.048, 2011.

[13] P. Coutteau, P. Sorgeloos, "The use of algal substitutes and the requirement for live algae in the hatchery and nursery rearing of bivalve molluscs: an international survey," *Journal of Shellfish Research*, vol.11, pp. 467-476, 1992.

[14] C. E. Epifanio, "Growth in bivalve molluscs: nutritional effects of two or more species of algae in diets fed to the American oyster *Crassostrea virginica* (Gmelin) and the hard clam *Mercenaria mercenaria* (L.)," *Aquaculture*, vol. 18, pp. 1-12, 1979.

[15] D. A. Kreeger, C. J. Langdon, "Effect of dietary protein content on growth of juvenile mussels, *Mytilus trossulus* (Gould 1850)," *Biological Bulletin*, vol. 185, pp. 123-139, 1993.

[16] J. K. Volkman, S. W. Jeffrey, P. D. Nichols, G. I. Rogers, C. D. Garland, "Fatty acid and lipid composition of 10 species of microalgae used in mariculture," *Journal of Experimental Marine Biology and Ecology*, vol. 128, pp. 219-240, 1989.

[17] G. A. Dunstan, J. K. Volkman, S. M. Barrett, J.-M. Leroi, S. W. Jeffrey, "Essential polyunsaturated fatty acids from 14 species of diatom (Bacillariophyceae)," *Phytochemistry*, vol. 35, pp. 155-161, 1994.

[18] R. M. Knuckey, M. R. Brown, S. M. Barrett, G. M. Hallegraeff, "Isolation of new nanoplanktonic diatom strains and their evaluation as diets for juvenile Pacific oysters (*Crassostrea gigas*)," *Aquaculture*, vol. 211, pp. 253-274, 2002.

[19] D. Chuntapa, S. Powtongsook, P. Menasveta, "Water quality control using *Spirulina platensis* in shrimp culture tanks," *Aquaculture*, vol. 220, pp. 355-366, 2003.

[20] G. D. Lio-Po, E. M. Leaño, M. M. D. Peñaranda, A. U. Villa- Franco, C. D. Sombito, N. G. Guanzon, "Antiluminous *Vibrio* factors associated with the 'green water' growout culture of the tiger shrimp *Penaeus monodon*," *Aquaculture*, vol. 250, pp. 1-7, 2005.

[21] L. Rodolfi, G. C. Zittelli, L. Barsanti, G. Rosati, M. R. Tredici, "Growth medium recycling in *Nannochloropsis* sp. mass cultivation," *Biomolecular Engineering*, vol. 20, pp. 243-248, 2003.

[22] A. Muller-Feuga, "Microalgae for aquaculture. The current global situation and future trends." In Richmond, A. (ed.), *Handbook of Microalgal Culture*. Blackwell, Oxford, 2004, pp. 352-364.

[23] A. Irianto, B. Austin, "Probiotics in aquaculture," *Journal of Fish Diseases*, vol. 25, pp. 633-642, 2002.

[24] P. Spolaore, C. Joannis-Cassan, E. Duran, A. Isambert, "Review: commercial applications of microalgae," *Journal of Bioscience and Bioengineering*, vol. 101, pp. 87-96, 2006.

[25] D. S. Nichols, "Prokaryotes and the input of polyunsaturated fatty acids to the marine food web," *FEMS Microbiology Letters*, vol. 219, pp. 1-7, 2003.

[26] C. Aragão, L. E. C. Conceição, M. T. Dinis, H.-J. Fyhn," Amino acid pools of rotifers and *Artemia* under different conditions: nutritional implications for fish larvae," *Aquaculture*, vol. 234, pp. 429-445, 2004.

[27] K. E. Apt, P. W. Behrens, "Commercial developments in microalgal biotechnology," *Journal of Phycology*, vol. 35, pp. 215-226, 1999.

[28] L. Waldenstedt, J. Inborr, I. Hansson, K. Elwinger, "Effects of astaxanthin-rich algal meal (*Haematococcus pluvalis*) on growth performance, faecal campylobacter and clostridial counts and tissue astaxanthin concentration of broiler chickens," *Animal Feed Science and Technology*, vol. 108, pp. 119-132, 2003.

[29] M. R. Brown, "The amino acid and sugar composition of 16 species of microalgae used in mariculture," *Journal of Experimental Marine Biology and Ecology*, vol. 145, pp. 79-99, 1991.

[30] M. R. Brown, G. A. Dunstan, S. W. Jeffrey, J. K., Volkman, S. M. Barrett, J. M. Leroi, "The influence of irradiance on the biochemical composition of the prymnesiophyte *Isochrysis* sp. (clone T-ISO)," *Journal of Phycology*, vol. 29, pp. 601-612, 1993.

[31] M. R. Brown, C. D. Garland, S. W. Jeffrey, I. D. Jameson, J. M. Leroi, "The gross and amino acid compositions of batch and semi-continuous cultures of *Isochrysis* sp. (clone T.ISO), *Pavlova lutheri* and *Nannochloropsis oculata*," *Journal of Applied Phycology*, vol. 5, pp. 285-296, 1993.

[32] J. Knauer, S. M. Barrett, J. K. Volkman, P. C. Southgate, "Assimilation of dietary phytosterols by Pacific oyster *Crassostrea gigas* spat," *Aquaculture Nutrition*, vol. 5, pp. 257-266, 1999.

[33] J. Fabregas, C. Herrero, "Marine microalgae as a potential source of minerals in fish diets," *Aquaculture*, vol. 51, pp. 237-243, 1986.

[34] S. M. Renaud, L. V. Thinh, D. L. Parry, "The gross composition and fatty acid composition of 18 species of tropical Australia microalgae for possible use in mariculture," *Aquaculture*, vol. 170, pp. 147-159, 1999.

[35] P. J. Harrison, P. A. Thompson, G. S. Calderwood, "Effects of nutrient and light limitation on the biochemical composition of phytoplankton," *Journal of Applied Phycology*, vol. 2, pp. 45-56, 1990.

[36] J. N. C. Whyte, N. Bourne, C. A. Hodgson, "Influence of algal diets on biochemical composition and energy reserves in *Patinopecten yessoensis* (Jay) larvae," *Aquaculture*, vol. 78, pp. 333-347, 1989.

[37] J. A. Mendiola, C. F. Torres, P. J. Martín-Alvarez, S. Santoyo, A. Toré, B. O. Arredondo, F. J. Señoráns, A. Cifuentes, E. Ibáñez, "Use of supercritical CO_2 to obtain extracts with antimicrobial activity from *Chaetoceros muelleri* microalga. A correlation with their lipidic content," *European Food Research and Technology*, vol. 224, pp. 505-510, 2007.

[38] K. Benkendorff, A. R. Davis, C. N. Rogers, J. B. Bremner, "Free fatty acids and sterols in the benthic spawn of aquatic molluscs, and their associated antimicrobial

properties," *Journal of Experimental Marine Biology and Ecology*, vol. 316, pp. 29-44, 2005.

[39] V. J. Smith, A. P. Desbois, E. A. Dyrynda, "Conventional and unconventional antimicrobials from fish, marine invertebrates and micro-algae," *Marine Drugs*, vol. 8, pp. 1213-1262, 2010.

[40] F. Jüttner, "Liberation of 5,8,11,14,17-eicosapentaenoic acid and other polyunsaturated fatty acids from lipids as a grazer defense reaction in epiphithic diatom biofilms," *Journal of Phycology*, vol. 37, pp. 744-755, 2001.

[41] G. d'Ippolito, S. Tucci, A. Cutignano, G. Romano, G. Cimino, A. Miralto, A. Fontana, "The role of complex lipids in the synthesis of bioactive aldehydes of the marine diatom *Skeletonema costatum*," *Biochimica et Biophysica Acta*, vol. 1686, pp. 100-107, 2004.

[42] M. Naviner, J.-P. Bergé, P. Durand, H. le Bris, "Antibacterial activity of the marine diatom *Skeletonema costatum* against aquacultural pathogens," *Aquaculture*, vol. 174, pp. 15-24, 1999.

[43] K. Das, J. Pradhan, P. Pattnaik, B. R. Samantaray, S. K. Samal, "Production of antibacterials from the freshwater alga *Euglena viridis* (Ehren)," *World Journal of Microbiology and Biotechnology*, vol. 21, pp. 45-50, 2005.

[44] M. A. Borowitzka, "Commercial production of microalgae: ponds, tanks, tubes and fermenters," *Journal of Biotechnology*, vol. 70, pp. 313–321, 1999.

[45] H. M. Amaro, A. C. Guedes, F. X. Malcata, "Advances and perspectives in using microalgae to produce biodiesel," *Applied Energy*, vol. 88, pp. 3402-3410, 2011.

[46] S. A. Scott, M. P. Davey, J. S. Dennis, I. Horst, C. J. Howe, D. J. Lea-Smith, "Biodiesel from algae: challenges and prospects," *Current Opinion in Biotechnology*, vol. 21, pp. 277-286, 2010.

[47] P. Soong, "Production and development of *Chlorella* and *Spirulina* in Taiwan." In: Shelef, G., Soeder, C. J. (Eds.), *Algae Biomass*. Elsevier, Amsterdam, 1980, pp. 97-113.

[48] L. J. Borowitzka, M. A. Borowitzka, "Industrial production: methods and economics." In: Cresswell, R. C., Rees, T. A. V., Shah, N. (Eds.), *Algal and Cyanobacterial Biotechnology*. Longman Scientific, London, 1989, pp. 294–316.

[49] A. Belay, "Mass culture of *Spirulina* outdoors—the Earthrise Farms experience". In: Vonshak, A. (Ed.), *Spirulina platensis* (*Arthrospira*): *Physiology, Cell-biology and Biotechnology*. Taylor & Francis, London, 1997, pp. 131-158.

[50] I. Setlík, S. Veladimir, I. Malek, "Dual purpose open circulation units for large scale culture of algae in temperate zones. I. Basic design considerations and scheme of pilot plant," *Algology Studies (Trebon)*, vol. 1, pp. 11, 1970.

[51] K. Kawaguchi, "Microalgae production systems in Asia." In: Shelef, G., Soeder, C. J. (Eds.), *Algae Biomass Production and Use*. Elsevier, Amsterdam, 1980, pp. 25-33.

[52] J. Kyle, R. M. Gladue, "Eicosapentaenoic acids and methods for their production," *World Patent*, vol. 9, pp. 114,427, 1991.

[53] J. Kyle, S. E. Reeb, V. J. Sicotte, "Dinoflagellate biomass, methods for its production, and compositions containing the same," *USA Patent* 5,711,983, 1998.

[54] A. P. Carvalho, L. A. Meireles, F. X. Malcata, "Microalgal reactors: a review of enclosed system designs and performances," *Biotechnology Progress*, vol. 22, pp. 1490-1506, 2006.

[55] C. Jiménez, B. R. Cossío, D. Labella, N. F. Xavier, "The feasibility of industrial production of *Spirulina* (*Arthrospira*) in southern Spain," *Aquaculture*, vol. 217, pp. 179-190, 2003.

[56] D. Chaumont, "Biotechnology of algal biomass production: a review of systems for outdoor mass culture," *Journal of Applied Phycology*, vol. 5, pp. 593-604, 1993.

[57] L. Terry, L. P. Raymond, "System design for the autotrophic production of microalgae," *Enzyme and Microbial Technology*, vol. 7, pp. 474-487, 1985.

[58] Y. Chisti, "Biodiesel from microalgae beats bioethanol," *Trends in Biotechnology*, vol. 26, pp. 126-131, 2008.

[59] L. Rodolfi, G. C. Zittelli, N. Bassi, G. Padovani, N. Biondi, G. Bonini, "Microalgae for oil: strain selection, induction of lipid synthesis and outdoor mass cultivation in a low-cost photobioreactor," *Biotechnology and Bioengineering*, vol. 102, pp. 100-112, 2008.

[60] U. Ugwu, H. Aoyagi, H. Uchiyama, "Photobioreactors for mass cultivation of algae," *Bioresource Technology*, vol. 99, pp. 4021-4028, 2008.

[61] Y. Chisti, "Biodiesel from microalgae," *Biotechnology Advances*, vol. 25, pp. 294-306, 2007.

[62] O. Pulz, "Photobioreactors: production systems for phototrophic microorganisms," *Applied Microbiology and Biotechnology*, vol. 57, pp. 287-293, 2001.

[63] B. Metting, "Biodiversity and application of microalgae," *Journal of Industrial Microbiology*, vol. 17, pp. 477-489, 1996.

[64] E. Molina-Grima, "Microalgae, mass culture methods." In M. C. Flickinger, & S. W. Drew (Eds.), *Encyclopedia of Bioprocess Technology: Fermentation, Biocatalysis and Bioseparation.* New York: Wiley, 1999, vol. 3, pp. 1753-1769.

[65] R. Tredici, "Bioreactors, photo." In Flickinger, M. C. & Drew, S. W. (Eds.), *Encyclopedia of Bioprocess Technology: Fermentation, Biocatalysis and Bioseparation*, vol. 1. New York, Wiley, 1999, pp. 395-419.

[66] G. Acién-Fernández, F. García-Camacho, J. A. Sánchez-Pérez, J. M. Fernández-Sevilla, E. Molina-Grima, "Modelling of biomass productivity in tubular photobioreactors for microalgal cultures: effects of dilution rate, tube diameter and solar irradiance," *Biotechnology and Bioengineering*, vol. 58, pp. 605-616, 1998.

[67] S. Aiba, "Growth kinetics of photosynthetic microorganisms," *Advances in Biochemical Engineering*, vol. 23, pp. 85-156, 1982.

[68] L. E. Erickson, H. Y. Lee, "Process analysis and design of algal growth system." In: Barclay, W. R. and McIntosh, R. P. (Eds), *Algal Biomass Technologies: an Interdisciplinary Perspective.* J. Cramer, Berlin-Stuttgart, 1986, pp. 197–206.

[69] S. L. Pirt, Y. K. Lee, M. R. Walach, M. W. Pirt, H. H. Balyuzi, M. J. Bazin, "A tubular bioreactor for photosynthetic production of biomass from carbon dioxide: design and performance," *Journal of Chemical Technology and Biotechnology*, vol. 33B, pp. 35-58, 1983.

[70] L. Rorrer, R. K. Mullikin, "Modeling and simulation of a tubular recycle photobioreactor for macroalgal cell suspension cultures," *Chemical Engineering Science*, vol. 54, pp. 3153-3162, 1999.

[71] C. Posten, "Design principles of photo-bioreactors for cultivation of microalgae," *Engineering in Life Sciences*, vol. 9, pp. 165-177, 2009.

[72] A. A. Tsygankov, "Laboratory scale photobioreactors," *Applied Biochemistry and Microbiology*, vol. 37, pp. 333-341, 2001.

[73] A. Borowitzka, "Microalgae as sources of pharmaceuticals and other biologically active compounds," *Journal of Applied Phycology*, vol. 7, pp. 3-15, 1995.

[74] A. Vonshak, "Tubular photobioreactors for algal mass production; prospects and achievements," *Israeli Journal of Aquaculture (Bamidgeh)*, vol. 44, pp. 151, 1992.

[75] M. Olaizola, "Commercial development of microalgal biotechnology: from the test tube to the marketplace," *Biomolecular Engineering*, vol. 20, pp. 459-466, 2003.

[76] E. Molina-Grima, F. G. Acién-Fernández, F. García-Camacho, F. Camacho-Rubio, Y. Chisti, "Scale-up of tubular photobioreactors," *Journal of Applied Phycology*, vol. 12, pp. 355-368, 2000.

[77] N. de Pauw, G. Persoone, "Micro-algae for aquaculture." In: Borowitzka, M. A., Borowitzka, L. J. (Eds.), *Micro-Algal Biotechnology*. Cambridge University Press, Cambridge, 1988, pp. 197-221.

[78] R. Benemann, W. J. Oswald, "Systems and economic analysis of microalgae ponds for conversion of CO_2 to biomass," *Final Report, Subcontract XK 4-04136- 06*, Pittsburgh Energy Technology Center Grant No. DE-FG22-93PC93204, p. 260, 1996.

[79] FAO, 2000. *Aquaculture Production 1992–1998*. FAO, Rome.

[80] I. Tzovenis, G. Triantaphyllidis, X. Naihong, E. Chatzinikolaou, K. Papadopoulou, G. Xouri, T. Tafas, "Cryopreservation of marine microalgae and potential toxicity of cryoprotectants to the primary steps of the aquacultural food chain," *Aquaculture*, vol. 230, pp. 457-473, 2004.

[81] M. A. Borowitzka, "Microalgae for aquaculture: opportunities and constraints," *Journal of Applied Phycology*, vol. 9, pp. 393-401, 1997.

[82] J. Knauer, P. C. Southgate, "A review of the nutritional requirements of bivalves and the development of alternative and artificial diets for bivalve aquaculture," *Reviews in Fisheries Science*, vol. 7, pp. 241–280, 1999.

[83] R. Robert, P. Trintignac, "Substitutes for live microalgae in mariculture: a review," *Aquatic Living Resources*, vol. 10, pp. 315–327, 1997.

[84] G. Chini-Zittelli, F. Lavista, A. Bastianini, L. Rodolfi, M. Vincenzini, M. R. Tredici, "Production of eicosapentaenoic acid by *Nannochloropsis* sp. cultures in outdoor tubular photobioreactors," *Journal of Biotechnology*, vol. 70, pp. 299-312, 1999.

[85] B. Spolaore-Robinson, T. M. Samocha, J. M. Fox, R. L. Gandy, D. A. McKee, "The use of inert artificial commercial food sources as replacements of traditional live food items in the culture of larval shrimp, *Farfantepenaeus aztecus*," *Aquaculture*, vol. 245, pp. 135-147, 2005.

[86] C. Seguineau, A. Laschi-Loquerie, J. Moral, J.-F. Samain, "Vitamin requirements in great scallop larvae," *Aquaculture International*, vol. 4, pp. 315-324, 1996.

[87] R. Brown, S. M. Barret, J. K. Volkman, S. P. Nearhos, J. Nell, G. L. Allan, "Biochemical composition of new yeasts and bacteria evaluated as food for bivalve aquaculture," *Aquaculture*, vol. 143, pp. 341-360, 1996.

[88] J. Langdon, E. T. Bolton, "A microparticulate diet for suspension-feeding bivalve mollusc, *Crassostrea virginica* (Gmelin)," *Journal of Experimental Marine Biology and Ecology*, vol. 89, pp. 239-258, 1984.

[89] P. Coutteau, M. Dravers, P. Leger, P. Sorgeloos, "Manipulated yeast diets and dried algae as a partial substitute for live algae in the juvenile rearing of the Manila clam *Tapes philippinarum* and the Pacific oyster *Crassostrea gigas*," *Special Publication of the European Aquatic Society Ghent*, Belgium, vol. 18, pp. 523-531, 1993.

[90] J. A. Nell, W. A. O'Connor, "The evaluation of fresh algae and stored algal concentrates as a food source for Sydney rock oyster, *Saccostrea commercialis* (Iredale and Roughley) larvae," *Aquaculture*, vol. 99, pp. 277-284, 1991.

[91] A. McCausland, M. R. Brown, S. M. Barrett, J. A. Diemar, M. P. Heasman, "Evaluation of live and pasted microalgae as supplementary food for juvenile Pacific oysters (*Crassostrea gigas*)," *Aquaculture and Research*, vol. 174, pp. 323-342, 1999.

[92] M. Heasman, J. Diemar, W. O'Connor, T. Sushames, L. Foulkes, "Development of extended shelf-life micro-algae concentrate diets harvested by centrifugation for bivalve molluscs—a summary," *Aquaculture and Research*, vol. 31, pp. 637–659, 2000.

[93] A. Muller-Feuga, R. Robert, C. Cahu, J. Robin, P. Divemach, "Use of microalgae in aquaculture." In: Støttrup, J. A., McEvoy, L. A., (Eds), *Live Feeds in Marine Aquaculture*. Blackwell, Oxford, 2003, pp. 253-299.

[94] E. Montaini, G. C. Zittelli, M. R. Tredici, E. M. Grima, J. M. F. Sevilla, J. A. S. Perez, "Long-term preservation of *Tetraselmis suecica*: influence of storage on viability and fatty acid profile," *Aquaculture*, vol. 134, pp. 81-90, 1995.

[95] A. Csordas, J.-K. Wang, "An integrated photobioreactor and foam fractionation unit for the growth and harvest of *Chaetoceros* spp. in open systems," *Aquacultural Engineering*, vol. 30, pp. 15-30, 2004.

[96] M. Millamena, E. J. Aujero, I. G. Borlongan, "Techniques on algae harvesting and preservation for use in culture as larval food," *Aquaculture Engineering*, vol. 9, pp. 295-304, 1990.

[97] E. Poelman, N. de Pauw, B. Jeurissen, "Potential of electrolytic flocculation for recovery of micro-algae," *Resources, Conservation* and *Recycling*, vol. 19, pp. 1-10, 1997.

[98] N. Rossingol, L. Vandanjon, P. Jaouen, F. Quéméneur, "Membrane technology for the continuous separation microalgae/culture medium: compared performances of cross-flow microfiltration and ultrafiltration," *Aquaculture Engineering*, vol. 20, pp. 191-208, 1999.

[99] E. Sandbank, "The utilization of microalgae as feed for fish," *Ergeb Limnology*, vol. 11, pp. 108-120, 1978.

[100] R. M. Knuckey, M. R. Brown, R. Robert, D. M. F. Frampton, "Production of microalgal concentrates by flocculation and their assessment as aquaculture feeds," *Aquacultural Engineering*, vol. 35, pp. 300-313, 2006.

[101] J. O. M. Karlson, M. Toner, "Long-term storage of tissues by cryopreservation: critical issues," *Biomaterials*, vol. 17, pp. 243-256, 1996.

[102] P. Mazur, "Freezing of living cells: mechanisms and implications," *American Journal of Physiology*, vol. 247, pp. 125-142, 1984.

[103] M. Fahy, "The relevance of cryoprotectant 'toxicity' to cryobiology," *Cryobiology*, vol. 23, pp. 1-13, 1986.

[104] M. Fahy, T. H. Lilley, H. Lindsell, M. J. Douglas, H. T. Meryman, "Cryoprotectant toxicity and cryoprotectant toxicity reduction: in search of molecular mechanisms," *Cryobiology*, vol. 27, pp. 247- 268, 1990.

[105] R. Taylor, R. L. Fletcher, "Cryopreservation of eukaryotic algae—a review of methodologies," *Journal of Applied Phycology*, vol. 10, pp. 481–501, 1999.

[106] W. Barclay, S. Zeller, "Nutritional enhancement of n-3 and n-6 fatty acids in rotifers and *Artemia nauplii* by feeding spray-dried *Schizochytrium* sp.," *Journal of the World Aquaculture Society*, vol. 27, pp. 314-322, 1996.

[107] C. Langdon, E. Önal, "Replacement of living microalgae with spray-dried diets for the marine mussel *Mytilus galloprovincialis*," *Aquaculture*, vol. 180, pp. 283-294, 1999.

[108] A. Dunstan, J. K. Volkman, S. W. Jeffrey, S. M. Barret, "Biochemical composition of microalgae from the green algal classes Chlorophyceae and Prasinophyceae," *Journal of Experimental Marine Biology and Ecology*, vol. 161, pp. 115-134, 1992.

[109] J. Dhont, G. van Stappen, *Live Feeds in Marine Aquaculture*, Blackwell Science, pp. 65-121. 2003.

[110] A. Elston, "Mollusc diseases, guide for the shellfish farmer," *Washington Sea Grant Program*, Seattle, WA, p. 73, 1990.

[111] P. Douillet, C. J. Langdon, "Effects of marine bacteria on the culture of axenic oyster *Crassostrea gigas* (Thunberg) larvae," *Biological Bulletin*, vol. 184, pp. 36-51, 1993.

[112] P. Douillet, C. J. Langdon, "Use of a probiotic for the culture of larvae of the Pacific oyster (*Crassostrea gigas* Thunberg)," *Aquaculture*, vol. 119, pp. 25-40, 1994.

[113] J. Donaldson, *Proceedings of US-Asia Workshop*, Honolulu, HA, January 28-31, 1991, The Oceanic Institute, HA, pp. 229-236.

[114] B. Zhou, W. Liu, W. Qu, C. K. Tseng, "Application of *Spirulina* mixed feed in the breeding of bay scallop,"*Bioresource Technology*, vol. 38, pp. 229-232, 1991.

[115] P. D. Nichols, D. G. Holdsworth, J. K. Volkman, M. Daintith, S. Allanson, "High incorporation of essential fatty acids by the rotifer *Brachionus plicatilis* fed on the prymnesiophyte alga *Pavlova lutheri*," *Australian Journal of Marine and Freshwater Research*, vol. 40, pp. 645-655, 1989.

[116] B. Gara, R. J. Shields, L. McEvoy, "Feeding strategies to achieve correct metamorphosis of Atlantic halibut, *Hippoglossus hippoglossus* L., using enriched *Artemia*," *Aquaculture Research*, vol. 29, pp. 935-948, 1998.

[117] C. Rodríguez, J. A. Pérez, P. Badía, M. S. Izquierdo, H. Fernández-Palacios, A. Lorenzo Hernández, "The n-3 highly unsaturated fatty acid requirements of gilthead seabream (*Sparus aurata* L.) larvae when using an appropriate DHA/EPA ratio in the diet," *Aquaculture*, vol. 169, pp. 9-23, 1998.

[118] M. R. Brown, S. Skabo, B. Wilkinson, "The enrichment and retention of ascorbic acid in rotifers fed with microalgal diets," *Aquaculture Nutrition*, vol. 4, pp. 151-156, 1998.

[119] G. Merchie, P. Lavens, P. Dhert, M. Dehasque, H. Nelis, A. de Leenheer, P. Sorgeloos, "Variation of ascorbic acid content in different live food organisms," *Aquaculture*, vol. 134, pp. 325-337, 1995.

[120] I. Rønnestad, S. Helland, Ø. Lie, "Feeding *Artemia* to larvae of Atlantic halibut (*Hippoglossus hippoglossus* L.) results in lower larval vitamin A content compared with feeding copepods," *Aquaculture*, vol. 165, pp. 159-164, 1998.

[121] C. S. Tamaru, R. Murashige, C.-S. Lee, "The paradox of using background phytoplankton during the larval culture of striped mullet, *Mugil cephalus* L.," *Aquaculture*, vol. 119, pp. 167-174, 1994.

Part 2

Genetics

5

Determination of Fish Origin by Using 16S rDNA Fingerprinting of Microbial Communities by PCR-DGGE: An Application on Fish from Different Tropical Origins

Didier Montet[1], Doan Duy Le Nguyen[1,2]
and Amenan Clementine Kouakou[1,3]
[1]CIRAD, UMR 95 Qualisud, TA B-95/16, Montpellier
[2]Can tho University, Faculty of Agriculture
[3]University of Abobo-Adjame, Laboratory of
Microbiology and Molecular Biology
[1]France
[2]Viet Nam
[3]Ivory Coast

1. Introduction

Food safety is now a compulsory issue for food imported to European Union. Bovine Spongiform Encephalitis, pathogens and avian influenza remain in the memories of European consumers. With similar scares occurring globally, the need for vigilance and strict monitoring is necessary. EU regulation 178/2002 imposed traceability to all food imported to EU. For a long time, food industry had simple traceability systems, but with the increasing implementation of current Good Manufacturing Practice, traceability systems have become more important in the production chain.

There are only a few analytical techniques that permit to trace food. In view of the difficulties of installing these documentary systems in developing country, and to follow the product during processing, we propose to identify and validate some pertinent biological markers which come from the environment of the fish to assure traceability of aquaculture product during international trade.

We proposed to trace the origin of fish by analysing in a global way the bacterial communities on the fish samples. The predominant bacterial flora would permit the determination of the capture area, production process or sanitary or hygienic conditions during post harvest operations (Montet et al., 2004; Le Nguyen et al., 2007, Montet, 2008).

Aquatic micro-organisms are known to be closely associated with the physiological status of fish. Numerous studies of the microbiota in fish captured from various geographical locations have been done (Grisez et al., 1997; Spanggaard et al., 2000; Al-Harbi and Uddin, 2003; Leesing, 2005). The bacterial communities of fish could be influence by water composition, temperature, weather conditions and farmer practices.

Separation of PCR products in DGGE is based on the decrease of the electrophoretic mobility of partially melted doubled-stranded DNA molecules in polyacrylamide gels containing a linear gradient of DNA denaturants like formamide and urea at 60°C. Molecules with different sequences will have a different melting behaviour and will stop migrating at different position in the gel (Muyzer et al., 1993; Leesing, 2005). PCR-DGGE has been already used to investigate several patterns of distribution of fish bacterial assemblages (Murray et al., 1996; Øvreas et al., 1997; Moeseneder et al., 1999; Riemann et al., 1999; Maiworé et al., 2009a, 2009b; Tatsadjieu et al., 2010) and was used by our team to study the bacteria on fresh water fish for their traceability (Le Nguyen et al., 2007 ; Montet et al., 2008).

A specific advantage of this technique is that it permits the analysis of both cultivable and non cultivable, anaerobic and aerobic bacteria and provides a rapid method to observe the changes in community structure in response to different environmental factors (Yang et al., 2001).

The purpose of our study is to apply the PCR-DGGE method for analyzing the bacteria in fish in order to create a technique to link bacterial communities to the geographical origin and avoid the individual analysis of each bacterial strain. The acquired band patterns for the bacterial species of different fish form Viet Nam were compared and analysed statistically to determine the fish origin. We give also an example of the following of the ecology of bacteria in a tropical and traditional fish fermentation form Ivory Cost, the Adjuevan.

2. Materials and methods

2.1 Fish sampling

Pangasius fish (*Pangasius hypophthalmus*) were collected in a unique pond in five aquaculture farms of five different districts from the South Vietnam namely Chau Phu, An Phu, Phu Tan, Chau Doc, Tan Chau of An Giang Province (Fig. 1). The samples were collected in two seasons in Vietnam: the rainy season (October 2005) and the dry season (February 2006). In each farm of each district, the samples were taken from the same pond and aseptically transferred to storage bags. The samples were maintained on crushed ice and transported to the laboratory. Then the skin, gills and intestines were aseptically removed from each fish specimen and put in separate sealed plastic bags, then kept frozen at -20°C prior to analysis.

Tilapia fish (*Oreochromis niloticus*) were collected in freshwater aquariums at 27°C at Cemagref-Cirad at Montpellier (France) in April and June 2010. Sample weights were 500 ± 10 g. After sampling with sterile gloves, fish were placed individually in sterile plastic bags and then transported in cooled containers refrigerated by crush ice to the laboratory (500m). Fermentation immediately followed the fish arrival.

2.2 Production of fermented fish "Adjuevan"

Adjuevan is a salted and fermented fish traditionally produced in the west coast of Ivory Coast at ambient temperature (28-30°C) following two traditional methods. First method of production took place in jars covered with plastics and stones for 5 days and second method

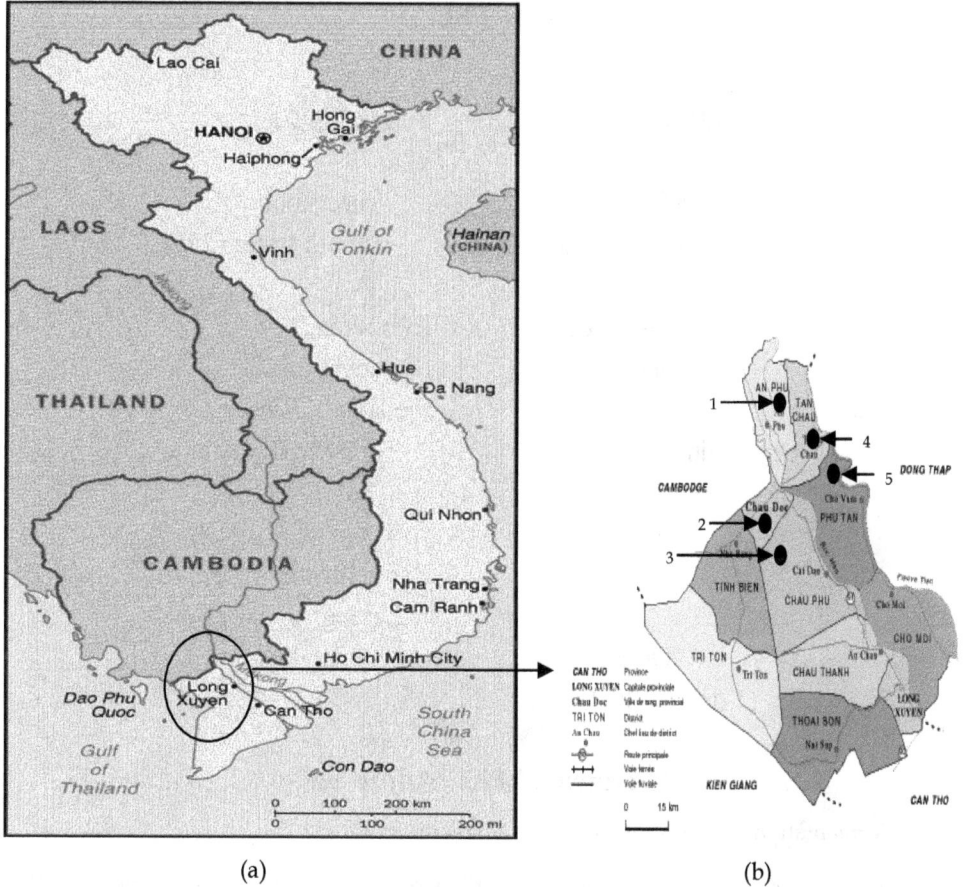

(a) (b)

Fig. 1. (a). The map of Viet nam (b) The expansion of An giang province with five different sampling locations: 1) An Phu; 2) Chau Doc; 3) Chau Phu; 4) Tan Chau; 5) Phu Tan.

followed the same fermentation process and then fish were dried on racks or nets for at least 10 days. We studied the changes in physicochemical characteristics and the dynamic of bacterial flora by PCR-DGGE on artificial Adjuevan made in our laboratory at 30°C, with different percentages of salts of 10%, 15% 20%, 25%, and 30% following the both methods.

Fermentation methods are described in Figure 2. About 1.5 kg of fish was gutted, washed and salt was added at the following concentrations 10%, 15%, 20%, 25%, 30% (w/w) after 24 h of maturation (stored at room temperature). For method 1, fish was salted, then wrapped in sterile plastic containers and arranged in different coolers. They are left to ferment for 5 days at ambient temperature (30°C) followed by drying in a ventilated dryer (at 30°C with minimal ventilation) for 24 h. For method 2, fish after been salted, was deposited on a sterile plastic surface then placed in sterile trays. Fermentation was done simultaneously as well as drying for 5 days at 30°C followed by ventilation for 24 h (Fig. 2). Fermentation experiments were done in triplicate.

Fresh   Fish

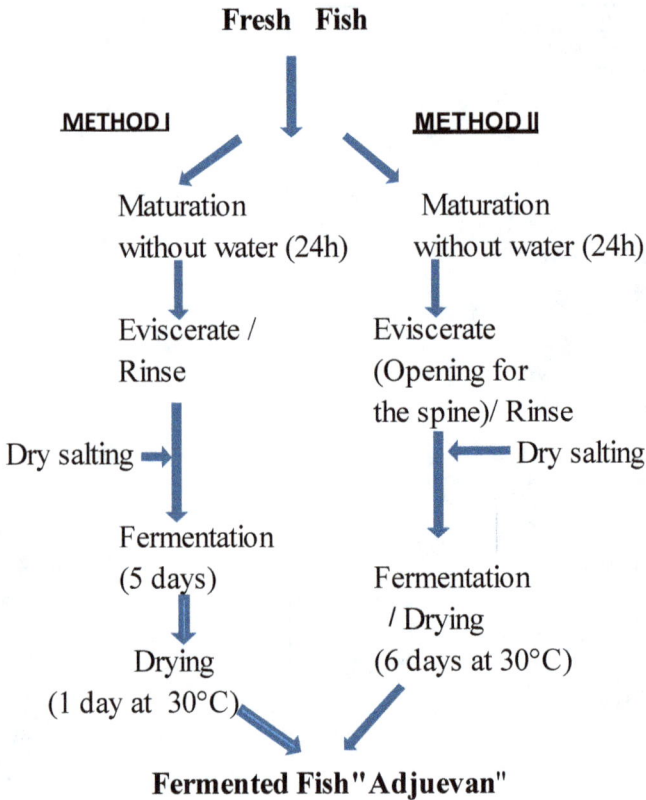

Fig. 2. Fermentation methods for Adjuevan at laboratory level

Fish samples were collected during fermentation primarily on fresh fish, then after 24 h followed by maturation and 1, 2, 3, 4 and 5 days during fermentation. They were put in aseptic sterile tubes, stored in a cooler filled with ice and transported immediately to laboratory for physicochemical analysis.

2.3 Total DNA extraction

DNA extraction was based on the methods of Ampe et al. (1999) and Leesing (2005) but modified and optimised. Around 2g each of gills, skin and intestine were homogenized for 3 min with vortexing after addition of 6 mL sterile peptone water (pH 7.0, Dickinson, France). Four 1.5-mL tubes containing the resulting suspension were then centrifuged at 10,000g for 10 min. 100 µL of lysis buffer TE (10 mM Tris-HCl; 1 mM EDTA; pH 8.0, Promega, France) and 100 µL of lysozyme solution (25 µg.µL^{-1}, Eurobio, France) and 50 µL of proteinase K solution (10 µg.µL^{-1}, Eurobio, France) were added to each pellet. Samples were vortexed for 1 min and incubated at 42°C for 30 min. Then 50µL of 20% SDS (Sodium Dodecyl Sulphate, Sigma, France) were added to each tube, and the tubes were incubated at 42°C for 10 min. 300 µL of MATAB (Mixed Alkyltrimethyl Ammonium Bromide, Sigma France) were added to each tube, and the tubes were incubated at 65°C for 10 min. The lysates were then purified by repeated extraction with 700 µL of phenol-chloroform-isoamyl alcohol (25:24:1, Carlo Erba, France), and the residual phenol was removed by extraction with an equal volume of chloroform-isoamyl alcohol (24:1). The DNA was precipitated with isopropanol, washed with 70% ethanol and then air dried at room temperature. Finally, the DNA was resuspended in 100 µL of ultra pure water and stored at - 20°C until analysis.

2.4 PCR-denaturing gradient gel electrophoresis (DGGE) analysis

The V3 variable region of bacterial 16S rDNA from fish was amplified using primers gc338f (5'CGCCCGCCGCGCGCGGCGGGCGGGGCGGGGGCACGGGGGGACTCCTACGGGAG GCAGCAG, Sigma, France) and 518r (5'-ATTACCGCGGCTGCTGG, Sigma, France) (Øvreas et al., 1997; Ampe et al., 1999; Leesing, 2005). A 40-bp GC-clamp (Sigma, France) was added to the forward primer in order to insure that the fragment of DNA will remain partially double-stranded and that the region screened is in the lowest melting domain (Sheffield et al., 1989). Each mixture (final volume 50 µL) contained about 100ng of template DNA, all the primers at 0.2µM, all the deoxyribonucleotide triphosphate (dNTPs) at 200µM, 1.5mM MgCl$_2$, 5µL of 10x of reaction *Tag* buffer (MgCl$_2$ free) (Promega, France) and 5U of Taq polymerase (Promega, France). In order to increase the specificity of amplification and to reduce the formation of spurious by-products, a "touchdown" PCR was performed according to the protocol of Díez et al. (2001). An initial denaturation at 94°C for 1 min and 10 touchdown cycles of denaturation at 94°C for 1 min, then annealing at 65°C (with the temperature decreasing 1°C per cycle) for 1 min, and extension at 72°C for 3 min, followed 20 cycles of 94°C for 1 min, 55°C for 1 min and 72°C for 3 min. During the last cycle, the extension step was increased to 10 min. Aliquots (5µL) of PCR products were analysed first by conventional electrophoresis in 2% (w/v) agarose gel with TAE 1X buffer (40 mM Tris-HCl pH 7.4, 20 mM sodium acetate, 1.0 mM Na$_2$-EDTA), stained with ethidium bromide (Sigma, France) 0.5 µg/mL in TAE 1X and quantified by using a standard (DNA mass ladder 100 bp, Promega, France).

The PCR products were analyzed by Denaturing Gradient Gel Electrophoresis (DGGE) by using a Bio-Rad DCodeTM universal mutation detection system (Bio-Rad Laboratories, Hercules, USA) and the procedure first described by Muyzer et al. (1993) and improved by Leesing (2005, 2011). Samples containing approximately equal amounts of PCR amplicons were loaded into 8% (wt/v) polyacrylamide gels (acrylamide/NN'-methylene bisacrylamide, 37.5:1, Promega, France) in 1X TAE buffer (40 mM Tris-HCl pH 7.4, 20 mM sodium acetate, 1.0mM Na$_2$-EDTA). All electrophoresis experiments were performed at 60°C using a denaturing gradient ranging from 30 to 60% (100% corresponded to 7M urea and 40% [v/v] formamide, Promega, France). The gels were electrophoresed at 20 V for 10 min and then at 180 V for 12h.

After electrophoresis, the gels were stained for 30 min with ethidium bromide and rinsed for 20 min in distilled water and then photographed on a UV transilluminator with the Gel Smart 7.3 system (Clara Vision, Les Ulis, France).

2.5 Image and statistical analysis

Individual lanes of the gel images were straightened and aligned using ImageQuant TL software version 2003 (Amesham Biosciences, USA). Banding patterns were standardized with the two reference patterns included in all gels which are the patterns of *Escherichia coli* DNA and *Lactobacillus plantarum* DNA. This software permitted to identify the bands and their relative position compared with the standard patterns.

In DGGE analysis, the generated banding pattern is considered as an "image" of all of the major bacterial species in the population. An individual discrete band refers to a unique "sequence type" or phylotype (Muyzer et al., 1995; van Hannen et al., 1999), which is treated as a discrete bacterial population. It is expected that PCR fragments generated from a single population will display an identical electrophoretic mobility in the analysis. This was confirmed by Kowalchuk et al. (1997) who showed that co-migrating bands generally corresponded to identical sequence.

The DGGE fingerprints were manually scored by the presence and absence of co-migrating bands, independent of intensity. Pairwise community similarities were quantified using the Dice similarity coefficient (S_D) (Heyndrickx et al., 1996).

$$S_D = 2 N_c / N_a + N_b$$

Where N_a represented the number of bands detected in the sample A, N_b represented the number of bands detected in the sample B, and N_c represented the numbers of bands common to both sample. Similarity index were expressed within a range of 0 (completely dissimilar) to 1.0 (perfect similarity). Dendograms were constructed using the Statistica version 6 software (StatSoft, France). Similarities in community structure were determined using the cluster analysis by the single linkage method with the Euclidean distance measure. Significant differences of bacterial communities of fish between seasons were determined by factorial correspondence analysis using the first 2 variances which described most of the variation in the data set.

3. Results

3.1 DGGE pattern of different locations in Viet Nam within the same sampling period

Fish were collected during the rainy season (October 2005) in 5 different districts of An Giang province, Viet Nam. PCR-DGGE fingerprinting of 5 replicates for each location revealed the presence of 8 to 12 bands of bacteria in the fish (Fig. 3). Some of the bands were common to all the different regions. The pattern obtained for the bacterial community for 5 replicates of the same pond of a unique farm in each district was totally similar among the same season (Fig. 4). We observed also high similarities on bacteria patterns for the samples from the same districts, as well as the neighbouring district where the water was supplied by the same branch of the Mekong River. Statistical analysis of the DGGE gel patterns for the 5 replicates of fish samples from 5 different districts of An Giang province harvested in the rainy season (25 samples), showed the community similarity among the different geographical locations where the fish samples were collected (Fig. 3). Two main clusters were observed at 70% similarities level (Fig. 3): 1. The first cluster included the samples from Chau Doc and Chau Phu districts; 2. The second cluster comprised the samples from An Phu, Tan Chau and Phu Tan districts. The bacterial communities of Chau Doc and Chau Phu districts were closely related, as well as the bacterial communities from An Phu, Tan Chau and Phu Tan districts.

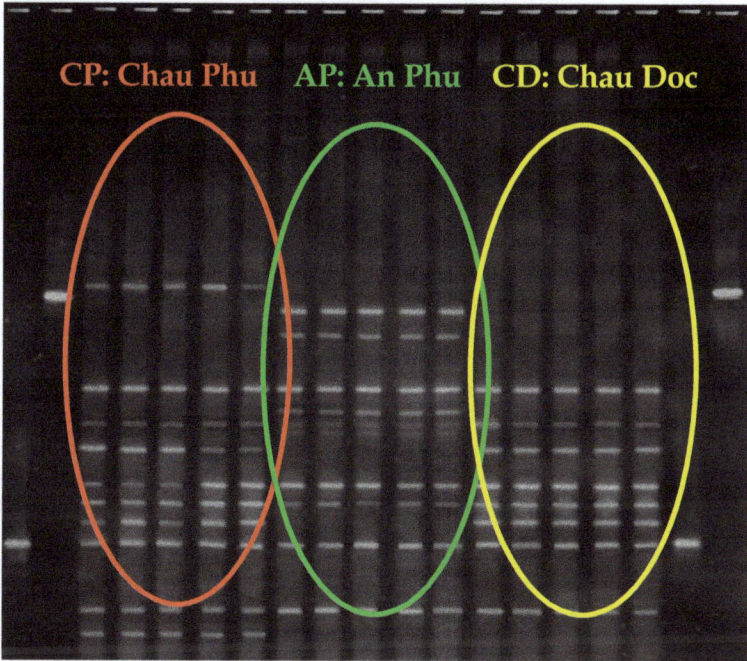

Fig. 3. PCR-DGGE 16S rDNA banding profiles of fish bacteria from 3 districts of An Giang province (five fish from the same pond in the same farm in each district), Viet Nam in rainy season : CP: Chau Phu district; AP: An Phu district; CD: Chau Doc district;

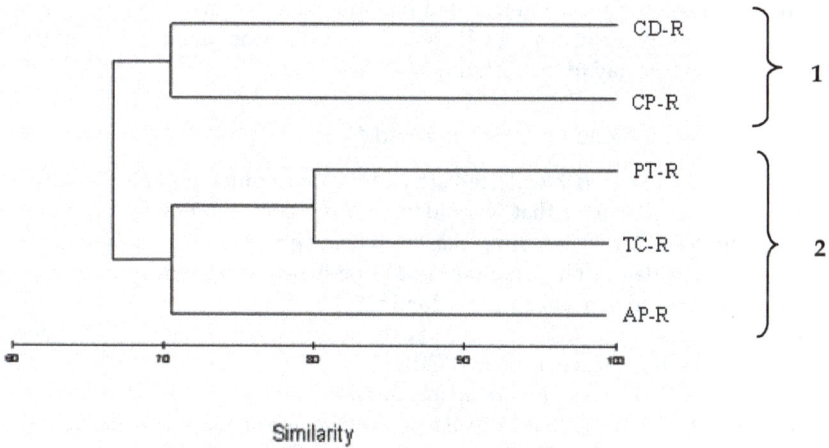

Fig. 4. Cluster analysis of 16S rDNA banding profiles for fish bacterial communities from 5 districts of An Giang province, Viet Nam in rainy season (R) 2006: CP: Chau Phu district; AP: An Phu district; CD: Chau Doc district; TC: Tan Chau district; PT: Phu Tan district.

3.2 DGGE pattern of fish within the same location in Viet Nam at two sampling seasons

Samples were taken at 8 months interval from the same ponds as previous experiments in the dry (February 2006) and rainy season (October 2005). Only four districts were studied for the dry season due to the closing down of Phu Tan pond. Differences on the DGGE band patterns of the same ponds at two different seasons can be clearly noted (Fig. 5). However, some dominant DNA bands could be observed that remained the same during the two seasons. Cluster analysis by Statistica software also showed that the bacterial communities of the same ponds were quite similar for the two seasons. Two main clusters were also observed at 75% similarities levels (Fig. 5): 1. The first cluster included samples from An Phu and Tan Chau districts; 2. The second cluster comprised all the samples from Chau Doc and Chau Phu districts. It was also shown that throughout the 2 seasons for the same location, the number of main bands observed on the gel was slightly different. For example, for the farm in An Phu district, 10 bands were observed in the dry season and 9 bands in the rainy season but only 7 bands are common and the calculation by Statistica showed that there was 75 % similarity between the two seasons. The similarities were higher for the other farms (88% for Than Chau, 84% for Chau Phu and 86 % for Chau Doc).

In addition, similarities of the bacterial communities from the 4 different geographical locations for two seasons were compared at 8 months interval by Factorial Correspondence Analysis (FCA). The first two variances described 76 % of the variation in the data set (Fig. 6). Four different groups of 4 different districts were observed regardless the seasons.

3.3 Microbiological changes during "Adjuevan" fermentation in Ivory Coast

Fish samples were collected between April and June 2010 at the pilot plant of Cemagref-Cirad. There were used to produce fermented fish following the two methods of Adjuevan fermentation used in Ivory Coast. PCR-DGGE analysis was done on samples taken successively from the first day to the fifth day of fermentation.

3.3.1 Microbiological changes following method 1

DGGE analyses conducted on fish collected after 24 h of maturation and before salting, gave 3 to 5 common bands of bacteria that were identified by sequencing as *Pseudomonas putida, Pseudomonas fluorescens, Aeromonas* spp., *Staphylococcus* spp. A similar DGGE profile was obtained at the second day of fermentation. But more bands appeared, from 9 to 21 bands composed of 94 % of positive gram bacteria dominated by *Staphylococcus* spp. (55%), *Bacillus* spp. (24 %), *Miccrococcus* spp. (13%) and 5% of aerobic mesophilic bacteria. Three percent of unidentified bacteria were revealed by PCR-DGGE in the third day of fermentation. *Staphylococcus xylosus, Staphylococcus auricularis, Staphylococcus piscifermentans, Staphylococcus saprophyticus, Micrococcus luteus, Bacillus subtilis, Bacillus licheniformis* and *Streptococcus* spp. were the genera identified (Fig.7). *Staphylococcus saprophyticus, Bacillus licheniformis, and Micrococcus luteus* were present in all samples. Some bacteria were common to some samples. However a loss of about 40% of bands of aerobic mesophilic bacteria and some genera of *Micrococcus* was observed at the fifth day of fermentation.

Determination of Fish Origin by Using 16S rDNA Fingerprinting of Microbial Communities by PCR-DGGE:
An Application on Fish from Different Tropical Origins

89

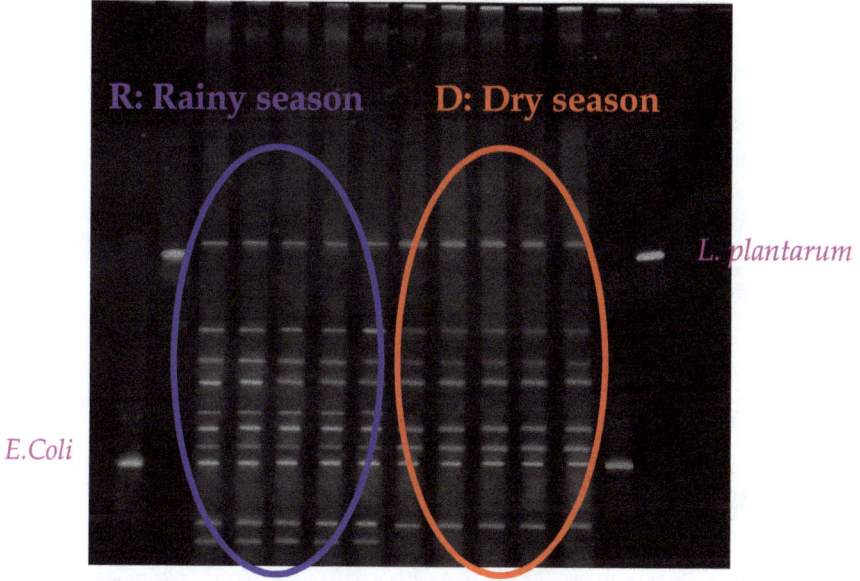

Fig. 5. PCR-DGGE 16S rDNA banding profiles of fish bacteria from one district of An Giang province, Viet Nam in the dry season (D) and rainy season (R)

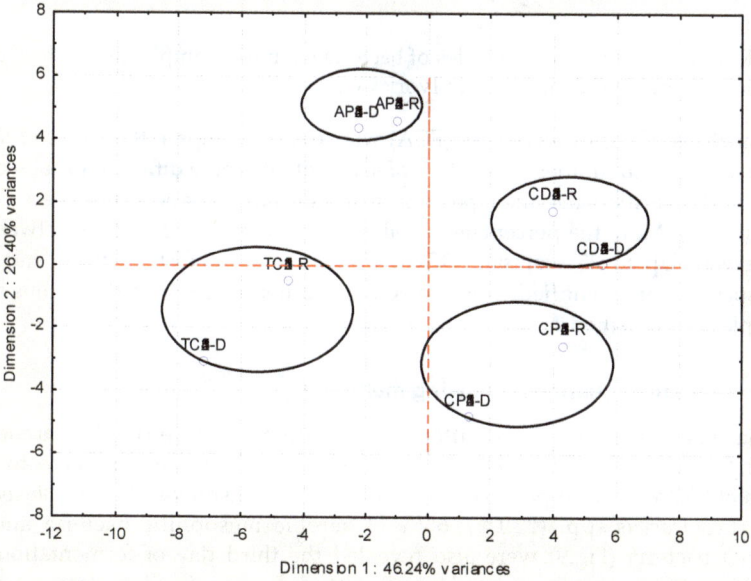

Fig. 6. Factorial variance analysis of 16S rDNA banding profiles of fish bacteria from 4 districts of An Giang province, Viet Nam in the dry season (D) and rainy season (R) 2006: CP: Chau Phu district; AP: An Phu district; CD: Chau Doc district; TC: Tan Chau district.

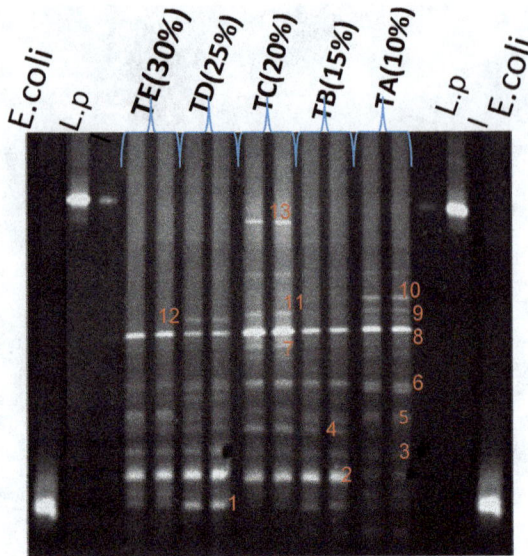

1: *P. putida*; 2:*B. licheformis*; 3, 5: *Bacillus spp.*; 4: *B. subtilis*; 6: *Micrococcus luteus*; 7: *S. Piscifermentans*; 8:*S. saprophyticus*; 9, 11: *S. xylosus*; 12: *Staphylococcus spp.*; 10: *S. auricularis*; 13: *Steptococcus spp.*. *(Abbreviations: B.; Bacillus; S, Staphylococcus. P, Pseudomonas).*

TA (10%): sample with 10% (w/w) of salt; TB (15%): sample with 15% (w/w) of salt; TC (20%): sample with 20% (w/w) of salt; TD (25%): sample with 25% (w/w) of salt; TE (30%): sample with 30% (w/w) of salt.

Fig. 7. PCR- DGGE of 16s rDNA profiles of bacteria from five samples at the third day of fermentation following the method 1 in Ivory Coast.

Factorial analysis of correspondences (FCA) was used to compare the levels of similarities between bacteria communities in function of salt content. Three different groups of samples were observed regardless the initial percentage of salt. Two variances described 75.1% of the variation (Fig. 8). More the percentage of salt was great, more differences between profiles were observed. Up to 25% of salt, DGGE profiles were almost identical and the same bacterial species were identified. This showed the influence of salt on the strains present on meat samples from method 1.

3.3.2 Microbiological changes following method 2

DGGE analyses conducted on fish after 24h of maturation gave two bands identified as *citrobacter freundii* and *Pseudomonas putida*, common to all samples. Five to 12 bands composed of 89.2% of gram positive bacteria with *Bacillus* spp. (49.8%), *Staphylococcus* spp. (17.3%), *Miccrococcus* spp. (22.1%), 6.7% of aerobic mesophilic bacteria and 4.1% of unidentified bacteria (Fig.9) were also revealed the third day of fermentation by PCR-DGGE. The genera were *Bacillus subtilis, Bacillus licheniformis, Staphylococcus auricularis, Staphylococcus piscifermentans,* and *Micrococcus* spp. *Pseudomonas putida, Bacillus licheniformis, Bacillus subtilis* and *Micrococcus* spp. were predominant in all samples during fermentation and persisted up to the end of fermentation. Samples D (25% salt) and E

Determination of Fish Origin by Using 16S rDNA Fingerprinting of Microbial Communities by PCR-DGGE:
An Application on Fish from Different Tropical Origins

91

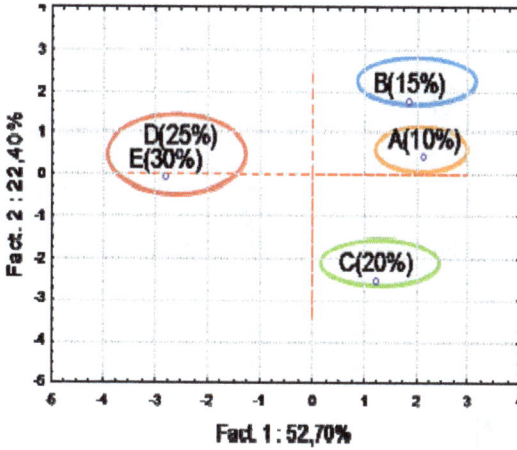

A (10%); B (15%); C (20%); D (25%); E (30%) five samples of fermented fish prepared with different percentages of salt.

Fig. 8. Factorial variance analysis of 16s rDNA banding profiles of bacteria from five samples at the third day of fermentation following method 1 in April 2010 in Ivory Coast.

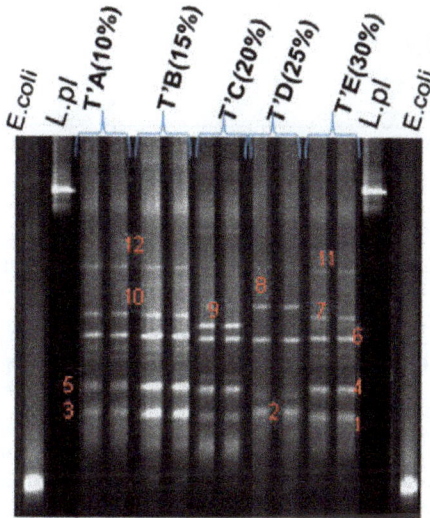

1: *Pseudomonas spp.*; .2: *P. fluorescens* ; 3: *P. putida*; 4,5: *B. licheformis*; 6: *B. subtilis*; 7,10: *S. piscifermentans*; 8: *Micrococcus spp.*; 9 : *Bacillus spp.*; 11,12: *S. auricularis*. (*Abbreviations: B.; Bacillus; S, Staphylococcus; P, Pseudomonas*).

T'A (10%): sample with 10% (w/w) of salt; T'B (15%): sample with 15% (w/w) of salt; T'C (20%): sample with 20% (w/w) of salt; T'D (25%): sample with 25% (w/w) of salt; T'E (30%): sample with 30% (w/w) of salt.

Fig. 9. PCR- DGGE 16s rDNA profiles of bacteria from five samples at the third day of fermentation following method 2 in Ivory Coast.

(30%) have common bands. A loss of only 10 % of bands was observed at the fifth day of fermentation.

Factorial analysis of correspondences (FCA) comparing the levels of similarities of bacterial communities at the third day of fermentation showed a variance described by 67.8% of the variation in the data set and a low difference between bacterial communities for all samples (Fig. 10). Overall flora varied very little from one sample to another but since 20% of salt, DGGE profiles were almost identical with the same bacterial species. All these results showed that the effect of salt was lower on the bacterial flora of method 2 than in method 1.

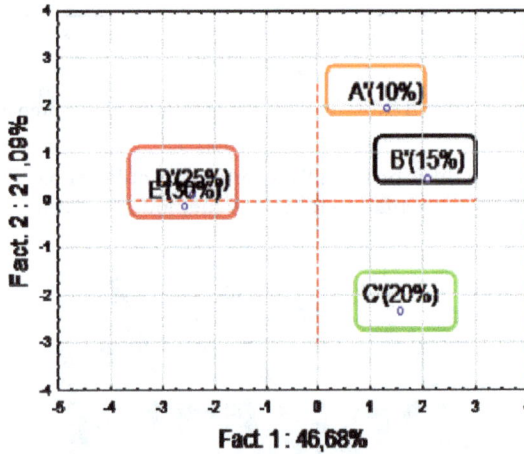

A' (10%); B' (15%); C' (20%); D' (25%); E' (30%) five samples of fermented fish prepared following different percentages of salt.

Fig. 10. Factorial variance analysis of 16s rDNA banding profiles of bacteria from five samples at the third day of fermentation following method 2 in June 2010 in Ivory Coast.

4. Discussion

Analysis of bacterial communities in fish samples has been often investigated using culture dependent methods and culture-independent methods by random amplified polymorphic DNA (RAPD) (Spanggaard et al., 2000). There are only a few published works that analyzed the bacterial communities in fish samples by PCR-DGGE methods (Spanggaard et al., 2000; Huber et al., 2004, Maiworé et al., 2009 a,b; Tatsadjieu et al., 2010).

In our study, we found that the band pattern of the bacterial communities isolated from fish obtained by PCR-DGGE was strongly linked to the microbial environment of the fish. The skin is in direct contact with the water, and the gills that filter the air from water that goes to the lung of fish is a good accumulator of the environmental bacteria. The intestine contains also a high amount of bacteria which is affected by the feeding habits of the fish.

The analysis of fish samples from different locations within the same period (rainy season) showed some significant differences in the migration patterns on the DGGE gel. However, the five replicates for each sampling location had statistically similar DGGE patterns throughout the study. The differences in the band profiles can be attributed to the differences of the feeding methods in between farms and the type of aquaculture system applied. The variations may also due to the water supply which can be affected by the pollution from urban life. Furthermore, the antibiotics needed to cure diseases and stress factors could also affect the microbial communities of the fish (Sarter et al., 2007). However, some common bands obtained by DGGE have been found in all the profiles within the same sampling periods and origin.

In fact, when seeing the different locations on the map of An Giang province on South Viet Nam, there is a separation of the Mekong River in two main branches which are then divided into many small canals. Chau Phu and Chau Doc are on the same branch in the west of the Mekong river and An Phu, Tan Chau and Phu Tan are on another branch in the east of the river. We could conclude that there were enough differences in the water quality and the environment of the fish to obtain a major effect on the bacterial ecology.

The study of the fish samples from the same locations in Viet Nam at two different seasons showed that there are some significant statistical variations in the DGGE bands profiles. This fact could result from the important variations due to the heavy rains in the rainy season which can greatly affect the salinity and the pH of the pond water. These factors will greatly affect the microbial communities of the fish which are dependent on the outside environment. These results suggested that the DGGE band profile of the bacterial community of the fish is unique and representative for a farm and a season. However, there was a relatively higher similarity between the samples of the same location across different sampling seasons than the samples from different locations. The DGGE profiles showed some dominant bands of the same locations which are present throughout the whole sampling time.

Concerning the fermented fish from Ivory Coast, our results showed that the salt content in the meat was influenced by the salt percentage in the brine and especially by the fermentation method used. It was also found, after sequencing DGGE bands, a strong dominance of different species of halophilic bacteria depending of the fermentation method. The variation of the bacterial flora of fermented fish was primarily influenced by the fermentation method and by the use of high salt content that conducted to at least 7% in meat. So the analysis of fermented fish bacteria communities by PCR–DGGE could be applied to differentiate the methods of fermentation.

In conclusion, PCR-DGGE technique could be applied to differentiate geographical location of fish production by using their bacterial community. This technique could be applied also to follow the influence of a process on the bacterial flora. Biological markers stayed stable for specific locations among the different seasons and that they showed sufficient statistical specificity per farm. This global technique is quicker (less than 24 h) than all of the classical microbial techniques and avoids the precise analysis of bacteria by biochemistry or molecular biology. Sequencing could be done directly from the DGGE gel extracts to identify bacteria. This method is a rapid analytical traceability tool for fish products and provides fish with a unique biological bar code.

5. References

Al Harbi, A.H., Uddin, N., 2003. Quantitative and qualitative studies on bacterial flora of hybrid tilapia (*Oreochromis niloticus x O. aureus*) cultured in earthen pond in Saudi Arabia. Aquaculture 34, 43-48.

Ampe, F., Omar, N.B., Moizan, C., Wacher, C., Guyot, J.P., 1999. Polyphasic study of the spatial distribution of microorganisms in Mexican pozol, a fermented maize dough, demonstrates the need for cultivation-independent methods to investigate traditional fermentations. Appl. and Environ. Microbiol. 65, 5464-5473.

Díez, B., Pedrós–Alió, C., Marsh, T.L., Massana, R., 2001. Application of denaturing gradient gel electrophoresis (DGGE) to study the diversity of marine picoeukaryotic assemblage and comparison of DGGE with other molecular techniques. Appl. and Environ. Microbiol. 67, 2942-2951.

Grisez, L., Reyniers, J., Verdonck, L., Swings, J., Ollevier, F., 1997. Dominant intestinal microbiota of sea breeam and sea bass larvae, from two hatcheries, during larval development. Aquaculture 155, 387-399.

Heyndrickx, M., Vauterin, L., Vandamme, P., Kersters, K., De Vos, P., 1996. Applicability of combined amplified ribosomal DNA restriction analysis (ARDRA) patterns in bacterial phylogeny and taxonomy. Journal of Microbiol. Methods 26, 247-259.

Huber, I., Spanggaard, B., Appel, K.F., Rossen, L., Nielsen, T., Gram, L., 2004. Phylogenetic analysis and in situ identification of the intestinal microbial community of rainbow trout (*Oncorhynchus mykiss*, Walbaum). Journal of Appl. Microbiol. 96, 117-132.

Kowalchuk, G.A., Stephen, J.R., de Boer, W., Prosser, J.I., Embley, T.M., Woldendorp, J.W., 1997. Analysis of ammonia-oxydizing bacteria of the beta subdivision of the class *Proteobacteria* in coastal sand dunes by denaturing gradient gel electrophoresis and sequencing of PCR amplified 16S ribosomal DNA fragments. Applied and Environ. Microbiol. 63, 1489-1497.

Leesing, R., 2005. Identification and validation of specific markers for traceability of aquaculture fish for import/export. PhD dissertation. University of Montpellier 2, France.

Leesing Ratanaporn, Dijoux Daniel, Le Nguyen Doan Duy, Loiseau Gérard, Ray Ramesh C, Montet Didier. Improvement of DNA Extraction and Electrophoresis Conditions for the PCR-DGGE Analysis of Bacterial Communities Associated to Two Aquaculture Fish Species. Dynamic Biochemistry, Process Biotechnology and Molecular Biology, In press July 2011.

Le Nguyen D.D., Ha Ngoc H., Dijoux D., Montet D., Loiseau G., 2008. Determination of fish origin by using 16S rDNA fingerprinting of bacterial communities by PCR- DGGE: application to Pangasius fish from Viet Nam. Food control. 19, 454-460.

Moeseneder, M.M., Arrieta, J.M., Muyzer, G., Winter, C., Herndl, G.J., 1999. Optimization of terminal-restriction length polymorphism analysis for complex marine bacterioplankton communities and comparison with denaturing gradient gel electrophoresis. Appl. and Environ. Microbiol. 65, 3518-3525.

Maiworé J., Tatsadjieu Ngouné L., Montet D., Loiseau G., Mbofung C.M., 2009a. Comparison of bacterial communities of Tilapia fish from Cameroon and Vietnam using PCR-DGGE. African Journal of Biotech. 8 (24), 7156-7163.

Maiworé J., Tatsadjieu Ngouné L., Montet D., Loiseau G., Mbofung C.M., 2009b. Comparison of bacterial communities of tilapia fish from Cameroon and Vietnam using PCR-DGGE (polymerase chain reaction-denaturing gradient gel electrophoresis). African journal of Biotechnol 8 (24), 7156-7163.

Montet, D., Leesing, R., Gemrot, F., Loiseau, G., 2004. Development of an efficient method for bacterial diversity analysis: Denaturing Gradient Gel Electrophoresis (DGGE). In: Seminar on Food Safety and International Trade, Bangkok, Thailand.

Montet, D., Le Nguyen D.D., El Sheikha A.F., Condur A., Métayer I., Loiseau G., 2008. Application of PCR-DGGE in determining food origin: Cases studies of fish and fruits. Greening the Food Chain 3 and 4; 3: Traceability: Tracking and tracing in the food chain. Aspects of Applied Biol 87, 11-22.

Murray, A.E., Hollibaugh, J.T, Orrego, C., 1996. Phylogenetic composition of bacterioplankton from two California estuaries compared by denaturing gradient gel electrophoresis of 16S rDNA fragments. Appl. Environ. Microbiol. 6, 2676-2680.

Muyzer, G., De Waal, E.C., Uitterlinden, A.G., 1993. Profiling of complex microbial populations by denaturing gradient gel electrophoresis analysis of polymerase chain reaction-amplified genes coding for 16S rRNA. Appl. Environ. Microbiol. 59, 695-700.

Muyzer, G., Teske, A., Wirsen, C.O., Jannasch, H.W., 1995. Phylogenetic relationships of Thiomicrospira species and their identification in deep-sea hydrothermal vent sample by denaturing gradient gel electrophoresis of 16S rDNA fragment. Archives of Microbiol. 164, 165-172.

Øvreas, L., Forney, L. Dae, F.L., Torsvik., V., 1997. Distribution of Bacterioplankton in Meromictic Lake Sælenvannet, as determined by Denaturing Gradient Gel Electrophoresis of PCR-Amplified gene fragment coding for 16S rRNA. Appl. and Environ. Microbiol. 63, 3367-3373.

Riemann, L., Steward, G.F., Fandino, L.B., Campbell, L., Landry, M.R., Azam, F., 1999. Bacterial community composition during two consecutive NE Monsoon periods in the Arabian Sea studied by denaturing gradient gel electrophoresis (DGGE) of rRNA genes. Deep-Sea Res. 46, 1791-1811.

Sheffield, V.C., Beck, J.S., Stone, E.M., Myers, R.M., 1989. Attachment of a 40 bp G+C rich sequence (GC-clamp) to genomic DNA fragments by polymerase chain reaction results in improved detection of single-base changes. Proceeding of the National Academy of Sciences of the United States of American 86, 232-236.

Sarter S., Hoang Nam Nguyen K., Le Thanh H., Lazard J., Montet D., 2007 Antibiotic resistance in Gram-negative bacteria isolated from farmed catfish. Food Control 18, 1391–1396.

Spanggaard, B., Huber, I., Nielsen, T.J., Nielsen, T., Appel, K., Gram, L., 2000. The microbiota of rainbow trout intestine: a comparison of traditional and molecular identification. Aquaculture 182, 1-15.

Tatsadjieu Ngouné L., Maiworé J., Hadjia B.M., Loiseau G., Montet D., Mbofung C.M., 2010. Study of the microbial diversity of *Oreochromis niloticus* of three lakes of Cameroon by PCR-DGGE: Application to the determination of the geographical origin. Food control 21, 673-678.

Van Hannen, E.J., Zwart, G., van Agterveld, M.P., Gons, H.J., Ebert, J., Laanbroek, H.J., 1999. Changes in bacterial and eukaryotic community structure after mass lysis of filamentous cyanobacteria associated with viruses. Appl. and Environ. Microbiol. 65, 795-801.

Yang, C.H., Crowley, D.E., Menge, J.A., 2001. 16S rDNA fingerprinting of rhizosphere bacterial communities associated with healthy and Phytophthora infected avocado roots. FEMS Microbiol. Ecol. 35, 129-136.

6

Validation of Endogenous Reference Genes for qPCR Quantification of Muscle Transcripts in Atlantic Cod Subjected to Different Photoperiod Regimes

Kazue Nagasawa, Carlo Lazado and Jorge M. O. Fernandes
Faculty of Biosciences and Aquaculture,
University of Nordland, Bodø
Norway

1. Introduction

Atlantic cod (*Gadus morhua*, L.) is a commercially important species worldwide and overfishing has contributed to a decline of wild stocks below sustainable levels. This has stimulated aquaculture production of this species, which has increased remarkably over the last decade to over 20,000 tonnes in 2008 (FAO). Nevertheless, cod farming still faces several production bottlenecks related to larval quality, nutrition, diseases and precocious sexual maturation. The early onset of sexual maturation at around two years in farmed conditions seriously restricts the profitability of the industry. Sexual maturation and the subsequent spawning result in loss of appetite, reduced feed conversion and increased mortality rate (Karlsen et al., 2006), which leads to an increase in the production time required to reach the desired harvest size.

Photoperiod manipulation, typified by continuous light illumination, has been used to delay sexual maturation to some extent in several aquaculture species, including Atlantic salmon (*Salmo salar*, L.) (Endal et al., 2000), European sea bass (*Dicentrachus labrax*, L.) (Begtashi et al., 2004) and Atlantic cod (Davie et al., 2003; Hansen et al., 2001; Norberg et al., 2004). The application of continuous light from the summer solstice prior to maturation is thought to mask the photoperiod signal that acts as a trigger for gonadal development and spawning (Davie et al., 2003). Taranger et al. (2006) have shown that gonadal maturation of cod kept in sea cages can be delayed by three to five months through application of continuous broad-spectrum light. In addition to inhibiting sexual maturation, photoperiod manipulation has a direct effect on somatic growth, particularly during juvenile stages (Davie et al., 2007; Taranger et al., 2006). In fact, short-term application of continuous light was found to induce a 5 to 9% increase in body weight when compared to cod reared under normal photoperiod conditions and significant differences can still be observed at harvesting size, nearly three years later (Imsland et al., 2007). In spite of its obvious relevance for the aquaculture industry, the molecular basis of this growth plasticity induced by light cues is not known. In order to better control the precocious sexual maturation of farmed cod, it is crucial to identify the transcriptional networks related to this phenomenon and to understand how they are influenced by photoperiod.

Muscle is the main tissue supporting fish growth. Teleost myogenesis is a complex phenomenon which involves a number of molecules regulating distinct phases of this process. The development and formation of muscle involves either hypertrophy (expansion of muscle fibre by absorption of myoblast nuclei) or hyperplasia (formation of fibres on the surface of an existing muscle fibre) (Johnston, 1999). The progression of muscle formation is associated with the sequential expression of key genes from the myogenic regulatory factors (MRFs) family, which include *myoblast differentiation 1 (myoD), myogenic factor 5 (myf-5), myogenin (myoG) and myogenic factor 6/myogenic regulatory factor 4 (myf-6/MRF4)* (Watabe, 1999). Another molecule of significant importance in muscle development is myosin heavy chain *(myhc)*, which serves as marker of muscle development in several studies (Johnston, 1999). *Myhc* genes code for a family of ATP-dependent motor proteins that are involved in muscle contraction (Ikeda et al., 2007). *Myhc* activity can be used to monitor fish growth, since most fish have a continuous hyperplasic growth throughout their lifespan and *myhc* is actively involved in muscle protein synthesis (Dhillon et al., 2009).

Quantification of transcript levels by real-time PCR (qPCR) is currently the method of choice, since it is reliable and sensitive enough to quantify even lowly expressed mRNAs in small amount of target tissues (Bustin, 2002). For example, in tiger pufferfish (*Takifugu rubripes*, Temminck & Schlegel) this technique has been used to validate suppression subtractive hybridization results (Fernandes et al., 2005), to examine how temperature affects expression of the growth-related genes *myoG* (Fernandes et al., 2006) and *forkhead box protein K1 (foxk1)* (Fernandes et al., 2007a) during embryonic development, and to examine differential regulation of splice variants of the master transcription factor *myoD1* (Fernandes et al., 2007b). In spite of its enormous potential, relative qPCR quantification has several pitfalls that must be carefully considered (Bustin and Nolan, 2004). In particular, selection of suitable reference genes with even expression in all samples is critical to normalise qPCR data and the use of non-validated reference genes can lead to erroneous conclusions that are biologically meaningless (Fernandes et al., 2008). It is a general consensus that a versatile reference gene stable under various experimental conditions does not exist. Before proceeding to quantifying the expression of a target gene, it is necessary to select the most appropriate reference genes for each species and tissue for a particular experimental setup. A sensible practice involves testing multiple genes for each experiment and using statistical applications to identify the best combination of the two or three most stable genes that will be used to normalise qPCR data (Andersen et al., 2004; Vandesompele et al., 2002).

The aim of the present research paper was to identify suitable reference genes for relative quantification by qPCR of growth- and maturation-related genes that may be affected by photoperiod manipulation. Five commonly used reference genes were evaluated, namely: *β-actin (actb), acidic ribosomal protein (arp), eukaryotic elongation factor 1a (eef1a), glyceraldehyde-3-phosphate dehydrogenase (gapdh)* and *ubiquitin (ubi)*. Their transcript levels in the above fast muscle samples were determined by qPCR using SYBR chemistry. GeNorm (Vandesompele et al., 2002) and NormFinder (Andersen et al., 2004) were used to evaluate expression stability of above candidate genes. In addition, to demonstrate the importance of using validated reference genes in qPCR analysis, *myhc* expression was examined. Here, we showed the impact of normalisation strategies (i.e., different individual candidate genes versus the normalisation factor from the two best validated reference genes) on *myhc* expression levels and the necessity of validation to select the most stable reference gene in each experimental plot.

2. Materials and methods

2.1 Photoperiod experiment and sampling

2.1.1 Fish husbandry

Atlantic cod juveniles, *Gadus morhua* L. with an initial size of 2.7 ± 0.8 g (mean ± standard deviation [SD], n=123) were divided into six tanks (250 m³) with an open flow system at a density of approximately 130 individuals per tank. Sea water was continuously supplied to each tank at 7.4 ± 0.4 °C. A commercial diet (Amber Neptun, Skretting AS, Stavanger, Norway) was provided daily by automatic belt feeders, at 5% (w/w) body weight of the fish and adjusted on a weekly basis.

2.1.2 Photoperiod experiment

Each group of fish in three tanks was either kept under continuous light (LL) or reared under a normal light regime (NL) that corresponded to natural environmental photoperiod conditions in Bodø (67°N), Norway during 6 months from January to July 2010. Day light time was recreated indoors using white light fluorescent tubes (Aura Light International AB, Karlskrona, Sweden) and controlled by a scheduled timer according to local sunrise and sunset times in Bodø. Light intensity on the central surface of each tank was 120 Lux.

2.1.3 Sample collection

The fish were killed by immersion in seawater containing 0.2 g·L⁻¹ tricaine methanesulfonate (MS222; Sigma, Oslo, Norway). Fast muscle samples were carefully dissected from the trunk area below the second dorsal fin from six fish at the start of the experiment and 0.5, 1, 7, 30, 60, 120 and 180 days thereafter. Tissues were snap-frozen in liquid nitrogen and stored at – 80 °C until RNA extraction.

2.1.4 Ethics statement

All procedures of fish rearing and tissue sampling were in accordance with the guidelines set by the National Animal Research Authority (Forsøksdyrutvalget, Norway).

2.2 Real-time PCR (qPCR)

2.2.1 Primer design

To validate expression stability of reference genes, five candidate genes (*β-actin* (*actb*), *acidic ribosomal protein* (*arp*), *elongation factor 1 alpha* (*eef1a*), *glyceraldehyde-3-phosphate dehydrogenase* (*gapdh*), *ubiquitin* (*ubi*) and *myosin heavy chain* (*myhc*)) were selected and analysed for qPCR validation.

2.2.2 RNA extraction and cDNA synthesis

Total RNA was extracted from the fast muscle samples above and used to synthesized cDNA as detailed elsewhere (Campos et al., 2010). Two micrograms of total RNA were used for cDNA synthesis by reverse transcription, following treatment with gDNA wipe out buffer (Qiagen, Nydalen, Sweden) to remove genomic DNA contamination.

2.2.3 qPCR amplification

qPCR reactions were conducted with the primer sets indicated on table 1. Quantification of transcripts were analysed by qPCR with SYBR Green chemistry (SYBR Green I Master, Roche) on a LightCycler® 480 (Roche) as previously described (Campos et al., 2010). Fifty-fold diluted muscle cDNA were run in duplicate, and minus reverse transcriptase and no template controls were included in the reactions. Thermocycling parameters were as follows: 95°C for 15 min, followed by 45 cycles of 15 s at 94°C, 20 s at 60°C and 20 s at 72°C. Five-point standard curves of a 2-fold dilution series (1:1, 1:2, 1:4, 1:8 and 1:16) were prepared from pooled RNA that was reverse transcribed as above. These dilution curves were used to calculate amplification efficiencies of the PCR reactions (Fernandes et al., 2006). Cycle threshold (C_t) values were determined by the LightCycler® 480 software with a fluorescence level arbitrarily set to 1.

Gene	GenBank	Sequence	Size (bp)	E (%)	Reference
actb	AJ555463	Fw: TGACCCTGAAGTACCCCATC Rv: TCTTCTCCCTGTTGGCTTTG	162	77	(Lilleeng et al., 2007)
arp	EX741373	Fw: TGATCCTCCACGACGATGAG Rv: CAGGGCCTTGGCGAAGA	113	86	(Olsvik et al., 2008)
eef1a	CO541820	Fw: CACTGAGGTGAAGTCCGTTG Rv: GGGGTCGTTCTTGCTGTCT	142	84	(Lilleeng et al., 2007)
gapdh	AY635584	Fw: GGTCGCAACCGCAAGGT Rv: TGACCGTTGAGCATTTCCTTCT	83	88	(Hall et al., 2006)
ubi	EX735613	Fw: GGCCGCAAAGATGCAGAT Rv: CTGGGCTCGACCTCAAGAGT	69	87	(Olsvik et al., 2008)
myhc	AY093703	Fw: CAGAAGCTATAAAAGGTGTCCG Rv: GCAGCCATTCTTCTTATCCTCCTC	86	81	(Koedijk et al., 2010)

Table 1. Primers used in this study. Primer sequences, Genbank accession numbers, amplicon sizes and PCR efficiencies are indicated.

2.3 Data analyses

2.3.1 Expression stability analyses

Raw qPCR data were converted to expression level using the above dilution curves. These were then analysed for expression stability using the statistical applications GeNorm (Vandesompele et al., 2002) and NormFinder (Andersen et al., 2004).

2.3.2 Statistical analyses

Differences in expression levels of *gapdh*, *actb*, *eef1a*, *arp*, *ubi* and *myhc* during the photoperiod manipulation experiment were examined by one-way ANOVA with Holm-Sidak post-hoc tests. ANOVA assumptions were checked prior to carrying out the analyses and when the data did not follow the Gaussian distribution or did not meet the equal variance requirements, a Kruskal-Wallis one-way ANOVA on ranks with Dunn's test for post-hoc comparisons was used instead. Statistical analyses were performed with the SigmaStat statistical package (Systat software, London, UK). In all cases, significance was set at $P < 0.05$.

Validation of Endogenous Reference Genes for qPCR Quantification of Muscle Transcripts in Atlantic
Cod Subjected to Different Photoperiod Regimes

101

3. Results and discussion

3.1 Validation of reference genes

Specificity of qPCR reactions was confirmed by melting curve analysis, which revealed a single dissociation peak for each gene. Global variation on expression profiles of the candidate reference genes can be observed in Fig.1. Mean C_t values of the candidate reference genes were 19.99, 18.36, 18.03, 17.76 for *actb*, *ubi*, *eef1a* and *arp*, respectively, whereas the median C_t value for *gapdh* (25.96) was above the range of values for the other four genes. Expression of *ubi* showed the least variation across samples, in contrast to *gapdh*, *actb* and *eef1a*.

Detailed expression of the individual reference genes during the 6-month photoperiod manipulation experiment is presented in Fig. 2, showing a differential expression trend between sampling points and photoperiod regimes. There were no significant differences ($P < 0.05$) on the overall expression amongst any of the candidate reference genes (*arp*: $P = 0.997$; *eef1*: $P = 0.735$; *ubi*: $P = 0.124$; *gapdh*: $P = 0.386$; *actb*: $P = 0.554$). It is also important to note that the transcript levels of *actb*, *ubi*, *eef1a* and *arp* were almost the same based on the close range of their C_t values and all showed a perceptible increase on their transcript levels at the last 3 sampling points. In contrast, transcript levels of *gapdh* were relatively lower compared with the other reference genes as characterized by having higher C_t values. There was an apparent difference on the expression of *actb* and *gapdh* between treatments.

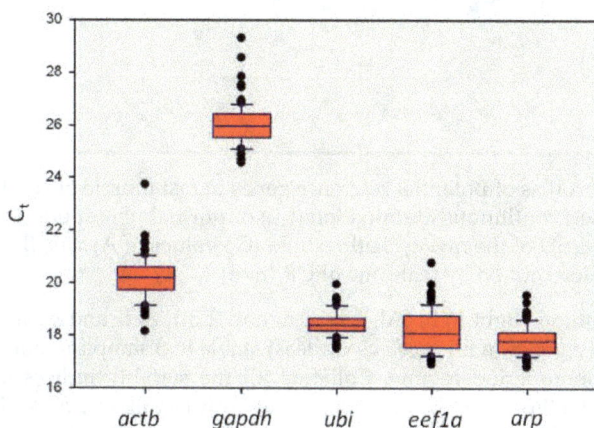

Fig. 1. Overall expression patterns of candidate reference genes in the muscle of Atlantic cod reared under different photoperiod regimes. Raw cycle threshold (C_t) qPCR data of individual reference genes in all samples (n=96) are represented as box-and-whisker plots. Median values are indicated by a solid line inside the boxes.

Expression stability indices of the candidate reference genes as assessed by geNorm varied with time and photoperiod regime (Table 2). *Arp* and *ubi* were identified as the most stable reference genes. Specifically, *arp* was the most stable in fish group under normal photoperiod (0 h, 6 h, 1 d, 1 w, 1 m and 6 m), while *ubi* was the most stable in the group

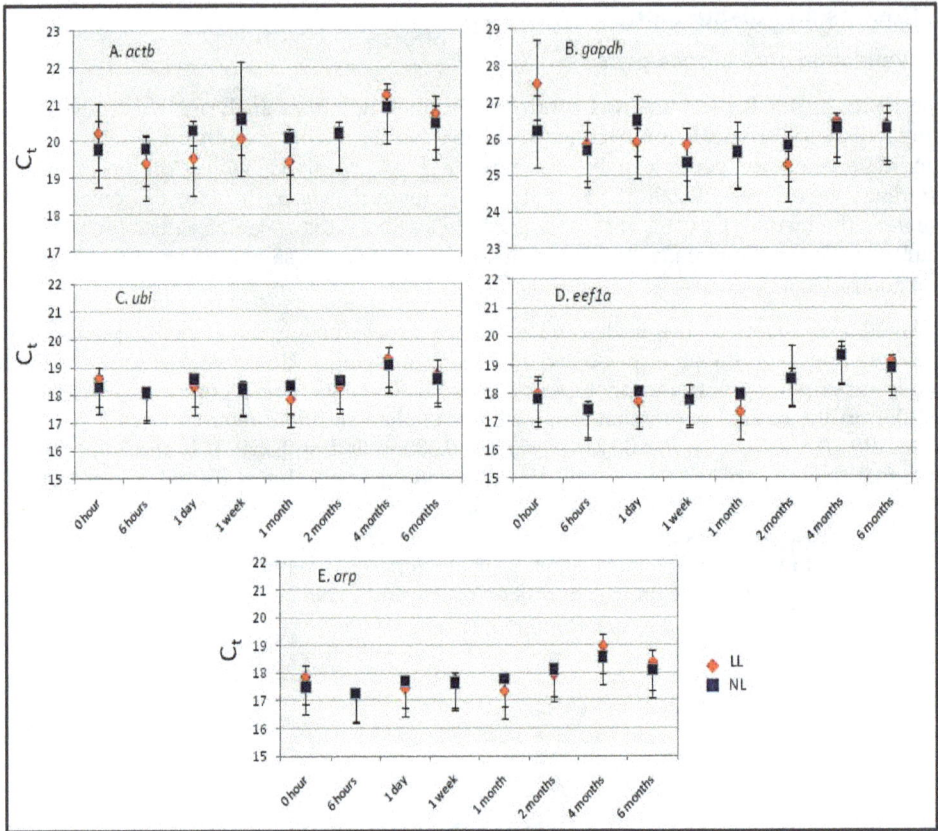

Fig. 2. Expression profiles of potential reference genes in fast muscle of cod kept under two photoperiod regimes: continuous illumination (LL) or normal photoperiod (NL). Data are presented as mean ± SD of the raw cycle threshold (C_t) values of A) *actb*, B) *gapdh*, C) *ubi*, D) *eef1a* and E) *arp* as determined by real-time qPCR (n=6).

reared under continuous light (6 h, 1 d, 1 w, 1 m and 2 m). *Actb* and *gapdh* were two of the least stable reference genes, with *gapdh* as the least stable in 3 sampling points (0 h, 6 h and 1 m) regardless of photoperiod regime. Collating all the stability indices of each reference gene, the order of stability from the most to the least was as follows: *arp* > *ubi* > *eef1a* > *actb* > *gapdh*. Pairwise comparisons revealed that *arp* and *ubi* were the best pair for two-gene normalisation with a joint stability value of 0.138 (Fig. 3A).

The validation software, NormFinder identified *arp* (0.084) and *gapdh* (0.259) as the most and least stable reference genes, respectively. The overall ranking reference genes from the most to the least stable was as follows: *arp* > *ubi* > *actb* > *eef1a* > *gapdh* (Fig. 3B). It was also determined that the best pair of candidate reference genes was *arp* and *ubi* with a joint stability value of 0.084. This pairwise result is similar to the result in geNorm that the best pair for two-gene normalisation was *arp* and *ubi*. It was also found that regardless of photoperiod regimes, *gapdh* was the least stable gene at 0 h, 6 h, 1 m and 4 m, corresponding broadly to the results obtained in geNorm.

		geNorm					NormFinder				
Time	Photoperiod	arp	eefla	ubi	actb	gapdh	arp	eefla	ubi	actb	gapdh
0 h	LL	0.389	0.425	0.401	0.445	0.862	0.023	0.031	0.025	0.001	0.342
	NL	0.372	0.389	0.381	0.464	0.602	0.027	0.016	0.034	0.049	0.145
6 h	LL	0.363	0.349	0.329	0.640	0.665	0.001	0.002	0.001	0.174	0.190
	NL	0.159	0.171	0.206	0.166	0.326	0	0.003	0.013	0.001	0.047
1 d	LL	0.182	0.189	0.153	0.258	0.195	0.007	0.007	0.001	0.026	0.010
	NL	0.298	0.357	0.313	0.356	0.602	0.010	0.025	0.013	0.013	0.160
1 w	LL	0.264	0.347	0.250	0.430	0.388	0.004	0.032	0.001	0.069	0.051
	NL	0.540	0.623	0.544	1.390	0.676	0.001	0.072	0.001	0.904	0.094
1 m	LL	0.287	0.277	0.275	0.367	0.605	0.014	0.009	0.009	0.020	0.164
	NL	0.230	0.298	0.231	0.295	0.542	0	0.022	0	0.013	0.134
2 m	LL	0.389	0.946	0.367	0.382	0.429	0.021	0.419	0.009	0.007	0.010
	NL	0.173	0.193	0.179	0.166	0.184	0.009	0.012	0.011	0.004	0.008
4 m	LL	0.210	0.225	0.239	0.224	0.321	0.008	0.010	0.016	0.006	0.044
	NL	0.188	0.239	0.186	0.316	0.313	0	0.011	0	0.039	0.039
6 m	LL	0.329	0.332	0.387	0.469	0.390	0.020	0.006	0.053	0.085	0.033
	NL	0.180	0.208	0.192	0.252	0.350	0	0.004	0.004	0.016	0.052

Table 2. Expression stability indices of the five reference genes, as determined by geNorm and NormFinder applications. Relative mRNA level were determined in fast muscle of cod kept under two photoperiod regimes: continuous illumination (LL) and normal photoperiod (NL). The most and the least stable reference genes are shaded in blue and red, respectively.

In order to obtain more robust conclusions it is advisable to do a parallel validation of reference gene stability using different alternative software applications, since there is no each method uses different mathematical models that can lead to different outputs. For example, NormFinder takes all candidate reference genes into account and ranks them with the intragroup and intergroup variation, whereas geNorm sequentially excludes the worst gene ending with two and ranks genes with the degree of similarity of expression. The overall results from NormFinder and geNorm applied to our data revealed that the most stable genes were arp and ubi and these two is the most suitable pair for two-gene normalisation. In previous qPCR studies in cod, it was also shown that arp and ubi were the most stable genes and could be used in studying wild populations of cod living in contaminated areas (Olsvik et al., 2008) and ontogeny in cod larvae (Sæle et al., 2009). To optimise the results in selecting the most suitable reference gene, candidate genes should belong to different biological pathways, so as to minimise errors associated with co-regulation. Co-regulation is still possible between ubi and arp, since they fall on the same biological pathway as important molecules in protein degradation and elongation step of protein synthesis, respectively. However, since these genes were validated by two applications generating similar results, it is fair to consider them as stable in this experimental setup. Both applications identified gapdh as the most unstable reference gene. It has been observed that gapdh is regulated under varying physiological conditions, which could render this gene inappropriate as reference gene (Olsvik et al., 2005). In a study in cod where candidate reference genes were evaluated during ontogeny with emphasis on the development of gastrointestinal tract, gapdh was also rated as one of the least suitable normalisation genes regardless of categorisation and analysis (Sæle et al., 2009). It is also mentioned that in cod, there are two isoforms of gapdh, one that is muscle-specific and the other as a brain-specific. Validation of reference genes in cod exposed to thermal stress revealed that gapdh was also the least favourable gene for normalisation (Aursnes et al., 2011).

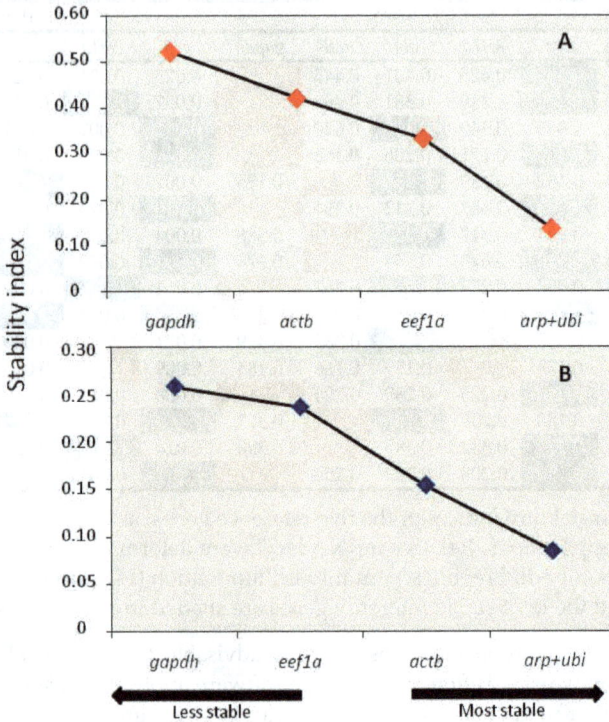

Fig. 3. Ranking of reference genes according to their expression stability in fast muscle of Atlantic cod reared under different photoperiod regimes. The average expression stability values were calculated with (A) geNorm and (B) NormFinder.

3.2 Influence of normalisation on photic-induced expression of *myhc* in cod muscle

Myosin is a ubiquitous eukaryotic motor that interacts with actin to generate the force for cellular movements as diverse as cytokinesis and muscle contraction (Cheney et al., 1993). This motor protein accounts for the majority of myofibrils, which themselves make up to two-thirds of muscle protein synthesis (Mommsen, 2001). For this reason, *myhc* has been used to study muscle growth and development in teleosts (Johnston, 2001). Environmental stimuli such as light influence most of the physiological processes in fish and muscle development is not an exception. The influence of photoperiod manipulation on the muscle physiology of Atlantic cod was assessed in this study by profiling the expression of this gene during a photoperiod manipulation experiment.

For comparison, raw expression data of *myhc* were normalised in two different ways: i) with the use of a two-gene normalisation factor from the most stable genes (*arp* and *ubi*) and, ii) with the least stable reference genes (Fig. 4). Using the best two-gene normalisation factor from geNorm, it was observed that from 0 h to 1 week the expression of *myhc* did not change significantly in either photoperiod regime. However, after a month of photoperiod manipulation, a significant difference was noted between treatments and the group exposed to normal photoperiod showed a significantly higher *myhc* expression than the group

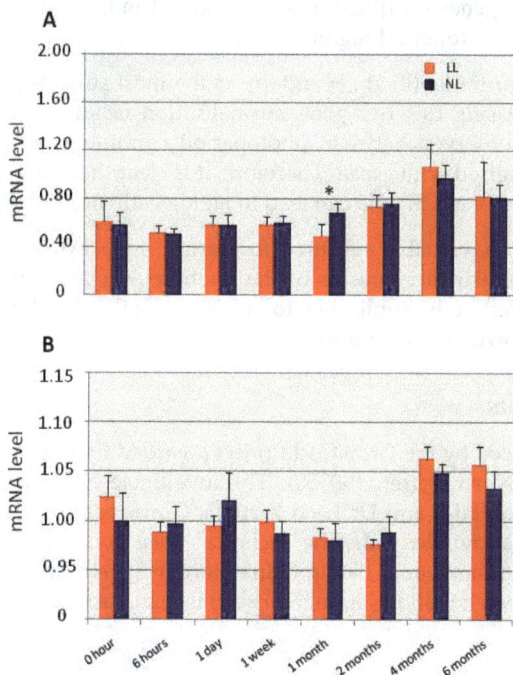

Fig. 4. Expression of *myhc* in the fast muscle of Atlantic cod subjected to different photoperiod regimes. A) Transcript levels of *myhc* gene normalised using *arp* and *ubi*, the best combination for a two-gene normalisation. B) Transcript levels of *myhc* gene normalised by the least stable reference gene, *gapdh* using $\Delta\Delta C_t$ method, Data are shown as mean ± SD of the normalised values (n=6). Asterisk (*) indicates that a significant difference was detected between photoperiod treatments ($P < 0.001$).

exposed to continuous light. Expression of *myhc* in the natural photoperiod group was approximately 40% higher than the expression in the continuous light group. From 2 to 6 months, *myhc* expression increased equally in both photoperiod regimes.

No significant difference was noted between light regimes throughout whole photoperiod manipulation experiment when *myhc* expression data were normalised using *gapdh*, the least stable reference gene. This stresses the importance of identifying suitable reference genes for a particular biological system, not only to draw robust conclusions but also to identify subtle and important differences in mRNA levels.

4. Conclusions

Though morphometric analysis is still an acceptable strategy in studying muscle growth in fish, molecular approaches have opened a new set of possibilities to study this phenomenon and to understand the role of key regulatory molecules in myogenesis. qPCR analysis is the most reliable method to quantify gene expression, provided that suitable reference genes are used for data normalisation. To the best of our knowledge, this report represents the first validation of reference genes for qPCR quantification of muscle transcripts in Atlantic cod reared under different photoperiod regimes.

NormFinder and geNorm identified *ubi* and *arp* as the most suitable gene pair to normalise our expression data. Using this two-gene normalisation factor, a 40% difference in *myhc* transcript levels was observed between photoperiod conditions, which was not detected when data were normalised with *gapdh*. Therefore, it is clear that using inadequate reference genes for normalisation of qPCR data can lead to biologically meaningless conclusions.

This study represents a valuable resource for future gene expression studies aimed at investigating the molecular mechanisms of the photic-plasticity of muscle development in Atlantic cod. Moreover, it is applicable to more general related topics in aquaculture research, including growth and nutrition.

5. Acknowledgments

This research was funded by the GrowCod Project awarded to Jorge M.O. Fernandes by the Research Council of Norway (ref. 190350). The invaluable help of Dr Alessia Giannetto (University of Messina, Italy) and Dr Lech Kirtiklis (University of Warmia and Mazury in Olsztyn, Poland) is acknowledged. The technical assistance of Marion Nilsen is also acknowledged. The authors would like to thank Bjørnar Eggen, Dalia Dahle and Hilde Ribe at Mørkvedbukta Research Station (Bodø, Norway) for their assistance with the photoperiod experiment.

6. References

Andersen, C.L., Jensen, J.L. & Orntoft, T.F. (2004). Normalisation of real-time quantitative reverse transcription-PCR data: A model-based variance estimation approach to identify genes suited for normalisation, applied to bladder and colon cancer data sets. *Cancer Research*, 64, 15, 5245-5250, ISSN 0008-5472

Aursnes, I.A., Rishovd, A.L., Karlsen, H.E. & Gjøen, T. (2011). Validation of reference genes for quantitative RT-qPCR studies of gene expression in Atlantic cod (*Gadus morhua l.*) during temperature stress. *BMC Research Notes*, 4, ISSN 1756-0500

Begtashi, I., Rodriguez, L., Moles, G., Zanuy, S. & Carrillo, M. (2004). Long-term exposure to continuous light inhibits precocity in juvenile male European sea bass (*Dicentrarchus labrax, L.*). I. Morphological aspects. *Aquaculture*, 241, 1-4, 539-559, ISSN 0044-8486

Bustin, S.A. (2002). Quantification of mRNA using real-time reverse transcription PCR (RT-PCR): trends and problems. *Journal of Molecular Endocrinology*, 29, 1, 23-39, ISSN 0952-5041

Bustin, S.A. & Nolan, T. (2004). Pitfalls of quantitative real-time reverse-transcription polymerase chain reaction. *Journal of Biomolecular Techniques*, 15, 3, 155-166, ISSN 1524-0215

Campos, C., Valente, L.M., Borges, P., Bizuayehu, T. & Fernandes, J.M. (2010). Dietary lipid levels have a remarkable impact on the expression of growth-related genes in Senegalese sole (*Solea senegalensis* Kaup). *The Journal of Experimental Biology*, 213, 2, 200-209, ISSN 1477-9145

Cheney, R.E., Riley, M.A. & Mooseker, M.S. (1993). Phylogenetic analysis of the myosin superfamily. *Cell Motility and the Cytoskeleton*, 24, 4, 215-223, ISSN 0886-1544

Davie, A., Porter, M.J.R. & Bromage, N.R. (2003). Photoperiod manipulation of maturation and growth of Atlantic cod (*Gadus morhua*). *Fish Physiology and Biochemistry*, 28, 1-4, 399-401, ISSN 0920-1742

Davie, A., Porter, M.J.R., Bromage, N.R. & Migaud, H. (2007). The role of seasonally altering photoperiod in regulating physiology in Atlantic cod (*Gadus morhua*). Part I. Sexual maturation. *Canadian Journal of Fisheries and Aquatic Sciences*, 64, 1, 84-97, ISSN 0706-652X

Dhillon, R.S., Esbaugh, A.J., Wang, Y.S. & Tufts, B.L. (2009). Characterization and expression of a myosin heavy-chain isoform in juvenile walleye *Sander vitreus*. *Journal of Fish Biology*, 75, 5, 1048-1062, ISSN 0022-1112

Endal, H.P., Taranger, G.L., Stefansson, S.O. & Hansen, T. (2000). Effects of continuous additional light on growth and sexual maturity in Atlantic salmon, *Salmo salar*, reared in sea cages. *Aquaculture*, 191, 4, 337-349, ISSN 0044-8486

FAO, http://www.fao.org/fishery/species/2218/en

Fernandes, J.M., Mackenzie, M.G., Elgar, G., Suzuki, Y., Watabe, S., Kinghorn, J.R. & Johnston, I.A. (2005). A genomic approach to reveal novel genes associated with myotube formation in the model teleost, *Takifugu rubripes*. *Physiological Genomics*, 22, 3, 327-338, ISSN 1531-2267

Fernandes, J.M., Mackenzie, M.G., Wright, P.A., Steele, S.L., Suzuki, Y., Kinghorn, J.R. & Johnston, I.A. (2006). Myogenin in model pufferfish species: Comparative genomic analysis and thermal plasticity of expression during early development. *Comparative Biochemistry and Physiology Part D: Genomics and Proteomics*, 1, 1, 35-45, ISSN 1878-0407

Fernandes, J.M., MacKenzie, M.G., Kinghorn, J.R. & Johnston, I.A. (2007a). FoxK1 splice variants show developmental stage-specific plasticity of expression with temperature in the tiger pufferfish. *Journal of Experimental Biology*, 210, 19, 3461-3472, ISSN 1477-9145.

Fernandes, J.M., Kinghorn, J.R. & Johnston, I.A. (2007b). Differential regulation of multiple alternatively spliced transcripts of MyoD. *Gene*, 391, 1-2, 178-185, ISSN 0378-1119

Fernandes, J.M., Mommens, M., Hagen, O., Babiak, I. & Solberg, C. (2008). Selection of suitable reference genes for real-time PCR studies of Atlantic halibut development. *Comparative Biochemistry and Physiology Part B: Biochemistry and Molecular Biology*, 150, 1, 23-32, ISSN 1096-4959

Hall, J.R., Short, C.E. & Driedzic, W.R. (2006). Sequence of Atlantic cod (*Gadus morhua*) GLUT4, GLUT2 and GPDH: Developmental stage expression, tissue expression and relationship to starvation-induced changes in blood glucose. *The Journal of Experimental Biology*, 209, Pt 22, 4490-4502, ISSN 0022-0949

Hansen, T., Karlsen, O., Taranger, G.L., Hemre, G.I., Holm, J.C. & Kjesbu, O.S. (2001). Growth, gonadal development and spawning time of Atlantic cod (*Gadus morhua*) reared under different photoperiods. *Aquaculture*, 203, 1-2, 51-67, ISSN 0044-8486

Ikeda, D., Ono, Y., Snell, P., Edwards, Y.J., Elgar, G. & Watabe, S. (2007). Divergent evolution of the myosin heavy chain gene family in fish and tetrapods: evidence from comparative genomic analysis. *Physiological Genomics*, 32, 1, 1-15, ISSN 1531-2267

Imsland, A.K., Foss, A., Koedijk, R., Folkvord, A., Stefansson, S.O. & Jonassen, T.M. (2007). Persistent growth effects of temperature and photoperiod in Atlantic cod *Gadus morhua*. *Journal of Fish Biology*, 71, 5, 1371-1382, ISSN 0022-1112

Johnston, I.A. (1999). Muscle development and growth: potential implications for flesh quality in fish. *Aquaculture*, 177, 1-4, 99-115, ISSN 0044-8486

Johnston, I.A. (2001). *Muscle development and growth*, Academic Press, ISBN 0123504422, San Diego, CA, USA

Karlsen, O., Norberg, B., Kjesbu, O.S. & Taranger, G.L. (2006). Effects of photoperiod and exercise on growth, liver size, and age at puberty in farmed Atlantic cod (*Gadus morhua* L.). *Ices Journal of Marine Science*, 63, 2, 355-364, ISSN 1054-3139

Koedijk, R.M., Le Francois, N.R., Blier, P.U., Foss, A., Folkvord, A., Ditlecadet, D., Lamarre, S.G., Stefansson, S.O. & Imsland, A.K. (2010). Ontogenetic effects of diet during early development on growth performance, myosin mRNA expression and metabolic enzyme activity in Atlantic cod juveniles reared at different salinities. *Comparative Biochemistry and Physiology Part A: Molecular & Integrative Physiology*, 156, 1, 102-109, ISSN 1531-4332

Lilleeng, E., Froystad, M.K., Vekterud, K., Valen, E.C. & Krogdahl, A. (2007). Comparison of intestinal gene expression in Atlantic cod (*Gadus morhua*) fed standard fish meal or soybean meal by means of suppression subtractive hybridization and real-time PCR. *Aquaculture*, 267, 1-4, 269-283, ISSN 0044-8486

Mommsen, T.P. (2001). Paradigms of growth in fish. *Comparative Biochemistry and Physiology Part B: Biochemistry & Molecular Biology*, 129, 2-3, 207-219, ISSN 10964959

Norberg, B., Brown, C.L., Halldorsson, O., Stensland, K. & Bjornsson, B.T. (2004). Photoperiod regulates the timing of sexual maturation, spawning, sex steroid and thyroid hormone profiles in the Atlantic cod (*Gadus morhua*). *Aquaculture*, 229, 1, 451-467, ISSN 0044-8486

Olsvik, P.A., Lie, K.K., Jordal, A.E.O., Nilsen, T.O. & Hordvik, I. (2005). Evaluation of potential reference genes in real-time RT-PCR studies of Atlantic salmon. *BMC Molecular Biology*, 6, ISSN 1471-2199

Olsvik, P.A., Softeland, L. & Lie, K.K. (2008). Selection of reference genes for qRT-PCR examination of wild populations of Atlantic cod *Gadus morhua*. *BMC Research Notes*, 1, 47, ISSN 1756-0500

Sæle, Ø., Nordgreen, A., Hamre, K. & Olsvik, P.A. (2009). Evaluation of candidate reference genes in Q-PCR studies of Atlantic cod (*Gadus morhua*) ontogeny, with emphasis on the gastrointestinal tract. *Comparative Biochemistry and Physiology Part B: Biochemistry & Molecular Biology*, 152, 1, 94-101, ISSN 1096-4959

Taranger, G.L., Aardal, L., Hansen, T. & Kjesbu, O.S. (2006). Continuous light delays sexual maturation and increases growth of Atlantic cod (*Gadus morhua* L.) in sea cages. *Ices Journal of Marine Science*, 63, 2, 365-375, ISSN 1054-3139

Vandesompele, J., De Preter, K., Pattyn, F., Poppe, B., Van Roy, N., De Paepe, A. & Speleman, F. (2002). Accurate normalisation of real-time quantitative RT-PCR data by geometric averaging of multiple internal control genes. *Genome Biology*, 3, 7, RESEARCH0034, ISSN 1465-6914

Watabe, S. (1999). Myogenic regulatory factors and muscle differentiation during ontogeny in fish. *Journal of Fish Biology*, 55, SUPPL. A, 1-18, ISSN 0022-1112

Mitochondrial DNA Variation as a Tool for Systematic Status Clarification of Commercial Species – The Case of Two High Commercial *Flexopecten* Forms in the Aegean Sea

Anastasia Imsiridou[1*], Nikoleta Karaiskou[2], Elena Aggelidou[1],
Vassilios Katsares[1] and Sofia Galinou-Mitsoudi[1]
[1]*Department of Fisheries and Aquaculture Technology,
Alexander Technological Educational Institute of
Thessaloniki, Nea Moudania, Halkidiki*
[2]*Department of Genetics, Development and Molecular Biology,
School of Biology, Aristotle University of
Thessaloniki, Thessaloniki, Macedonia*
Greece

1. Introduction

The determination and the identification of species constitute some of the first basic steps for biodiversity monitoring and conservation (Dayrat, 2005). Species identification is usually carried out by taxonomists, but in many cases it is restricted by the lack of available morphological characters. This is the case for two *Flexopecten* taxa: *Flexopecten glaber* L. and *F. proteus* Dillwyn ex Solander ms. The smooth scallop *F. glaber* is an edible bivalve, with a maximum length of 8.66 cm (Pisor & Poppe, 2008). It is an epibenthic species of soft and/or hard substrate and its habitat is characterized as muddy and sandy, with organic detritus. The species has a fast growth rate. Its reproduction takes place mainly during late summer, while the gonad activity can be observed all the year. It inhabits between 5 and 900 m depth or more (even 1600 m), in Mediterranean and Black Seas (Poppe & Goto, 1993) and its distribution appears in Figure 1.

The expansion of *F. proteus* is controversial, since some authors reported restricted distribution of the species to Adriatic Sea (Pountiers, 1987). However, this form has been reported as a commercial species in the area of Thessaloniki Gulf (North Western Aegean Sea), where it coexists with *F. glaber* (Zenetos, 1996; Galinou-Mitsoudi & Sinis, 2000). According to these authors, the occurrence ratio of *F. glaber* to *F. proteus* in three random fisheries samples of 99, 130 and 120 individuals in total was 0.74, 1.28 and 6.50, respectively. Besides Thessaloniki Gulf, *F. proteus* is found in Ionian Sea (Amvrakikos and Korinthiakos Gulfs) and also in Aegean Sea (Pagasitikos Gulf , N. Evoikos Gulf , Saronikos Gulf , Lesvos Island and Lemnos Island) (Koutsoubas et al., 2007; Zenetos, 1996) (Fig. 1).

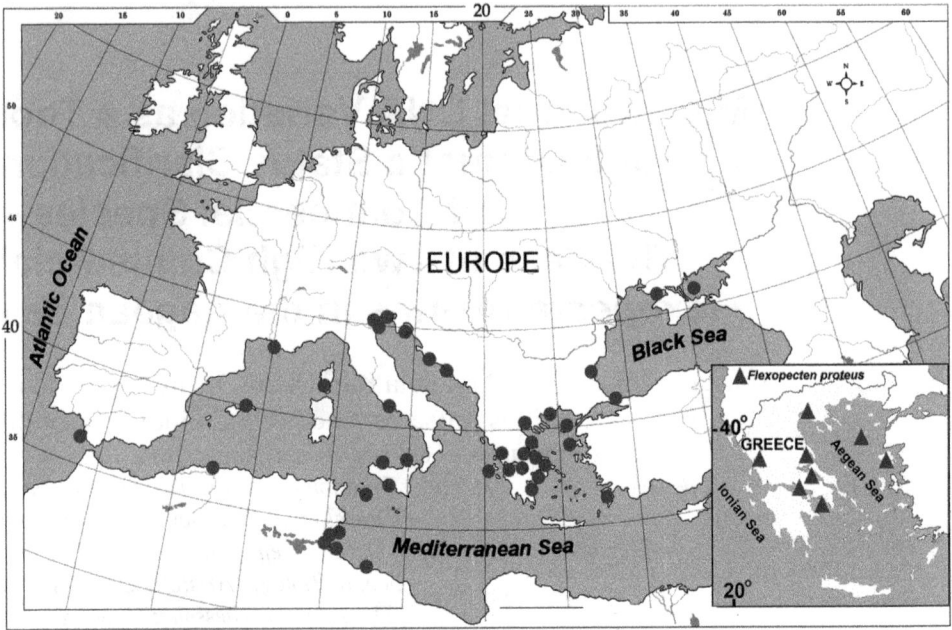

Fig. 1. *F. glaber* distribution in Mediterranean and Black Sea. In inlaid map of Greece the *F. proteus* distribution is denoted, according to Zenetos (1996).

These commercial scallops are major target species due to their high demand in the national and European market. Nevertheless, the species seems to be overexploited, as the annual Hellenic production from over 20 t between mid 1970s and late 1980s, was reduced to 5 t from the late 1980s to 1990s (Koutsoubas et al., 2007). However, *Flexopecten* scallops fishing is forbidden in the Thessaloniki Gulf since 2002, due to the presence of high concentrations of heavy metals (Cd), while the stocks in the central Aegean Sea have rather collapsed in 2003 (Koutsoubas et al., 2007).

The two forms of *F. glaber* and *F. proteus* are also exploitable in Adriatic Sea (Pujolar et al., 2010). Besides the fishing, *Flexopecten* scallops having high growth rate, high demand and good price/kg, are also potential species for aquaculture.

Mitochondrial DNA (mtDNA) has been shown to be useful for analyzing relationships among bivalve populations (Hare & Avise, 1996; Rawson & Hilbish, 1998; Katsares et al., 2008), due to its maternal inheritance, its haploid nature and the fact that its effective population size is the one fourth of that of the nuclear DNA. The invertebrate mitochondrial 16S rDNA gene has been largely used for studying the population structure and the levels of genetic variation in many Pectinidae taxa (Kong et al., 2003; Mahidol et al., 2007; Yuan et al., 2009; Pujolar et al., 2010). As the Aegean populations of *F. glaber* and *F. proteus* are the main commercial stocks of these bivalves in Greece, the main objective of our study was to conduct a survey on the genetic resources of the two taxa using mitochondrial 16S rDNA nucleotide sequences.

Chlamys proteus or *F. proteus* Dillwyn has been considered since long, as a species of the bivalve family Pectinidae (Cossignani et al., 1992; Delamotte & Vardala–Theodorou, 1994; Doneddu & Trainito, 2005; Repetto et al., 2005; Zenetos et al., 2005). The majority of these sources mention that *F. proteus* is another similar species to *F. glaber*. The main systematic characteristic for the recognition of the two pectinids, is the difference in the number of major and dominated ribs which are 9-12 in *F. glaber*, but only five in *F. proteus* (Fig. 2). On the other hand, according to Poppe & Goto (1993), CLEMAM & ERMS, *F. proteus* is accepted as a subspecies or a form of *F. glaber*. Also, Raines & Poppe (2006) underline that *F. proteus* "is treated only as a synonyme" of *F. glaber*.

Flexopecten

F. proteus   F. glaber

Fig. 2. Fished scallops in the study area of the two *Flexopecten* forms *proteus* and *glaber*, with epibiots on their left valves

Many biologists have argued that the future of descriptive taxonomy will depend on successfully embracing new techniques. Many ideas have been proposed and much progress has been achieved by using molecular data that provide a complementary approach to discriminate species separated by subtle morphological characters (Knowlton, 1993; Avise, 1994; Chan & Chu, 1996; Sarver et al., 1998; Mathews et al., 2002; Goetze, 2003). The invertebrate mtDNA 16S rDNA gene has been relatively well studied due to availability of universal primers (Kocher et al., 1989; Palumbi, 1996) and was used to resolve taxonomic problems in the family Mytilidae (Rawson & Hilbish, 1995), Veneridae (Canapa et al., 1996) and Pectinidae (Canapa et al., 2000a; Saavedra & Pena, 2004; Pujolar et al., 2010).

Taking into account all the existed considerations and the uncertain systematic status of these two common *Flexopecten* "forms", another aim of our work was an attempt to solve the identification problem of the two Aegean taxa, using the sequence analysis of the mtDNA

16S rDNA gene. A parallel and similar study (Pujolar et al., 2010), was also made for the Adriatic populations of the *Flexopecten* complex.

2. Methods and tools

2.1 Sampling

Specimens were collected from three different regions of the Gulf of Thessaloniki, northern Aegean Sea (Fig. 3). In total, 78 adult individuals of *F. glaber* and 57 adult individuals of *F. proteus* from the three different locations, were collected (Table 1). Specimens were classified according to the number of dominant ribs.

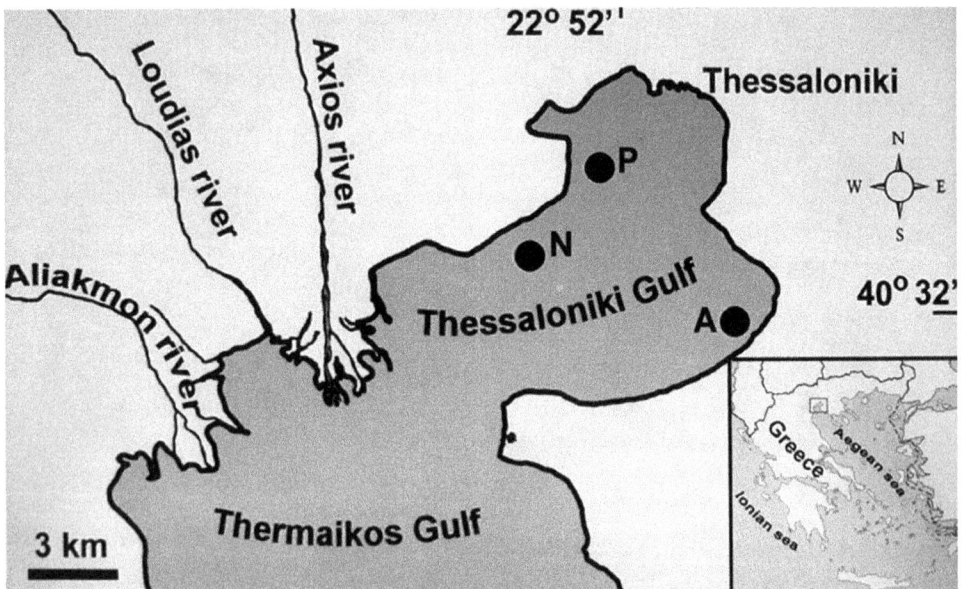

Fig. 3. Sampling sites of the analyzed population for both *F. glaber* and *F. proteus*. A: Airport; P: Paliomana; N: Naziki.

2.2 DNA extraction and PCR amplification

Total DNA was extracted from the anterior adductor muscle according to Hillis et al. (1996). A universal primer set (Palumbi, 1996) was used for the amplification of the 16S rDNA gene in both *F. glaber* and *F. proteus*. The reaction mixture contained template DNA (approximately 100 ng), 1X PCR buffer, 2.2 mM MgCl$_2$, 20 pmol of each primer, 0.25 mM of each dNTP and 0.5 U of Promega polymerase. Amplification was started at 94°C for 3 min, followed by 31 cycles at 94°C for 50 s, 50°C for 50 s, 72°C for 50 s and a final extension at 72°C for 5 min.

Electrophoresis of 3 µl of the PCR product was performed in 1XTBE buffer for 1 h at 150 V, in 1.5% agarose gel containing 0.5 µg/ml ethidium bromide. The size of the PCR products

Mitochondrial DNA Variation as a Tool for Systematic Status Clarification of Commercial Species –
The Case of Two High Commercial Flexopecten Forms in the Aegean Sea

113

was checked against a 100 bp DNA ladder and was approximately 500 bp for both taxa. The resulting DNA fragments were visualized by UV transilumination and photographed.

2.3 DNA sequencing

A sequencing analysis on a 3730Xl DNA Analyzer (Applied Biosystems) was followed using both forward and reverse primers for crosschecking. DNA sequences were deposited to GenBank (accession numbers GU320272 – GU320288; HM 627014 – HM627051).

2.4 Data analysis

The nucleotide sequences of all individuals were aligned using the Clustal X software (Thompson et al., 1997) and the BioEdit software (Hall, 1999), set to default parameters and corrected by eye. Mt DNA 16S rDNA diversity was estimated by the number of haplotypes, haplotype frequency, haplotype diversity (h) and nucleotide diversity (π), via Arlequin version 3.5 (Excoffier & Lischer, 2010). Pairwise genetic distances among haplotypes were computed based on the 2-parameter Kimura distance model (Kimura, 1980) using MEGA software (Tamura et al., 2007; Kumar et al., 2008). The obtained pairwise genetic distances were used to construct a Neighbor-Joining tree with the same software. The species *Aequipecten opercularis* (GenBank: AM494412) was used as outgroup and two relevant ingroups (*Argopecten irradians*: GU119971; *Chlamys multistriata*: FN667665) were also included in the analysis. Pairwise sample differentiation was assessed using the exact test described by Raymond & Rousset (1995), included in the Arlequin package. Analysis of molecular variance was conducted also with the same package (Arlequin version 3.5; Excoffier & Lischer, 2010), to determine the genetic differentiation of populations.

To test whether the populations underwent recent demographic population growth we calculated the mismatch distribution in Arlequin package, which assumes an infinite sites model of selectively neutral nucleotide substitutions and assesses significance via coalescent simulations of a large, neutrally evolving population of constant size (Slatkin & Hudson, 1991). Finally, Tajima's D (Tajima, 1989) was used to examine the selective neutrality of mitochondrial fragment and it was also used in mismatch distribution, because significant negative D-values are what the hypothesis of population growth predicts (Bertorelle & Slatkin, 1995).

3. Results

A total of 436 base pairs of the 16S rRNA gene fragment were successfully sequenced for 135 individuals from both taxa, and 50 polymorphic sites were identified (Appendix 1). The transition/transversion rate ratios were 2.312 for purines and 6.283 for pyrimidines while the overall transition/transversion bias was 1.7. No species specific positions were detected that could discriminate the two species. Fifty one haplotypes were detected among all samples (37 for *F. glaber* and 32 for *F. proteus*) and their frequencies are given in Appendix 1. Haplotype 5 (20% in total) was shared among all samples of both taxa. Haplotype 1 revealed a higher percentage of appearance among samples (25.2%) and it was dominant in all samples apart from Paliomana population of *F. proteus* (Appendix 1). Haplotypes 3, 10 and 15 were present in three samples in total of the same and different taxa, whereas haplotypes 6, 19 and 21 were shared between two samples of different taxa. Haplotypes 8, 45 and 50 were observed twice in the same population and all the other haplotypes were unique.

Haplotype diversity, nucleotide diversity, number of polymorphic sites and number of haplotypes are given also in Appendix 1. The highest values of haplotype and nucleotide diversity were revealed in the population of *F. proteus* from Paliomana (0.9744 and 0.0078 respectively) while the lowest values of these indices were observed in the sample of *F. glaber* from Airport (0.7793 and 0.0027). Mean haplotype and nucleotide diversity for *F. glaber* populations were h=0.854 and π=0.004 respectively, whereas for *F. proteus* were h=0.892 and π=0.005.

Values of genetic distances between pairs of populations are given in Table 1 and ranged from 0.003 (between pA and gA) to 0.008 (between pP and pN). Genetic distance between the two taxa was estimated in a value of D=0.005. Pairwise genetic distances among haplotypes were used to construct a neighbour joining tree. As it can be seen by Figure 4, no clustering of haplotypes corresponded to specific taxon and/or sampling site was detected. All internal nodes are supported by relatively low bootstrap values (≤67%).

	gA	gP	gN	pA	pP	pN
gA		[0.001]	[0.002]	[0.001]	[0.002]	[0.001]
gP	0.005		[0.002]	[0.001]	[0.002]	[0.002]
gN	0.005	0.006		[0.002]	[0.002]	[0.002]
pA	0.003	0.005	0.005		[0.002]	[0.001]
pP	0.006	0.008	0.006	0.007		[0.002]
pN	0.004	0.006	0.007	0.004	0.008	

Table 1. Genetic distances among the studied population samples, based on 16S rDNA sequences and the 2-parameter Kimura model. SE values are shown in brackets. gA: *F. glaber* from Airport; gP: *F. glaber* from Paliomana; gN: *F. glaber* from Naziki; pA: *F. proteus* from Airport; pP: *F. proteus* from Paliomana; pN: *F. proteus* from Naziki.

The analysis of the partitioning of the haplotype diversity indicated that the majority of the genetic variation (96.98%) was distributed within populations (Table 2) and only a percentage of 4.53% could be attributed to variation among populations within groups. AMOVA with two groups (i.e. *F. glaber* versus *F. proteus*) revealed a low F_{ST} value of 0.0295 and showed that only a 1.52% of the genetic variation occurred among groups. Pairwise exact test (Raymond & Rousset, 1995) for samples of both taxa showed no population differentiation with all *P* values>0.05 (*P*=1.000 for all the estimates). When pooling together

Source of variation	Df	Variance components	Percentage of variation
Among groups	1	-0.006	-1.52
Among populations within groups	4	0.020	4.53
Within populations	129	0.430	96.98
Total	134	0.444	100

Table 2. Analysis of molecular variance, in populations of both taxa. Df: degrees of freedom. The fixation indices are: F_{CT}=- 0.01518; F_{SC}=0.03017; and F_{ST}=0.04467.

Mitochondrial DNA Variation as a Tool for Systematic Status Clarification of Commercial Species –
The Case of Two High Commercial Flexopecten Forms in the Aegean Sea

115

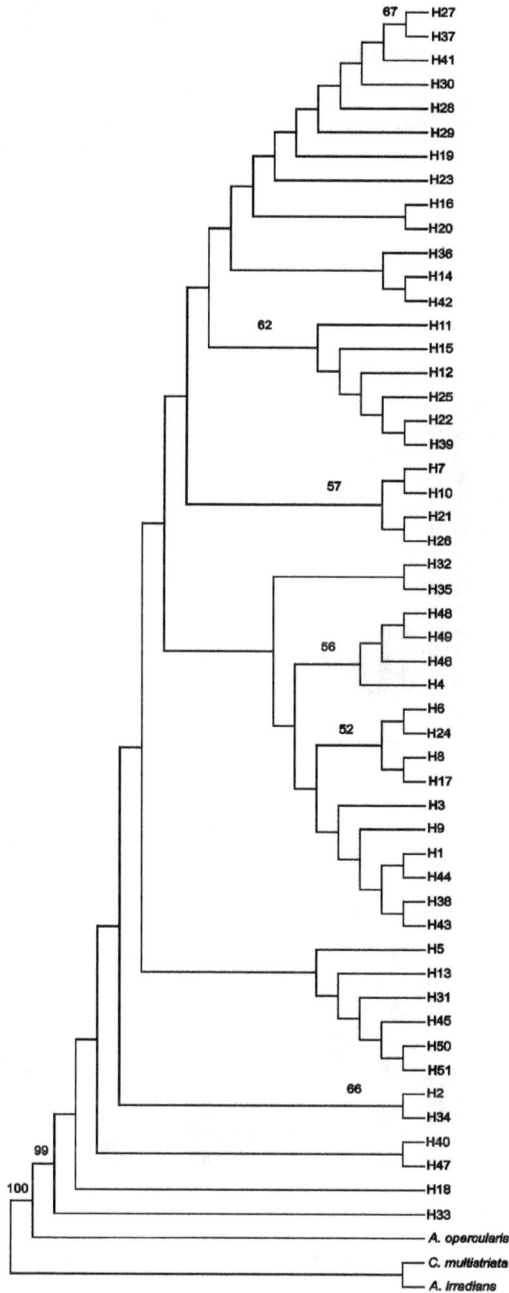

Fig. 4. Phylogenetic tree of the 51 haplotypes of *F. glaber* and *F. proteus* taxa recovered from 16S rRNA sequences, estimated by the Neighbor-joining method. Only bootstrap values based on 100 replications higher than 50% are displayed.

the three sampling sites of *F. glaber* and *F. proteus* respectively, the exact test of population differentiation was still not statistically significant (*P* values=0.20).

Mismatch distribution was calculated for all studied populations of both taxa (Fig. 5). All populations followed unimodal distribution (i.e a bell-shaped distribution) that it is assumed to be the signature of population expansion occurred probably after a bottleneck event. However, Tajima's *D*-values were not significantly negative for any of the studied populations (p=0.05–0.1), thus rejecting the hypothesis for population growth.

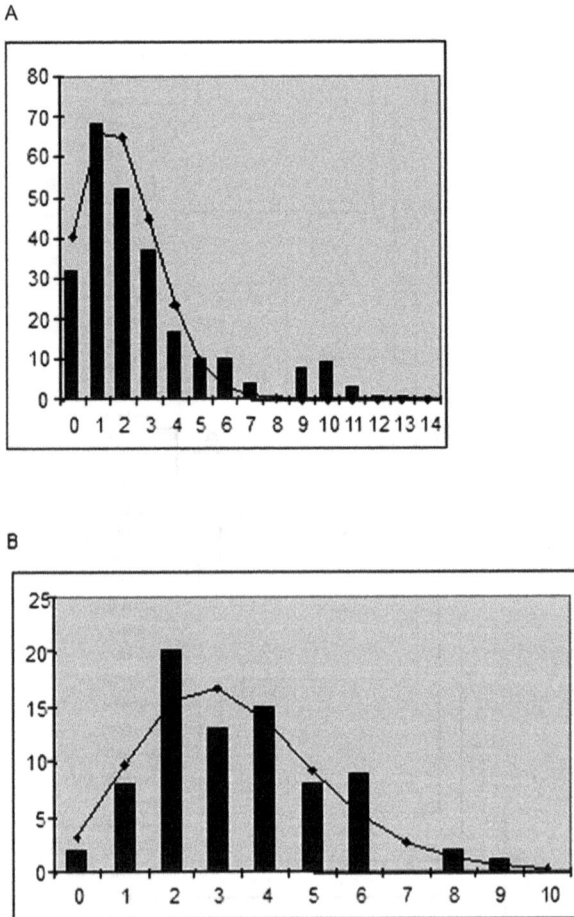

A

B

Fig. 5. Mismatch distributions for A) *F. glaber* and B) *F. proteus*. All studied samples followed the same unimodal distribution and thus graphs depict frequencies of pairwise differences for one of the populations of each taxon. The observed frequency of pairwise nucleotide differences among sequences is represented by black bars, and expected frequencies under a model of sudden population expansion are represented by continuous line.

Mitochondrial DNA Variation as a Tool for Systematic Status Clarification of Commercial Species –
The Case of Two High Commercial Flexopecten Forms in the Aegean Sea

117

4. Discussion

4.1 Levels of population variation

Low levels of intraspecific variation reported by many authors (Palumbi, 1996; Stepien et al., 1999; Saavedra & Pena, 2004; Pujolar et al., 2010) indicate that the mt 16S rDNA gene is highly conserved in many taxa, apparently due its functional role in protein assembly. However, it was not the case in the present study. Values of nucleotide diversity found for *F. glaber* (π=0.004) and *F. proteus* (π=0.005) are higher than the values reported for Adriatic populations of the same taxa (π=0.001 and π=0.000 respectively) (Pujolar et al., 2010). Also our values are higher than those estimated for other Pectinidae taxa as *Chlamys nobilis* (π =0.0012; Yuan et al., 2009), *C. farreri* (π=0.0035; Kong et al., 2003), *Pecten jacobaeus* (π=0.0015), *P. novaezelandiae* (π=0.0020), *P. fumatus* (π=0.000) and *P. maximus* (π=0.0039) (Saavedra & Pena, 2004).

The levels of intrapopulation variation for *F. glaber* and *F. proteus*, revealed by sequences of the mt DNA 16S rDNA region in the present study are also relatively high. Values of haplotype diversity ranged from 0.7793 to 0.9744 for populations of both taxa, are much higher than the values of 0.000 and 0.400 reported for the Adriatic populations (Pujolar et al., 2010) and higher than the values of 0.5356 and 0.7328 reported for two wild populations of *C. nobilis*, using sequences of the same region (Yuan et al., 2009). Also the highest value of π revealed in our study (0.0078) exceeds much the value of π=0.0012 observed in *C. nobilis* populations (Yuan et al., 2009). Kong et al. (2003) reported lower to similar values of π (0.00175 to 0.00591) and similar values of h (0.750 to 1.000) for different populations of Zhikong scallop *C. farreri*, based on the sequencing analysis of 16S ribosomal gene. Also, values of nucleotide and haplotype diversity observed in our samples are considerably higher than the values observed for different populations of the Asian moon scallop *Amusium pleuronectes* (0.000 to 0.0017 for π; 0.000 to 0.0511 for h) (Mahidol et al., 2007b).

The demographic analysis revealed that all the studied samples of both taxa were probably stable over time. Pujolar et al. (2010) who have studied the demographic history of the Adriatic *Flexopecten* complex using three mitochondrial and one nuclear gene proposed that the lack of large demogarphic expansion since the last glaciations, suggests that colonization must have been conducted by a substantial number of individuals that occupied the new habitat relatively fast. Demographic expansion does not appear either as a correct explanation of the observed genetic variation in our samples. A possible explanation for high values of intrapopulation indices for *F. glaber* and *F. proteus* could be the large effective population sizes as genetic variability is abundant. It is common knowledge that large *Ne* results to large amounts of neutral variability in populations.

An alternative hypothesis to explain this finding is that there is more than one species in the data set, but the character used to identify it is not valid. In other words, there may be more than one evolutionary entity, so we are artificially lumping more than one species in the analysis. One way to sort this out was to include other taxa in the analysis and make a phylogenetic analysis of the haplotypes, to see if they form unique phylogenetic species. There may simply be cryptic species to be considered or perhaps the morphotype doesn't correspond to the phylotype, which would certainly not be the first time in molluscs (Knowlton, 2000). The phylogenetic analysis of haplotypes revealed no clustering corresponding to each putative species (*F. glaber* and *F. proteus*). Also, the taxa *A. irradians* and *C. multistriata* formed a separate group in the Neighbour – Joining tree.

The decreased intrapopopulation genetic diversity can have serious consequences on the survival and reproduction of the species (Soule, 1980). Nevertheless, in the marine invertebrates reduction of genetic variation is expressed mainly with the loss of rare haplotypes despite with the total reduction of heterozygosity in a population (Gosling, 2003). Our samples are characterized from high DNA variability and the existence of some rare haplotypes, a result which indicates the good population status for both taxa. As the fishing of the Aegean *F. glaber* and *F. proteus* stocks is forbidden since 2002 because of heavy metals (Cd) presence in high concentrations (Koutsoubas et al., 2007), our data are fully justified.

Despite the high degree of genetic variation found within populations, the results of the statistical analysis of the 16S rDNA haplotypes indicated that little geographic structure was present among populations. The genetics of marine species with pelagic larval development has often been characterized by low genetic variation among populations, a pattern driven by high dispersal capabilities and large scale oceanic mixing (Reichow & Smith 2001; Rivera et al., 2004). After spawning and external fertilization, developing larvae spend a variable period of time as part of the plankton, which can be passively drifted by water currents (Seed, 1969; Cho et al., 2007). Similarly, gene flow among the populations of the present study seems not to be significantly blocked in this relatively small area of the Gulf of Thessaloniki. Consequently, our results support the general observation that marine invertebrates possessing a planktonic larval stage have a high genetic relatedness due to increased potential for larval dispersal (Crisp, 1978).

4.2 Systematic status of the two taxa

The presence of *F. proteus* in the Aegean Sea does not support the restricted gene flow or the recent origin of this form, reported by Pulojar et al. (2010). It seems that the high dispersal of the larval stage has provoked the expansion of this morph outside of the Adriatic Sea or *F. proteus* has appeared for the first time in other Mediterranean region, apart from Adriatic Sea.

The phylogenetic position of *Chlamys glabra* (now *F. glaber*) is under discussion. Previous phylogenetic data (Canappa et al., 2000a, 2000b; Barucca et al., 2004; Liu et al., 2007) support the idea that *C. glabra* seems to belong to a completely distinct genus from *Chlamys*, in agreement with some malacologists who have included it in the genus *Flexopecten* (Vaught, 1989; Rombouts, 1991). According to these studies *C. glabra* is clustered together with *Aequipecten opercularis* and not with other species of the genus *Chlamys* i. e. *C. islandica*, *C. varia* and *C. farreri*.

Later on Saavedra & Pena (2006) used the updated latin name *F. glaber* instead of *C. glabra* in a phylogenetic analysis for American scallops, and they also found that taxa *F. glaber* and *A. opercularis* clustered together. There has been one effort for identification of *C. glabra* larvae from other bivalve species larvae, with a method based on PCR - SSCP combined with sequencing of partial 18S rDNA region, but it was not successful (Livi et al., 2006). In the latest study of the two Adriatic *Flexopecten* forms using three mitochondrial and one nuclear gene, Pujolar et al. (2010) suggest that *F. glaber* and *F. proteus* are the same species as evidenced by both putative species appearing mixed in all genetic trees, with no clustering according to species.

Mitochondrial DNA Variation as a Tool for Systematic Status Clarification of Commercial Species –
The Case of Two High Commercial Flexopecten Forms in the Aegean Sea

119

The present study tried to resolve the two Aegean taxa relationship using a mitochondrial gene. Sequencing analysis of 16S rDNA revealed no taxon specific positions that could discriminate between different species. Additionally, two highly frequent and six less frequent haplotypes are shared between the two taxa, while no clustering in different groups was detected. This is an evidence of interbreeding between *F. glaber* and *F. proteus*, suggesting that they are conspecific taxa. Similarly, Wilding et al. (1999) found no high frequency private haplotypes within *P. maximus* and *P. jacobaeus* after a mtDNA PCR - RFLP analysis, and this fact together with the low genetic divergence demonstrated that the two species are also capable of hybridization. Later on, Saavedra & Pena (2004), in a phylogenetic analysis of *P. maximus* and *P. jacobaeus* noticed that the two most common haplotypes appeared in both taxa and this supports the held view of conspecificity between the two European taxa.

Another evidence of conspecific taxa was the levels of genetic distance. The genetic distance found between the two taxa (D=0.005) does not suggest genetic differences between two species, and shown to be of a similar magnitude to intraspecific than interspecific values. This value is considerably lower than expected for congeneric species (Liu et al., 1998; Stepien et al., 1999; Saavedra & Pena 2004; Yu et al., 2004; Mahidol et al., 2007a). Our data cannot even justify the classification of *F. proteus* as a subspecies of *F. glaber*. According to Therriault et al. (2004), the previous classification of *Dreissena bugensis* as a subspecies of *D. rostriformis* should be revised because genetic distance between them was relatively low (0.003–0.004; a value similar to our results), suggesting that these two taxa constitute a single species.

The existence of a single species in our study is also reinforced by the exact test of population differentiation. When pooling together the three sampling sites of *F. glaber* and *F. proteus* respectively, no statistically significant population differentiation was detected. So, our findings support the aspect that *F. proteus* is only a form or simply a synonyme of *F. glaber*.

In Bivalve, differences in shell morphology are often the result of phenotypic plasticity, so that many past descriptions of species were unjustified (reviewed by Knowlton, 2000). Consequently, the number of genetic studies that have resulted in the synonimization of species is probably higher for bivalves than most other marine invertebrate phyla. Our study is one example of numerous cases where distinctions between sympatric taxa based on shell morphology were not supported by genetic data (Wilding et al., 1999; reviewed by Knowlton, 2000). Present data reinforce also the validity of the mtDNA markers for clarifying uncertain systematic relationships among taxa.

A global DNA-based barcode identification system that is applicable to all animal species provides a universal tool for the identification of different species. The barcode system is based on sequence diversity in a single gene region (a section of the mitochondrial DNA cytochrome c oxidase I gene, COI). When the reference sequence library is in place, new specimens can be identified by comparing their DNA barcode sequences against this barcode reference library. Herbert et al. (2004a, 2004b) have demonstrated that the COI region is appropriate for discriminating between closely related species and these results have prompted international efforts to accelerate the process of cryptic species identification (Herbert et al., 2003). A similar survey based on the barcode identification system, could be done in the future for the two *Flexopecten* forms, in order to confirm the existence of a single species.

5. Appendix

```
                     Variable nucleotide sites
          1 1111112222 2222222223 3333333333 3344444444
5555556790 1555990111 2335666781 3333333333 9900013333
8567890628 7244785678 9450035343 2699836693 7912603456
TGTCTTTTCT TGCAAATGGG GGTTGAATCT TTTTGTCTTT GCGTTTATTG
```

Haplotype	Variable nucleotide sites	gA(30)	gP(23)	gN(25)	pA(18)	pP(13)	pN(26)	Total
H1	`TGTCTTTTCT TGCAAATGGG GGTTGAATCT TTTTGTCTTT GCGTTTATTG`	12	7	7	3	5	–	34
H2	`.......... ..G....... .......... .......... ...T......`	1	–	–	–	–	–	1
H3	`.......... .......... .......... .......... .....A....`	1	–	1	–	2	–	4
H4	`.......... .....C.... .......... .......... ..T.......`	1	–	–	–	–	–	1
H5	`.......... .......... .......... .......... ..T.......`	8	1	4	2	8	5	28
H6	`......C... .......... .......... .......... ..........`	1	–	–	–	1	–	2
H7	`.......... .......... C......... .......... ..A.......`	1	–	–	–	–	–	1
H8	`.......... ....G..... .......... .......... ..........`	2	–	–	–	–	–	2
H9	`.......... ...C...... .......... .......... ..........`	1	–	–	–	–	–	1
H10	`.......... .......... C......... .......... ..T.......`	2	1	1	–	–	1	5
H11	`.....G.... .......... C......... .......... ..T.......`	–	1	–	–	–	–	1
H12	`C......... .....G.... .......... .......... ..T....A..`	–	1	–	–	–	–	1
H13	`.......... .....G.... .......... .......... ..T......A`	–	1	–	–	–	–	1
H14	`..CA...... .....G.... ...C...... .......... ..T...A...`	–	1	–	–	–	–	1
H15	`.......... .....G.TTC .C........ .......... ..T.......`	–	3	1	1	–	–	5
H16	`.......... .G........ CA.A...... .......... ...C......`	–	1	–	–	–	–	1
H17	`......C... .......... .G........ .......... ..T....G..`	–	1	–	–	–	–	1
H18	`.......... .....G.... .C........ .......... ..........`	–	1	–	–	–	–	1
H19	`.......... .....G.... .......... .......... ......G...`	–	7	–	2	–	–	9
H20	`.......... .....G.... .......... C......... ..TC......`	–	1	–	–	–	–	1
H21	`.......... .....G.... .......... .A........ ..T.......`	–	–	–	2	–	–	2
H22	`.......... .....G.... A......... .......... ..T...G...`	–	1	–	–	–	–	1
H23	`.......... .....G.... ..A....... .......... ..........`	–	1	–	–	–	–	1
H24	`......C... .....G.... C......... .......... ..T.......`	–	1	–	–	–	–	1
H25	`.......... .....G.... ...T...... .......... ..T.......`	–	1	–	–	–	–	1
H26	`.....C.G.. .......... C......... .......... ..........`	–	1	–	–	–	–	1
H27	`C......... .....G.... .......... C......... ..........`	–	1	–	–	–	–	1
H28	`.......... .....G.... .......... ..C....... ..........`	–	1	–	–	–	–	1
H29	`.......... .....G.... .......... .......... ......G...`	–	1	–	–	–	–	1
H30	`.......... .....G.... .......... ...T...... ..........`	–	–	1	–	–	–	1
H31	`.......... .......... .......... .......... ...T.....T`	–	–	–	–	–	1	1
H32	`.......... .......... .......... ..C....... ...T......`	–	–	1	–	–	–	1
H33	`.......... .......... .......... ...C...... ...T......`	–	–	1	–	–	–	1
H34	`.......... ..G....... .......... .......... ...T......`	–	–	1	–	–	–	1
H35	`.......... .......... .......... ..C....... ...T...G..`	–	–	1	–	–	–	1

Mitochondrial DNA Variation as a Tool for Systematic Status Clarification of Commercial Species –
The Case of Two High Commercial Flexopecten Forms in the Aegean Sea

121

	gA	gP	gN	pA	pP	pN
No. polymorphic sites	9	21	15	8	16	15
No. haplotypes	10	12	15	9	11	12
Haplotype diversity (h)	0.7793	0.8735	0.9100	0.8235	0.9744	0.8769
Nucleotide diversity (m)	0.0027	0.0065	0.0042	0.0032	0.0078	0.0053

Appendix 1. Polymorphic sites, haplotype frequencies and genetic diversity indices, in all three studied populations of both taxa. The nucleotides in each position are given in comparison to haplotype H1. For all haplotypes variable sites are indicated, while identity is given by dots. In brackets it is the sample size. gA: *F. glaber* from Airport; gP: *F. glaber* from Paliomana; gN: *F. glaber* from Naziki; pA: *F. proteus* from Airport; pP: *F. proteus* from Paliomana; pN: *F. proteus* from Naziki.

6. Conclusion

We developed a DNA methodology based on PCR amplification and sequencing analysis of the mitochondrial 16S rDNA gene, in order to discriminate two different taxa of the family Pectinidae - *Flexopecten glaber* and *Flexopecten proteus* - and for a population study of both taxa. Fifty polymorphic nucleotide sites and fifty one different haplotypes were revealed in total. The majority of the genetic variation (96.98%) was distributed within populations. The levels of haplotype and nucleotide diversity were relatively high for both taxa, probably due to the large effective population sizes and the good population status. No species specific position was found and the two highly frequent haplotypes are shared between taxa, suggesting possible hybridization between them. Genetic distance between *F. glaber* and *F. proteus* was D = 0.005 and no genetic differentiation was revealed among taxa, when pooling together samples from different localities. Our results show that *F. glaber* and *F. proteus* are conspecific, which is in agreement with the latest classification.

7. Acknowledgements

The authors are indebted to Mr. Gregory Tselepidis, for his help in preparing the paper. The financial support for this study was provided by the Alexander Technological Educational Institute of Thessaloniki and is gratefully acknowledged.

8. References

Avise, J.C. (1994). *Molecular markers, Natural history and Evolution*. Chapman & Hall, ISBN 9780412037818, New York

Barucca, M.; Olmo, E; Schiaparelli, S. & Canapa, A. (2004). Molecular phylogeny of the family Pectinidae (Mollusca: Bivalvia) based on mitochondrial 16S and 12S rRNA genes. *Molecular Phylogenetics and Evolution*, Vol.31, No.1 (April 2004), pp. 89-95

Bertorelle, G. & Slatkin, M. (1995). The number of segregating sites in expanding human populations, with implications for estimates of demographic parameters. *Molecular Biology and Evolution*, Vol.12, No.5 (July 1995), pp. 887-892

Canapa, A.; Marota, I.; Rollo, F. & Olmo, E. (1996). Phylogenetic analysis of Veneridae (Bivalvia): comparison of molecular and palaeontological data. *Journal of Molecular Evolution*, Vol.43, No.5 (November 1996), pp. 517–522

Canapa, A.; Barucca, M.; Marinelli, A. & Olmo, E. (2000a). Molecular data from the 16S rRNA gene for the phylogeny of Pectinidae (Mollusca: Bivalvia). *Journal of Molecular Evolution*, Vol.50, No.1 (January 2000), pp. 93–97

Canapa, A.; Barucca, M.; Caputo, V.; Marinelli, A.; Cerioni, P.N.; Olmo, E.; Capriglione, T. & Odierna, G. (2000b). A molecular analysis of the systematics of three Antarctic bivalves. *Italian Journal of Zoology*, Vol.67, S1, pp. 127-132

Chan, T.Y. & Chu, K.H. 1996. On the different forms of *Panulirus longipes femoristriga* (von Martens, 1872) (Crustacea: Decapoda: Palinuridae), with description of a new species. *Journal of Natural History*, Vol.30, No.3, pp. 367-387

Cho, E.S.; Joung, C.G.; Sohn, S.G.; Kim, C.D. & Han, S.J. (2007). Population genetic structure of the ark shell Scapharca broughtonii Schrenck from Korea, China, and Russia based on COI gene sequences. *Marine Biotechnology*, Vol.9, No.2 (March 2007), pp. 203-216

CLEMAM, Taxonomic Database on European Marine Molusca.
(http://www.somali.asso.fr/clemam/index.php) (5/2010)

Cossignani, T.; Cossignani, V.; Di Nisio, A. & Passamonti, M. (1992). *Atlante delle conchiglie del medio Adriatico*. L' Informatore Piceno, ISBN 8886070004, Ancona

Crisp, D.J. (1978). Genetic consequences of different reproductive strategies in marine invertebrates, In: *Marine Organisms: Genetics, Ecology and Evolution*, Battaglia, B. & Beardmore, J.A., (Eds.), 257–273, Plenum Press, ISBN 978-0306400209, New York

Dayrat, B. (2005). Towards integrative taxonomy. *Biological Journal of the Linnean Society of London*, Vol.85, No.3, (July 2005), pp. 407–415

Delamotte, M. & Vardala–Theodorou, E. (1994). *Shells from the Greek Seas*. The Goulandris National History Museum, ISBN 9608519381, Athens

Doneddu, M. & Trainito, E. (2005). *Conchiglie del Mediterraneo*. Il Castello, ISBN 9788865200582, Trezzano sul Naviglio, Italy

ERMS. European Register of Marine Species (http://www.marbef.org) (5/2010)

Excoffier, L. & Lischer, H.E.L. (2010). Arlequin suite ver 3.5: A new series of programs to perform population genetics analyses under Linux and Windows. *Molecular Ecology Resources*, Vol.10, No.3, (May 2010), pp. 564-567

Galinou-Mitsoudi, S. & Sinis, A.I. (2000). Influence of exploitation on to natural bivalve populations in Thessaloniki gulf, *Proceedings of the 9th Panhellenic Ichtyhologists Conference*, pp. 21-24, Mesologi, Greece, January 20-23, 2000

Goetze, E. (2003). Cryptic speciation on the high seas; global phylogenetics of the copepod family Eucalanidae. *Proceedings of the Royal Society of London B*, Vol.270, No.1531, (November 2003), pp. 2321-2331

Gosling, E. (2003). Genetics in Aquaculture, In: *Bivalve, Molluscs. Biology, Ecology and Culture*, Gosling, E. (Ed.), 333-369, Fishing New Books, Blackwell Science, ISBN 978-0-85238-234-9, USA

Hall, T.A. (1999). BioEdit: a user-friendly biological sequence alignment editor and analysis program for Windows 95/98/NT. *Nucleic Acids Symposium Series*, Vol.41, No.41, pp. 95-98, ISSN 02613166

Hare, M.P. & Avise, J.C. (1996). Molecular genetic analysis of a stepped multilocus cline in the American oyster (*Crassostrea virginica*). *Evolution*, Vol.50, No.6, (December 1996), pp. 2305-2315

Hebert, P.D.N.; Cywinska, A.; Ball, S.L. & Ward, J.R. (2003). Biological identifications through DNA barcodes. *Proceedings of the Royal Society of London, Series B*, Vol.270, No.1512, (February 2003), pp. 313–321

Hebert, P.D.N.; Penton, E.H.; Burns, J.M.; Janzen, D.H. & Hallwachs, W. (2004a). Ten species in one: DNA barcoding reveals cryptic species in the neotropical skipper butterfly *Astraptes fulgerator*. *Proceedings of the National Academy of Sciences USA*, Vol.101, No.41, (October 2004), pp. 14812–14817

Hebert, P.D.N.; Stoeckle, M.Y.; Zemlak, T.S. & Francis, C.M. (2004b) Identification of birds through DNA barcodes. *Public Library of Science*, Vol.2, No.10, (September 2004), 1657–1663

Hillis, D.M.; Moritz, C. & Mable, B.K. (1996). *Molecular Systematics*. Sinauer Associates, Inc., Sunderland, ISBN 0878932828 9780878932825, MA, USA

Katsares, V.; Tsiora, A.; Galinou – Mitsoudi, S. & Imsiridou, A. (2008). Genetic structure of the endangered species *Pinna nobilis* Linnaeus, 1758 (Mollusca: Bivalvia) inferred from mtDNA sequences. *Biologia*, Vol.63, No.3, (June 2008), pp. 412 – 417

Kimura, M. (1980). A simple method for estimating evolutionary rate of base substitution through comparative studies of nucleotide sequences. *Journal of Molecular Evolution*, Vol.16, No.2, (June 1980), pp. 111-120

Knowlton, N. (1993). Sibling species in the sea. *Annual Review of Ecology, Evolution and Systematics*, Vol.24, (November 1993), pp. 189-216

Knowlton, N. (2000). Molecular genetic analyses of species boundaries in the sea. *Hydrobiologia*, Vol.420, No.1, (February 2000), pp. 73-90

Kocher, T.D.; Thomas, W.K.; Meyer, A.; Edwards, S.V.; Paabo, S.; Villablanca, F.X. & Wilson, A.C. (1989). Dynamics of mitochondrial DNA evolution in animals: Amplification and sequencing with conserved primers. *Proceedings of the National Academy of Sciences of USA*, Vol.86, No.16, (August 1989), pp. 6196-6200

Kong, X.; Yu, Z.; Liu, Y. & Chen, L. (2003). Intraspecific genetic variation in mitochondrial 16S ribosomal gene of Zhikong scallop *Chlamys farreri*. *Journal of Shellfish Research*, Vol.22, No.3, (December 2003), pp. 655-660

Koutsoubas, D.; Galinou-Mitsoudi, S.; Katsanevakis, S.; Leontarakis, P.; Metaxatos, A. & Zenetos, A. (2007). Bivalve and gastropod mollusks of commercial interest for human consumption in the Hellenic Seas, In: *State of Hellenic Fisheries*, Papaconstantinou, C.; Zenetos, A.; Vassilopoulou, V. & Tserpes, G., (Eds.), 70-84, HCMR publications, ISBN 9789609805414, Athens, Greece

Kumar, S.; Dudley, J.; Nei, M. & Tamura, K. (2008). MEGA: A biologist-centric software for evolutionary analysis of DNA and protein sequences. *Briefings in Bioinformatics*, Vol.9, No.4, (July 2008), pp. 299-306

Liu, B.; Dai, J. & Yu, Z. (1998). The investigation of populations by RAPD markers on oyster, Crassostrea talienwhanesis. *Journal of Ocean University of Qingdao*, Vol.28, pp. 82-88

Liu, B.; Dong, B.; Xiang, J. & Wang, Z. (2007). The phylogeny of native and exotic scallops cultured in China based on 16S rDNA sequences. *Journal of Ocean University of Qingdao*, Vol.25, pp. 85-90

Livi, S.; Cordisco, C.; Damiani, C.; Romanelli, M. & Crosetti, D. (2006). Identification of bivalve species at an early developmental stage through PCR-SSCP and sequence analysis of partial 18S rDNA. *Marine Biology*, Vol.149, No.5, (August 2006), pp. 1149-1161

Mahidol, C.; Na-Nakorn, U.; Sukmanomon, S.; Yoosuk, W.; Taniguchi, N. & Nguyen, T.T.T. 2007a. Phylogenetic relationships among nine scallop species (Bivalvia: Pectinidae) inferred from nucleotide sequences of one mitochondrial and three nuclear regions. *Journal of Shellfish Research*, Vol.26, No.1, (April 2007), pp. 25-32

Mahidol, C.; Na-Nakorn, U.; Sukmanomon, S.; Taniguchi, N. & Nguyen, T.T.T. 2007b. Mitochondrial DNA diversity of the Asian moon scallop, *Amusium pleuronectes* (Pectinidae) in Thailand. *Marine Biotechnology*, Vol.9, No.3, (May 2007), pp. 352-359

Mathews, L.M.; Schubart, C.D.; Neigel, J.E. & Felder, D.L. (2002). Genetic, ecological and behavioural divergence between two sibling snapping shrimp species (Crustacea: Decapoda: *Alpheus*). *Molecular Ecology*, Vol.11, No.8, (August 2002), pp. 1427-1437

Palumbi, S.R. (1996). Nucleic acids II: The polymerase chain reaction, In: *Molecular Systematics*, Hillis, D.M.; Moritz, C. & Mable, B.K. (Eds.), 205-247, Sinauer Associates, Inc., Sunderland, ISBN 0878932828 9780878932825, MA, USA

Pisor, D. & Poppe, G. (2008). *Registry of World Record Size Shells*. Conchbooks & Conchology, ISBN 9783939767121, Germany

Poppe, G.T. & Goto, Y. (1993). *European seashells, Vol. II*. ConchBooks Hachenheim, ISBN 3925919104, Germany

Poutiers, J.M. (1987). Bivalves, In: *Fiches FAO d'identification des espèces pour les besoins de la pêche. Méditerranée et Mer Noire. Zone de pêche 37*, Fischer, W.; Schneider, M. & Bauchot, M.L., (Eds.), 369-514, FAO, Rome, Italy

Pujolar, J.M.; Marčeta, T.; Saavedra, C.; Bressan, M. & Zane, L. (2010). Inferring the demographic history of the Adriatic *Flexopecten* complex. *Molecular Phylogenetics and Evolution*, Vol.57, No.2, (November 2010), pp. 942-947

Repetto, G.; Orlando, F. & Arduino, G. (2005). *Conchiglie del Mediterraneo*. Amici del Museo "Federico Eusebio", Alba, Italy

Raines, B.K. & Poppe, G.T. (2006). *A conchological Iconography: the family Pectinidae*. ConchBooks, Hackenheim, Germany

Rawson, P.D. & Hilbish, T.J. (1995). Evolutionary relationships among the male and female mitochondrial DNA lineages in the *Mytilus edulis* species complex. *Molecular Biology and Evolution*, Vol.12, No.5, (July 2005), pp. 893-901

Rawson, P.D. & Hilbish, T.J. (1998). Asymmetric introgression of mitochondrial DNA among European populations of blue mussels (*Mytilus* spp.). *Evolution*, Vol.52, No.1, (February 2008), pp. 100-108

Raymond, M. & Rousset, F. (1995). An exact test for population differentiation. *Evolution*, Vol.49, No.6, (December 1995), pp. 1280-1283

Reichow, D. & Smith, M.J. (2001). Microsatellites reveal high levels of gene flow among populations of the California squid *Loligo opalescens*. *Molecular Ecology*, Vol.10, No.5, (May 2001), pp. 1101-1109

Rivera, M.A.J.; Kelley, C.D. & Roderick, G.K. (2004). Subtle population genetic structure in the Hawaiian grouper, *Epinephelus quernus* (Serranidae) as revealed by mitochondrial DNA analysis. *Biological Journal of the Linnean Society. Linnean Society of London*, Vol.81, No.3, (March 2004), pp. 449-468

Rombouts, A. (1991). *Guidebook to Pecten shells: recent Pectinidae and Propeamussiidae of the world*. Oegstgeest Universal Book Service/Dr. W. Backhuys, ISBN 1863330267, The Netherlands

Saavedra, C. & Pena, J.B. (2004). Phylogenetic relationships of European and Australasian king scallops (*Pecten* spp.) based on partial 16S ribosomal RNA gene sequences. *Aquaculture*, Vol.235, No.1-4, (June 2004), pp. 153-166

Saavedra, C. & Pena, J.B. (2006). Phylogenetics of American scallops (Bivalvia: Pectinidae) based on partial 16S and 12S ribosomal RNA gene sequences. *Marine Biology*, Vol.150, No.1, (October 2006), pp. 111-119

Sarver, S.K.; Silberman, J.D. & Walsh, P.J. (1998). Mitochondrial DNA sequence evidence supporting the recognition of two subspecies or species of the Florida spiny lobster *Panilurus argus. Journal of Crustacean Biology*, Vol.18, No.1, (February 1998), pp. 177-186

Seed, R. (1969). The ecology of *Mytilus edulis* L. (Lamelibrabchiata) on exposed rocky shores. 1. Breeding and settlement. *Oecologia*, Vol.3, No.3-4, (September 1969), pp. 277–316

Slatkin, M. & Hudson, R.R. (1991). Pairwise comparisons of mitochondrial DNA sequences in stable and exponential growth populations. *Genetics*, Vol.129, No.2, (October 1991), pp. 555–562

Soule, M.E. (1980). Thresholds for survival: maintaining fitness and evolutionary potential, In: *Conservation biology: an evolutionary-ecological perspective*, Soule, M.E. & Wilcox, B.A., (Eds.), 151-170, Sinauer Association, ISBN 978-0878938001, Sunderland, Massachusetts

Stepien, C.A.; Hubers, A.N. & Skidmore, J.L. (1999). Diagnostic genetic markers and evolutionary relationships among invasive Dreissenoid and Corbiculoid Bivalves in North America: phylogenetic signal from mitochondrial 16S rDNA. *Molecular Phylogenetics and Evolution*, Vol.13, No.1, (October 1999), pp. 31-49

Tajima, F. (1989). Statistical method for testing the neutral mutation hypothesis by DNA polymorphism. *Genetics*, Vol.123, No.3, (November 1989), pp. 585–595

Tamura, K.; Dudley, J.; Nei, M. & Kumar, S. (2007). MEGA4: Molecular Evolutionary Genetics Analysis (MEGA) software version 4.0. *Molecular Biology and Evolution*, Vol.24, No.8, (August 2007), pp. 1596-1599

Therriault, T.W.; Docker, M.F.; Orlova, M.I.; Heath, D.D. & MacIsaac, H.J. (2004). Molecular resolution of the family Dreissenidae (Mollusca: Bivalvia) with emphasis on Ponto-Caspian species, including first report of Mytilopsis leucophaeata in the Black Sea basin. *Molecular Phylogenetics and Evolution*, Vol.30, No.3, (March 2004), pp. 479-489

Thompson, J.D.; Gibson, T.J.; Plewniak, F.; Jeanmougin, F. & Higgins, D.G. (1997). The CLUSTAL X windows interface: Flexible strategies for multiple alignments aided by quality analysis tool. *Nucleic Acids Research*, Vol.25, No.24, (December 2007), pp. 4876-4882

Vaught, K.C. (1989). *A classification of the living mollusca.* American Malacologists, ISBN 9780915826223, Melbourne, FL

Wilding, C.S.; Beaumont, A.R. & Latchford, J.W. (1999). Are *Pecten maximus* and *Pecten jacobaeus* different species? *Journal of the Marine Biological Association of the United Kingdom*, Vol. 79, No.5, (September 2000), pp. 949-952

Yu, X.Y.; Mao, Y.; Wang, M.F.; Zhou, L. & Gui, J.F. (2004). Genetic heterogeneity analysis and RAPD marker detection among four forms of *Atrina pectinata* Linnaeus. *Journal of Shellfish Research*, Vol.23, No.1, (April 2004), pp. 165-171

Yuan, T.; He, M. & Huang, L. (2009). Intraspecific genetic variation in mitochondrial 16S rRNA and COI genes in domestic and wild populations of Huaguizhikong scallop *Chlamys nobilis* Reeve. *Aquaculture*, Vol.289, No.1-2, (April 2009), pp. 19-25

Zenetos, A. (1996). *Fauna Graeciae. VII. The marine Bivalvia (Mollusca) of Greece.* N.C.M.R., ISBN 960-85952-0-7, Athens

Zenetos, A.; Vardala-Theodorou, E. & Alexandrakis, C. (2005). Update of the marine Bivalvia Mollusca checklist in Greek waters. *Journal of the Marine Biological Association of the United Kingdom*, Vol.85, No.4, (June 2005), pp. 993-998

Genomics and Genome Sequencing: Benefits for Finfish Aquaculture

Nicole L. Quinn[1], Alejandro P. Gutierrez[1],
Ben F. Koop[2] and William S. Davidson[1]
*[1]Department of Molecular Biology and Biochemistry,
Simon Fraser University, Burnaby, British Columbia*
*[2]Department of Biology, University of Victoria,
Victoria, British Columbia*
Canada

1. Introduction

1.1 What is genomics?

The central dogma of molecular biology states that DNA is transcribed to RNA, which is translated into proteins (Crick, 1970), the molecules that facilitate all biological functions. A genome comprises all of an organism's DNA, or hereditary information. The field of genomics is the study of whole genomes. Genomics has been defined as "a branch of biotechnology concerned with applying the techniques of genetics and molecular biology to the genetic mapping and DNA sequencing of sets of genes or the complete genomes of selected organisms" (Mirriam-Webster Dictionary). Specifically, whereas genetics is the study of a single gene, or a few genes in isolation, genomics examines all of the genes, as well as the non-coding elements (i.e., regions that do not encode proteins or RNA components of the cell), within the DNA of a genome. Although the term genomics was first used in 1987, the field is relatively new and is growing rapidly as new technologies for exploring genomes emerge. The applications of genomics are vast, spanning realms such as medicine, industry and ecology, with implications for global issues including cancer diagnosis, prevention and treatment, alternative energy sources, agriculture, conservation and sustainable development. Furthermore, given that the fundamental basis of genomics—DNA—is common across all living organisms, genomics stands to bridge the gaps between these fields of study, with synergistic results as multi-disciplinary approaches to answering biological questions are developed.

1.2 Genomics in aquaculture: Phenotypic vs. genotypic selection

In traditional selective breeding practices, individuals showing desirable phenotypic, or visible characteristics are bred to produce new strains of plants or animals that exhibit features such as faster growth, greater size, increased overall robustness, or improved aesthetic appeal. Farmers, through thousands of years of following the mantra "breed the best to the best and hope for the best" (quotation attributed to American Thoroughbred and

Standardbred breeder, John E. Madden), have drastically changed natural populations, as exemplified by the now hundreds of breeds of the domestic dog, or differing cattle strains for dairy or beef production. However, these phenotype-based breeding tactics come with inherent drawbacks. Specifically, they are draining of resources and time, as they often rely on trial and error, they require vast numbers of individuals exhibiting extensive phenotypic variation, and often numerous generations of repetitive selective breeding are required to see an effect, which is especially difficult for species with long generation times and for which extensive parental investment is needed. Furthermore, often animals need to be sacrificed to detect or measure morphological characteristics such as flesh color and tissue quality, and these, as well as disease challenged (i.e., to determine immune function or disease resistance) animals cannot be bred. Finally, most characteristics of interest to aquaculture are complex traits – i.e., those that are governed by numerous genes and complex interactions between multiple pathways, or those for which morphological variation is based on environmental cues, and it is very difficult to select for multiple complex traits together based on phenotypic information alone. Thus, the amount of time and the costs associated with traditional, phenotypic approaches to selective breeding are large.

In contrast to phenotype-based selection, which directly selects for a given heritable trait, in genotype-based selection, or marker-assisted selection (MAS), traits are indirectly selected for based on variability at the DNA level. More specifically, the genomes of breeding populations are screened for multiple markers, or DNA tags, that are associated with genes of interest that work together to produce a desirable phenotype for a complex trait. This genetic screening can be accomplished using high-throughput genomics systems that incorporate tools such as polymerase chain reaction (PCR) or DNA sequence-based screening.

MAS has been used extensively for the genetic improvement of cultivated plant cultivars, such as wheat (Gupta et al., 2008), corn (Tuberosa et al., 2002) and soy (Kim et al., 2010), as well as cultured stocks (broodstocks) in agriculture species such as swine (Liu et al., 2007) and cattle (Veerkamp & Beerda 2007). As the availability of genomics resources has increased for aquatic species, more research is being done to improve broodstocks for many key aquaculture species worldwide. Specifically, genomics, and the use of genomic tools, enables one to examine the differences and similarities among organisms at the genotypic (DNA) level, as opposed to more traditional broad-scaled phenotype-based (appearance-based) approaches. This very fine-scaled perspective, looking at differences in genes (alleles) as well as variability at non-coding loci, or positions within the genome, means that complex traits that are driven by more than one gene or pathway can be broken down into their components. That is, differences at the individual, family, population, species and even the organism level can be assessed in very fine detail. Such insight into the genetic factors that drive complex traits can facilitate the development of effective and efficient breeding methods, which have far-reaching implications for the aquaculture industry.

2. Atlantic salmon: A model aquaculture species for MAS

The holy grail of any genomics program for a species is a whole genome sequence that is well assembled and annotated. The advantages of this are many, and are discussed in further detail in subsequent sections, but are mainly centered around two things: 1) the wealth of data that is produced, including the full gene repertoire with additional information such as gene location and copy number, and 2) the ability of the sequenced genome to act as a reference genome, both for the sequenced species itself (i.e., such that the genomes of additional individuals can be easily re-sequenced using the original as a reference for assembly), as well as to provide information for other, closely related species. Currently, however, even with the great advances in sequencing technology that have come to light in recent years, obtaining a whole genome sequence remains an extremely difficult, costly and time-consuming undertaking. This is particularly true for fish species simply due to the evolutionary age of fish and the more than 20,000 extant species (Nelson, 2006), factors that make the fish genomes diverse and complex and complicate sequencing. Indeed, only five fish genomes have been reported to date (medaka, *Oryzias latipes*; tiger pufferfish, *Takifugu rubripes*; green spotted pufferfish, *Tetraodon nigriviridis*; zebrafish, *Danio rerio* and stickleback, *Gasterosteus aculeatus*), although more are underway. Their sequences, as well as those for the sequenced genomes of other organisms are publically available within the Ensembl database (www.ensembl.org). These fish species were chosen for their abilities to act as model sequences for genetics research, rather than for their utility for aquaculture. Specifically, the medaka and zebrafish genomes were sequenced to provide model organisms for studying developmental biology, while the stickleback genome serves as a model for studying adaptive evolution, and the two pufferfish represent the smallest known vertebrate genomes. Figure 1 illustrates the phylogenetic relationships among these species as well as some key aquaculture species, and shows that the full spectrum of teleosts is not represented by the genome sequences that are currently available.

When a whole genome sequence is not available, there are numerous genomics resources and tools that can be developed which, particularly when used in combination, can provide extensive insight into a genome and can be used for applications such as MAS, both for the species of interest, and for other closely-related species. In the following sections, we will use the example of Atlantic salmon (*Salmo salar*) and its genomics program to describe these tools, their development and utility for aquaculture. Atlantic salmon is a particularly good model fish species for genomics because it is a major aquaculture species, with approximately 1.5 million tonnes produced worldwide in 2009 (Food and Agriculture Organization [FAO] Fishery Statistic, 2009), and like other salmonids, is of substantial environmental, economic and social importance. In addition, there are extensive genomics resources for Atlantic salmon, which were developed using standard methods and approaches that are applicable to other genomes. Furthermore, a strong argument has been made for obtaining the full genome sequence for Atlantic salmon, a project that is currently in progress (Davidson et al., 2010). Specifically, aside from the merits of the species itself, there are no salmonid species yet sequenced, and thus Atlantic salmon can serve as a reference salmonid genome, providing extensive opportunities for cross-referencing, or comparative synteny analyses with other salmonids, particularly the Pacific salmon (*Oncorhynchus* sp.), rainbow trout (*Oncorhynchus mykiss*) and Arctic charr (*Salvelinus alpinus*).

Atlantic salmon also provides an example of some of the challenges that face fish genomics in general. The common ancestor of salmonids underwent a whole genome duplication event between 20 and 120 million years ago (Allendorf & Thorgaard, 1984; Ohno, 1970). Thus, whereas there are usually two copies of each gene within a genome, Atlantic salmon have four, and the duplicate copies are evolving into genes with new functions or non-coding DNA. The genome duplication also increased the size and the repetitiveness of the genome. These characteristics, combined with the lack of a closely related guide sequence, mean that sequencing and assembling the Atlantic salmon genome are extremely challenging.

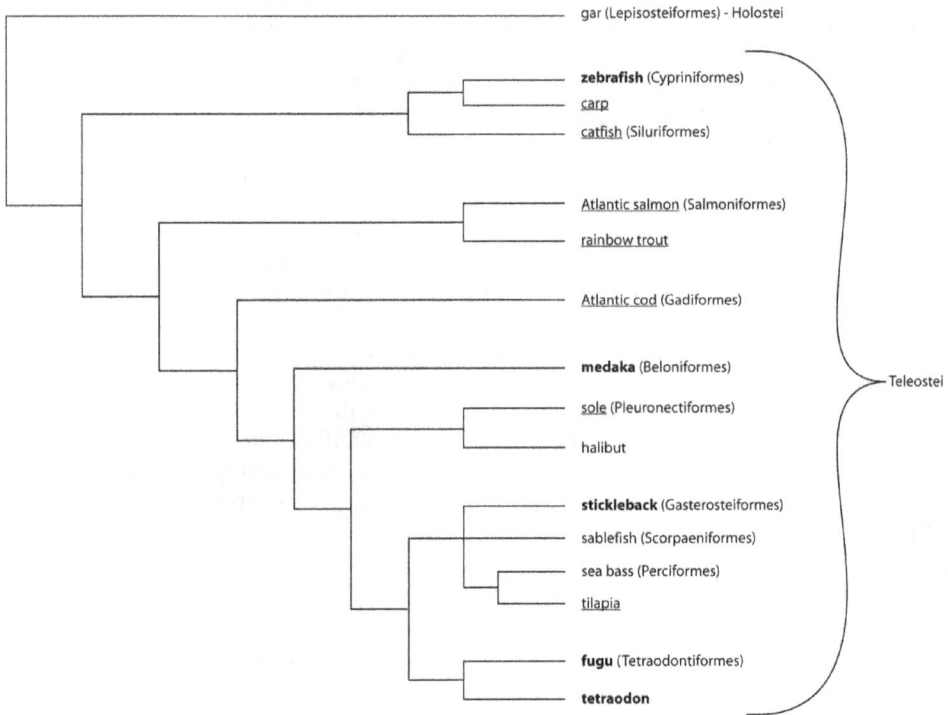

Fig. 1. Schematic representation of the phylogenetic relationships among fish species. Species in bold have publically available full genome sequences, while those that are underlined are currently being sequenced. Note that the full spectrum of teleosts is not represented by the genome sequences that are currently available. Species are listed by their common names with orders in parentheses. Gar is used as an outgroup.

3. Genomics resources

3.1 ESTs

An expressed sequence tag (EST) is a sequenced section of a transcribed, or expressed gene. ESTs are developed by isolating the mRNA, i.e., the transcribed DNA, from a tissue, which is then reverse-transcribed into cDNA (complementary DNA). Thus, an EST library is the full repertoire—essentially an inventory—of expressed genes within a genome in a given tissue at a given time. More than 499,000 ESTs have been identified for Atlantic salmon, the sequences of which are available publicly in an online database (Rise et al., 2004; Koop et al., 2008) as well as within the NCBI database.

Perhaps the most common use of ESTs is to study gene expression (described below). Other uses of ESTs include the identification of molecular markers, or DNA tags, which are described in detail in the subsequent section. EST-based markers have unique applications because they occur in expressed genes, as opposed to non-coding DNA. Thus, they can provide insight into allelic differences between individuals or populations, thereby facilitating an understanding of the relationship between phenotypes and particular genes.

3.2 Genetic markers

Genetic markers are short sequences of DNA that are associated with a specific location, or locus, on a chromosome. To be useful, a genetic marker must be sufficiently variable (polymorphic) among individuals or populations. Markers are valuable tools in molecular biology and they have many practical uses. For example, they are used to examine ancestry and to define discrete populations, as well as for forensic work, paternity testing and tissue typing for transplants given that the genotypes of the markers are inherited and thus an individual's combination of marker genotypes (commonly called a "DNA fingerprint") is unique.

3.2.1 Microsatellite markers

Satellite markers are a group of DNA markers that are characterized by containing tandemly repeated sequences of nucleotides (e.g., ...ACACACAC....). Depending on their size, which is a function of the number of repeated bases in a repeat unit and the number of tandemly arrayed units, satellite markers are classed as microsatellites (short repeat unit) or minisatellites (long repeat unit). Microsatellite markers are heavily utilized in genomics. They act as DNA anchors or tags within a genome because: 1) like all molecular markers, their genotypes are heritable, 2) the repeat units make them more vulnerable to mutation than other types of DNA sequence (e.g., coding sequence), which means they are variable at all levels, from individual to species, and generally have several alleles, 3) their locations (loci) within the chromosome can be determined relative to one another, and 4) their small size (vs. minisatellites, which can be very long) facilitates ease of detection using PCR. The development of a suite of microsatellite markers is often one of the first steps in a genomics program for a species of interest.

The first microsatellites characterized for Atlantic salmon came from genomic libraries (McConnell et al. 1995; O'Reilly et al., 1998; Sanchez et al., 1996; Slettan et al., 1995). These were classified as "anonymous" or "Type II" markers as they were not associated with a particular gene. The mining of large EST databases allowed the identification of Type I microsatellites, which could be related directly to a gene (Anti Vasemagi et al., 2005; Ng et al., 2005,). More recently, the sequencing of the ends of Bacterial Artificial Chromosome (BAC) clones has provided a rich source of microsatellites (Danzmann et al., 2008; Phillips et al., 2009). Microsatellites have been used extensively to examine the population structure of Atlantic salmon (e.g., King et al., 2001) as well as for assigning parentage and relatedness of individuals in a mixed family stock (Norris et al., 2000), which precludes the necessity of rearing families in individual tanks or using physical tags. Many of the microsatellite markers that have been developed for one species of salmonid can be used in others (e.g., McGowan et al., 2004; Presa & Guyomard, 1996; Scribner et al., 1996), which enables genomic comparisons among salmonids such as Atlantic salmon, brown trout (*Salmo trutta*), rainbow trout and Arctic charr (Danzmann et al., 2005, 2008; Woram et al., 2004). Most aquaculture species have a set of microsatellites now, and the list is too extensive to reference here.

3.2.2 Single Nucleotide Polymorphisms (SNPs)

A single nucleotide polymorphism (SNP) (pronounced *snip*) is a sequence change in a single nucleotide – A, T, C or G - that occurs with sufficient frequency in a population to serve as a genetic marker. SNPs can be within genes or within non-coding regions of the genome. Because they are usually bi-allelic (i.e., exhibit only two alleles, such as an A or a T at a given locus), they are easily identified and studied. The development of novel sequencing technologies has made it possible to obtain thousands SNP markers for genotyping (see Molecular Ecology Resources (2011) 11: Supplement 1, for several recent articles on this subject). Moreover, new genotyping technologies enable the simultaneous analysis of a great number of these markers, which allows the construction of high density genetic maps (Goddard & Hayes, 2009). These techniques are now being widely used in species of agricultural importance and have led to the development of numerous dense SNP arrays, which are tools that are used to screen entire genomes for the genotypes of known SNPs (Matukumalli et al., 2009; Muir et al., 2008; Ramos et al., 2009).

There have been several attempts to identify SNPs in individual genes in Atlantic salmon (Ryynanen & Primmer, 2006) and other salmonids (Smith et al., 2005), but these were small-scale efforts. It was not until large EST databases became available that an intensive search was undertaken for SNPs that could be tied to actual genes (Hayes et al., 2007). Re-sequencing BAC-end sequences (described below) was another approach taken to identify SNPs for Atlantic salmon (Lorenz et al., 2010), but this process was highly labor intensive relative to the information gained. A next generation sequencing strategy (Van Tassell et al., 2008) has also been used to identify SNPs in Atlantic salmon; however, the repetitive nature of the genome made it difficult to distinguish between putative SNPs arising from duplicated regions and those from a unique locus. Recently a custom SNP-array (iSelect by Illumina) containing approximately 6,500 SNP markers was developed for Atlantic salmon. This has been used to search for generic genetic differences between farmed and wild Atlantic salmon (Karlsson et al., 2011) and to assess the level of heterozygosity in Tasmanian Atlantic salmon (Dominik et al., 2010). In addition, the SNP-array was used to construct a relatively dense SNP-based linkage map (described below) (Lien et al., 2011). For other aquaculture species the availability of SNPs is still somewhat limited, with only a large number of SNPs being reported for catfish (*Ictalurus* sp.), (Wang et al., 2010) rainbow trout (Sanchez et al., 2009) and Atlantic cod (*Gadus morhua*) (Hubert et al., 2010; Moen et al., 2008b).

3.3 Linkage maps

One of the most useful applications of genetic markers is the generation of a genetic map, also known as a linkage map, of an organism's genome. Specifically, markers that are located physically close to one another on a chromosome, or 'linked', tend to be inherited together. By determining the frequency that alleles at different genetic loci are inherited together (i.e., the degree of linkage between makers) one can estimate the recombination distances that separate markers (i.e., the relative positions of markers within a genome). With sufficient markers (i.e., a "dense" linkage map), the number of linkage groups should correspond to the number of different chromosomes comprising a genome. Linkage maps are extremely useful tools in genomics and form the foundation for QTL mapping (described below).

The first linkage maps for Atlantic salmon were constructed using relatively few microsatellite markers (50–60) (Gilbey et al., 2004; Moen et al., 2004). The number of microsatellite markers was expanded to approximately 2,000 thanks to BAC-end sequencing, and a relatively dense genetic map covering the 29 pairs of chromosomes in the European strain of Atlantic salmon was built (Danzmann et al., 2008). Additional maps were constructed using SNPs derived from ESTs (Moen et al., 2008a) and BAC-end sequences (Lorenz et al., 2010), and these were incorporated into a dense SNP map comprising ~5,600 SNPs (Lien et al., 2011). In terms of other salmonids, linkage maps have been constructed for rainbow trout (Guyomard et al., 2006; Nichols et al., 2003; Rexroad et al., 2005; Sakamoto et al., 2000), brown trout (Gharbi et al., 2006), coho salmon (*Oncorhynchus kisutch*) (McClelland & Naish, 2008), and Arctic charr (Woram et al., 2004). Other aquaculture species of interest for which linkage maps have been developed include catfish (Kucuktas et al., 2009; Waldbieser et al., 2001), Japanese flounder (Coimbra et al., 2003) European sea bass (Chistiakov et al., 2005), tilapia (Lee et al., 2005), common carp (Sun & Liang, 2004), olive flounder (Kang et al., 2008), Atlantic cod (Hubert et al., 2010; Moen et al., 2009b) and the list is expanding.

3.4 Bacterial Artificial Chromosome (BAC) libraries

3.4.1 BACs and BAC libraries

A bacterial artificial chromosome (BAC) clone is comprised of a special plasmid and a large segment of genomic DNA from the species of interest. Specifically, the entire genome of a selected individual is cut into pieces of approximately 150,000–300,000 base pairs, each of which is inserted into the plasmid, thereby generating thousands of closed, circular "hybrid DNA constructs". Thus, a BAC library refers to all of the BACs, which in combination comprise the entire genome, for a representative individual of a species. Each BAC is then inserted ("transformed") into live *E. coli* cells; thus, the foreign (non-bacterial) DNA can be replicated as if it were a bacterial chromosome using the cellular machinery found within the bacterial cell. Ideally there should be redundancy of the segments of the genome represented in a BAC library such that overlaps of individual clones occur. The ability to identify which BACs share a segment of the genome enables a network called a physical map to be constructed. Therefore, the physical map represents the genome, and has traditionally been used to assist with sequence assembly following whole genome sequencing. BAC libraries also facilitate the identification of target sequences or regions, as the library can be "screened" for genetic markers or other sequence-specific probes. An Atlantic salmon BAC library comprising approximately 300,000 clones and providing an 18-fold coverage of the genome is available (Thorsen et al., 2005), and it was used to generate a physical map (Ng et al., 2005). BAC libraries have been constructed for several aquaculture species including rainbow trout (Palti et al., 2009), catfish (Xu et al., 2007), common carp (Li et al., 2011b) and barramundi (*Lates calcarifer*) (Xia et al., 2010).

3.4.2 BAC-end sequencing

Once a BAC library has been generated, typically the next step is to select a subset of BACs for end sequencing. Because the sequence of the plasmid vector within the BAC is known, the bacterial sequences at the junctions of the genomic DNA and vector DNA can act as sequence primers. Thereby, the first 500–1,000 nucleotides (based on Sanger sequencing) of the genomic DNA inserts—i.e., the BAC-ends—can be determined. End-sequencing a large subset of BACs can provide extensive insight into the full genome sequence. For example, the 207,869 BAC-end sequences for the Atlantic salmon BAC library cover approximately 3.5% of the whole genome sequence. This glance into a genome is very powerful, as it can provide information about the complexity of the genome (i.e., repeat content), and the BAC-end sequences can be a source of molecular markers (as described above for Atlantic salmon (Danzmann et al., 2008; Lorenz et al., 2010)). Finally, the BAC-end sequences can be used for comparative synteny analyses by aligning them against other, fully sequenced genomes, which can provide insight into the gene content of the BACs, thus providing a partial, putative annotation of segments of the genome (Sarropoulou et al., 2007). This is particularly powerful if the two genomes being compared are closely related, as one can act as a reference genome (e.g., common carp and zebrafish, which are both Cypriniformes) (Xu et al., 2011). Even when there is no obvious reference genome there is value in this exercise as it often suggests syntenic regions among phylogenetically distantly related species, and can lead to candidate genes being suggested for specific traits (see below for an example with ISA resistance in the MAS section). The Atlantic salmon BAC end sequences have been made available in a publicly accessible searchable database (www.asalbase.org) along with comparative genomic information for four of the five published fish genomes. Similar work is ongoing with aquaculture species such as catfish (Liu et al., 2009), rainbow trout (Genet et al., 2011) and sea bream (Kuhl et al., 2011).

3.4.3 Integration of physical and linkage maps and chromosomes

Once both a physical map (i.e., a BAC library is assembled into overlapping, contiguous BACs), and a linkage map (generated using the relationships between variable markers) exist for a species, the next step is to integrate the two maps. This procedure is relatively straight forward if the linkage map contains many markers from BAC-ends, as illustrated by the integration of the Atlantic salmon physical and linkage maps (Danzmann et al., 2008). However, this is seldom the case and considerable effort is normally required to carry out this process (e.g., see preliminary data for rainbow trout) (Palti et al., 2011). The integration of the linkage and physical maps then allows BACs containing particular markers to be selected for fluorescent *in situ* hybridization on chromosomes so that the each linkage group can be assigned to a particular chromosome. This information is now available for several salmonid species including Atlantic salmon (Phillips et al., 2009) and rainbow trout (Phillips et al., 2006).

3.5 Quantitative trait loci (QTL)

Generally, the characteristics that are desirable for selection are complex, or 'quantitative', which means that they are not simply governed by a single gene, but rather are controlled by a suite of genes and regulatory factors that interact to produce a phenotypic characteristic. The resulting phenotype usually falls somewhere on a spectrum. This is not always the case; traits such as genetic disorders can be determined by a single gene (e.g., cystic fibrosis and Duchenne muscular dystrophy), but most phenotypic traits associated with health, growth, meat quality and the capacity for utilizing new and ecologically sustainable food items are complex traits that are strongly influenced by the environment, and thus show a range of phenotypes.

A deep biological understanding of phenotypic patterns, ranging from intracellular molecular signatures to whole individual physiological, morphological and behavioral characteristics of genes of commercial importance is synonymous with an understanding of the functioning of the underlying gene regulatory networks and how these are molded by the environment. This calls for the identification of genes and characterization of their functional repertoires in various systemic settings. Specifically, the first step in understanding the mechanisms that govern complex traits is to determine the heritability of the trait – i.e., the extent to which it is determined genetically and by the environment. Heritability estimates vary tremendously depending on the population, the experimental design and the researchers. Garcia de Leaniz et al. provide the best available review of heritability estimates for Atlantic salmon (Garcia de Leaniz et al., 2007). What is clear is that the genes do play a role in the variation detected in the majority of economically important production traits.

The two most common approaches to identify genes associated with complex traits are to examine candidate genes, based on previous knowledge of a biochemical process, or by mapping quantitative trait loci (QTL). We will not discuss the candidate gene approach here, but rather concentrate on QTL analysis. QTL are regions of the genome that contain, or are linked to, genes that contribute to a particular trait (Lynch & Walsh, 1998). Note that the term "QTL" does not refer to a particular gene, but rather its putative location in the genome. This association, or linkage, is determined using the genetic markers within a linkage map, and how alleles at these markers co-segregate with the trait in question. Thus, finding functional markers to assist in MAS is largely dependent on the variability and polymorphic level of the species. In most agricultural examples, there are highly inbreed strains that make MAS usable. When a species is highly variable, such as the case with fish and most marine species, the QTL markers have to be very close to the causal variant to be usable in MAS breeding studies. The more variable the species, the closer the marker has to be to be used in selective breeding. Given sufficiently high linkage, the markers that define QTL can feed directly into marker assisted selection (MAS). The number, positions and the magnitude of the QTL affecting a trait are determined by statistical associations between marker alleles or genotypes and particular trait phenotypes. QTL analysis, therefore, relies heavily on robust, dense linkage maps as well as large families that are accurately pedigreed and carefully described with respect to phenotype. Indeed, a limiting factor for effective QTL analysis is the difficulty in characterizing "phenomes", the full set of phenotypes of an individual (Houle et al., 2010).

Growth is one of the most economically important traits for aquaculture species. It is no surprise therefore, that several QTL have been identified for body weight and condition factor in Atlantic salmon (Baranski et al., 2010; Boulding et al., 2008; Moghadam et al., 2007; Reid et al., 2005) rainbow trout (Haidle et al., 2008; Martyniuk et al., 2003; O'Malley et al., 2003; Wringe et al., 2010), coho salmon (McClelland & Naish, 2010; O'Malley et al., 2010) and Arctic charr (Moghadam et al., 2007; Kuttner et al., 2011). QTL for several other traits in salmonids have been mapped, including upper temperature tolerance (Jackson et al., 1998; Perry et al., 2001; Somorjai et al., 2003), spawning time (O'Malley et al., 2003), developmental rate (Nichols et al., 2008), and resistance to pathogens (Gheyas et al., 2010; Gilbey et al., 2006; Houston et al. 2008, 2009, 2010; Jones et al., 2002; Moen et al. 2007, 2009a). Given that many of the microsatellite markers derived from one salmonid species amplify the DNA from other salmonid species, it has been relatively straightforward to carry out comparative analyses of the rainbow trout, Arctic charr and Atlantic salmon genomes (Danzmann et al., 2005).

Most of the QTL studies described above were based on a limited number of microsatellite markers; thus, many regions of the genome were sparsely represented. Therefore, it was not possible to obtain precise and complete information about the number and locations of most of these QTL, and to identify specific genes and the underlying genetic variation, or to obtain markers with sufficiently robust associations with the QTL to warrant their integration into MAS programs. Cost has been the main factor limiting the number of microsatellites that could reasonably be used. However, with the advent of SNP arrays, this problem has largely been overcome for agriculture animals such as cattle, pigs and chicken (Groenen et al., 2009; Matukumalli et al., 2009; Ramos et al., 2009) and great progress is being made for some aquaculture species such as Atlantic salmon (Kent M et al., in preparation).

3.6 Expression profiling (microarrays and qPCR)

The concept of gene expression was introduced above in the discussion on expressed sequence tags, or ESTs and EST libraries, which are assemblages of all small sequences of all the expressed genes in a tissue at a particular time point. The term "expression profiling" refers to the examination of the genes expressed in an individual, or group of individuals, under specific conditions. This is based on the fact that gene expression varies with genotype (i.e., the suite of genes and their alleles that are present within an individual's genome) and environment, which cues the expression of particular genes as well as the extent of expression. Expression profiling studies generally compare groups (treatment groups) of individuals that differ in some way, such as genetic background, or environmental surroundings. Thereby, one can examine which genes are relatively over- and under-expressed across treatment groups using statistical analyses, thus gaining insight into the genes that govern particular traits that differ between the subjects.

There are two approaches to studying gene expression that are now standard: real-time PCR (also called quantitative real time PCR or qPCR, or sometimes abbreviated qrtPCR), and microarray analysis. The former is used to examine the expression of a single gene, or a few genes at a time. The technique follows the general principles of regular PCR, in which DNA (in this case, having been reverse-transcribed from RNA, which is present in amounts that are directly proportional to the extent of gene expression) is amplified from a template strand, with the key difference that the amount of product is measured in real time. This is accomplished in one of two ways: 1) the DNA components in the reaction mixture are tagged with dyes that fluoresce when integrated into the DNA molecule, and the level of fluorescence (i.e., corresponding to the amount of product) is measured after each PCR cycle, or 2) sequence-specific probes are similarly tagged, and fluoresce when hybridized to the complementary DNA target, thus providing a direct measurement of the amount of target present after each PCR cycle. The benefits of qPCR are that the process is gene-specific and the results are relatively accurate and reproducible. The cons are that only a small number of genes within a genome can be examined at once, and that sequence-specific PCR primers must be designed, which is dependent on knowing the sequence of the gene *a priori*. This is further complicated when examining gene families or duplicated genes, as it is often not feasible to design primers specific for each gene, a problem of particular relevance when there is no genome sequence for the study organism.

Whereas qPCR targets single genes at a time, DNA microarrays are specifically designed to examine the gene expression at thousands of genes at once. A microarray (also referred to as a "chip") is comprised of thousands of single-stranded DNA spots, or "features" attached in known locations to a surface, such as a glass slide, with each spot corresponding to a gene. To analyze gene expression using a microarray, cDNA (again, reverse-transcribed from RNA) that has been fluorescently tagged is washed over the slide, and complementary strands will hybridize to corresponding spots on the microarray. Thus, when the level of fluorescence is measured, the intensity of each spot will correspond to extent of the expression of that particular gene. Statistical tests are then used to determine the relative expression of each gene on the microarray for the individual/tissue/environmental condition being tested. This method is extremely powerful given the number of genes that can be assessed at once. For example, several microarrays have been designed using the Atlantic salmon EST library, including an initial 3.7K (i.e., 3700 spots) array (Rise et al., 2004), followed by 16K (von Schalburg et al., 2005) and a 32K cDNA (Koop et al., 2008) arrays, as well as two recent 44K oligonucleotide arrays. These resources, which are collectively referred to as the GRASP (Genomic Research on All Salmon Project) arrays, have been used extensively for gene expression studies throughout the salmonid genomics world. For example, recent studies have used the GRASP arrays to look for genetic markers of immune responses in Atlantic salmon (LeBlanc et al., 2010), as well as of thermal stress in rainbow trout (Lewis et al., 2010) and Arctic charr (Quinn et al., 2011a, 2011b), and to predict spawning failure in wild sockeye populations (Miller et al., 2011).

Despite their utility, however, microarrays (particularly those based on cDNA clones) have inherent drawbacks, such that they are resource intensive and highly susceptible to technical variations as well as statistical pitfalls (Martin et al., 2007). The latter may result in Type I errors (false positive results), a particular concern when using a large microarray, as the false-positive rate increases as the number of spots increases. Thus, often microarray data are confirmed, or validated, by selecting a few genes to test using qPCR. Another challenge is the potential for cross-hybridization between similar transcripts, which is a problem of particular concern for duplicated genomes, or for large gene families, such as that of the hemoglobin genes, which contains multiple members of high sequence similarity (Quinn et al., 2010). This can mute out evidence of differential expression between individual transcripts, and instead, produce the general trends for all highly similar genes on the microarray. In addition, this phenomenon may cause information to be lost if the elevated expression of a small number of family members is spread among all similar genes on the microarray such that no spots meet the statistical filters assigned. The use of transcript-specific oligonucleotide arrays may overcome some of these challenges. For these reasons, profiling using microarrays is sometimes described as an exploratory tool, for which the data must be interpreted with a caution, or which must be followed up or correlated with other types of analysis (Lewis et al., 2010; Quinn et al., 2011a).

3.7 Bioinformatics and genomics databases

The advancement of genomics technologies has resulted in the generation of vast amounts of biological data. This has fueled the need for computational systems and techniques to manage such data as well as for data analysis, the development of which has led to the advent of the field of bioinformatics. Specifically, bioinformatics refers to the application of computer science and information technology to biology to increase the understanding of biological processes. Bioinformatics and genomics now go hand in hand; indeed, often the bottlenecks for obtaining and interpreting genomics results occur in the bionformatics steps, as high throughput genomics technologies produce more data than the available computational tools can handle. Individuals specializing both in biology and computer science are becoming increasingly in demand. For genomics, the general goals of bioinformatics include the development of computational tools for data management (e.g., databases and interactive online tools), data manipulation (map assembly, sequence/genome assembly, protein structure alignment) and data mining (e.g., pattern recognition, genome annotation, sequence comparisons, predictions of gene expression and genome-wide association studies). Given that the fundamental goals of genomics research are shared across all genomes, standard computational tools can often be used for numerous genomes, requiring only minor tweaking and manipulating to accommodate the specific resources for the species of interest.

The concept of "open data", or public accessibility is generally adopted for all genomics projects, and most genomics data are posted online in publically accessible databases. Indeed, it is a requirement for publication in most peer-reviewed academic journals that sequences as well as gene expression data be made available online before any results can be considered for publication within the journal. Besides preventing redundant work and enabling research groups to build off of the work of others, open access data allows the potential for cross-species comparisons, which is particularly true among the Salmonidae. That is, online, public databases such as those managed by NCBI (National Center for Biotechnology Information), Ensembl (Ensembl Genome Browser), and GEO (Gene Expression Omnibus) provide an opportunity to make predictions about the genes in closely related species for which there are few genomic resources available (e.g., Arctic charr). Furthermore, for most species for which there is a large genomics program, a species-specific database will be generated. For example, the Atlantic salmon database, publically available at www.asalbase.org, comprises both the linkage and physical maps, as well as all of the BAC-end sequences and marker information. Also included are tools that enable comparative synteny analyses between Atlantic salmon and the other sequenced species. The next step for Asalbase will be integrating the sequence scaffolds for the whole genome sequence. Finally, the last decade saw many research labs pack away their pipettes and centrifuges and replace them with large computer networks and powerful servers, as the wealth of genomics data available online now allows entire experiments, such as comparative synteny analyses, evolutionary studies and even gene expression meta-analyses to be conducted "*in silico*", or with computers alone. This not only greatly improves the speed of research, but also reduces the cost and number of live specimens, or research animals required.

4. MAS in aquaculture

4.1 Examples of QTL-based markers for aquaculture

Over the last decade breeders have begun to include molecular genetics in their strategies for genetically improving salmonids for aquaculture production efficiency. Markers that exhibit strong associations with QTL that play a substantial role in a desirable trait can be integrated into MAS programs. That is, rather than using trial and error to cross individuals that exhibit phenotypic traits of interest, breeders can purposely select individuals that exhibit the marker associated with the desirable QTL. This fine-scale, targeted approach to artificial selection has the potential to be substantially more precise, efficient and effective than traditional, phenotype based approaches.

One example of an economically important trait for which there is a QTL is Infectious Salmon Anemia (ISA) resistance in Atlantic salmon. Previous studies determined that there is a genetic component to ISA resistance in Atlantic salmon (Gjøen & Bentsen 1997, Ødegård et al., 2007) and a QTL for this trait was mapped (Moen et al., 2004). A more recent study confirmed this QTL and showed that it explains 6% of the phenotypic variation and 32-47% of the additive genetic variation (Moen et al., 2007). A comparative genomics analysis of the most likely QTL region suggested several candidate genes for ISA resistance (Li et al., 2011a). These genes remain to be validated based on the association of polymorphisms within them and the trait. However, it is anticipated that a combination of fine mapping and genomic sequence information will identify the causative allele that confers resistance and that this will be incorporated into a MAS breeding program.

Another excellent example of the potential of MAS involves resistance to a pathogen that causes Infectious Pancreatic Necrosis (IPN) in Atlantic salmon. A major QTL for IPN resistance in Atlantic salmon has been identified and confirmed (Gheyas et al., 2010; Houston et al., 2008; Moen et al., 2009a). A four-marker haplotype has been identified that strongly segregates with IPN resistance in Norwegian Atlantic salmon broodstock, providing a solid framework for linkage-based MAS in this population (Moen et al., 2009a).

These results support the prediction that selective breeding for resistance ISA and IPN (caused by viruses) as well as furunculosis (caused by a bacterium) should be possible given the heritabilities for these traits (Kjoglum et al., 2008). However, the speed at which selective breeding for these traits can be performed efficiently and effectively will depend on the ability to identify the underlying genetic basis for the resistance, and this requires a reference sequence for Atlantic salmon. Other traits of interest for QTL identification for MAS in aquaculture include tissue quality, the ability to thrive on commercial diets (i.e., increased vegetable oil content, pathway analysis for vaccine development, and ability/suitability for high-density rearing conditions. Overall, integrating MAS into broodstock development programs can provide an effective and efficient, science-based means of minimizing animal stress while maximizing growth, health and sustainability

4.2 Challenges facing MAS for aquaculture

Currently, given the lack of fully sequenced genomes for aquaculture species, genomics-based selection, or MAS is fully dependent on a having a suite of well-developed genomics tools for the genome of the species of interest. In addition to the cost associated with the development of these tools, the complex nature of fish genomes combined with the large number of diverse fish species of interest for aquaculture complicates this goal. However, the availability of fully sequenced reference genomes for a few key species groups would go a long way to fill in the major under-represented gaps in the fish phylogeny. As illustrated in Figure 1, there is no reference genome from a closely related species for either the

Salmoniformes or the Gadiformes. However, work on the genomes of Atlantic salmon (Davidson et al., 2010) and Atlantic cod (Johansen et al., 2009) is in progress, and the results are eagerly anticipated by the aquaculture community that breeds these and related species.

5. Role of whole genome sequencing (WGS) in genomics

5.1 Benefits of whole genome sequences

The availability of a complete genome sequence for Atlantic salmon will have a major impact on all sectors of the international salmonid community. For the aquaculture industry it will provide a complete suite of genetic markers for the identification of the genes and alleles responsible for production traits. Companies will be able to develop tailored broodstock using nucleotide or allele assisted selection rather than the more general marker assisted selection. In conjunction with traditional breeding practices, this approach promises rapid gains that will make companies who embrace this technology more competitive.

5.2 Methods and approaches to whole genome sequencing

The conventional approach to sequencing whole genomes is to break the genome into small parts, and sequence each part using Sanger sequencing (reviewed in Hutchison, 2007). Although this approach remains the gold standard for sequence and assembly quality, limitations with respect to cost, labor-intensiveness and speed have fueled the demand for new approaches to DNA sequencing. In recent years, several novel high-throughput sequencing platforms, or 'next generation' sequencing approaches have entered the market (for a review, see Metzker, 2010). Most of these are targeted to the goal of re-sequencing an entire human genome for <$1,000, and their capabilities with respect to sequencing whole genomes *de novo* (i.e., for the first time, and without a reference sequence for comparative assembly), remain unknown. This is complicated by the diverse, highly repetitive and overall complex nature of the fish genomes given that, these technologies are currently limited by their sequence read lengths and difficulties with specific types of repetitive sequences. Generally, the next generation sequencing platforms are capable of producing vast amounts of sequence data in very short periods of time for significantly lower cost than traditional Sanger sequencing, but the bottleneck lies in developing the computational tools to handle the data, and ultimately, to assemble the short sequence reads into contiguous (gap-free) stretches of DNA sequence when no reference genome is available (Quinn et al., 2008).

Aside from their potential contribution to whole genome sequencing, next generation sequencing technologies have countless implications for genomics. One of the most common uses for next generation sequencing in human genomics is re-sequencing, i.e., sequencing the genomes of individuals for which there is a full reference sequence. This could be applied to aquaculture to look for allelic variations between individuals of the same species that show differences in characteristics of interest (e.g., disease resistance or growth). Similarly, next generation sequencing has been used for "transcriptome sequencing" (also referred to as "transcriptomics") i.e., sequencing the entire repertoire of transcribed RNA of a species. This is a form of expression profiling, as the relative number of transcripts of genes can be compared against one another, thus providing insight into the expression levels of the genes at a given time. This approach is being used much more extensively as the cost of next generation sequencing technologies decreases. Finally, next generation sequencing provides an efficient and effective means identifying molecular markers, including microsatellites, SNPs (Chan, 2009) and RAD (Restriction site Associated DNA) markers, which are markers (e.g., SNPs) contained within RAD tags, or the sequences immediately flanking specific DNA sites that are targeted by restriction enzymes. The use of next generation sequencing technologies to sequence RAD tags to identify RAD markers is referred to as RADseq, and its popularity is increasing quickly given its utility for identifying individual-specific markers, and thus to generate an indexed marker library for an individual organism. Therefore, as the capabilities and availability of next generation technologies continue to advance, we are sure to see drastic changes in the approaches to genomics centered around high-throughput, next generation technologies as well as bioinformatics-based analyses. These advances will provide vast opportunities for integrating genomics into aquaculture breeding programs.

6. Conclusion

Genomics has become an increasingly powerful field with far-reaching implications from medicine to pharmacology to industry to ecology. When applied to aquaculture, genomics has vast potential to improve factors such as production, efficiency, animal welfare, sustainability, reducing inbreeding, and increasing the profitability of the industry. The current genomics tools for fish, and specifically for aquaculture species, have already enabled drastic improvements to characteristics such as growth and disease resistance, and show great promise for future developments; however, their utility remains limited by the lack of whole genome sequences for a few key species, which could provide reference sequences for other aquaculture species. As these sequences, particularly those of Atlantic salmon, rainbow trout and Atlantic cod, become available over the next several years, which will certainly correspond to further advances in technology and computational tools, the potential for genomics to facilitate major developments in aquaculture is enormous.

7. Glossary of terms

Alleles: Alternative forms (two or more) of a gene, located at a specific position on a chromosome.

Annotation: Identification of the genomic position of intron-exon boundaries, regulatory sequences, repeats, gene names and protein products within a DNA sequence.

Bioinformatics: refers to the application of computer science and information technology to developing computationally intensive techniques to increase the understanding of biological processes.

Complex trait: Trait influenced by more than one genetic and/or environmental factor.

cDNA: Complementary DNA, which is DNA synthesized from a mature mRNA template by reverse transcription.

DNA sequence: Linear representation of a DNA strand by the presence of the four nucleotide bases (A, T, C, G).

Ensembl Genome Browser: A joint project between EMBL-EBI and the Wellcome Trust Sanger Institute to develop a software system that produces and maintains automatic annotation on selected eukaryotic genomes (www. ensembl.org).

EST: Expressed sequence tag, which is a short sequence obtained from one shot sequencing of cDNA, corresponding to a fragment (~500 bp) of an expressed gene.

Gene: Name given to stretches of DNA that code for a specific protein for a specific characteristic or function. It represents the heredity unit in living organisms.

Gene Expression Omibus (GEO): A database repository of high-throughput gene expression data and hybridization arrays, chips and microarrays.

Generation time: The average interval between the birth of an individual and the birth of its offspring.

Genome: Is the entirety of an organism's hereditary information. It is encoded either in DNA or RNA (some viruses).

Genomics: Discipline in genetics concerning the study of the genomes of organisms.

Genotype: The total set of alleles possessed by an organism that determines a specific characteristic or trait. It comprises the entire complex of genes inherited from both parents.

Heritability: Proportion of phenotypic variation in a population that is due to genetic variation between individuals.

High-throughput: Processes usually performed via increased levels of automation and robotics. In sequencing: involves the application of rapid sequencing technology at the scale of whole genomes.

Locus: (pl. loci) Specific location of a gene or DNA sequence on a chromosome.

Marker-assisted selection (MAS): The use of DNA markers linked to traits of interest to assist in the selection of individuals for breeding purposes.

Molecular marker: Specific fragments of DNA that can be identified within the whole genome. These can be associated with the position of a particular gene or the inheritance of a particular characteristic.

Molecular pathway: Series of molecular processes that are connected by their intermediates such that the products of one process may trigger or participate in the other.

Morphology: The visible form and structure of living organisms.

mRNA: Molecule of RNA encoding a specific protein product. mRNA is transcribed from a DNA template in the cell nucleus then moves to the cytoplasm where it is translated by the ribosomes.

NCBI: The National Center for Biotechnology Information, which houses the world's biggest sequence datasets in GenBank and an index of biomedical research articles in PubMed.

Non-coding: Components of an organism's genome (DNA sequences) that do not encode for protein sequences.

Phenotype: Observable physical or biochemical characteristics of an organism, as determined by both genetic makeup and environmental influences.

Polymerase Chain Reaction (PCR): Scientific technique designed to amplify a DNA fragment across several orders of magnitude, generating thousands to millions of copies of a particular DNA sequence

Quantitative trait locus (QTL): Stretch of DNA containing or linked to genes that underlie a quantitative trait of interest.

RAD: Restriction site Associated DNA. RAD tags are the sequences that immediately flank a restriction enzyme site. RAD markers refer to sequence variations identified within RAD tags that can be used as molecular markers (e.g., SNPs).

Reference genome: Nucleic acid sequence database, assembled into a whole genome and representative of a species' genetic code. Typically used as a guide on which new genomes are built.

Restriction enzyme: An enzyme that cuts DNA at a specific recognition site/sequence, referred to as a restriction site.

RNA: Ribonucleic acid. Single stranded molecule transcribed from one strand of DNA. Unlike DNA, the sugar in RNA is ribose and one of the four bases, T (thymine) is replaced by U (uracil). There are several types of RNA mainly involved in protein synthesis (mRNA, tRNA, rRNA).

Sequence assembly: Refers to aligning and merging sequence fragments into a longer sequence.

Synteny: Physical co-localization of genetic loci on the same chromosome within an individual or species.

Transcription: Process involved in the synthesis of mRNA from a DNA template, catalyzed by RNA polymerase.

Translation: Process in which the messenger RNA (mRNA) produced by transcription is decoded by the ribosome to produce a specific amino acid chain, or polypeptide, that will later fold into an active protein.

Whole genome sequence: The complete DNA sequence of the genome of an organism.

8. References

Allendorf, F.W. & Thorgaard, G.H. (1984). Tetraploidy and the Evolution of Salmonid Fishes. in Evolutionary Genetics of Fishes, Plenum Press, pp. 55, ISBN 978-0306415203, New York, USA.

Anti Vasemagi, Jan Nilsson, Craig R. Primmer (2005). Seventy-five EST-linked Atlantic Salmon (*Salmo salar* L.) Microsatellite Markers and their Cross-amplification in Five Salmonid Species. Molecular Ecology Notes, Vol. 5, No. 2, pp. 282-288, ISSN 1471-8286.

Baranski, M., Moen, T. & Vage, D.I. (2010). Mapping of Quantitative Trait Loci for Flesh Colour and Growth Traits in Atlantic Salmon (*Salmo salar*). Genetics, Selection, Evolution, Vol. 42, No. 1, pp. 17. ISSN 1297-9686.

Boulding, E.G., Culling, M., Glebe, B., Berg, P.R., Lien, S. & Moen, T. (2008). Conservation Genomics of Atlantic Salmon: SNPs Associated with QTLs for Adaptive Traits in Parr from Four Trans-Atlantic Backcrosses. Heredity, Vol. 101, No. 4, pp. 381-391, ISSN 0018-067X.

Chan, E.Y. (2009). Next-Generation Sequencing Methods: Impact of Sequencing Accuracy on SNP Discovery. In Single Nucleotide Polymorphisms, Humana Press, pp. 95-111, ISBN 978-1-60327-411-1.

Chistiakov, D.A., Hellemans, B., Haley, C.S., Law, A.S., Tsigenopoulos, C.S., Kotoulas, G., Bertotto, D., Libertini, A. & Volckaert, F.A. (2005). A Microsatellite Linkage Map of the European Sea Bass *Dicentrarchus labrax* L. Genetics, Vol. 170, No. 4, pp. 1821-1826, ISSN 0016-6731.

Coimbra, M.R.M., Kobayashi, K., Koretsugu, S., Hasegawa, O., Ohara, E., Ozaki, A., Sakamoto, T., Naruse, K. & Okamoto, N. (2003). A genetic Linkage Map of the Japanese Flounder, *Paralichthys olivaceus*. Aquaculture, Vol. 220, No. 1-4, pp. 203, ISSN 0044-8486.

Crick, F. (1970). Central Dogma of Molecular Biology. Nature, Vol. 227, No. 5258, pp. 561-563, ISSN 0028-0836.

Danzmann, R.G., Cairney, M., Davidson, W.S., Ferguson, M.M., Gharbi, K., Guyomard, R., Holm, L.E., Leder, E., Okamoto, N., Ozaki, A., Rexroad, C.E.,3rd, Sakamoto, T., Taggart, J.B. & Woram, R.A.(2005). A Comparative Analysis of the Rainbow Trout Genome with 2 Other Species of Fish (Arctic charr and Atlantic salmon) Within the Tetraploid Derivative Salmonidae Family (subfamily: Salmoninae). Genome, Vol. 48, No. 6, pp. 1037-1051, ISSN 0831-2796.

Danzmann, R.G., Davidson, E.A., Ferguson, M.M., Gharbi, K., Koop, B.F., Hoyheim, B., Lien, S., Lubieniecki, K.P., Moghadam, H.K., Park, J., Phillips, R.B. & Davidson, W.S. (2008). Distribution of Ancestral Proto-Actinopterygian Chromosome Arms Within the Genomes of 4R-derivative Salmonid Fishes (Rainbow trout and Atlantic salmon). BMC Genomics, Vol. 9, No.1, ISSN 1471-2164.

Davidson, W.S., Koop, B.F., Jones, S.J., Iturra, P., Vidal, R., Maass, A., Jonassen, I., Lien, S. & Omholt, S.W. (2010). Sequencing the Genome of the Atlantic Salmon (*Salmo salar*). Genome Biology, Vol. 11, No. 9, pp. 403, ISSN 1465-6906.

Dominik, S., Henshall, J.M., Kube, P.D., King, H., Lien, S., Kent, M.P. & Elliott, N.G.(2010). Evaluation of an Atlantic Salmon SNP Chip as a Genomic Tool for the Application in a Tasmanian Atlantic Salmon (*Salmo salar*) Breeding Population. Aquaculture, Vol. 308, Suppl. 1, pp. S56, ISSN 0044-8486.

Garcia de Leaniz, C., Fleming, I.A., Einum, S., Verspoor, E., Jordan, W.C., Consuegra, S., Aubin-Horth, N., Lajus, D., Letcher, B.H., Youngson, A.F., Webb, J.H., VÃ‚llestad, L.A., Villanueva, B., Ferguson, A. & Quinn, T.P. (2007). A Critical Review of Adaptive Genetic Variation in Atlantic Salmon: Implications for Conservation. Biological Reviews, Vol. 82, No. 2, pp. 173-211, ISSN 1469-185X.

Genet, C., Dehais, P., Palti, Y., Gao, G., Gavory, F., Wincker, P., Quillet, E. & Boussaha, M. (2011). Analysis of BAC-end Sequences in Rainbow Trout: Content

Characterization and Assessment of Synteny Between Trout and Other Fish Genomes. BMC Genomics, Vol. 12, No. 1, ISSN 1471-2164.

"genomics" Mirriam-Webster.com, Mirriam-Webster, 2011. Web. 27 June 2011.

Gharbi, K., Gautier, A., Danzmann, R.G., Gharbi, S., Sakamoto, T., Hoyheim, B., Taggart, J.B., Cairney, M., Powell, R., Krieg, F., Okamoto, N., Ferguson, M.M., Holm, L.E. & Guyomard, R. (2006). A Linkage Map for Brown Trout (Salmo trutta): Chromosome Homeologies and Comparative Genome Organization With Other Salmonid Fish. Genetics, Vol. 172, No. 4, pp. 2405-2419, ISSN 0016-6731.

Gheyas, A.A., Haley, C.S., Guy, D.R., Hamilton, A., Tinch, A.E., Mota-Velasco, J.C. & Woolliams, J.A. (2010). Effect of a Major QTL Affecting IPN Resistance on Production Traits in Atlantic Salmon. Animal Genetics, Vol. 41, No. 6, pp. 666-668, ISSN 1365-2052.

Gilbey, J., Verspoor, E., McLay, A. & Houlihan, D. (2004). A Microsatellite Linkage Map for Atlantic Salmon (Salmo salar). Animal Genetics, Vol. 35, No. 2, pp. 98-105, ISSN 1365-2052.

Gilbey, J., Verspoor, E., Mo, T.A., Sterud, E., Olstad, K., Hytterod, S., Jones, C. & Noble, L. (2006). Identification of Genetic Markers Associated with Gyrodactylus salaris Resistance in Atlantic Salmon Salmo salar. Diseases of Aquatic Organisms, Vol. 71, No. 2, pp. 119-129, ISSN 0177-5103.

Gjøen, H.M. & Bentsen, H.B. (1997). Past, Present, and Future of Genetic Improvement in Salmon Aquaculture. ICES Journal of Marine Science: Journal du Conseil, Vol. 54, No. 6, pp. 1009-1014, ISSN 1095-9289.

Goddard, M.E. & Hayes, B.J. (2009). Mapping Genes for Complex Traits in Domestic Animals and Their use in Breeding Programmes. Nature Reviews. Genetics, Vol. 10, No. 6, pp. 381-391, ISSN 1471-0056.

Groenen, M.A., Wahlberg, P., Foglio, M., Cheng, H.H., Megens, H.J., Crooijmans, R.P., Besnier, F., Lathrop, M., Muir, W.M., Wong, G.K., Gut, I. & Andersson, L. (2009). A High-density SNP-based Linkage Map of the Chicken Genome Reveals Sequence Features Correlated with Recombination Rate. Genome Research, Vol. 19, No. 3, pp. 510-519, ISSN 1088-9051.

Gupta, P.K., Mir, R.R., Mohan, A. & Kumar, J. (2008). Wheat Genomics: Present Status and Future Prospects. International Journal of Plant Genomics, Vol. 2008, pp.36, ISSN 1687-5389.

Guyomard, R., Mauger, S., Tabet-Canale, K., Martineau, S., Genet, C., Krieg, F. & Quillet, E. (2006). A Type I and Type II Microsatellite Linkage Map of Rainbow Trout (Oncorhynchus mykiss) with Presumptive Coverage of All Chromosome Arms. BMC Genomics, Vol. 7, No. 1, ISSN 1471-2164.

Haidle, L., Janssen, J., Gharbi, K., Moghadam, H., Ferguson, M. & Danzmann, R. (2008). Determination of Quantitative Trait Loci (QTL) for Early Maturation in Rainbow Trout (Oncorhynchus mykiss). Marine Biotechnology, Vol. 10, No. 5, pp. 579-592, ISSN 1436-2228.

Hayes, B.J., Nilsen, K., Berg, P.R., Grindflek, E. & Lien, S. (2007). SNP Detection Exploiting Multiple Sources of Redundancy in Large EST Collections Improves Validation Rates. Bioinformatics, Vol. 23, No. 13, pp. 1692-1693, ISSN 1367-4803.

Houle, D., Govindaraju, D.R. & Omholt, S. (2010). Phenomics: The Next Challenge. Nature Reviews. Genetics, Vol. 11, No. 12, pp. 855-866, ISSN 1471-0056.

Houston, R.D., Bishop, S.C., Hamilton, A., Guy, D.R., Tinch, A.E., Taggart, J.B., Derayat, A., McAndrew, B.J. & Haley, C.S. (2009). Detection of QTL Affecting Harvest Traits in a Commercial Atlantic Salmon Population. Animal Genetics, Vol. 40, No. 5, pp. 753-755, ISSN 1365-2052.

Houston, R.D., Gheyas, A., Hamilton, A., Guy, D.R., Tinch, A.E., Taggart, J.B., McAndrew, B.J., Haley, C.S. & Bishop, S.C. (2008). Detection and Confirmation of a Major QTL Affecting Resistance to Infectious Pancreatic Necrosis (IPN) in Atlantic Salmon (Salmo salar). Developments in Biologicals, Vol. 132, pp. 199-204, ISSN 1424-6074.

Houston, R.D., Haley, C.S., Hamilton, A., Guy, D.R., Mota-Velasco, J.C., Gheyas, A.A., Tinch, A.E., Taggart, J.B., Bron, J.E., Starkey, W.G., McAndrew, B.J., Verner-Jeffreys, D.W., Paley, R.K., Rimmer, G.S., Tew, I.J. & Bishop, S.C. (2010). The susceptibility of Atlantic Salmon Fry to FreshwaterI pancreatic Necrosis is Largely Explained by a Major QTL. Heredity, Vol. 105, No. 3, pp. 318-327, ISSN 0018-067X.

Houston, R.D., Haley, C.S., Hamilton, A., Guy, D.R., Tinch, A.E., Taggart, J.B., McAndrew, B.J. & Bishop, S.C. (2008). Major Quantitative Trait Loci Affect Resistance to Infectious Pancreatic Necrosis in Atlantic Salmon (Salmo salar). Genetics, Vol. 178, No. 2, pp. 1109-1115, ISSN 0016-6731.

Hubert, S., Higgins, B., Borza, T. & Bowman, S. (2010). Development of a SNP Resource and a Genetic Linkage Map for Atlantic Cod (Gadus morhua). BMC Genomics, Vol. 11, No. 1, ISSN 1471-2164.

Hutchison, C.A.,3rd (2007). DNA Sequencing: Bench to Bedside and Beyond. Nucleic Acids Research, Vol. 35, No. 18, pp. 6227-6237, ISSN 1362-4962.

Johansen, S.D., Coucheron, D.H., Andreassen, M., Karlsen, B.O., Furmanek, T., Jorgensen, T.E., Emblem, A., Breines, R., Nordeide, J.T., Moum, T., Nederbragt, A.J., Stenseth, N.C. & Jakobsen, K.S. (2009). Large-scale Sequence Analyses of Atlantic Cod. New Biotechnology, Vol. 25, No. 5, pp. 263-271, ISSN 1871-6784.

Jones, C.S., Lockyer, A.E., Verspoor, E., Secombes, C.J. & Noble, L.R. (2002). Towards Selective Breeding of Atlantic Salmon for Sea Louse Resistance: Approaches to Identify Trait Markers. Pest Management Science, Vol. 58, No. 6, pp. 559-568, ISSN 1526-4998.

Kang, J.H., Kim, W.J. & Lee, W.J. (2008). Genetic Linkage Map of Olive Flounder, Paralichthys olivaceus. International Journal of Biological Sciences, Vol. 4, No. 3, pp. 143-149, ISSN 1449-2288.

Karlsson, S., Moen, T., Lien, S., Glover, K.A. & Hindar, K. (2011). Generic Genetic Differences Between Farmed and Wild Atlantic Salmon Identified from a 7K SNP-chip. Molecular Ecology Resources, Vol. 11, Suppl. 1, pp. 247-253, ISSN 1755-098X.

Kim, D.H., Kim, K.H., Van, K., Kim, M.Y. & Lee, S.H. (2010). Fine Mapping of a Resistance Gene to Bacterial Leaf Pustule in Soybean. Theoretical and Applied Genetics.Theoretische und Angewandte Genetik, Vol. 120, No. 7, pp. 1443-1450, ISSN 0040-5752.

King, T.L., Kalinowski, S.T., Schill, W.B., Spidle, A.P. & Lubinski, B.A. (2001). Population Structure of Atlantic Salmon (Salmo salar L.): A Range-wide Perspective from

Microsatellite DNA Variation. Molecular Ecology, Vol. 10, No. 4, pp. 807-821, ISSN 0962-1083.

Kjoglum, S., Larsen, S., Bakke, H.G. & Grimholt, U. (2008). The Effect of Specific MHC Class I and Class II Combinations on Resistance to Furunculosis in Atlantic Salmon (*Salmo salar*). Scandinavian Journal of Immunology, Vol. 67, No. 2, pp. 160-168, ISSN 0300-9475.

Koop, B.F., von Schalburg, K.R., Leong, J., Walker, N., Lieph, R., Cooper, G.A., Robb, A., Beetz-Sargent, M., Holt, R.A., Moore, R., Brahmbhatt, S., Rosner, J., Rexroad, C.E.,3rd, McGowan, C.R. & Davidson, W.S. (2008). A Salmonid EST Genomic Study: Genes, Duplications, Phylogeny and Microarrays. BMC Genomics, Vol. 9, No. 1, ISSN 1471-2164.

Kucuktas, H., Wang, S., Li, P., He, C., Xu, P., Sha, Z., Liu, H., Jiang, Y., Baoprasertkul, P., Somridhivej, B., Wang, Y., Abernathy, J., Guo, X., Liu, L., Muir, W. & Liu, Z. (2009). Construction of Genetic Linkage Maps and Comparative Genome Analysis of Catfish Using Gene-associated Markers. Genetics, Vol. 181, No. 4, pp. 1649-1660, ISSN 0016-6731.

Kuhl, H., Sarropoulou, E., Tine, M., Kotoulas, G., Magoulas, A. & Reinhardt, R. (2011). A Comparative BAC Map for the Gilthead Sea Bream (*Sparus aurata* L.). Journal of Biomedicine & Biotechnology, Vol. 2011, 7 pp, ISSN 1110-7243

Kuttner, E., Moghadam, H.K., Skulason, S., Danzmann, R.G. & Ferguson, M.M. (2011). Genetic Architecture of Body Weight, Condition Factor and Age of Sexual Maturation in Icelandic Arctic Charr (*Salvelinus alpinus*). Molecular Genetics and Genomics, Vol. 286, No. 1, pp. 67-79, ISSN 1617-4623.

LeBlanc, F., Laflamme, M. & Gagne, N. (2010). Genetic Markers of the Immune Response of Atlantic Salmon (*Salmo salar*) to Infectious Salmon Anemia Virus (ISAV). Fish & Shellfish Immunology, Vol. 29, No. 2, pp. 217-232, ISSN 1050-4648.

Lee, B.Y., Lee, W.J., Streelman, J.T., Carleton, K.L., Howe, A.E., Hulata, G., Slettan, A., Stern, J.E., Terai, Y. & Kocher, T.D. (2005). A Second-generation Genetic Linkage Map of Tilapia (*Oreochromis spp.*). Genetics, Vol. 170, No. 1, pp. 237-244, ISSN 0016-6731.

Lewis, J.M., Hori, T.S., Rise, M.L., Walsh, P.J. & Currie, S. (2010). Transcriptome Responses to Heat Stress in the Nucleated Red Blood Cells of the Rainbow Trout (*Oncorhynchus mykiss*). Physiological Genomics, Vol. 42, No. 3, pp. 361-373, ISSN 1094-8341.

Li, J., Boroevich, K.A., Koop, B.F. & Davidson, W.S. (2011a). Comparative Genomics Identifies Candidate Genes for Infectious Salmon Anemia (ISA) Resistance in Atlantic Salmon (*Salmo salar*). Marine Biotechnology, Vol. 13, No. 2, pp. 232-241, ISSN 1436-2228.

Li, Y., Xu, P., Zhao, Z., Wang, J., Zhang, Y. & Sun, X.W. (2011b). Construction and Characterization of the BAC Library for Common Carp *Cyprinus Carpio* L. And Establishment of Microsynteny with Zebrafish *Danio Rerio*. Marine Biotechnology, Vol. 13, No. 4, pp. 706-712, ISSN 1436-2228.

Lien, S., Gidskehaug, L., Moen, T., Hayes, B.J., Berg, P.R., Davidson, W.S., Omholt, S.W., & Kent, M.P. (2011). A dense SNP-based linkage map for Atlantic salmon (*Salmo salar*) reveals extended chromosome homeologies and striking difference in sex-specific recombination rates. BMC Genomics (in press).

Liu, G., Jennen, D.G., Tholen, E., Juengst, H., Kleinwachter, T., Holker, M., Tesfaye, D., Un, G., Schreinemachers, H.J., Murani, E., Ponsuksili, S., Kim, J.J., Schellander, K. & Wimmers, K.(2007). A Genome Scan Reveals QTL for Growth, Fatness, Leanness and Meat Quality in a Duroc-Pietrain Resource Population. Animal Genetics, Vol. 38, No. 3, pp. 241-252, ISSN 1365-2052.

Liu, H., Jiang, Y., Wang, S., Ninwichian, P., Somridhivej, B., Xu, P., Abernathy, J., Kucuktas, H. & Liu, Z. (2009). Comparative Analysis of Catfish BAC end Sequences with the Zebrafish Genome. BMC Genomics, Vol. 10, No. 1, ISSN 1471-2164.

Lorenz, S., Brenna-Hansen, S., Moen, T., Roseth, A., Davidson, W.S., Omholt, S.W. & Lien, S. (2010). BAC-based Upgrading and Physical Integration of a Genetic SNP Map in Atlantic Salmon. Animal Genetics, Vol. 41, No. 1, pp. 48-54, ISSN 1365-2052.

Lynch, M. & Walsh, B. (1998), Genetics and Analysis of Quantitative Traits, Sinauer Associates Inc., ISBN 0-87893-481-2, Sunderland, USA.

Martin, L.J., Woo, J.G., Avery, C.L., Chen, H.S., North, K.E., Au, K., Broet, P., Dalmasso, C., Guedj, M., Holmans, P., Huang, B., Kuo, P.H., Lam, A.C., Li, H., Manning, A., Nikolov, I., Sinha, R., Shi, J., Song, K., Tabangin, M., Tang, R. & Yamada, R. (2007). Multiple Testing in the Genomics Era: Findings From Genetic Analysis Workshop 15, Group 15. Genetic Epidemiology, Vol. 31, Suppl. 1, pp. S124-131, ISSN 0741-0395.

Martyniuk, C.J., Perry, G.M.L., Mogahadam, H.K., Ferguson, M.M. & Danzmann, R.G. (2003). The Genetic Architecture of Correlations Among Growth-related Traits and Male Age at Maturation in Rainbow Trout. Journal of Fish Biology, Vol. 63, No. 3, pp. 746-764, ISSN 1095-8649.

Matukumalli, L.K., Lawley, C.T., Schnabel, R.D., Taylor, J.F., Allan, M.F., Heaton, M.P., O'Connell, J., Moore, S.S., Smith, T.P., Sonstegard, T.S. & Van Tassell, C.P. (2009). Development and Characterization of a High Density SNP Genotyping Assay for Cattle. PloS One, Vol. 4, No. 4, pp. e5350, ISSN 1932-6203.

McClelland, E.K. & Naish, K.A. (2010). Quantitative Trait Locus Analysis of Hatch Timing, Weight, Length and Growth Rate in Coho Salmon, Oncorhynchus kisutch. Heredity, Vol. 105, No. 6, pp. 562-573, ISSN 0018-067X.

McClelland, E.K. & Naish, K.A. (2008). A Genetic Linkage Map for Coho Salmon (Oncorhynchus kisutch). Animal Genetics, Vol. 39, No. 2, pp. 169-179, ISSN 1365-2052.

McConnell, S.K., O'Reilly, P., Hamilton, L., Wright, J.M. & Bentzen, P. (1995). Polymorphic Microsatellite Loci from Atlantic Salmon (Salmo salar): Genetic Differentiation of North American and European Populations. Canadian Journal of Fisheries and Aquatic Sciences, Vol. 52, No. 9, pp. 1863-1872, ISSN 1205-7533.

McGowan, C.R., Davidson, E.A., Woram, R.A., Danzmann, R.G., Ferguson, M.M. & Davidson, W.S. (2004). Ten Polymorphic Microsatellite Markers from Arctic Charr (Salvelinus alpinus): Linkage Analysis and Amplification in Other Salmonids. Animal Genetics, Vol. 35, No. 6, pp. 479-481, ISSN 1365-2052.

Metzker, M.L. (2010). Sequencing Technologies - The Next Generation. Nature Reviews. Genetics, Vol. 11, No. 1, pp. 31-46, ISSN 1471-0056.

Miller, K.M., Li, S., Kaukinen, K.H., Ginther, N., Hammill, E., Curtis, J.M., Patterson, D.A., Sierocinski, T., Donnison, L., Pavlidis, P., Hinch, S.G., Hruska, K.A., Cooke, S.J.,

English, K.K. & Farrell, A.P. (2011) "Genomic Signatures Predict Migration and Spawning Failure in Wild Canadian Salmon. Science, Vol. 331, No. 6014, pp. 214-217, ISSN 0036-8075.

Moen, T., Baranski, M., Sonesson, A.K. & Kjoglum, S. (2009a), "Confirmation and Fine-mapping of a Major QTL for Resistance to Infectious Pancreatic Necrosis in Atlantic Salmon (Salmo salar): Population-level Associations Between Markers and Trait", BMC Genomics, Vol.10, No.1, ISSN 1471-2164.

Moen, T., Delghandi, M., Wesmajervi, M., Westgaard, J. & Fjalestad, K. (2009b), "A SNP/Microsatellite Genetic Linkage Map of the Atlantic Cod (Gadus morhua)", Animal Genetics, Vol. 40, No.6, pp. 993-996, ISSN 1365-2052.

Moen, T., Fjalestad, K.T., Munck, H. & Gomez-Raya, L. (2004). A Multistage Testing Strategy for Detection of Quantitative Trait Loci Affecting Disease Resistance in Atlantic Salmon. Genetics, Vol. 167, No. 2, pp. 851-858, ISSN 0016-6731.

Moen, T., Hayes, B., Baranski, M., Berg, P.R., Kjoglum, S., Koop, B.F., Davidson, W.S., Omholt, S.W. & Lien, S. (2008a). A Linkage Map of the Atlantic Salmon (Salmo salar) Based on EST-derived SNP Markers. BMC Genomics, Vol. 9, No.1, ISSN 1471-2164.

Moen, T., Hayes, B., Nilsen, F., Delghandi, M., Fjalestad, K.T., Fevolden, S.E., Berg, P.R. & Lien, S. (2008b). Identification and Characterisation of Novel SNP Markers in Atlantic Cod: Evidence for Directional Selection. BMC Genetics, Vol. 9, No.1, ISSN 1471-2156.

Moen, T., Hoyheim, B., Munck, H. & Gomez-Raya, L. (2004). A Linkage Map of Atlantic Salmon (Salmo salar) Reveals an Uncommonly Large Difference in Recombination Rate Between the Sexes. Animal Genetics, Vol. 35, No. 2, pp. 81-92, ISSN 1365-2052.

Moen, T., Sonesson, A.K., Hayes, B., Lien, S., Munck, H. & Meuwissen, T.H. (2007). Mapping of a Quantitative Trait Locus for Resistance Against Infectious Salmon Anaemia in Atlantic Salmon (Salmo salar): Comparing Survival Analysis with Analysis on Affected/Resistant Data. BMC Genetics, Vol. 8, No.1, ISSN 1471-2156.

Moghadam, H.K., Poissant, J., Fotherby, H., Haidle, L., Ferguson, M.M. & Danzmann, R.G. (2007). Quantitative Trait Loci for Body Weight, Condition Factor and Age at Sexual Maturation in Arctic Charr (Salvelinus alpinus): Comparative Analysis with Rainbow Trout (Oncorhynchus mykiss) and Atlantic Salmon (Salmo salar). Molecular Genetics and Genomics, Vol. 277, No. 6, pp. 647-661, ISSN 1617-4623.

Muir, W.M., Wong, G.K., Zhang, Y., Wang, J., Groenen, M.A., Crooijmans, R.P., Megens, H.J., Zhang, H., Okimoto, R., Vereijken, A., Jungerius, A., Albers, G.A., Lawley, C.T., Delany, M.E., MacEachern, S. & Cheng, H.H. (2008). Genome-wide Assessment of Worldwide Chicken SNP Genetic Diversity Indicates Significant Absence of Rare Alleles in Commercial Breeds. Proceedings of the National Academy of Sciences of the United States of America, Vol. 105, No. 45, pp. 17312-17317, ISSN 0027-8424.

Nelson, J.S. (2006). Fishes of the World, John Wiley and Sons Inc., ISBN 978-0-471-25031-9, New Jersey, USA.

Ng, S.H., Artieri, C.G., Bosdet, I.E., Chiu, R., Danzmann, R.G., Davidson, W.S., Ferguson, M.M., Fjell, C.D., Hoyheim, B., Jones, S.J., de Jong, P.J., Koop, B.F., Krzywinski, M.I., Lubieniecki, K., Marra, M.A., Mitchell, L.A., Mathewson, C., Osoegawa, K., Parisotto, S.E., Phillips, R.B., Rise, M.L., von Schalburg, K.R., Schein, J.E., Shin, H.,

Siddiqui, A., Thorsen, J., Wye, N., Yang, G. & Zhu, B. (2005). A Physical Map of the Genome of Atlantic Salmon, *Salmo salar*. Genomics, Vol. 86, No. 4, pp. 396-404, ISSN 0888-7543.

Nichols, K.M., Young, W.P., Danzmann, R.G., Robison, B.D., Rexroad, C., Noakes, M., Phillips, R.B., Bentzen, P., Spies, I., Knudsen, K., Allendorf, F.W., Cunningham, B.M., Brunelli, J., Zhang, H., Ristow, S., Drew, R., Brown, K.H., Wheeler, P.A. & Thorgaard, G.H. (2003). A Consolidated Linkage Map for Rainbow Trout (*Oncorhynchus mykiss*). Animal Genetics, Vol. 34, No. 2, pp. 102-115, ISSN 0268-9146.

Nichols, K., Felip-Edo, A., Wheeler, P. & Thorgaard, G. (2008). The Genetic Basis of Smoltification-related Traits in *Oncorhynchus mykiss*. Genetics, Vol. 179, No. 3, pp. 1559-1575, ISSN 0016-6731.

Norris, A.T., Bradley, D.G. & Cunningham, E.P. (2000). Parentage and Relatedness Determination in Farmed Atlantic Salmon (*Salmo salar*) Using Microsatellite Markers. Aquaculture, Vol. 182, No. 1-2, pp. 73, ISSN 0044-8486.

Ødegård, J., Olesen, I., Gjerde, B. & Klemetsdal, G. (2007). Positive Genetic Correlation Between Resistance to Bacterial (Furunculosis) and Viral (Infectious Salmon Anaemia) Diseases in Farmed Atlantic Salmon (*Salmo salar*). Aquaculture, Vol. 271, No. 1-4, pp. 173, ISSN 0044-8486.

Ohno, S. (1970). Evolution by Gene Duplication, Springer-Verlag, ISBN 0-04-575015-7, New York, USA.

O'Malley, K.G., McClelland, E.K. & Naish, K.A. (2010). Clock Genes Localize to Quantitative Trait Loci for Stage-specific Growth in Juvenile Coho Salmon, *Oncorhynchus kisutch*. The Journal of Heredity, Vol. 101, No. 5, pp. 628-632, ISSN 0022-1503.

O'Malley, K.G., Sakamoto, T., Danzmann, R.G. & Ferguson, M.M. (2003). Quantitative Trait Loci for Spawning Date and Body Weight in Rainbow Trout: Testing for Conserved Effects Across Ancestrally Duplicated Chromosomes. Journal of Heredity, Vol. 94, No. 4, pp. 273-284, ISSN 1465-7333.

O'Reilly, P.T., Herbinger, C. & Wright, J.M. (1998). Analysis of Parentage Determination in Atlantic Salmon (*Salmo salar*) Using Microsatellites. Animal Genetics, Vol. 29, No. 5, pp. 363-370, ISSN 1365-2052.

Palti, Y., Genet, C., Luo, M.C., Charlet, A., Gao, G., Hu, Y., Castano-Sanchez, C., Tabet-Canale, K., Krieg, F., Yao, J., Vallejo, R.L. & Rexroad, C.E.,3rd (2011). A First Generation Integrated Map of the Rainbow Trout Genome. BMC Genomics, Vol. 12, No. 1, ISSN 1471-2164.

Palti, Y., Luo, M.C., Hu, Y., Genet, C., You, F.M., Vallejo, R.L., Thorgaard, G.H., Wheeler, P.A. & Rexroad, C.E.,3rd (2009). A First Generation BAC-based Physical Map of the Rainbow Trout Genome. BMC Genomics, Vol. 10, No. 1, ISSN 1471-2164.

Perry, G.M., Danzmann, R.G., Ferguson, M.M. & Gibson, J.P. (2001). Quantitative Trait Loci for Upper Thermal Tolerance in Outbred Strains of Rainbow Trout (*Oncorhynchus mykiss*). Heredity, Vol. 86, No. 3, pp. 333-341, ISSN 0018-067X.

Phillips, R.B., Keatley, K.A., Morasch, M.R., Ventura, A.B., Lubieniecki, K.P., Koop, B.F., Danzmann, R.G. & Davidson, W.S. (2009). Assignment of Atlantic salmon (Salmo salar) Linkage Groups to Specific Chromosomes: Conservation of Large Syntenic

Blocks Corresponding to Whole Chromosome Arms in Rainbow Trout (*Oncorhynchus mykiss*). BMC Genetics, Vol. 10, No. 1, ISSN 1471-2156.

Phillips, R.B., Nichols, K.M., DeKoning, J.J., Morasch, M.R., Keatley, K.A., Rexroad, C.,3rd, Gahr, S.A., Danzmann, R.G., Drew, R.E. & Thorgaard, G.H. (2006). Assignment of Rainbow Trout Linkage Groups to Specific Chromosomes. Genetics, Vol. 174, No. 3, pp. 1661-1670, ISSN 0016-6731.

Presa, P. & Guyomard, R. (1996). Conservation of Microsatellites in Three Species of Salmonids. Journal of Fish Biology, Vol. 49, No. 6, pp. 1326-1329, ISSN 1095-8649.

Quinn, N.L., Boroevich, K.A., Lubieniecki, K.P., Chow, W., Davidson, E.A., Phillips, R.B., Koop, B.F. & Davidson, W.S. (2010). Genomic Organization and Evolution of the Atlantic Salmon Hemoglobin Repertoire. BMC Genomics, Vol. 11, No.1, ISSN 1471-2164.

Quinn, N.L., Levenkova, N., Chow, W., Bouffard, P., Boroevich, K.A., Knight, J.R., Jarvie, T.P., Lubieniecki, K.P., Desany, B.A., Koop, B.F., Harkins, T.T. & Davidson, W.S. (2008). Assessing the Feasibility of GS FLX Pyrosequencing for Sequencing the Atlantic Salmon Genome. BMC Genomics, Vol. 9, No. 1, ISSN 1471-2164.

Quinn, N.L., McGowan, C.R., Cooper, G.A., Koop, B.F. & Davidson, W.S. (2011a). Identification of Genes Associated with Heat Tolerance in Arctic Charr Exposed to Acute Thermal Stress. Physiological Genomics, Vol. 43, No. 11, pp 685-96, ISSN 1094-8341.

Quinn, N.L., McGowan, C.R., Cooper, G.A., Koop, B.F. & Davidson, W.S. (2011b). Ribosomal Genes and Heat Shock Proteins as Putative Markers for Chronic, Sub-lethal Heat Stress in Arctic Charr: Applications for Aquaculture and Wild Fish. Physiological Genomics, Jul 12 2011 [epub ahead of print].

Ramos, A.M., Crooijmans, R.P., Affara, N.A., Amaral, A.J., Archibald, A.L., Beever, J.E., Bendixen, C., Churcher, C., Clark, R., Dehais, P., Hansen, M.S., Hedegaard, J., Hu, Z.L., Kerstens, H.H., Law, A.S., Megens, H.J., Milan, D., Nonneman, D.J., Rohrer, G.A., Rothschild, M.F., Smith, T.P., Schnabel, R.D., Van Tassell, C.P., Taylor, J.F., Wiedmann, R.T., Schook, L.B. & Groenen, M.A. (2009). Design of a High Density SNP Genotyping Assay in the Pig Using SNPs Identified and Characterized by Next Generation Sequencing Technology. PloS One, Vol. 4, No. 8, pp. e6524, ISSN 1932-6203.

Reid, D.P., Szanto, A., Glebe, B., Danzmann, R.G. & Ferguson, M.M. (2005). QTL for Body Weight and Condition Factor in Atlantic Salmon (Salmo salar): Comparative Analysis with Rainbow Trout (*Oncorhynchus mykiss*) and Arctic Charr (*Salvelinus alpinus*). Heredity, Vol. 94, No. 2, pp. 166-172, ISSN 0018-067X.

Rexroad, C.E.,3rd, Rodriguez, M.F., Coulibaly, I., Gharbi, K., Danzmann, R.G., Dekoning, J., Phillips, R. & Palti, Y. (2005). Comparative Mapping of Expressed Sequence Tags Containing Microsatellites in Rainbow Trout (*Oncorhynchus mykiss*). BMC Genomics, Vol. 6, No.1, ISSN 1471-2164.

Rise, M.L., von Schalburg, K.R., Brown, G.D., Mawer, M.A., Devlin, R.H., Kuipers, N., Busby, M., Beetz-Sargent, M., Alberto, R., Gibbs, A.R., Hunt, P., Shukin, R., Zeznik, J.A., Nelson, C., Jones, S.R., Smailus, D.E., Jones, S.J., Schein, J.E., Marra, M.A., Butterfield, Y.S., Stott, J.M., Ng, S.H., Davidson, W.S. & Koop, B.F. (2004). Development and Application of a Salmonid EST Database and cDNA Microarray:

Data Mining and Interspecific Hybridization Characteristics. Genome Research, Vol. 14, No. 3, pp. 478-490, ISSN 1088-9051.

Ryynanen, H.J. & Primmer, C.R. (2006). Single Nucleotide Polymorphism (SNP) Discovery in Duplicated Genomes: Intron-primed Exon-crossing (IPEC) As a Strategy for Avoiding Amplification of Duplicated Loci in Atlantic Salmon (*Salmo salar*) and Other Salmonid Fishes. BMC Genomics, Vol. 7, No.1, ISSN 1471-2164.

Sakamoto, T., Danzmann, R.G., Gharbi, K., Howard, P., Ozaki, A., Khoo, S.K., Woram, R.A., Okamoto, N., Ferguson, M.M., Holm, L.E., Guyomard, R. & Hoyheim, B. (2000). A Microsatellite Linkage Map of Rainbow Trout (*Oncorhynchus mykiss*) Characterized by Large Sex-specific Differences in Recombination Rates. Genetics, Vol. 155, No. 3, pp. 1331-1345, ISSN 0016-6731.

Sanchez, C.C., Smith, T.P., Wiedmann, R.T., Vallejo, R.L., Salem, M., Yao, J. & Rexroad, C.E.,3rd (2009). Single Nucleotide Polymorphism Discovery in Rainbow Trout by Deep Sequencing of a Reduced Representation Library. BMC Genomics, Vol. 10, No.1, ISSN 1471-2164.

Sanchez, J.A., Clabby, C., Ramos, D., Blanco, G., Flavin, F., Vazquez, E. & Powell, R. (1996). Protein and Microsatellite Single Locus Variability in *Salmo Salar* L. (Atlantic salmon). Heredity, Vol. 77, No. 4, pp. 423-432, ISSN 0018-067X.

Sarropoulou, E., Franch, R., Louro, B., Power, D.M., Bargelloni, L., Magoulas, A., Senger, F., Tsalavouta, M., Patarnello, T., Galibert, F., Kotoulas, G. & Geisler, R. (2007). A Gene-based Radiation Hybrid Map of the Gilthead Sea Bream *Sparus aurata* Refines and Exploits Conserved Synteny with *Tetraodon nigroviridis*. BMC Genomics, Vol. 8, No.1, ISSN 1471-2164.

Scribner, K.T., Gust, J.R. & Fields, R.L. (1996). Isolation and Characterization of Novel Salmon Microsatellite Loci: Cross-species Amplification and Population Genetic Applications. Canadian Journal of Fisheries and Aquatic Sciences, Vol. 53, No. 4, pp. 833-841, ISSN 1205-7533.

Slettan, A., Olsaker, I. & Lie, O. (1995). A Polymorphic Dinucleotide Repeat Microsatellite in Atlantic Salmon, *Salmo salar* (SSOSL436). Animal Genetics, Vol. 26, No. 5, pp. 368, ISSN 1365-2052.

Smith, C.T., Elfstrom, C.M., Seeb, L.W. & Seeb, J.E. (2005). Use of Sequence Data From Rainbow Trout and Atlantic Salmon for SNP Detection in Pacific Salmon. Molecular Ecology, Vol. 14, No. 13, pp. 4193-4203, ISSN 0962-1083.

Somorjai, I.M., Danzmann, R.G. & Ferguson, M.M. (2003). Distribution of Temperature Tolerance Quantitative Trait Loci in Arctic Charr (*Salvelinus alpinus*) and Inferred Homologies in Rainbow Trout (*Oncorhynchus mykiss*). Genetics, Vol. 165, No. 3, pp. 1443-1456, ISSN 0016-6731.

Sun, X. & Liang, L. (2004). A Genetic Linkage Map of Common Carp (*Cyprinus carpio* L.) and Mapping of a Locus Associated with Cold Tolerance. Aquaculture, Vol. 238, No. 1-4, pp. 165, ISSN 0044-8486.

Thorsen, J., Zhu, B., Frengen, E., Osoegawa, K., de Jong, P.J., Koop, B.F., Davidson, W.S. & Hoyheim, B. (2005). A Highly Redundant BAC Library of Atlantic Salmon (*Salmo salar*): An Important Tool for Salmon Projects. BMC Genomics, Vol. 6, No. 1, ISSN 1471-2164.

Jackson, T.R, Ferguson, M.M, Danzmann, R.G, Fishback, A.G, Ihssen, P.E, O'Connell, M. & Crease, T.J.(1998). Identification of Two QTL Influencing Upper Temperature Tolerance in Three Rainbow Trout (*Oncorhynchus mykiss*) Half-sib Families. Heredity, Vol. 80, No. 2, pp. 143-151, ISSN 0018-067X.

Tuberosa, R., Salvi, S., Sanguineti, M.C., Landi, P., Maccaferri, M. & Conti, S. (2002). Mapping QTLs Regulating Morpho-physiological Traits and Yield: Case Studies, Shortcomings and Perspectives in Drought-stressed Maize. Annals of Botany, Vol. 89 Spec No, pp. 941-963, ISSN 0305-7364.

Van Tassell, C.P., Smith, T.P., Matukumalli, L.K., Taylor, J.F., Schnabel, R.D., Lawley, C.T., Haudenschild, C.D., Moore, S.S., Warren, W.C. & Sonstegard, T.S.(2008). SNP Discovery and Allele Frequency Estimation by Deep Sequencing of Reduced Representation Libraries. Nature Methods, Vol. 5, No. 3, pp. 247-252, ISSN 1548-7091.

Veerkamp, R.F. & Beerda, B. (2007). Genetics and Genomics to Improve Fertility in High Producing Dairy Cows. Theriogenology, Vol. 68, Suppl. 1, pp. S266-273, ISSN 0093-691X.

von Schalburg, K.R., Rise, M.L., Cooper, G.A., Brown, G.D., Gibbs, A.R., Nelson, C.C., Davidson, W.S. & Koop, B.F. (2005). Fish and Chips: Various Methodologies Demonstrate Utility of a 16,006-gene Salmonid Microarray. BMC Genomics, Vol. 6, No. 1, ISSN 1471-2164.

Waldbieser, G.C., Bosworth, B.G., Nonneman, D.J. & Wolters, W.R. (2001). A Microsatellite-Based Genetic Linkage Map for Channel Catfish, *Ictalurus punctatus*. Genetics, Vol. 158, No. 2, pp. 727-734, ISSN 0016-6731.

Wang, S., Peatman, E., Abernathy, J., Waldbieser, G., Lindquist, E., Richardson, P., Lucas, S., Wang, M., Li, P., Thimmapuram, J., Liu, L., Vullaganti, D., Kucuktas, H., Murdock, C., Small, B.C., Wilson, M., Liu, H., Jiang, Y., Lee, Y., Chen, F., Lu, J., Wang, W., Xu, P., Somridhivej, B., Baoprasertkul, P., Quilang, J., Sha, Z., Bao, B., Wang, Y., Wang, Q., Takano, T., Nandi, S., Liu, S., Wong, L., Kaltenboeck, L., Quiniou, S., Bengten, E., Miller, N., Trant, J., Rokhsar, D., Liu, Z. & Catfish Genome Consortium (2010). Assembly of 500,000 Inter-specific Catfish Expressed Sequence Tags and Large Scale Gene-associated Marker Development for Whole Genome Association Studies. Genome Biology, Vol. 11, No. 1, pp. R8, ISSN 1465-6906.

Woram, R.A., McGowan, C., Stout, J.A., Gharbi, K., Ferguson, M.M., Hoyheim, B., Davidson, E.A., Davidson, W.S., Rexroad, C. & Danzmann, R.G. (2004). A Genetic Linkage Map for Arctic Char (Salvelinus alpinus): Evidence for Higher Recombination Rates and Segregation Distortion in Hybrid Versus Pure Strain Mapping Parents. Genome, Vol. 47, No. 2, pp. 304-315, ISSN 0831-2796.

Wringe, B., Devlin, R., Ferguson, M., Moghadam, H., Sakhrani, D. & Danzmann, R. (2010). Growth-related Quantitative Trait Loci in Domestic and Wild Rainbow Trout (*Oncorhynchus mykiss*). BMC Genetics, Vol. 11, No. 1, ISSN 1471-2156.

Xia, J.H., Feng, F., Lin, G., Wang, C.M. & Yue, G.H. (2010). A First Generation BAC-based Physical Map of the Asian Seabass (*Lates calcarifer*). PloS One, Vol. 5, No. 8, pp. e11974, ISSN 1932-6203.

Xu, P., Li, J., Li, Y., Cui, R., Wang, J., Wang, J., Zhang, Y., Zhao, Z. & Sun, X. (2011). Genomic Insight Into the Common Carp (Cyprinus carpio) Genome by Sequencing Analysis of BAC-end Sequences. BMC Genomics, Vol. 12, No. 1, ISSN 1471-2164.

Xu, P., Wang, S., Liu, L., Thorsen, J., Kucuktas, H. & Liu, Z. (2007). A BAC-based Physical Map of the Channel Catfish Genome. Genomics, Vol. 90, No. 3, pp. 380, ISSN 0888-7543.

Part 3

Post-Harvest

Novel Approach for Controlling Lipid Oxidation and Melanosis in Aquacultured Fish and Crustaceans: Application of Edible Mushroom (*Flammulina velutipes*) Extract *In Vivo*

Angel Balisi Encarnacion, Huynh Nguyen Duy Bao
Reiko Nagasaka and Toshiaki Ohshima
Tokyo University of Marine Science and Technology
Japan

1. Introduction

The demand for fish and fishery products in the global market has been increasing with increase in the world population. This has necessitated the introduction of more aquaculture technologies in the fisheries industry in order to meet such demand. Data from the Food and Agriculture Organization (FAO, 2006) shows that the world aquaculture contribution to the global supplies of fish, crustaceans, molluscs and other aquatic animals continues to grow, increasing from 3.9, 27.1, and 32.4% of total production by weight in 1970, 2000, and 2004, respectively. It has been reported that aquaculture continues to grow more rapidly than all other animal food-producing sectors. Furthermore, the FAO (2006) reported that the aquaculture sector has grown at an average rate of 8.8% per year since 1970, compared with only 1.2% for capture fisheries and 2.8% for terrestrial farmed meat production systems over the same period worldwide. Notably, production from aquaculture has greatly outpaced population growth, with per capita supply from aquaculture increasing from 0.7 kg in 1970 to 7.1 kg in 2004, representing an average annual growth rate of 7.1%.

However, the generally acknowledged limitations of production from aquaculture and capture fisheries, coupled with the widening gap between the supply of and demand for fish for human consumption, reaffirms that postharvest technology is a very important component of this industry. Undoubtedly, postharvest losses are an unacceptable waste given our scarce natural resources.

Postharvest losses of fish occur in various forms during handling, processing, and preservation. Significant economic losses occur when spoilage of fish and crustaceans decrease market value or the product needs to be reprocessed, thereby increasing the cost of the finished product. Improper handling and processing methods can also reduce nutrient levels and conversion of large quantities of fish catches into fish meal for animal feeds can be considered under certain conditions as a "loss" for human food security. According to FAO (2006), fish losses caused by spoilage are estimated to be 10 to 12 million tonnes per year, accounting for approximately 10% of the total production from capture fisheries and aquaculture. Thus, the postharvest losses in fisheries can be among the highest for all the commodities in the entire food production system. An appropriate preservation method can significantly reduce these losses, particularly those incurred during the handling, processing, distribution, and marketing of fishery products.

2. Postharvest technology for fisheries

Our fisheries resources have not only been a source of food for people but are also economically important because they are a source of livelihood for many. However, the potential for harvesting more products is now very limited because of overexploitation and the demand for food protein is increasing worldwide. Applying effective postharvest technology could reduce gap between production and demand and result in effective utilization of aquatic resources. This includes the different processes and techniques used during the postharvest handling, processing, and marketing of aquatic products.

2.1 Methods for handling and preserving fresh seafood products

Seafood products are among the world`s most perishable commodities, and their spoilage begins soon after the death of the fish. Improper handling could hasten spoilage, resulting in gradual development of undesirable qualities in seafood products. Thus, effective handling and preservation techniques need to be used to address the problems of postharvest losses.

Many developments have taken place in postharvest technology. Freezing at sea was a major development in the advance nations in the 1950`s along with the bulk chilling in chilled and refrigerated seawater systems on board fishing vessels and no new major technologies have been recently introduced but only a consolidation and anticipated development of current ones (Hermes, 1998). The industry tends to be more perceptive of the necessity of satisfying the requirements and demands of consumers, particularly with regards to the safety and quality of seafood products.

2.1.1 Chilling

Chilling is most commonly used in the industry to keep seafood fresh after harvest. Chilling is used to reduce the temperature of seafood products to some point below (-2 to -4°C for superchilling) or above (0 to 5°C) the freezing point of water (Hermes, 1998). The methods used for chilling include wet icing and use of chilled seawater, ice slurry, refrigerated air, dry ice, and gel ice mats. Icing is the most common and useful way of chilling the fish catch and is affected by direct contact between the melted ice and the fish. This requires sufficient ice and proper arrangement of ice and the product to allow rapid cooling. The use of chilled seawater or slush ice involves seawater and ice. The amount of ice used depends on the initial temperature of the water and fish, the size of the container and the quality of its insulation, and the length of the trip. Refrigerated air has been used in some big commercial boats, wherein chilled air is circulated by a finned evaporator and a fan situated at the end of the cold room. Cooling of fish through dry ice (solid carbon dioxide) is affected by evaporation of the ice, but dry ice should not be used in direct contact with the product in order to avoid cold burns (Shoemaker, 1991 as cited by Hermes, 1998). This method is normally used for air shipment. However, these chilling methods do not stop spoilage but slow it down considerably; therefore, they are used only to delay spoilage of the fish catch but not for longer storage of the product.

2.1.2 Freezing

Freezing seafood becomes essential when other methods of preserving fish such as chilling are unsuitable for longer storage. Under proper conditions, it can keep the product frozen for several months without considerably changing its quality. Freezing is a method of stopping, either partly or entirely, the deteriorative activities of microorganisms and enzymes.

Microorganisms stop multiplying at approximately around -10°C and below, and the enzymatic activity is generally controlled when the temperature is reduced below the freezing point, to approximately -1°C (Cornell, 1995 as cited by Hermes, 1998). The water in the flesh begins to freeze at temperatures ranging from -1°C to -3 °C, and most of the water is converted into ice during freezing. At -5°C, approximately 75% of the water in fish muscles is frozen.

In the fishing industry, there are 3 basic methods available for freezing seafood products, namely, air-blast freezing, contact or plate freezing, and spray or immersion freezing. Air-blast freezing involves the continuous flow of cold air over the product. Uniform freezing is attained only if the temperature and speed of the air over the product is constant. This type of freezing system is very versatile, and is thus very useful in producing individual quick frozen (IQF) products for crustaceans, fish fillets, and value-added products. In contrast, contact or plate freezing allows the product to come into direct contact with hallow freezing metal plates, through which a cold fluid is passed. It generally used for freezing products such as whole fish, fillets, shrimps, and other seafood products into blocks. In the spray- or immersion-freezing method, the product comes into direct contact with the fluid refrigerant, such as liquid nitrogen. The method is normally used for producing very high value and specialized IQF products. The choice regarding freezing systems will depend upon cost, function, and feasibility because this method of preservation can be expensive.

2.1.3 Preservatives

Various problems are encountered with seafood products even when they are kept in cold storage. Frozen tuna becomes dark brown during cold storage because of oxidation of hemoglobin in the blood and myoglobin (Mb) in the meat. In crustaceans, browning or blackening of frozen shrimps and prawns is also observed. These problems prompted the food industry to look for ways to control quality deterioration in seafood products, and the use of preservatives has been one of the immediate responses.

The addition of antioxidants to products is one of the most widely studied methods for controlling discoloration and lipid oxidation in fish meat and other meat products. Antioxidants have many modes of action which include sequestering catalytic metal ions, decreasing oxygen concentration, quenching singlet oxygen and superoxide anion, decomposing primary oxidation products to non-volatile compounds, preventing first-chain initiation by scavenging initially generated radicals and chain breaking. The chain-breaking mechanism has been studied for several antioxidants (Roginsky & Lissi, 2005). In this mechanism, the antioxidant donates a hydrogen atom to a lipid peroxyl radical and forms an antioxidant radical, the antioxidant radical subsequently either combines with other lipid peroxyl radical or another antioxidant radical to terminate the reaction.

Several categories of antioxidants can be used for food applications. However, selecting antioxidants for foods is a major concern in the industry because of strict regulations. Generally, the antioxidants involved in food additives must be effective at low doses, must not affect sensory flavor, and must not toxic. Therefore, natural antioxidants are generally preferred for food applications.

3. Lipid oxidation in seafood products

Lipids are one of the important structural and functional components of foods. They provide energy to humans and essential nutrients such as eicosapentaenoic acid; docosahexaenoic acid; and fat-soluble vitamins including, vitamins A, D, E, and K. Lipids are generally defined as "fatty acids and their derivatives, and substances related biosynthetically or functionally to these components" (Christie, 1987). They have been known to significantly affect food quality even though they constitute a minor component of food. Lipids not only impart flavor, odor, texture, and color to foods but also contribute to the feeling of satiety and help in making food products palatable.

However, constant exposure of lipids, particularly unsaturated fatty acids, to air could adversely affect food quality. The susceptibility of lipids to oxidation is one of the main causes of quality deterioration in various types of fresh food products as well as in processed foods. Lipid oxidation is a perennial problem for both the food industry and the consumers. Lipid oxidation has been believed to be one of the factors limiting the shelf life of foods, particularly that of many complex products (Jacobsen, 1999). It is a complex process whereby lipids, particularly polyunsaturated fatty acids, are degraded via free radical formation, causing the deterioration of flavor, texture, color, aroma, taste, consistency, nutritional benefits, and to some extent, the safety of foodstuffs for human consumption. Notably, consumers` preferences for foods are being influenced by such factors. Thus, lipid oxidation is a decisive factor in the useful processing and storage of food products.

Various effects of lipid oxidation on food properties are briefly summarized in Figure 1. Lipid oxidation itself is primarily the formation of reactive compounds like hydroperoxides (HPO) and peroxy radicals. The primary products very often undergo further reactions to form more stable compounds such as hydroxy acids or epoxides. It has been reported that compounds like hydroxy acids can contribute to bitter taste (Grosch et al., 1992). In complex food systems, the interaction of lipid HPO and secondary oxidation products with proteins and other components significantly impact oxidative and flavor stability and texture during processing, cooking, and storage (Erickson, 1992). Oxidized lipids can react with amines, amino acids, and proteins to form brown macromolecular products (Frankel, 1998). Color formation is known to be primarily influenced by the degree of fatty acid unsaturation, water activity, oxygen pressure, and the presence of phenolic compounds (Pan, 2004).

Some of the known factors that promote or inhibit lipid oxidation in foods are shown in Table 1. Metal, metallo-proteins, and enzymes are important factors affecting lipid oxidation in raw materials. Water activity, lipid interactions, proteins, and sugars are important elements affecting the food quality of processed foods.

Novel Approach for Controlling Lipid Oxidation and Melanosis in Aquacultured Fish and Crustaceans:
Application of Edible Mushroom (Flammulina velutipes) Extract In Vivo

161

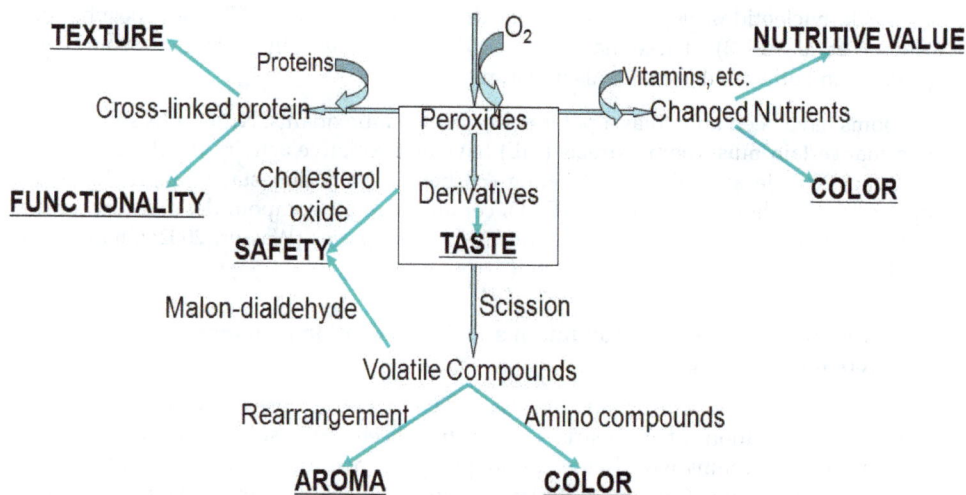

Fig. 1. Reactions of polyunsaturated fatty acids leading to quality and nutritional changes in foods (Erickson, 1992).

Type of lipid fatty acid	Non-polar, polar, fatty acids, sterols, terpenes, chain length, unsaturation, cis-trans isomers, free or bound
Catalysts	Enzymes, heme compounds, trace metals
State and access of oxygen	Triplet, singlet or radical, package
Light	Frequency, intensity, sensitizers
Temperature	Denaturating, non-denaturating
pH	Dissociation, denaturing
Inhibitors	Antioxidants, chelators, enzymes, enzyme inhibition or activation

Table 1. Lipid oxidation factors in food (Frankel, 1998).

3.1 Control of lipid oxidation by natural antioxidants

Fish tissue contains many endogenous antioxidants that can contribute to stabilizing the natural color of fish meat by acting against either lipid oxidation or Mb oxidation in the meat. These antioxidants may act directly or indirectly to inhibit the initiation and propagation steps of lipid oxidation. Antioxidants that interfere with the propagation step by converting free radicals to stable compounds are generally referred to as primary antioxidants, and those interfering with the initiation step are considered as secondary antioxidants or preventive inhibitors. Secondary antioxidants include both oxygen scavengers and chelators.

The endogenous antioxidants found in aquatic food products include the tocopherols (Syvaoja & Salminen, 1985), ubiquinol (Petillo et al., 1998), carotenoids (Miki, 1991), ascorbate, organic acids, glutathione peroxidase (Watanabe et al., 1996), peroxide dismutase (Aksnes & Njaa, 1981), catalase, peroxidases (Kanner & Kinsella, 1983), ferroxidases (Kanner et al., 1988), nucleotides, peptides, amino acids (Boldyrev et al., 1987), and phospholipids (Ohshima et al., 1993). These natural antioxidants are usually extracted from fruits, vegetables, and other edible materials in nature.

Mushrooms have been known as a potential source of antioxidants. Numerous studies have shown that certain mushroom extracts (ME) have antioxidative activity *in vitro* and *in vivo* (Mau et al., 2002; Jang et al., 2004; Cheung & Cheung, 2005; Elmastas et al., 2007). Active compounds, including ergothioneine (ERT), certain phenolic compounds, and saccharides, have been found in extracts from different mushroom species (Wasser, 2002; Quang et al., 2006; Dubost et al., 2007).

3.2 Prevention of fish meat discoloration and lipid oxidation by dietary supplementation with ME

Mushrooms are widely cultivated because of their known uses and health benefits. However, accumulation of industrial waste has been an issue with the increasing production of mushrooms worldwide. Although spent medium is still underutilized, it is normally used as compost or raw material for extracting soluble sugars (Makishima et al., 2006). Among the mushroom species that are cultivated, *Flammulina velutipes* has been known as a medicinal mushroom and has also been cultured as an edible mushroom on a large scale in Japan and other Asian countries (Wasser, 2002).

Jang et al. (2004) reported the antioxidative properties of the *F. velutipes* extracts against the oxidation of cod liver oil in oil-in-water emulsions. In addition, water extracts from this mushroom have been reported to inhibit the oxidation of oxymyoglobin (MbO_2) isolated from cattle meats (Ashida & Sato, 2005). The activity of phenolic compounds depends on their structure and is relative to the number and location of the hydroxyl (OH) groups involved (Dziedric & Hudson, 1984). Recently, ERT is another potent antioxidant which usually exists in mushrooms. It has been known as a powerful scavenger of hydroxyl radicals ($\cdot OH$) and an inhibitor of $\cdot OH$ generation from hydrogen peroxide; $\cdot OH$ generation is catalyzed by iron or copper ions (Akanmu et al., 1991). Arduini et al. (1990) showed that ferrylmyoglobin (ferrylMb), which was formed when deoxymyoglobin (deoxyMb) and metmyoglobin (metMb) were exposed to hydrogen peroxide, was reduced to metMb in the presence of ERT.

Bao et al. (2008) found that *F. velutipes* contains 300 ug/g ERT and can prevent lipid oxidation as well as stabilize the color of beef and fish meats during low-temperature storage. In addition, Bao et al. (2009a) evaluated the antioxidative activity of hydrophilic extract prepared from solid waste medium obtained during *F. velutipes* cultivation with regard to value-added utilization of the waste for recovery of bioactive compounds. The extract contained 341 ug/mL ERT and was used for stabilizing the lipids and the color of the dark meat of two year-old yellowtail, *Seriola quinqueradiata* through feeding prior to harvest (Bao et al., 2009a).

The DPPH radical scavenging activity of the extracts correlated with their ERT content is shown in Figure 2. In this study, the effective content of ERT was 0.3 ug in the extract in which the DPPH radicals were scavenged by 50% of the original.

Changes in the HPO content of the yellowtail dark muscle during chilled storage are shown in Figure 3. The HPO content in the dark muscle of the control yellowtail increased with

Novel Approach for Controlling Lipid Oxidation and Melanosis in Aquacultured Fish and Crustaceans:
Application of Edible Mushroom (Flammulina velutipes) Extract In Vivo
163

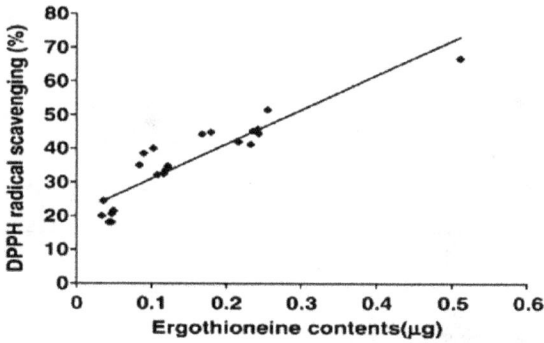

Fig. 2. DPPH radical scavenging activity of the mushroom (*Flammulina velutipes*) extract.
Results are presented in terms of mean ± standard deviation ($n = 3$) (Bao et al., 2009a).

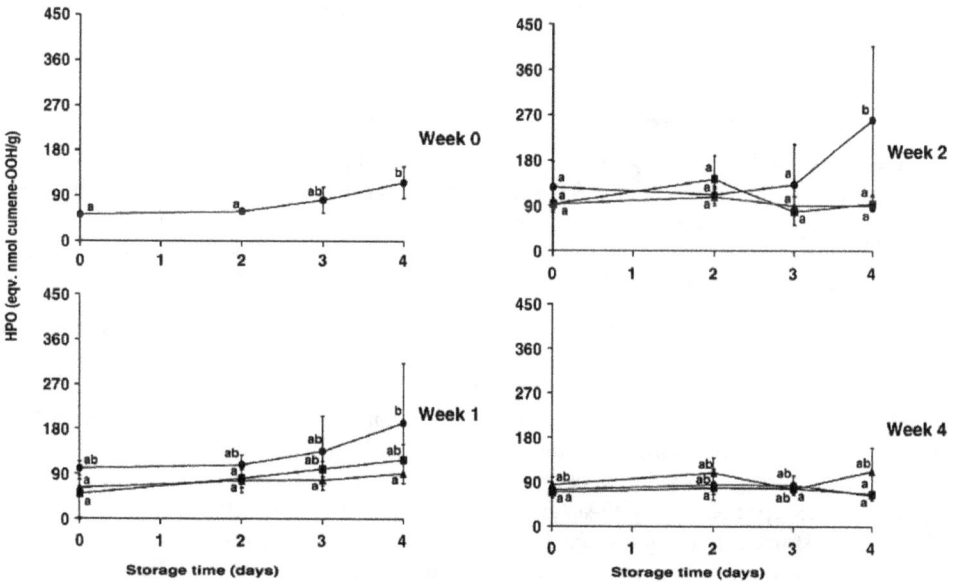

Fig. 3. Changes in the total lipid hydroperoxide level of yellowtail (*Seriola quinqueradiata*)
dark muscle during chilled storage. (●), fed a diet without mushroom extract; (■), fed
with diet containing 1% concentrated mushroom extract; (▲), fed with diet containing 10%
mushroom extract. Data are presented in terms of mean ± standard deviation (n = 5). The
values with different superscript letters represent significant difference ($P < 0.05$) (Bao et al.,
2009a).

prolongation of storage time, whereas HPO accumulation in the ME-supplemented fish was
significantly suppressed. This effect is believed to be because of the radical scavenging
behavior of ERT (Franzoni et al., 2006). ERT is also known to not only protect organs against
lipid peroxidation but also conserve endogenous glutathione and α-tocopherol (Deiana et
al., 2004). The same study also showed that feeding yellowtails with the ME remarkably
delayed metMb formation in the dark muscle during chilled storage, as shown in Figure 4.

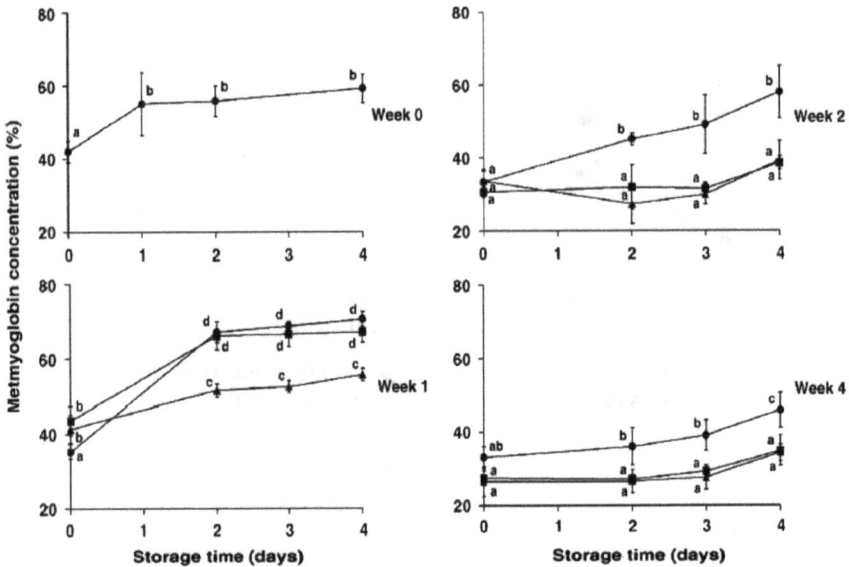

Fig. 4. Changes in the metmyoglobin content of yellowtail (*Seriola quinqueradiata*) dark muscle during chilled storage. (●), fed a diet without mushroom extract; (■), fed with diet containing 1% concentrated mushroom extract; (▲), fed with diet containing 10% mushroom extract. Data are presented in terms of mean ± standard deviation (n = 5). The values with different superscript letters represent significant difference (P <0.05) (Bao et al., 2009a).

Because ERT conserves endogenous antioxidants such as glutathione and α-tocopherol, it enhances the color stability of fish meat during postharvest handling and chilled storage. In addition, discoloration of meats is known to result from interaction between Mb and lipid oxidation (Renerre, 1990). Mb exists in 3 forms in fresh meat: deoxyMb, oxyMb, and metMb. metMB is an undesirable form not only because of its brown color but also because of its catalytic effect during the oxidation of unsaturated lipids (Love & Pearson, 1971). Grunwald & Richards (2006) reported that the reaction between metMb and lipid HPO generates ferrylMb, which can abstract a hydrogen atom from lipid (LH) to form an alkyl radical (L·). In the presence of oxygen, the alkyl radical then generates a peroxyl radical (LOO·), which can in turn abstract a hydrogen atom from another unoxidized lipid, resulting in the production of lipid HPO. The lipid HPOs generated are continuously involved in the redox reactions of Mb to generate radicals. Once underway, the alkoxyl radical (LO·) formed rearranges to generate an epoxyradical (epoxyL·), which reacts with molecular oxygen to form LOO·. The epoxyL· can also react with ferrylMB to form metMB, which continuously plays a role as an intermediate for oxidizing lipids in meat. In this case, the antioxidants that are present in the ME and accumulated in the meat could have acted as chain-breaking antioxidants. The radicals generated in meat from the interaction between Mb and lipid oxidation could be scavenged by the antioxidants, and the ferrylMb produced from this process could be reduced by donating electrons from the thiol groups of the antioxidants (Romero et al., 1992).

The proposed mechanism for ERT-delayed oxidation of lipid and Mb in minced big eye tuna (*Thunnus obesus*) meat to which *F. velutipes* extract has been added is shown in Figure 5.

Novel Approach for Controlling Lipid Oxidation and Melanosis in Aquacultured Fish and Crustaceans:
Application of Edible Mushroom (Flammulina velutipes) Extract In Vivo

165

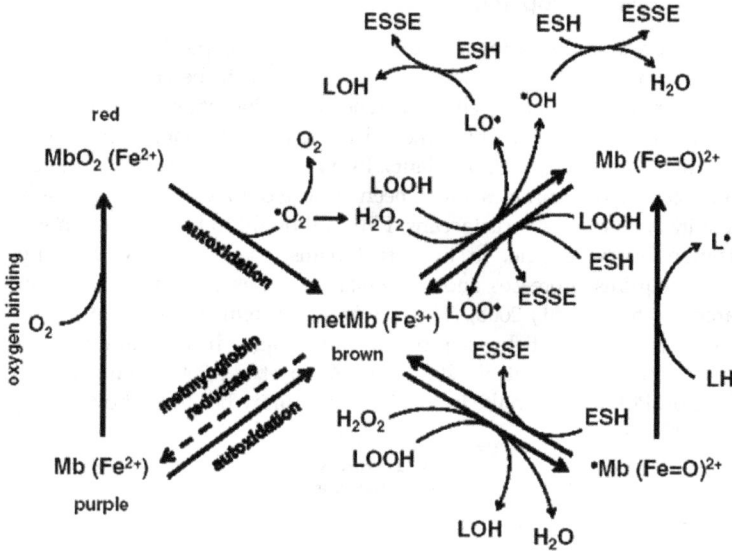

ESH, ergothioneine; ESSH, dimer of ergothioneine; LO, lipid; LO•, alcoxyl radical, LOOH, lipid hydroperoxide, LOO, Peroxy radical; Mb, myoglobin; Mb(Fe=O)$^{2+}$, ferryl myoglobin; •Mb(Fe=O)$^{2+}$, ferryl myoglobin radical; MbO$_2$, oxymyoglobin; Mb·, myoglobin radical; metMb, metmyoglobin.

Fig. 5. Proposed mechanism for ERT-delayed oxidation of lipid and myoglobin in minced big eye tuna *(Thunnus obesus)* meat to which *F. velutipes* extract has been added (Bao et al., 2009b).

Thus, the delay in metMb formation in the dark muscle of yellowtail fed with ME might result from both suppression of lipid oxidation and conservation of endogenous antioxidants. The *in vivo* study of Bao et al. (2009a) was the first model to show that dietary supplementation with ME can be used on an industrial scale for preventing discoloration and lipid oxidation in the dark muscle of yellowtail.

4. Melanosis in crustaceans

The development of melanosis or blackspot during the postharvest period of crustaceans is a well-known postmortem phenomenon attributed to the polymerization of phenol into an insoluble black pigment, melanin. Phenol polymerization is mainly initiated by the action of an enzymatic complex called polyphenoloxidase (PPO; also known as tyrosinase). Severe blackspot formations can cause tremendous economic losses because of the high value commanded by these aquatic products in the marketplace (Kim et al., 2002). Its prevention has been a challenge to the industry, especially for food scientists.

Intensive studies on PPO in crustaceans could help understand the mechanisms underlying the development of melanosis in crustaceans; the properties of PPO; their substrates and inhibitors; and the physical, biological and chemical factors that affect each of these parameters. Complete understanding of this phenomenon and the mechanisms underlying it may help provide a scientific approach to prevent melanosis and slow its rate, thus extending the shelf life and acceptability of the product.

4.1 PPO in melanosis development

PPO plays an important physiological role in crustaceans, particularly in the process of sclerotization. The melanization cascade has been reported to be closely associated with the occurrence of factors that stimulate cellular defence by aiding phagocytosis and encapsulation reactions (Cerenius et al., 2008). Thus, melanization is an important immune response in crustaceans, similar to that observed in plants. In plants, the compounds produced as a result of the polymerization of quinones have been reported to exhibit both antibacterial and antifungal activities, whereas in crustaceans, PPO is thought to be involved in wound healing and sclerotization of the cuticle. In live crustaceans, the activation of prophenoloxidase (proPO) to PPO requires proteases and microbial activators such as polysaccharide-binding proteins (García-Carreño et al., 2008). This activation system plays an important role in the primary immune response, cuticle sclerotization, and injury healing in crustaceans (Buda & Shafer 2005; Martínez-Álvarez et al., 2005; José-Pablo et al., 2009). Simple diagram on the induction of the proPO cascade in the invertebrate immunity has been shown in Figure 6.

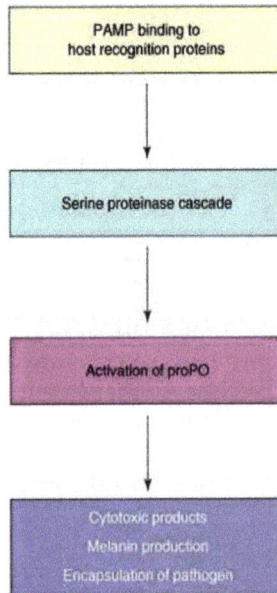

Fig. 6. Induction of the prophenoloxidase (proPO) cascade in the invertebrate immunity. Pathogen-associated molecular patterns (PAMPs; e.g., peptidoglycans, lipopolysaccharides, and β-1,3-glucans) are bound by host recognition proteins. This initiates a serine proteinase cascade that leads to the conversion of zymogenic proPO into catalytically active phenoloxidase and ultimately results in the generation of cytotoxic products and encapsulation of the pathogen (Cerenius et al., 2008).

Postharvest PPO-catalyzed blackening of the shell in crustaceans adversely affects both quality and the consumers' acceptance of these products. Crustaceans such as lobsters, shrimps, and crabs are extremely vulnerable to enzymatic blackening or melanosis. Although the occurrence of melanosis in these aquatic products does not necessarily mean that they are unfit for human consumption, consumers tend to be selective regarding these products, mainly because browning in the carapace connotes spoilage; thus, melanosis decreases the product's market value.

Enzymatic browning or melanosis takes place in the presence of oxygen when tyrosinase and their polyphenolic substrates are mixed. Tyrosinase catalyzes 2 fundamental reactions: 1) hydroxylation at the *o*-position adjacent to an existing hydroxyl group of phenolic compounds (monophenolase activity) and 2) oxidation of diphenol to *o*-benzoquinones (diphenolase activity).

The subunits of tyrosinase have been reported to differ with respect to chemical, physical and kinetic properties, which were believed to be responsible for the relative affinities of the enzymes for both mono- and diphenolic substrates (Kim et al., 2002). The aforementioned considerations have led to many studies on the molecular mechanism underlying the monophenolase and diphenolase activity of tyrosinase. Many studies have been performed on the monophenolase activity of tyrosinase, based on 3 forms of the enzyme. The monophenolase cycle is shown in Figure 7, wherein monophenol reacts with the oxy form and binds at the axial position of one of the coppers ions of this oxy form. It has been reported that rearrangement through the trigonal bipyramidal intermediate leads to binding of peroxide and thereby generates a coordinated *o*-diphenol that is oxidized to *o*-quinone, resulting in a deoxy form ready for dioxygen binding. In the diphenolase cycle, both the oxy and met forms react with *o*-diphenol, oxidizing it to *o*-quinone. However, monophenol can compete with *o*-

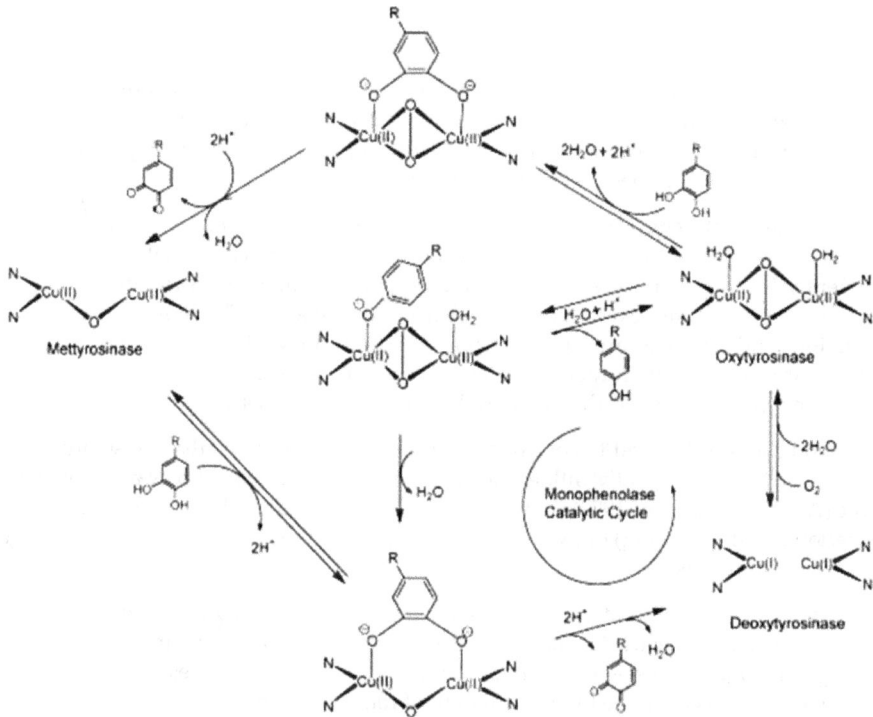

Fig. 7. Catalytic cycles of the hydroxylation of monophenol and oxidation of *o*-dipehol to *o*-quinone by tyrosinase (Kim and Uyama, 2005).

diphenol for binding to the met form site, thereby inhibiting its reduction (Kim & Uyama, 2005). Diphenolases have received much attention because of their high catalytic rate and their association with quinone formation that in turn leads to melanin production.

4.2 Control of melanosis

The fundamental step in melanosis is the transformation of an *o*-diphenol such as 3,4-dihydroxyphenylalenine (L-DOPA) to the corresponding *o*-quinone, which undergoes further oxidation to form brown or black pigments (Martinez & Whitaker, 1995). *o*-Quinones are known to be powerful electrophiles that can be attacked by water, other polyphenols, amino acids, peptides, and proteins, leading to Michael-type products (Mayer et al., 1990).

Many studies have been performed on controlling or inhibiting the activity of PPO in foods, and various mechanisms and techniques have been proposed and developed over the years to mitigate concerns regarding undesirable products of enzyme activity. This involves elimination of one or more of the important components involved in the enzymatic reaction, such as oxygen, copper, substrate, and the enzyme itself. There are also known compounds that react with the product of enzyme activity, thus inhibiting the formation of colored compounds. Many other techniques are also being applied to prevent melanosis in foods, such as processing methods, different types and kinds of inhibitors, and a molecular approach to control PPO, as presented in the following section.

4.2.1 Processing

Traditionally, heating is one of the most popular method to destroy microorganism and to inactivate enzymes. The catalytic activity of PPO is inhibited at temperatures ranging from 70°C to 90°C (Vamos-Vigyoso, 1981). Lee et al. (1988) reported that blanching green beans at temperatures above 82 °C inactivated its enzymes, including PPO.

Similar to heat treatment, low-temperature treat can also help control enzyme-catalyzed reactions. At low temperatures, the kinetic energy of the reactant molecules decreases, subsequently leading to a decrease in the mobility and reaction necessary for the formation of enzyme-substrate complexes and their products (Kim et al., 2002). Thus, low-temperature preservation and storage are used during distribution and retailing to control the development of melanosis in food products. Temperatures as low as -18°C can be used to inactivate enzymes. However, this changes the physical attributes of the products.

Dehydration can also be used to control enzyme catalyzed reactions that cause browning in foods. Generally, water greatly influences enzyme activity because it acts as a solvent or a reactant (Ashie & Simpson, 1996). Water activity can be controlled by physical drying (i.e., dehydration or freeze drying) or by chemical methods (i.e., addition of water-binding agents such as sugars and salts).

Irradiation is another method used for controlling enzymatic browning in foods. It uses ionizing radiation to inactivate microorganisms as well as enzymes. Irradiation is being increasingly recognized as a method for reducing postharvest losses, maintaining food quality, as well as ensuring the hygienic quality of food products.

The rearrangement or destruction of noncovalent bonds such as hydrogen bonds, hydrophobic interactions, and ionic bonds of the tertiary protein structure can lead to enzyme denaturation. This can be achieved by high-pressure treatment that reduces molecular spacing and increases interchain reactions. However, this varies depending the enzymes; the nature of the substrate; and the temperature and duration of high-pressure processing (Cheftel, 1992).

Other processing methods that have been developed and used to control the enzymatic browning of food products included the treatment of food with supercritical carbon dioxide, ultrafiltration, or ultrasonication (Kim et al., 2002). Treatment with supercritical carbon dioxide inactivates the enzyme through carbonic acid production, which causes a significant pH reduction. In contrast, ultrafiltration removes PPO during the process, whereas ultrasonication causes severe shear stresses, thereby promoting enzyme denaturation.

4.2.2 Inhibitors

Melanosis can be inhibited by targeting the enzyme, the substrate, and/or the reaction products involved in the process. Inhibition that targets the enzyme includes metal chelators and compounds that consist of carboxylic acids of the benzoic and cinnamic series. These compounds can be competitive owing to their structural similarity with phenolic substrates. Inhibition that targets the substrate can be performed by removing either oxygen or the phenolic substrate from the reaction. Inhibition targeting the products involves compounds that can react with o-quinones to produce a colorless addition product.

These PPO inhibitors can be classified into 6 categories that include: 1) reducing agents such as sulfites, ascorbic acid and its analogs, cysteine, and glutathione; 2) chelating agents such as phosphates, EDTA, and organic acids; 3) acidulants such as citric acid and phosphoric acid; 4) enzyme inhibitors such as aromatic carboxylic acids, aliphatic alcohols, anions, peptides, and substituted resorcinols; 5) treatments with enzymes such as oxygenases, o-methyltransferase and proteases; and 6) complexing agents such as cyclodextrins (McEvily et al., 1992). To date, there have been many studies on the different compounds used to inhibit melanosis in foods.

4.2.3 Molecular approach

The nature of PPO and its role in the immune system of plants and crustaceans have been extensively reviewed. One possible approach to reduce the activity of PPO and the resultant enzymatic browning reactions is to characterize and inactivate the PPO coding genes. This can be done by generating antisense RNAs specific for PPO.

Antisense genes have been successfully used for altering plant processes and improving crops. It involves blocking gene expression of the plant enzymes involved in a certain process. The process is based on blocking the flow of information from DNA via RNA to protein by introducing an RNA strand complementary to the sequence of the target mRNA. The antisense approach involves the insertion of a gene or a significant part of it into the cell in a reverse orientation. This approach has been used to increase the shelf life of fruits (Fray & Grierson, 1993), and commercial applications of this technology now include alteration of flower color, viral resistance induction, and fruit ripening. It has been reported that the lack of bruising sensitivity in transgenic potatoes without any side effects opens up the possibility of preventing melanosis in wide varieties of food crops and even to crustaceans, without the use of any physical and chemical treatments. Thus, by using this antisense technology, it is possible to develop fruits, vegetables and crustaceans that are resistant to enzymatic browning.

4.3 Application of ERT-rich edible mushroom (*Flammulina velutipes*) extract for controlling melanosis

As mentioned earlier, many technologies and techniques have been developed for the inhibition of melanosis in food products, and some more new approaches are under study. However, these alternatives must be evaluated in terms of their impact on the overall food quality, effectiveness, cost, and regulatory status. Because of health concerns, only a number of inhibitors have been accepted for food application, based on the basis of government health regulations.

A relatively new approach involves the design and development of another application of ERT-rich edible mushroom (*F. velutipes*) extract for inhibiting melanosis in commercially important crustaceans.

4.3.1 Dietary supplementation

The extract from the edible mushroom *F. velutipes* has been found to significantly inhibit mushroom PPO activity; prevent browning in apples; and delay melanosis in shrimps which usually develops during storage (Jang et al., 2002; 2003). These observations suggest that the mushroom extract contains certain compounds contributing to such actions. In section 3, it has been mentioned that ME containing ERT at the level of 3.03 ± 0.07 mg/mL has been reported to show remarkable DPPH radical scavenging activity and suppress lipid oxidation in bigeye tuna meat (Bao et al., 2008). ME have been reported to stabilize the fresh color of tuna meat during ice storage (Bao et al., 2009a). These results strongly suggest that ERT could be one of the major compounds in the ME that effectively inhibited mushroom PPO activity, apple browning, and melanosis in shrimps in the previous studies (Jang et al., 2002; 2003).

The *in vivo* application of L-ERT as an antioxidant in rats (Deiana et al., 2004) and the effects of feeding fish with ERT-rich ME have been studied; however, before this study, no study had been performed to determine the efficacy of extracts that contains this compound, when fed to crustaceans to control melanosis. Therefore, a study was performed on the use of a hydrophilic extract prepared from the fruiting body of *F. velutipes* for controlling melanosis and lipid oxidation in cultured kuruma shrimps (*Marsupenaeus japonicus*) *in vivo*.

In a study performed by Encarnacion et al. (2010), the hot water ME containing 2.05 ± 0.24 mg/mL ERT remarkably inhibited mushroom PPO. The PPO activity is usually found to be higher in the carapace area during postmortem of crustaceans (José-Pablo et al., 2009) and this study showed that the development of melanosis in the carapace area of the shrimps during ice storage was relatively controlled in the ME-fed group than in the control (Fig. 8). The accumulation of ERT from the ME added in the diet during feeding could have contributed to this effect.

In the same study, the PPO activities of hemolymph from shrimps fed a diet containing the ME was significantly lower than that of the control group, and the expression of the proPO genes in the hemocytes of shrimps fed with the ME was relatively lower than that in the

Novel Approach for Controlling Lipid Oxidation and Melanosis in Aquacultured Fish and Crustaceans:
Application of Edible Mushroom (Flammulina velutipes) Extract In Vivo

171

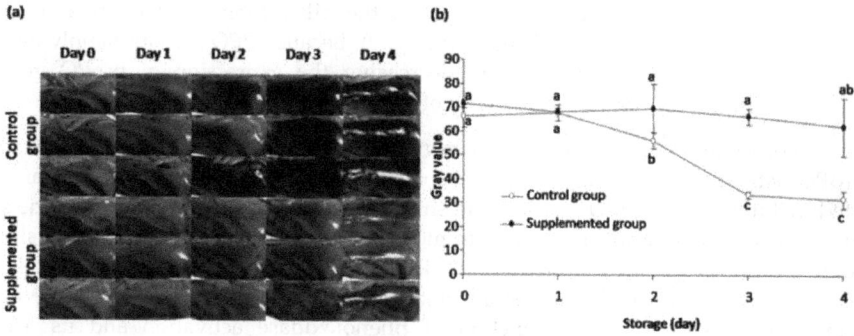

Fig. 8. Development of melanosis (a) and changes in the mean gray values of the carapace area (b) of kuruma shrimps (*Marsupenaeus japonicus*) during ice storage. Results are presented interms of mean±standard deviation ($n = 3$). The values with different superscript letters represent significant difference between groups for different storage periods ($p < 0.05$) (Encarnacion et al., 2010).

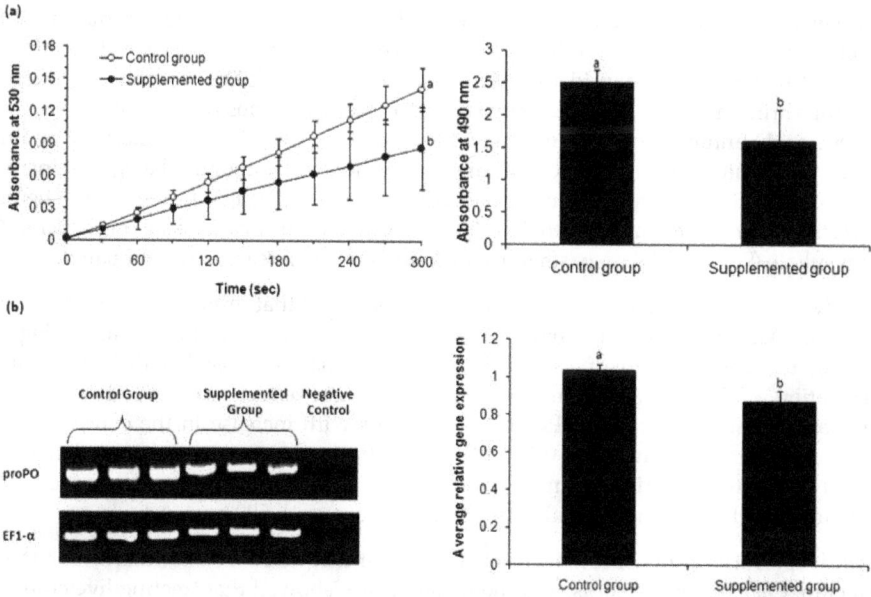

Fig. 9. (a) Enzymatic oxidation of catechol and L-DOPA (b) in the hemolymph of kuruma shrimps (*Marsupenaeus japonicus*) after 7 days of feeding. Results are presented in terms of mean ± standard deviation ($n = 4$). The values with different superscript letters represent significant difference ($p < 0.05$) at the end of the reaction period for each assay. (b) Gene expression analysis of prophenoloxidase (proPO) transcripts in the hemocytes of hemolymphs of kuruma shrimp (*M. japonicus*) after 7 days of feeding (representative gel and quantitative analyses of band intensities obtained using the ImageJ software. Elongation factor1-α gene (EF1-α) was used as the standard for computing relative gene expression level for each sample band. Results are presented in terms of mean ± standard deviation ($n = 3$). The values with different superscript letters represent significant difference ($p < 0.05$) (Encarnacion et al., 2010).

hemocytes of the control samples, showing that the ME inhibited PPO activation in the hemolymphs of the supplement-fed shrimps (Fig. 9). Because PPO activity mainly depends on the activation of the proPO system, decreasing the expression of proPO genes in hemocytes consequently reduces the activity of PPO.

The results of a study performed by Amparyup et al. (2009) indicated the significant role of the proPO activating system as one of the major immune responses in shrimps and also reported that inhibiting proPO activation in shrimps could cause death. However, the study did not show any significant differences in mortality between the supplement-fed shrimps the control samples. A study by Leclerc et al. (2006) reported that microbial infections in *Drosophila* did not require the activation of the proPO system for survival thus, raising questions regarding the precise function of phenoloxidase activation and its level of participation in the immune defense system of invertebrates because they use 2 broad but interacting categories of defense responses against pathogens, namely, the cellular and hormonal responses.

4.3.2 Immersion technique

Because feeding shrimps with MEs is time consuming and requires tedious technical work, the possibility of immersing live full-grown crustaceans in an ME solution for inhibiting postharvest melanosis was considered. Martínez-Álvarez et al. (2005) reported the efficacy of immersing shrimp in 0.05% hexylresorcinol (HR) for preventing postharvest melanosis. *In vivo* treatment with antimelanosic agents may be more effective than postmortem treatment because of greater absorption of antimelanosic agents through the gills, increased accumulation in the hemolymph, and enhanced distribution throughout the body. Therefore, the effects of *in vivo* treatment with *F. velutipes* ME, sodium ascorbate (ascorbic acid, AA), sodium sulfate (SS), and HR on postharvest melanosis in crustaceans were compared.

The study conducted by Encarnacion et al. (2011a) showed that immersing live *M. japonicus* shrimps in ME effectively inhibited postharvest melanosis in the shrimps (Fig. 10). Moreover, the development of melanosis in the carapace decreased with increase in the concentration of ME in the immersion solution. These effects may be due to greater absorption and accumulation of ERT in the shrimps with increase in the concentration of ME in the immersion solution. At 0.5 % ME concentration in pure seawater, melanosis was effectively inhibited in shrimp samples after 2 days of ice storage; the effect was the same as that obtained with 500 ppm HR and better than obtained with 500 ppm AA.

The study performed by Encarnacion et al. (2011b) also used this technique for the red queen crab, *Chionoecetes japonicus,* and the result clearly showed that treating live crabs with purified seawater containing 1.0% ME effectively prevented melanosis during 4 days of ice storage; and the effect obtained the same as that with 500 ppm SS or with 500 ppm HR solutions (Fig. 11).

ERT accumulation in shrimp tissue may inhibit PPO activity in the carapace during the postharvest period. Because thiols, such as ERT, are powerful nucleophiles that can chelate Zn^{2+} and Cu^{2+} (Hanlon, 1971; Park et al., 2006), the mechanism underlying the inhibitory effect of ERT could be due to its Cu^{2+} chelating activity. The mechanism of ERT could also be attributed to its Cu^{2+} chelating activity; thus, melanosis in the shrimps immersed in the ME solution was inhibited.

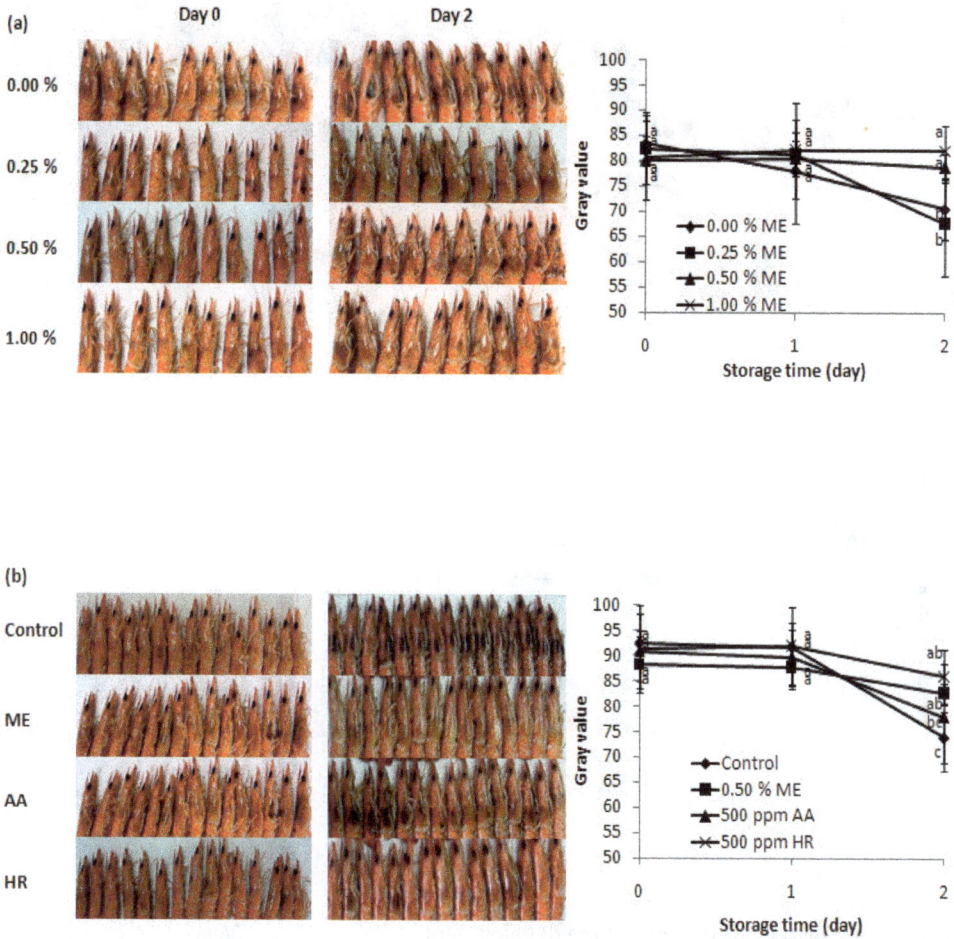

Fig. 10. Development of melanosis and changes in the mean gray value of the carapace of
Marsupenaeus japonicus shrimps before and after ice storage: (a) dose-dependent effect of
ME, (b) comparison with other antimelanosic compounds. Results are presented in terms of
mean ± standard deviation ($n = 3$). The superscript letters above each data point represent
statistically significant differences ($p < 0.05$) (Encarnacion et al., 2011a).

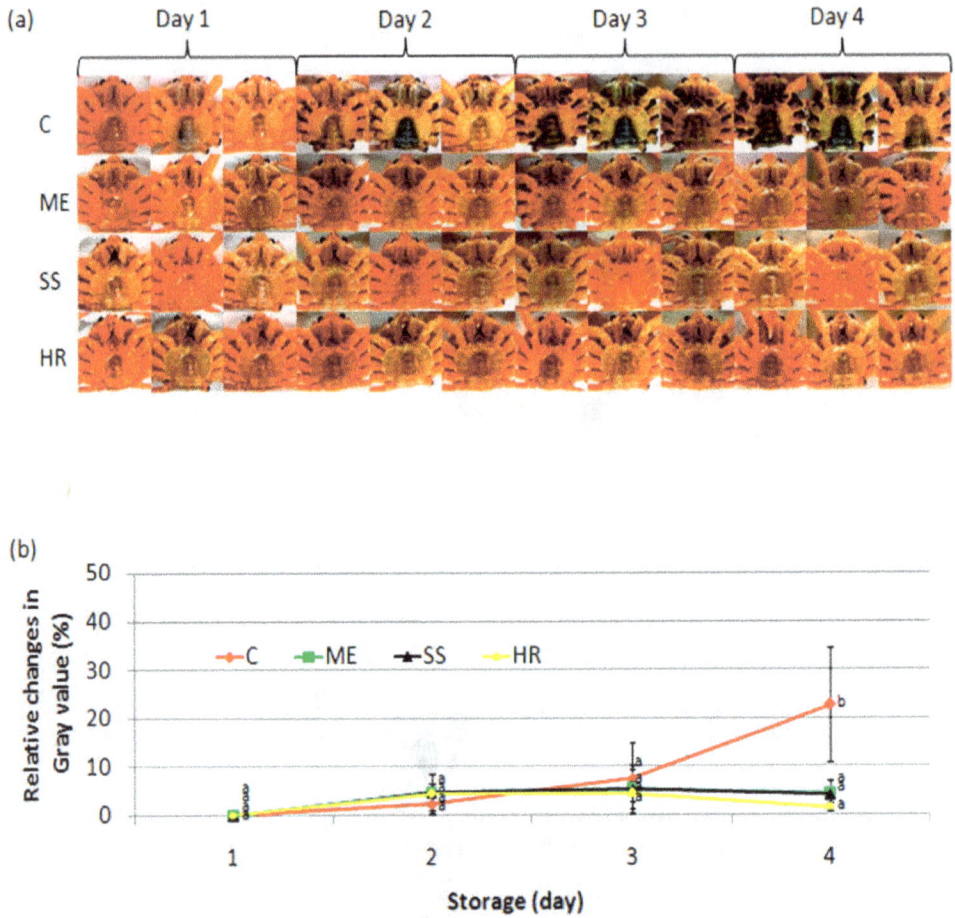

Fig. 11. (a) Digital photographs of the development of melanosis in *Chionoecetes japonicus* crabs pretreated with purified seawater (C), 1.0% mushroom extract (ME), 500 ppm sodium sulphite (SS), and 500 ppm 4-hexyl-1,3-benzenediol (HR) solutions during ice storage. (b) Relative changes in the mean gray value of the carapace of *C. japonicus* during ice storage, analyzed using the ImageJ software. Results are presented in terms of mean (SD) (n = 3). The superscript letters above each data point represent statistically significant differences ($p <$ 0.05) (Encarnacion et al., 2011b).

In addition to inhibiting PPO activity, ME also suppressed proPO gene expression in the hemolymph of *M. japonicus* (Fig. 12) and *C. japonicus* (Fig. 13). The same result was obtained in the feeding trial. Because immersing shrimp in ME significantly reduced PPO activity in

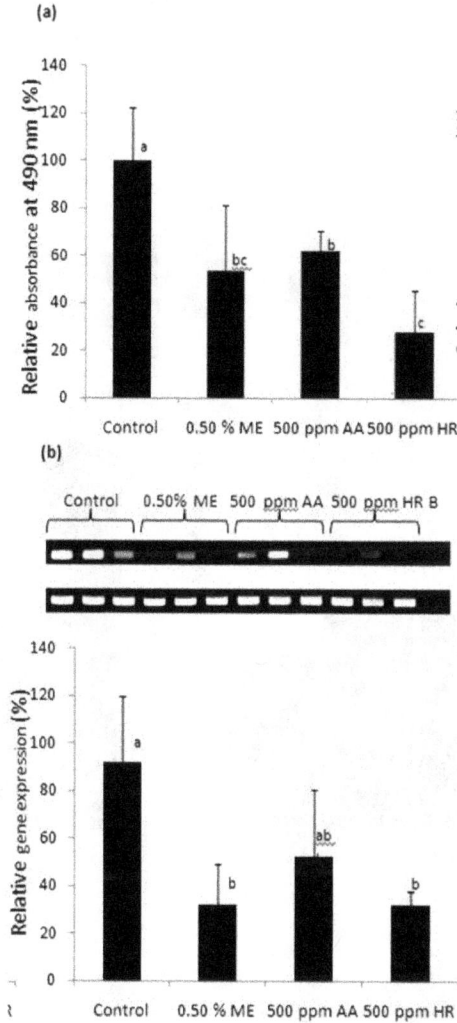

Fig. 12. (a) Enzymatic oxidation of L-DOPA in the hemolymph of kuruma shrimps (*M. japonicus*) immersed in control or treatment solutions. In each graph, the average absorbances of samples are shown relative to the average absorbance of the control (defined as 100%). Results are presented in terms of mean ± SD ($n = 5$). (b) Gene expression analysis of prophenoloxidase (proPO) transcripts in the hemocytes of *M. japonicus* shrimp immersed in control or treatment solutions. The top and bottom panels show representative gels and quantitative analyses of the band intensities obtained using the ImageJ software, respectively. The expression level of the elongation factor 1α (EF1-α) gene was used as the reference for calculating the relative expression level for each band. B denotes a blank sample (negative control). Results are presented in terms of mean ± SD ($n = 3$). The superscript letters above each bar represent statistically significant differences ($p < 0.05$) (Encarnacion et al., 2011a).

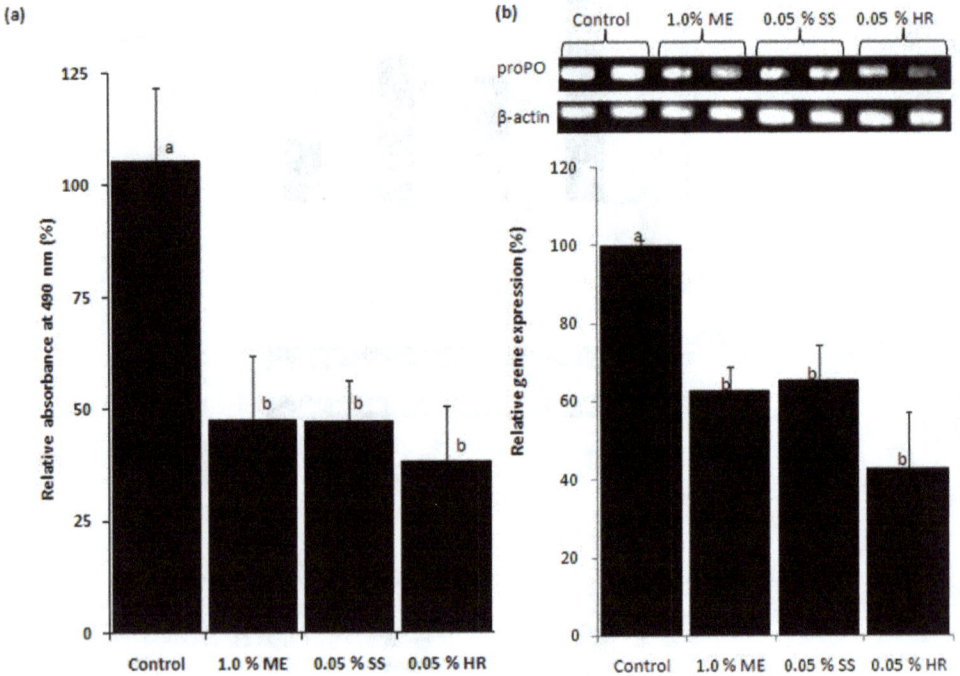

Fig. 13. (a) Enzymatic oxidation of L-DOPA in the hemolymph of red queen crabs (*Chionoecetes japonicus*) immersed in control or treatment solutions. In each graph, the average absorbances of samples are shown relative to the average absorbance of the control (defined as 100%). Results are presented in terms of mean ± SD ($n = 5$). (b) Gene expression analysis of prophenoloxidase (proPO) in the hemocytes of *C. japonicus* immersed in purified seawater (Control), 1.0% mushroom extract (ME), 500 ppm sodium sulphite (SS), or 500 ppm 4-hexyl-1,3-benzenediol (HR) solutions, performed using reverse transcription–polymerase chain reaction. (b) Quantitative analysis of the band intensities obtained using the ImageJ software. The expression level of the β-actin gene was used as the reference for carculating the relative expression level for each band. Results are presented in terms of mean (SD). The superscript letters above each bar represent statistically significant differences (p < 0.05) (Encarnacion et al., 2011b).

the hemolymph, it is possible that the ME containing ERT inhibited PPO activation. The observation that proPO gene expression in the hemocytes of ME-treated shrimps was lower than that in control shrimps supports this hypothesis. Thus, decreasing the expression of proPO reduces the amount that can be proteolytically activated into PPO (Adachi et al., 2003).

The efficacy of ME in inhibiting postharvest melanosis is comparable to the efficacy of existing antimelanosic compounds, such as HR, AA and SS in this regard. HR forms an inactive complex with PPO that is incapable of catalyzing melanosis development (Guererro-Beltran et al., 2005). Therefore, PPO activity significantly decreases in the presence of HR (Encarnacion et al., 2010; 2011). HR possesses antimicrobial activity in addition to its antimelanosic activity. The antimicrobial activity of HR (Martínez-Álvarez et al., 2005) may be related to the decrease in proPO gene expression because invading microorganisms trigger a proteolytic cascade culminating in the release of active PPO (Leclerc et al., 2006). Because AA did not significantly affect proPO gene expression, its mechanism of inhibiting postharvest melanosis is most likely to be due to its reducing power. The slight melanosis that occurred in AA-immersed shrimps after 2 days of ice storage may be due to the oxidation of AA, which prevented it from reducing o-quinones, thereby allowing melanin development (Guererro-Beltran et al., 2005). Sulfites are known to inhibit both enzymatic and nonenzymatic browning reactions. Bisulphite is reported to inhibit melanosis by 2 mechanisms: 1) by reacting with intermediate quinones in the melanosis reaction that forms sulfoquinones; and 2) by irreversibly reacting with PPO, causing complete inactivation (Ferrer et al., 1989). Therefore, PPO activity also significantly decreases in the presence of SS. SS and HR posses antimicrobial activities in addition to their antimelanosic activities, SS and HR are also antimicrobial agents. The antimicrobial activity of SS (McFeeters et al., 2004; Martínez-Álvarez et al., 2005) may also be related to the decrease in proPO gene expression.

4.3.3 Mechanism underlying the inhibitory effect of the extract on melanosis

The mushroom hot water ME remarkably inhibited mushroom PPO. Commercial L-ERT had the same effect on mushroom PPO. Mushroom PPO activity remarkably declined with increase in the concentrations of L-ERT (Fig. 14). PPO inhibition by the ME was also depended on L-ERT concentration. These results suggest that extracts containing ERT as one of their active compounds potently inhibit mushroom PPO activity.

In vitro experiments performed by Encarnacion et al. (2010) also showed that the transcript expression of the proPO genes in the HLS was lower than that in the L-ERT- and *p*-amidinophenyl methanesulfonyl fluoride hydrochloride (*p*-APMSF)-treated HLSs (Fig. 15). The PPO activity in the L-ERT-treated HLS was also remarkably low. ERT could have been involved in inhibiting transcriptional factors in the cascade system, leading to a decrease in proPO gene expression. Maeda et al. (2007) reported that in human melanosis, hydroperoxy traxastane-type triterpene decreased the protein levels of PPO and its related proteins in B16 melanoma cells because of inhibition of the transcription factor melanocyte-type isoform of the microphthalmia-associated transcription factor, leading to a decrease in the PPO gene and its related genes. In the case of crustaceans, it is generally believed that the enzyme responsible for the activation of proPO to PPO is a serine protease, or commonly called as proPO-activating enzyme (PPAE). This enzyme is believed to be involved in the final step in

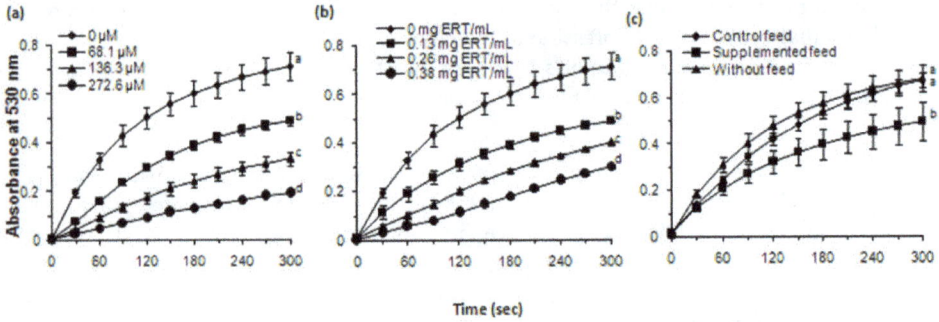

Fig. 14. Inhibitory effects of commercial L-ergothioneine (a), mushroom (*F. velutipes*) extract (b), and residues of diets used in the feeding trial (c) on the activity of mushroom polyphenoloxidase. Results are presented in terms of mean ± standard deviation ($n=3$). The values with different superscript letters represent significant difference ($p<0.05$) at the end of the reaction period (Encarnacion et al., 2010).

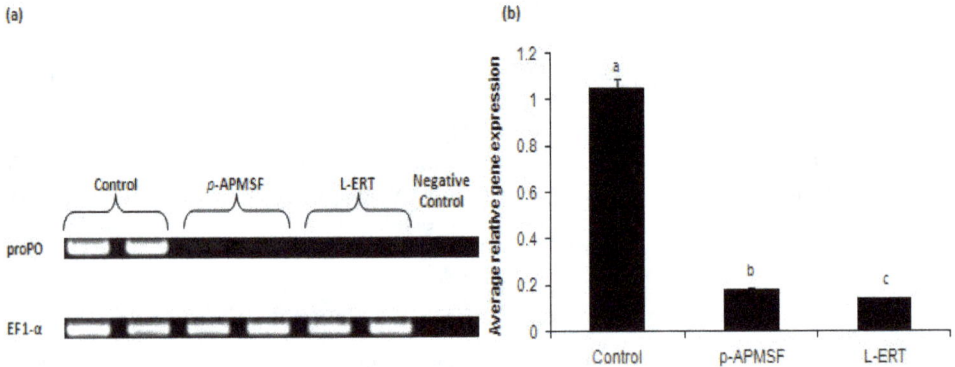

Fig. 15. Inhibition of the activation of the proPO system in hemocyte lysate supernatant by commercial L-ergothioneine (L-ERT) and *p*-amidinophenyl methanesulfonyl fluoride hydrochloride (*p*-APMSF). Polyphenoloxidase activity (a) and peptidase activity (b). 7-amino-4-methylcoumarin (AMC) was used as the standard solution in the peptidase activity assay. The final concentration of each inhibitor in the reaction system was 1 mM. Results are presented in terms of mean ± standard deviation (n = 3). The values with different superscript letters represent significant difference between groups for different storage periods ($p < 0.05$) (Encarnacion et al., 2010).

the proPO cascade leading to PPO activation and is itself tightly regulated (Buda & Shafer, 2005). The results of the peptidase activity assay indicated that L-ERT inhibited serine protease activity but that this effect was not as strong as that of *p*-APMSF (Fig. 16b). However, the presence of L-ERT could possibly affect the overall activation of the proPO system because it inhibited PPO activity (Fig.16a). At least 4 mechanisms underlying PPAE regulation have been recognized: gene induction, activation by another protease, requirement for noncatalytic serine protease homologues as cofactors, and inactivation by serine protease inhibitors (Wang & Jiang, 2004). Thus, other proPO activating factors such as

Novel Approach for Controlling Lipid Oxidation and Melanosis in Aquacultured Fish and Crustaceans:
Application of Edible Mushroom (Flammulina velutipes) Extract In Vivo
179

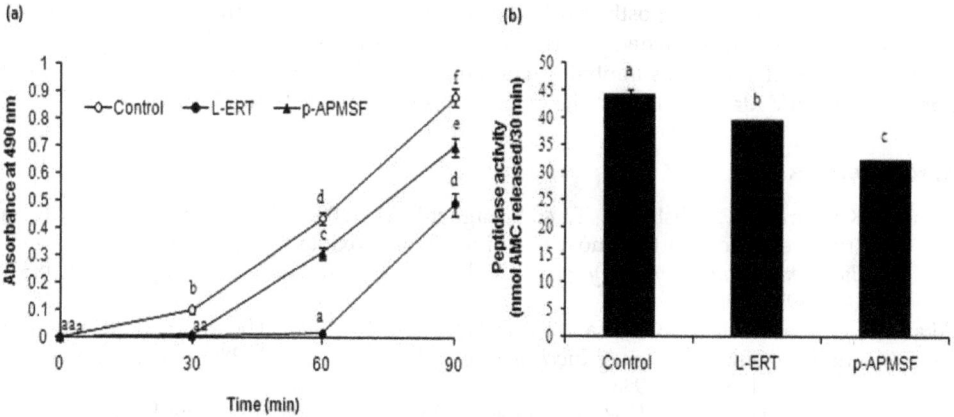

Fig. 16. Gene expression analysis of proPO transcripts in hemocyte lysate supernatant treated with commercial L-ergothioneine (L-ERT) and p-amidinophenyl methanesulfonyl fluoride hydrochloride (p-APMSF). Representative gel (a) and quantitative analyses of band intensities obtained using the ImageJ software (b). Elongation factor1-α gene (EF1-α) was used as the standard for computing the relative gene expression level for each sample band. The final concentration of the inhibitor in the reaction system was 1mM (Encarnacion et al., 2010).

serine protease homologues and other proteases could have been inhibited by L-ERT. The inhibition of PPO activity in the hemolymphs of the supplement-fed shrimp supports this phenomenon. Moreover, accumulation of ERT in the shrimp muscles by feeding or in the hemolymph by the immersion technique could directly inhibit PPO activity in the carapace during the postharvest period. The thiol (SH) group, present in compounds such as ERT, is a powerful nucleophile, that tends to chelate Zn^{2+} and Cu^{2+} (Park et al., 2006). The latent PPO activity of hemocyanin, a copper-binding protein, in whiteleg shrimp Penaeus vannamei has been shown to be involved in postmortem melanosis. The mechanism underlying the inhibitory effect of ERT could also be attributed to its Cu^{2+} chelating activity, leading to the inhibition of melanosis in ME-fed and ME-immersed shrimps and crabs.

5. Summary

Mushroom trimmings are cost-effective and scalable source of ME and contain significant amounts of ERT, which has been proven to have antioxidative and melanosis-inhibiting properties. Yellowtail fish S. quinqueradiata that were fed a diet, including ME showed significantly reduced myoglobin and lipid oxidation in dark muscles during chilled storage. The results of the feeding trial are also supported by data from similar trials on other aquacultured fish species such as jack mackerel and rainbow trout (data not shown). Immersion of live M. japonicus shrimps and C. japonicus crabs in an F. velutipes ME solution containing a significant amount of ERT effectively inhibited postharvest melanosis in shrimps and crabs, and this result was comparable to those obtained using melanosis-inhibiting agents such as HR, AA and SS. In vitro experiments with exogenous ERT confirmed that it inhibits PPO activity and proPO gene expression in shrimp HLS. Thus, ME containing ERT is a promising natural alternative to the synthetic melanosis-inhibiting agents used to prevent postharvest melanosis in shrimps and other crustaceans. Furthermore, the use of biowaste as the raw material for extracting and producing antioxidants or of melanosis-inhibiting compounds such as ERT for food and industrial applications could also help reduce the growing problem of agricultural waste.

6. References

Adachi, K.; Hirata, T.; Nishioka, T. & Sakaguchi, M. (2003). Hemocyte components in crustaceans convert hemocyanin into phenoloxidase-like enzyme. *Comparative Biochemistry and Physiology Part B*, Vol. 134, No. 1, (January 2003), pp. 135-41, ISSN 1096-4959

Akanmu, D.; Cecchini, R.; Aruoma, O. I. & Halliwell, B. (1991). The antioxidant action of ergothioneine. *Archives of Biochemistry and Biophysics*, Vol. 288, No. 1, (July 1991), pp. 10-16, ISSN 0003-9861

Aksnes, A. & Njaa, L. R. (1981). Catalase, glutathione peroxidase and superoxidase dismutase in different fish species. *Comparative Biochemistry and Physiology Part B*, Vol. 69, No.4, pp. 893-896, ISSN 1096-4959

Amparyup, P.; Charoensapsri, W. & Tassanakajon, A. (2009). Two prophenoloxidases are important for the survival of *Vibrio harveyi* challenged shrimp *Penaeus monodon*. *Developmental and Comparative Immunology*, Vol. 33, No. 2, (February 2009), pp. 247-256, ISSN 0145-305X

Arduini, A.; Eddy, L. & Hocstein, P. (1990). The reduction of ferryl myoglobin by ergothioneine: A novel function for ergothioneine. *Archives of Biochemistry and Biophysics*, Vol. 281, No. 1, (August 1990), pp. 41-43, ISSN 0003-9861

Ashida, S. & Sato, M. (2005). Screening of edible plants for reducing activity by monitoring their effects on the oxidation of oxymyoglobin. *Food Science and Technology Research*, Vol. 11, No. 3, (October 2005), pp. 349-354, ISSN 1344-6606

Ashie, I.N.A. & Simpson, B.K. (1996). α;2 Macroglobulin inhibition of endogenous proteases in fish muscle. *Journal of Food Science*, Vol. 61, No. 2, (March, 1996) pp. 357-361, ISSN 1750-3841

Bao, H.N.D.; Ushio, H. & Ohshima, T. (2008). Antioxidative activity and antidiscoloration efficacy of ergothionine in mushroom (*Flammulina velutipes*) extract added to beef and fish meats. *Journal of Agricultural and Food Chemistry*, Vol. 56, No. 21, (October 2008), pp. 10032-10040, ISSN 0021-8561

Bao, H.N.D.; Ushio, H. & Ohshima, T. (2009a). Antioxidative activities of mushroom (*Flammulina velutipes*) extract added to bigeye tuna meat: dose-dependent efficacy and comparison with other biological antioxidants. *Journal of Food Science*, Vol. 74, No. 2, (March 2009) pp. 162-169, ISSN 1750-3841

Bao, H.N.D.; Shinomiya, Y.; Ikeda, H. & Ohshima, T. (2009b). Preventing discoloration and lipid oxidation in dark muscle of yellowtail by feeding an extract prepared from mushroom (*Flammulina velutipes*) cultured medium. *Aquaculture*, Vol. 295, No. 4, (October 2009), pp. 243-249, ISSN 0044-8486

Boldyrev, A. A.; Dupin, A. M.; Bumin, A. Y.; Balbizhaev, M. A. & Severin, S. E. (1987). The antioxidative properties of carnosine, a natural histidine containing dipeptide. *Biochemistry International*, Vol. 15, No. 6, (June 1987), pp. 1105-1113, ISSN 0158-5231

Buda, E.S. & Shafer, T.H. (2005). Expression of a serine proteinase homolog prophenoloxidase-activating factor from the blue crab, *Callinectes sapidus*. *Comparative Biochemistry and Physiology Part B*, Vol. 140, No. 4, (April 2005), pp. 521-531 ISSN 1096-4959

Cerenius, L.; Lee, B.L. & Söderhäll, K. (2008). The proPO-pros and cons for its role in invertebrate immunity. *Trends in Immunology*, Vol. 29, No.6, (June 2008), pp. 263-271, ISSN 1471-4906

Christie, W. W. (1987). Lipids: their structures and occurrence, In: *High Performance Liquid Chromatography and Lipids. A Practical Guide*, W. W. Christie (Ed.), 1-20, Paragon Press, ISBN 008-03421-2-4, Oxford, USA

Cheftel, J. C. (1992). High-pressure, microbial inactivation and food preservation. *Food Science and Technology International*, Vol. 1, No. 2-3, (August 1992), pp. 75-90, ISSN 1532-1738

Cheung, L. M. & Cheung, P. C. K. (2005). Mushroom extracts with antioxidant activity against lipid peroxidation. *Food Chemistry*, Vol. 89, No. 3, (February 2005), pp. 403-409, ISSN 0308-8146

Deiana, M.; Rosa, A.; Casu, V.; Piga, R.; Dessí, M. & Aruoma, O.I. (2004). L-ergothioneine modulates oxidative damage in the kidney and liver of rats in vivo: studies upon the profile of polyunsaturated fatty acids. *Clinical Nutrition*, Vol. 23, No. 2, (April 2004), pp. 183-193, ISSN 0261-5614

Dubost, N.J.; Beelman, R.; Peterson, D. & Royse, D. (2006). Identification and quantification of ergothionine in cultivated mushrooms using liquid chromatography-mass spectroscopy. *International Journal of Medicinal Mushroom*, Vol. 8, No.3, (July 2006), pp. 215-222, ISSN 1521-9437

Dubost, N. J.; Ou, B. & Beelman, R. B. (2007). Quantification of polyphenols and ergothioneine in cultivated mushrooms and correlation to total antioxidant capacity. *Food Chemistry*, Vol. 105, No. 2, (February 2007), pp. 727- 735, ISSN 0308-8146

Dziedric, S. Z. & Hudson, B. J. F. (1984). Phenolic acids and related compounds as antioxidants for edible oils. *Food Chemistry*, Vol. 14, No. 1, (January 1984), pp. 45-51, ISSN 0308-8146

Elmastas, M.; Isildak, O.; Turkekul, I. & Temur, N. (2007). Determination of antioxidant activity and antioxidant compounds in wild edible mushrooms. *Journal of Food Composition Analysis*, Vol. 20, No. 3-4, (May 2007), pp. 337-345, ISSN 0889-1575

Encarnacion, A. B.; Fagutao, F.; Hirono, I.; Ushio, H. & Ohshima, T. (2010) Effects of ergothioneine from mushrooms (*Flammulina velutipes*) on melanosis and lipid oxidation of kuruma shrimp (*Marsupenaeus japonicus*). *Journal of Agricultural and Food Chemistry*, Vol. 58, No. 4, (January 2010), pp. 2577-2585, ISSN 0021-8561

Encarnacion, A. B.; Fagutao, F.; Hirayama, J.; Tereyama, M.; Hirono, I. & Ohshima T. (2011a). Ergothioneine from edible mushroom (*Flammulina velutipes*) as a potent inhibitor of melanosis in kuruma shrimp (*Marsupenaeus japonicus*). *Journal of Food Science*, Vol. 1, No. 76, (January/February 2011), pp. 52-58, ISSN 1750-3841

Encarnacion, A. B.; Fagutao, F.; Shozen, K.; Hirono, I., & Ohshima, T. (2011b) Biochemical intervention of ergothioneine-rich edible mushroom (*Flammulina velutipes*) extract inhibits melanosis in the crab (*Chionoecetes japonicus*). *Food Chemistry*, Vol. 127, No. 4, (August 2011), pp. 1594-1599, ISSN 0308-8146

Erickson, M.C. (1992). Changes in lipid oxidation during cooking of refrigerated minced channel catfish muscle. In: *Lipid oxidation in food*, A.J. St. Angelo (Ed.), 344-351, American Chemical Society, ISBN 084-12246-1-7, USA

Ferrer, O.; Otwell, W. & Marshall, M. (1989). Effect of bisulfate on lobster shell phenoloxidase. *Journal of Food Science*. Vol, 54, No. 2, (March 1989), pp. 478-480, ISSN 1750-3841

Food and Agriculture Organization (FAO) of the United Nations. (2006). Aquaculture, In: *The State of World Fisheries and Aquaculture 2006*, 11.05.2011, Available from: http://www.fao.org/docrep/009/A0699e/A0699e00.htm

Frankel, E.N. (1998). Antioxidants, In: *Lipids Oxidation*, E. N. Frankel, (Ed), 209-258, The Oily Press, ISBN 095-14171-9-3, Dundee, Scothland

Franzoni, F.; Colognato, R.; Galetta, F.; Laurenza, I.; Barsotti, M.; Stefano, R. D.; Bocchetti, R.; Regoli, F.; Carpi, A.; Balbarini, A.; Migliore, L. & Santoro, G. (2006). An *in vitro* study on the free radical scavenging capacity of ergothioneine comparison with reduced glutathione, uric acid and trolox. *Biomedicine and Pharmacotherapy*, Vol. 60, No. 8, (September 2006), pp. 45-47, ISSN 0753-3322

Fray, R.G. & Grierson D. (1993). Molecular genetics of tomato fruit ripening. *Trends in Genetics*, Vol. 9, No.12, (December 1993), pp. 438-443, ISSN 168-9525

Fu, H; Shieh, D. & Ho, C. (2002). Antioxidant and free radical scavenging activities of edible mushrooms. *Journal Food Lipids* Vol.9, No. 1, (March 2002), pp. 35-46, ISSN 1065-7258

García-Carreño, F.L.; Cota, K. & Navarrete del Toro, M.A. (2008). Phenoloxidase activity of hemocyanin in whiteleg shrimp *Penaeus vannamei*: conversion, characterization of catalytic properties, and role in postmortem melanosis. *Journal of Agricultural and Food Chemistry*, Vol. 56, No. 15, (July 2008), pp. 6454-6459, ISSN 0021-8561

Grosch, W.; Konopka, U. C. & Guth, H. (1992). Characterization of off-flavors by aroma extract dilution analysis, In: *Lipid oxidation in food*, A.J. St. Angelo (Ed), 266-279, American Chemical Society, ISBN 084-12246-1-7, USA

Grunwald, E. W. & Richarsd, M. P. (2006). Mechanisms of heme protein-mediated lipid oxidation using hemoglobin and myoglobin variants in raw and heated washed muscle. *Journal of Agricultural and Food Chemistry*, Vol. 54, No. 21, (September 2006), pp. 8271- 8280, ISSN 0021-8561

Guererro-Beltran, J.; Swanson, B. & Barbosa-Cánovas, G. (2005). Inhibition of polyphenoloxidase in mango puree with 4-hexylresorcinol, cysteine and ascorbic acid. *LWT-Food Science and Technology*, Vol. 38, No.6, (September 2005) pp. 625-630, ISSN 0023-6438

Hanlon, D. (1971). Interaction of Ergothioneine with metal ions and metalloenzymes. *Journal of Medicinal Chemistry*, Vol. 14, No. 1, (January 1971). pp. 1084-1087, ISSN 0023-6438

Hermes, J.E. (1998). *Fish Processing in the Tropics*, Tawid Publications, ISBN 971-91395-7-9, Quezon City, Philippines.

Jacobsen, C. (1999). Sensory impact of lipid oxidation in complex food systems. *Lipids*, Vol. 101, No. 5, (May 1999), pp. 484-490, ISSN 0024-4201

Jang, M.S.; Sanada, A.; Ushio, H.; Tanaka, M. & Ohshima, T. (2002). Inhibitory effect of 'enokitake' mushroom extracts on polyphenol oxidase and prevention of apple browning. *LWT-Food Science and Technology*, Vol. 35, No. 8, (December 2002), pp. 697-702, ISSN 0023-6438

Jang, M.S.; Sanada, A.; Ushio, H.; Tanaka, M. & Ohshima, T. (2003). Inhibitory effect of enokitake extract on melanosis of shrimp. *Fisheries Science*, Vol. 69, No. 2, (April 2003), pp. 379-384, ISSN 0919-9268

Jang, M.S.; Eun, J.B.; Ushio H. & Ohshima T. (2004). Antioxidative properties of mushroom (*Flammulina velutipes*) on the oxidation of cod liver oil in emulsion. *Food Science Biotechnology*, Vol. 13, No. 2, (June 2004), pp. 215-218, ISSN 1226-7708

José-Pablo, Z.; Martínez-Álvarez, O.; Montero, P. & Gómez-Guillén, M. (2009). Characterization and tissue distribution of polyphenoloxidase of deepwater pink shrimp (*Parapenaeus longirostris*). *Food Chemistry*, Vol. 112, No. 1, (January, 2009), pp. 104-111, ISSN 0308-8146

Kanner, J. & Kinsella, J. E. (1983). Initiation of Lipid oxidation by a peroxidase/hydrogen peroxide/halide system. *Lipids*, Vol. 18, No. 3, (May 1983), pp. 204-210, ISSN 0024-4201

Kanner, J.; Sofer, F.; Harel, S. & Doll, L. (1988). Antioxidant activity of ceruloplasmin in muscle membrane and *in situ* lipid peroxidation. *Journal of Agricultural and Food Chemistry*, Vol. 36, No. 3, (May 1988), pp. 415-417, ISSN 0021-8561

Kim, J.; Marshall, M.R. & Wei, C. (2002). Polyphenoloxidase, In: *Seafood Enzymes: Utilization and Influence on Postharvest Seafood Quality*, N. Haard, & B. Simpson (Ed.), 271-315, Marcel Dekker Inc., ISBN 082-47032-6-X, New York, USA

Kim, Y. J. & Uyama H. (2005). Tyrosinase inhibitors from natural and synthetic sources: structure, inhibition mechanism and perspective for the future. *Cellular and Molecular Life Science*, Vol. 62, No. 15, (August 2005), pp. 1707-23, ISSN 1420-682X

Leclerc, V.; Pelte, N.; El Chamy, L.; Martinelli, C.; Ligoxygakis, P.; Hoffmann, J.A. & Reichhart, J.M. (2006). Prophenoloxidase activation is not required for survival to microbial infections in *Drosophila*. *EMBO reports*, Vol. 7, No. 1, (July 2006), pp. 231-235, ISSN 1469-3178

Lee, C.Y.; Smith, N.L. & Hawbecker, D.E. (1988). Enzyme activity and quality of frozen green beans as affected by blanching and storage. *Journal of Food Quality*, Vol. 11, No. 4, (November, 1988), pp. 278-287, ISSN 0146-9428

Love, J. D. & Pearson, A. M. (1971). Lipid oxidation in meat and meat products- A review. *Journal of American Oil Chemist Society*, Vol. 48, No. 9, (September, 1971), pp. 547-549, ISSN 0003-021X

Maeda, K.; Naitou, T.; Umishio, K.; Fukuhara, T. & Motoyama, A. (2007). A novel melanin inhibitor: hydroperoxy traxastane-type triterpene from flowers of *Arnica montana*. *Biological and Pharmaceutical Bulletin*, Vol. 30, No. 5, (May 2007), pp. 873-879, ISSN 0918-6158

Makishima, S.; Nozaki, K.; Mizuno, M.; Netsu, E.; Shinji, K.; Shibayama, T.; Kanda, T. & Amano, Y. (2006). Recovery of soluble sugars from waste medium for Enokitake (*Flammulina velutipes*) mushroom cultivation with hydrothermal reaction and enzyme digestion. *Journal of Applied Glycoscience*, Vol. 53, No. 4, (October 2006), pp. 261-266, ISSN 1344-7882

Martínez-Álvarez, O.; López-Caballero, M.E; Montero, P. & Gómez-Guillén, M. (2005). A 4-hexylresorcinol-based formulation to prevent melanosis and microbial growth in chilled tiger prawns (*Marsupenaeus japonicas*) from aquaculture. *Journal of Food Science*, Vol. 70, No. 9, (November 2005), pp. 415-422, ISSN 1750-3841

Martinez, M. & Whitaker, J. (1995). The biochemistry and control of enzymatic browning. *Trends in Food Science and Technology*, Vol. 6, No. 6, (June 1995), pp. 195-200, ISSN 0924-2244

Mau, J. L.; Lin, H. C. & Song, S. F. (2002). Antioxidant properties of several specialty mushrooms. *Food Research International*, Vol. 35, No. 6, (November 2002), pp. 519-526, ISSN 0963-9969

Mayer, M. P.; Beyer, P. & Kleinig, H. (1990). Quinone compounds are able to replace molecular oxygen as terminal electron acceptor in phytoene desaturation in chromoplasts of *Narcissus pseudonarcissus* L. *European Journal of Biochemistry*, Vol. 191, No. 2, (July 1990), pp. 359-363, ISSN 0014-2956

McEvily, A.J.; Iyengar, R. & Otwell, W.S. (1992). Inhibition of enzymatic browning in foods and beverages. *Critical Review on Food Science and Nutrition*, Vol. 32, No. 3, (October 1992), pp. 253-273, ISSN 1040-8398

McFeeters, R.; Barrangou, L.; Barish, A. & Morrison, S. (2004). Rapid softening of acidified peppers: effect of oxygen and sulphite. *Journal of Agricultural and Food Chemistry*, Vol. 52, No. 14, (June 2004), pp. 4554-4557, ISSN 0021-8561

Miki, W. (1991). Biological functions and activities of animal carotenoids. *Pure and Applied Chemistry*, Vol. 63, No. 1, (May 1990), pp. 141-146, ISSN 0033-4545

Ohshima, T.; Fujita, Y. & Koizumi, C. (1993). Oxidative stability of sardine and mackerel lipids with reference to synergism between lipids and α-tocopherol . *Journal of American Oil Chemist Society*, Vol. 70, No. 3, (March 1993), pp. 269-276, ISSN 000-021X

Pan, X.; Hideki, U. & Ohshima, T. (2004). Photo-oxidation of lipids impregnated on the surface of dried seaweed (*Porphyra yezoenosis* Ueda): Hydroperoxide distribution. *Journal of American Oil Chemist Society*, Vol. 81, No. 8, (August 2004), pp. 765-771, ISSN 003-021X

Park, Y.D; Lyou, Y.J.; Hahn, H.S.; Hahn M.J. & Yang, J.M. (2006) Complex inhibition of tyrosinase by thiol-composed Cu^{2+} chelators: a clue for designing whitening agents. *Journal of Biomolecular Structure and Dynamics*, Vol. 24, No. 2, (October 2006), pp. 131-137, ISSN 0739-110

Petillo, D.; Hultin, H. O.; Krzynowek, J. & Autio, W. R. (1998). Kinetics of antioxidant loss in mackerel light and dark muscle. *Journal of Agricultural Food Chemistry*, Vol. 46, No.10, (October 1998), pp. 4128-4137, ISSN 0021-8561

Quang, D. N.; Hashimoto, T. & Asakawa, Y. (2006). Inedible mushroom: A good source of biologically active substances. *The Chemistry Record*, Vol. 6, No. 2, (February 2006) pp. 79-99, ISSN 1527-8999

Renerre, M. (1990). Review: Factors involved in the discoloration of beef meat. *International Journal of Food Science and Technology*, Vol. 25, No. 6, (December 1990), pp. 613-630, ISSN 1365-2621

Roginsky, V. & Lissi, E. A. (2005). Review of methods to determine chain-breaking antioxidant in food. *Food Chemistry*, Vol. 92, No. 2, (September 2005), pp. 235-254, ISSN 0308-8146

Romero, F. T.; Ordonez, I.; Arduini, A. & Cadenas, E. (1992). The reactivity of thiols and disulfides with different redox states of myoglobin. Redox and addition reactions and formation of thiyl radial intermediates. *Journal of Biology and Chemistry*, Vol. 267, No. 3, (January 1992), pp. 1680-1688, ISSN 0021-9258

Syvaoja, E. L. & Salminen, K. (1985). Tocopherols and tocotrienols in finfish foods: Fish and fish products. *Journal of American Oil Chemist Society*, Vol. 62, No. 8, (August 1995), pp. 1245-1248, ISSN0021-9258

Vamos-Vigyaso, L. (1981). Polyphenol oxidase and peroxidasein fruits and vegetables. *Critical Review on Food Science and Nutrition*, Vol. 14, No. 1 (January 1981), pp. 44-50, ISSN 1040-8398

Wang, Y. & Jiang, H. (2004). Purification and characterization of *Manduca sexta* serpin-6: a serine proteinase inhibitor that selectively inhibits polyphenoloxidase-activating proteinase-3. *Insect Biochemistry and Molecular Biology*, Vol. 34, No. 4, (April 2004), pp. 387-395, ISSN 0965-1748

Wasser, S. P. (2002). Medicinal mushrooms as a source of anti tumor and immunomodulating polysaccharides. *Applied Microbiology and Biotechnology*, Vol. 60, No. 3, (March 2002), pp. 258-274, ISSN 0175-7598

Watanabe, F.; Goto, M.; Abe, K. & Nakano, Y. (1996). Glutathione peroxidase activity during storage of fish muscle. *Journal of Food Science*, Vol. 61, No. 4, (July 1996), pp. 734-735, ISSN 1750-3841

Part 4

Culture Techniques and Management

10

Improving Larval Culture and Rearing Techniques on Common Snook (*Centropomus undecimalis*)

Carlos Yanes Roca[1] and Kevan L. Main[2]
[1]*Stirling University*
[2]*Mote Marine Laboratory*
[1]*UK*
[2]*USA*

1. Introduction

The common snook or *Centropomus undecimalis* (Bloch) is a diadromus, stenothermic, euryhaline, estuarine-dependent species found in the tropical and sub-tropical western Atlantic Ocean from about 34° N to about 25° S latitude (Howells *et al.*, 1990). The snook physiology is characterized by a distinct lateral line, high divided dorsal fin, sloping forehead, a large mouth, a protruding lower jaw and a yellow pelvic fin (Fore & Schmidt, 1973).

Partial genetic isolation occurs between Florida's Atlantic and Gulf Coast stocks (Tringali & Bert, 1996). Snook are protandric hermaphrodites: some males develop into females between 1 and 7 years of age, having a maximum 20-year lifespan. Females are generally larger than males of the same age, at the same time it is unusual to find females smaller than 500 mm in fork length. Snook growth rates are highly variable. For instance, Atlantic Coast fish grow more quickly and to a larger size than do fish on the Gulf Coast (Taylor *et al.* 2000).

Common snook form the basis of important fisheries throughout their range due to their sporting and culinary attributes (Tucker *et al.*, 1985; Matlock & Osburn, 1987). Numbers of common snook have declined over recent years due to shoreline development, fishing pressure, and loss of coastal habitats. As a result, common snook were designated as a game fish restricted to recreational harvest only. Depletion of some Florida stocks during the late 1970's and the early 1980's (Bruger & Haddad, 1986) resulted in common snook being declared a species of special concern, and they are now protected by strict regulations enacted by the Florida legislature.

The ultimate objective of hatchery-based production of common snook in Florida is the supply of high quality animals to restore declining stocks and enhance local populations. The quality of the juveniles, environmental conditions and releasing techniques are all involved in the success of restocking programs (Tsukamoto 1993). The objective of larval rearing is to mass-produce high-quality and healthy juvenile fish. The management of both the rearing environment and feeding regime are the most important aspects of this activity.

To improve larval rearing techniques, a good understanding of larval morphology, behaviour, live food and artificial diet requirements, and environmental conditions is fundamental (Liao et al., 2001).

Collection of data on the conditions required for spawning, larval rearing, and release of common snook into marine and freshwater systems began in 1974 at the Florida Game and Fresh Water Fish Commission (Ager *et al.*, 1978; Shafland and Koehl, 1980, Chapman et al., 1978). Information on the development of laboratory reared larvae and juveniles (Lau and Shafland, 1982) and on the lower lethal temperature (15°C) for juveniles (Howells *et al.*, 1990) was also documented. These studies described the basic common snook biology and the principles for captive rearing. Although research on this species in Florida and Texas was carried out during the 1970's, 1980's and 1990's, there are still a number of gaps in our understanding of the requirements for successful larval rearing and broodstock management.

1.1 Snook larval culture

Techniques of larviculture have gradually been developed from simply collecting the stocking material in the wild to using modern, advanced facilities for complete larviculture practices (Liao *et al.*, 2001). Common snook culture research in Florida, Texas, Mexico and Brazil has primarily relied on collection of fertilized eggs from mature wild fish and more recently at Mote Marine Laboratory on captive broodstock production. At this time, snook larviculture practices are still under-development when compared to many other marine fish species, such as red drum and cobia. This paper is focused on larval rearing during the first 14 days after hatching using wild, strip spawned eggs. The techniques investigated include the design of a larviculture system, diet requirements of larvae, and system management.

1.1.1 Importance of temperature on embryonic and larval development

Nearly every aspect of early fish development is affected by temperature (i.e., fertilization, hatching, first feeding) (Alderdice and Velsen, 1978; Heggberget and Wallace, 1984; Brännäns, 1987; Crisp, 1988; Kane, 1988; Jensen *et al.*, 1989; Beacham and Murray, 1990; Blaxter, 1992). Other aspects affected by temperature are the yolk conversion efficiency as demonstrated in salmonid embryos (Heming, 1982; Heming and Buddington, 1988; Marr, 1996; Peterson & Martin-Robichaud, 1995) and in stripped bass (Peterson *et al.*, 1996). Also larval size and fitness at the end of the endogenous feeding period are directly affected by temperature (Peterson *et al.*, 1977, 1996, Baynes and Howell, 1996). Therefore, temperature has a key controlling effect on metabolic processes through thermal dependence on enzymatic activity (Brett, 1970; Rombough, 1988; Blaxter, 1992).

1.1.2 Larval stocking densities

One of the key aspects of successful large-scale production is determining the optimum larval stocking densities. For several species of fish, such as sea bass (*Dicentrarchus labrax*) or sea-bream (*Sparus aurata*), optimum culture densities are well known. The optimal stocking density varies between species depending on the behavioural and physical characteristics (Tagawa *et al.*, 1997, 2004; Kaji *et al.*, 1999; Hernandez-Cruz *et al.*, 1994). Larval density studies for common snook have not been conducted, although some work has been done on fat snook (*Centropomus parallelus*) evaluating the effect of larval and juvenile densities on growth (Cerqueira *et al.*, 1995).

1.1.3 Prey density

Prey density (Werner and Blaxter, 1981) is one of the factors affecting feeding efficiency and consequently larval growth and survival under culture conditions. Enhancing feeding efficiency, at first feeding, can reduce the risk of starvation during the first days of development (Peña *et al.*, 2004). It has also been shown that foraging success increases with prey density (Wyatt, 1972; Laurence, 1974, 1978; Houde and Schekter, 1980; Munk and Kiørboe, 1985) until an asymptote is reached (Houde and Schekter, 1980; Klumpp and Von Westernhagen, 1986). Feeding levels (e.g. rotifer densities) must be tailored to the needs and consumption rates of the larvae at different ages so that food is not wasted, larvae are not underfed, and rearing water is not fouled. The usefulness of food to larvae at particular stages may be measured by food intake, growth and survival (Duray, *et al.*, 1996).

1.2 First feeding

One of the key restrictions in larval rearing is first feeding at early stages of development. This is a major bottleneck for larval culture, due primarily to their small size and often poorly developed digestive system (Person Le Ruyet, et al., 1993). Many marine fish larvae require motile prey organisms (Pedersen *et al.*, 1987; Pedersen and Hjelmeland, 1988). Visual skill is not only important for feeding but also for orientation, schooling and eluding predators (Blaxter, 1986, Batty 1987). Larval survival clearly depends on their ability to feed successfully (Heath, 1992). During the endo-exotrophic phase (Mani-Ponset *et al.*, 1996), larvae utilize nutrients from both yolk sac and their surrounding environment. This phase starts soon after hatching, especially in larvae with a small yolk sac (Calzada *et al.*, 1998). This first feeding phase is critical for larval survival; therefore, successful synchronization between exhaustion of endogenous reserves and first feeding must occur.

Larval mouth size at first feeding is also an important factor for larval survival. The mouth size of first-feeding larvae mechanically restricts the size of the food particles that can be ingested. In general, mouth size is correlated with body size, which in turn is influenced by egg diameter and the period of endogenous feeding (i.e., yolk sac consumption period). For example, Atlantic salmon eggs are usually at least four times larger than Gilthead sea bream eggs and consequently on hatching yield large salmon larvae with large yolk sac supplies (i.e., sufficient endogenous feed reserves for the first three weeks of their development). Whereas first-feeding Gilthead sea bream larvae are very small with limited yolk sac reserves, and consequently can only feed endogenously for about three days (Jones & Houde, 1981).

1.2.1 Background phytoplankton ('Green water')

Most marine fish larvae are visual feeders and feeding success of larvae at various developmental stages depends on the provision of suitable food, the rearing environment, and on the visibility and adequate density of the prey (Ina *et al.*, 1979, Hunter, 1980). Publications on the rearing of marine fish larvae indicate that phytoplankton cultures enhance survival rates (May, 1971; Al-Abdul-Elah, 1984; Hernandez-Cruz, et al., 1994; Marliave, 1994). Furthermore, several papers have discussed the beneficial effect of adding microalgae to larval rearing tanks in order to improve larval growth and survival (Howell 1979; Scott & Middleton 1979; Jones & Houde, 1981; Bromley & Howell 1983; Vasquez-Yeomans, et al., 1990; Naas, et al., 1992; Hernandez-Cruz et al., 1994; Marliave 1994; Tamaru, et al., 1994). These papers discuss the effect of micro-algae on the nutritional and behavioural aspects of fish larvae. Some fish larvae take up substantial amounts of micro-algae during the initial days after hatching (Van der Meeren, 1991; Reitan, et al., 1991) which maybe used as a food source. In recent years, the benefits of culturing larvae in `green water' is considered to be optical rather than nutritional to fish larvae (Marliave 1994).

1.2.2 Rotifers and their nutritional value

Live food organisms are an important food source for the first feeding of early larval stages. The most widely used starter live-food organism in fish larviculture is the marine rotifer *Brachionus plicatilis*. The successful development of commercial fish farms in the Mediterranean has been made possible by several improvements in production techniques for rotifers (Candreva et al., 1996; Dehasque et al., 1998). Rotifers are an ideal link in the food chain for different stages of fish and shrimp larvae. Rotatoria (=Rotifera) belong to the smallest metazoan of which over 1000 species have been described, 90% of which inhabit freshwater habitats. They seldom reach 2 mm in body length. Males have reduced sizes and are less developed than females; some measuring only 60 µm. The body of all species consists of a constant number of cells, with various *Brachionus* species containing approximately 1000 cells, which should not be considered as single identities, but as a plasma area. Growth of the animal is achieved by plasma volume increase and not by cell division. The epidermis contains a densely packed layer of keratin-like proteins and is called the lorica. The shape of the lorica and the profile of the spines and ornaments allow determination of different species and morphotypes. A rotifer body is differentiated into three distinct parts consisting of the head, trunk and foot. The head carries the rotatory organ or corona, which is easily recognized by its annular ciliation and is the characteristic that led to the name Rotatoria (bearing wheels). The retractable corona assures locomotion and a whirling water movement, for the uptake of small food particles (mainly algae and detritus). The trunk contains the digestive tract, the excretory system and the genital organs. A characteristic organ of rotifers is the mastax (a calcified apparatus in the mouth region) that is very effective in grinding ingested particles. The foot is a ring-type retractable structure without segmentation ending in one or four toes. *B. plicatilis,* a cosmopolitan inhabitant of inland saline and coastal brackish waters. It has a lorica length of 100 to 340 µm, with the lorica ending in 6 occipital spines (Fukusho, 1989).

The nutritional value of *B. plicatilis* is dependent on the nutritional value of its food source, which can influence its suitability as a starter feed for marine larvae, and is determined by the concentrations of highly unsaturated fatty acids ((n-3) HUFA), such as docosahexaenoic acid (DHA, 22:6(n-3)) and eicosapentaenoic acid (EPA, 20:5(n- 3)). Low dietary HUFA levels can lead to high mortality in fish larviculture. Koven et al. (1990) suggested that HUFAs function as essential components of bio-membranes, and that their levels in the tissue phospholipid fraction are associated with larval growth. Rainuzzo et al. (1997) emphasized the importance of DHA in the development of neural tissues such as brain and retina, considering that the larval head constitutes a significant part of the body mass, and that predatory fish larvae rely on vision to capture their food. Sorgeloos *et al.*, (1988) reported a strong correlation between dietary EPA content and survival, and between DHA and growth of Asian sea bass larvae. Watanabe (1993) concluded that DHA and EPA increased survival and growth of several marine fish larvae. At the same time, Kanazawa (1993) observed that high DHA levels increased the tolerance of red sea-bream larvae to various stressful conditions.

1.2.3 Copepods

The suitability of copepods as live prey for marine fish larvae is now well established, but their use in aquaculture remains sporadic. Although of lower nutritional value, the relative ease of production of rotifers and *Artemia* nauplii continues to ensure their predominance. Studies in the literature have highlighted differences in the levels and ratios of fatty acids, lipid classes and pigments between copepods and traditional live prey used in hatcheries. Such differences are important for fish larval nutrition, as previously mentioned. The consequences of poor nutrition during fish larval development can result in deformities or malpigmentation, and in some cases may be less obvious, such as effects on temperature tolerance or growth during later life stages. (Støttrup, 2000). Rearing the larvae of most marine fish species requires provision of live prey for variable periods from the onset of exogenous feeding. A common feature of these species is the production of small pelagic eggs. Larvae generally hatch at an early stage in their development of the digestive system as well as the development of organs critical for successful feeding, such as vision and motor development (Støttrup,. 2000). Effects such as tolerance to low temperatures during the juvenile stage have been shown to be related to the larval diet (enriched vs non-enriched *Artemia*) (Howell, 1994), which were not detectable during or at the end of the larval stage. Several studies have shown that rearing marine fish on natural zooplankton can ameliorate these nutritional deficiencies (Nellen, 1981)

1.2.4 Artificial microparticulate diets

A number of studies were carried out to find satisfactory, formulated diets that would substitute for natural live food (rotifers, *Artemia* sp.) in larval rearing of various fish species (Lazo, et al., 2000; Yufera, et al., 1999; Dabrowski, et al., 2003; Takeuchi, et al., 2003). Feeds used as first food during fish larval development must be fine-grained, acceptable, digestible and utilized for body protein/lipid synthesis by the larvae (Ostaszewska, *et al.*, 2005). They should also include the optimal composition of nutrients to achieve high survival and growth rate, and correct development (metamorphosis) of fish. Simultaneously with the efforts on feed formulation, studies of digestion physiology in the gastrointestinal system development in fish larvae must be performed (Ostaszewska, *et al.*, 2005). Ontogenesis, differentiation and development of functions of all organs are genetically determined. However, fish larvae are able to adapt, within some limits, to variable environmental and feeding conditions (Webb, 1999).

1.2.5 Snook aquaculture

Collection of data on the conditions required for spawning, larval rearing, and release into marine and freshwater systems began in 1974 at the Florida Game and Fresh Water Fish

Commission (Ager *et al.*, 1978; Shafland and Koehl, 1980, Chapman et al., 1978). These studies provided information on the lower lethal temperature (15°C) for juveniles (Howells *et al.*, 1990) and preliminary developmental results for laboratory reared larvae and juveniles (Lau and Shafland, 1982). These studies described basic common snook biology and the principles for captive rearing. The early studies on this species in Florida and Texas conducted in the 1970's, 1980's and 1990's, were unable to identify the appropriate culture requirements to support captive spawning and larval rearing of common snook.

1.2.6 Objectives

The objectives of the series of experiments reported here were to improve larval survival of common snook during the first 14 days after hatching.

The main aims were to:

1. Investigate the influence of temperature on hatch rate
2. Establish the effect of egg stocking density on larval survival and growth
3. Determine the influence of flow rate on larval survival
4. Investigate the effect of the green water technique on larval survival and growth
5. Determine the influence of rotifer density on larval survival
6. Investigate effect of alternative live food species on larval survival and growth
7. Evaluate the acceptance of micro-diet feeding by larval snook

2. Materials and methods

All the larval rearing experiments reported here took place at the Mote Marine Laboratory facilities, located in Sarasota, Florida. During the first two years experiments were carried out in aquaculture systems located at Mote Marine Laboratory. The studies carried out in the later two years were conducted in the new aquaculture systems located at Mote Aquaculture Research Park (MAP).

Artificial seawater (Instant Ocean ®) was used at both locations; however, conditions at Mote Marine Laboratory were not ideal because the systems were located under a building that had poor ventilation, limited lighting, and a lack of temperature control. At MAP, experiments were conducted in a variety of tank systems equipped with state-of-art filtration, and in isolated and temperature controlled experimental rooms.

2.1 Live culture

The suitability of three live food types (microalgae, rotifers, copepods) were investigated in parallel. Each live culture was maintained in separate rooms using water from a different reservoir, in order to avoid any possibility of cross-contamination.

2.1.1 Microalgae

The main microalgae used in the snook trials was *Nannochloropsis occulata*; this is non motile, green coloured cell with no flagella. It is a small, elliptical cell, 4-6 µm in diameter, with few distinguishing features. The chloroplast usually occupies much of the cell. Cells tend to float in culture and stay in suspension without aeration. This organism is placed in a separate division from Nannochloris because of its lacks of chlorophyll b. These algae are a popular food source for rotifers and filter feeders.

N. occulata was used to feed the rotifer cultures and to create a green water environment in the larval systems. The procedure used for its culture was the classical batch culture method, which consists of inoculating culture tubes with low density of algae cells. After two weeks, test tube cultures were transferred into 250 ml flasks and later (1 week) into a larger 19 L carboy culture vessels. After a week, a 100-liter cone shape transparent tank was inoculated with a full carboy. The culture was kept running with four 200 L transparent fiberglass tanks during the experiment's duration to ensure reliable microalgae production. All cultures were exposed to 24 hour white light condition (1000 lux), water temperature was kept at 29 °C, and had constant aeration.

N. oculata paste was also obtained from Reed Mariculture. This paste is a highly concentrated media (68 billion/ml) of *N. oculata* that was kept frozen until the day before it was used. The paste was used to reduce the time involved in batch culture of live algae and to test the difference between live and frozen paste algae as a food source for rotifers and for creating a green water environment.

2.1.2 Rotifers

Four different types of rotifers were used in the snook larval rearing trials. *Brachionus rotundiformis* or small (S-type) rotifers and *B. plicatilis* or large (L-type) rotifers, which can be clearly distinguished by their morphological characteristics: the lorica length of the L-type ranges from 130 to 340 μm (average 239 μm), and in the S-type ranges from 100 to 210 μm (average 160 μm). Moreover, the lorica of the S-type has pointed spines, while the L-type has obtuse angled spines. Two other types of rotifers were used. The SS type rotifer (Super small rotifers) ranges between 100-120 μm, which are preferred for the first feeding of fish larvae with small mouth openings (rabbitfish, groupers, and other fish with mouth openings less than 100 μm at first feeding). Those rotifers, however, are not genetically isolated from S-strains, but are smaller than common S-strains (Person Le Ruyet, *et al.*, 1993). The last strain was an SS rotifers from the University of Ghent, Belgium, that were genetically modified to resist warmer temperatures (above 30 °C).

All the rotifers strains were cultured using a batch culture method. Batch cultivation, due to its simplicity, is probably the most common type of rotifer production in marine fish hatcheries (Fukusho, 1989; Nagata and Hirata, 1986; Snell, 1991). The culture strategy consists of either the maintenance of a constant culture volume with an increasing rotifer density or the maintenance of a constant rotifer density by increasing the culture volume. In batch culture, a total harvest of the rotifers is done with part of the rotifers being used as food for fish larvae and part used as inoculum for the next culture (Hirata, 1980; Lubzens, 1987). All the rotifers were fed *N. oculata*.

2.1.3 Copepods

The calanoid copepod *Acartia tonsa* (*Acanthacartia*) was cultured for some of the feeding experiments. This species was chosen due its small size (80-100μm), nutritional value and availability. The copepods were cultured at Florida State University (Tallahassee) and eggs were sent every two weeks on ice (4°C) in 100 ml flasks. Once in the lab, eggs were refrigerated at 4°C until they were needed. Copepod eggs were taken out the 100 ml flask and placed in a 500 ml transparent flask with seawater at 35 ppt and 28°C under a 12 hours light:dark period. No aeration was needed during the 48 hours hatching period. After hatch, *A. tonsa* were fed to the snook larvae. Feeding densities varied depending on the experiment.

2.2 Larval rearing systems

All the experiments were conducted using two independent experimental systems: small and large microcosm systems.

2.2.1 Small microcosm system

The small microcosm system (System A) was a self-contained recirculating system (Rana, 1986), which allowed several different experiments to be run at the same time, with the appropriate replication.. The system was made of transparent plexiglass, and contained 48 2-L tanks (Figure 1), The system dimensions were 1.25 m in length by 70 cm in width by 20 cm in height. The individual tank dimensions (Figure 1) were 12x10x20 cm, with an opening to allow water exchange 17.5 cm from the bottom, which was covered by a mesh screen (75 µm). In addition to the larval tanks, the system included two sumps, a 140 liter sump (filled with biofiltration beads) and a 120 liter sump (with a fluidized bed and a carbon filter). Air stones, were placed in both sump and a pure oxygen ceramic stone was also in the second sump, in order to keep the dissolved oxygen levels between 8 to 10 mg/L. Flow rates within the larval tanks was individually regulated through a drip valve. Flow rates varied between tanks depending on the experimental requirements. The tank recirculating system was based on an overflowing system, with a drip valve on the inflow and a 75 µm mesh rectangular opening for tank outflow. For the first 3 days after hatching, a transparent

Fig. 1. System A (Microcosms) and individual tank dimensions

plastic separator was placed close to the outflow to avoid egg and larvae impingement on the mesh to reduce possible mortalities. All the tanks outflows drain to a common canal through the UV light filter and into the sumps.

Daily 10% water exchange was conducted. Water quality was checked three times a day (every five hours) for temperature, salinity, dissolved oxygen and pH. Nitrite, ammonia and nitrate were checked on a weekly basis. During the feeding experiments, residual prey counts were taken prior to feeding. Tanks were fed 1, 2, 3 and 4 times a day depending on the experiment.

2.2.2 Large microcosm system

The large microcosm system (System B) was built in 2005 to provide additional replicated experimental tanks. The water volume in the System B tanks (6L) was 3 times the water volume of the System A tanks (2L) and was used to conduct simultaneous trials to compare the influence of increased water volume on larval survival. System B tanks were placed in a green bottom fiberglass raceway, where individuals tanks where maintained in a water bath (Figure 2). Twelve (6 l) tanks shared the same filtration system, which included a fluidized bed, a moving bed bioreactor, UV, and a protein skimmer. The system had two 300-litre sumps under the raceway, where the filtration system was set up. Air stones were placed in each tank to keep dissolved oxygen at desired levels. Water heaters were placed in the raceways to maintain constant temperature.

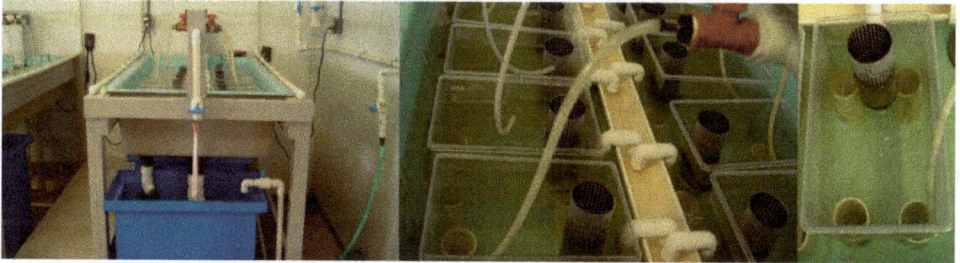

Fig. 2. Larval rearing System B. Full system and individual rearing tanks (left to right)

Inflow water was regulated individually per tank, and outflow water passed through a 75 μm mesh standpipe, draining into a common drain channel (Figure 2, middle picture) leading to the first sump. Slight aeration was also supplied to the individual tanks. A 25% water exchange was carried out every week. Like in the microcosms, water quality (temperature, salinity, dissolved oxygen and pH) was checked three times a day (every five hours). Nitrite, ammonia and nitrate were monitored weekly.

2.3 Experimental methodology

After collecting common snook eggs in the field, eggs were fertilized and transported to the laboratory. Eggs were stocked in experimental systems within 2 to 3 hours after fertilization. Eggs were stocked at a salinity of 35 ppt, temperature of 28°C (except for the temperature experiments), 9-10 mg/l of dissolved oxygen and pH of 7.9. During the first two years, larvae were fed 3 days after hatching; during the last two years, feeding started 2 days after hatching. All specimens that survived past Day 14 after hatching were sacrificed, for total length and myomere height measurements.

2.3.1 Influence of temperature on hatch rate

Eggs were stocked in 2 L PVC floating containers, similar containers were at the same time floating in a 300 L raceway tank. Three raceway tanks were used, each one of them had a different temperature (23, 28 and 30°C). Each raceway tank had three 2 L PVC floating container and each container was stocked with 100 snook eggs per litre. Aeration was removed before stocking. Water quality parameters were maintained within acceptable values. Twenty hours after fertilization, containers were removed from the raceways and percent hatch was determined (see the formula below). This experiment was replicated four times to obtain reliable results.

Hatch rate= Total number hatched larvae / Total number of eggs stocked *100

2.3.2 Effect of egg stocking density on larval survival and growth

Nine tanks were stocked with three different egg densities (three replicates per density). The densities used were: high density (375 eggs/L), medium density (200 eggs/L) and low density (200 eggs/L). All tanks were fed S type rotifers three times a day at a concentration of 30 rot/ml. The tanks were harvested fourteen days after hatch (DAH) and larvae were counted and measured (total length and myomere height). This experiment was run in both systems A and B.

2.3.3 Influence of flow rate on larval survival

Common snook eggs were stocked at a density of 200 eggs/L and exposed to three different flow rates (no flow, slow flow (10 ml/min) and high flow (30 ml/min)). Each flow treatment had 3 replicates per day. All tanks were fed three times a day with SS type rotifer at a concentration of 30 rotifers/ml. System A tanks (27 tanks) were stocked, nine per flow treatment. Larvae were harvested on 3, 6 and 10 DAH to determine survival.

2.3.4 Effect of background phytoplankton in the water (Green water technique) on larval survival and growth

Snook were stocked at 200 eggs/L in ten system A tanks and the flow rate was set at 15 ml/min. Five tanks were stocked with *Nannochloropsis oculata* at 1000/ml (green water) and five tanks without (clear water). All tanks were fed SS type rotifers three times a day with a concentration of 30 rot/ml. The experiment was terminated at 14 DAH and all larvae were counted to establish larval survival and measured (total length and myomere height) for growth.

2.3.5 Influence of rotifer density on larval survival

The effect of rotifer density on larval survival was evaluated in 15 System A and System B tanks. Three SS type rotifers densities were used with: 5 rotifers/ml, 15 rotifers/ml and 30 rotifers/ml. All the tanks were stocked at the same time with 200 eggs/L. Residual rotifer counts were taken before every feeding and feeding amounts were adjusted, in order to maintain the same rotifer concentration throughout the experiment. All larvae were harvested 14 DAH and counted to calculate survival.

2.3.6 Effect of alternative live food species on larval survival and growth

To investigate the effect of alternative live food species on larval survival and growth, four diets were given to the larvae: a) 100% rotifers, b) 75% rotifers and 25% copepods, c) 50% rotifers and 50% copepods, and d) 75% copepods and 25% rotifers. All the tanks were stocked at the same time and fed 30 prey/ml; the number of each prey (rotifer or copepod) was determined by the above-mentioned percentages. Twelve tanks were used in System A and System B for this experiment and there were three tanks per experiment diet. Flow rate was maintained at 10 ml/min during the 14 day trial. Total length and myomere height were taken at 14 DAH and stomach contents were examined.

2.3.7 Acceptance of micro-diet feeding by larval snook

A total of 21 (2L PVC) floating tanks were placed in the raceway and each tank was stocked with 300 eggs/L. The experiment was run twice for a period of 7 days, from day 2 after hatching till day 8 after hatching. Seven tanks were fed one of the following three diets: a) a SS type strain (150 μm) rotifer, at a density of 30 rot/ml, fed three times a day, b) a 100 μm artificial micro-diet, fed twice a day, c) a 150 μm artificial micro-diet fed twice a day. Each day three tanks were taken out and larvae stomach contents were examined under a microscope.

3. Results

3.1 Influence of temperature on hatch rate

The influence of incubation temperature on hatch rate was investigated to establish the optimal temperature for incubating snook eggs. Embryos exposed to 23°C water had the lowest mean hatch rate of 5.9% (range 0.8 to 11.4%) (Figure 3). The highest hatch rates of 23.5% occurred at 28°C (range 21.7% to 27%). No significant difference in hatching rate ($p > 0.05$) was found between the 28°C and 30°C treatment, but a significant difference ($p < 0.05$) was found when 28°C was compared to 26°C and 23°C.

3.2 Effect of egg stocking density on larval survival

The effect of egg stocking density on larval survival at 14 DAH is presented in (Figure 4). At a stocking density of 375 eggs/1 (High density) per tank, the mean survival from three tanks was 0.6%. Tanks stocked with a medium density (200 eggs/1) had a similar mean survival percentage with 0.5%. The highest survival was obtained at the low density stocking (100 eggs/1) with a mean overall 1.2% survival by 14 DAH (Figure 4). No significant difference ($p > 0.05$) was obtained between the high and medium density treatments, but a significant difference ($p < 0.05$) was obtained between the low density treatment and the other two.

3.3 Influence of flow rate on larval survival

Snook larvae were exposed to three different flow rates. Those fish exposed to no flow during the first 10 days (Figure 5) had a mean survival rate of 45% by 3 DAH, survival then

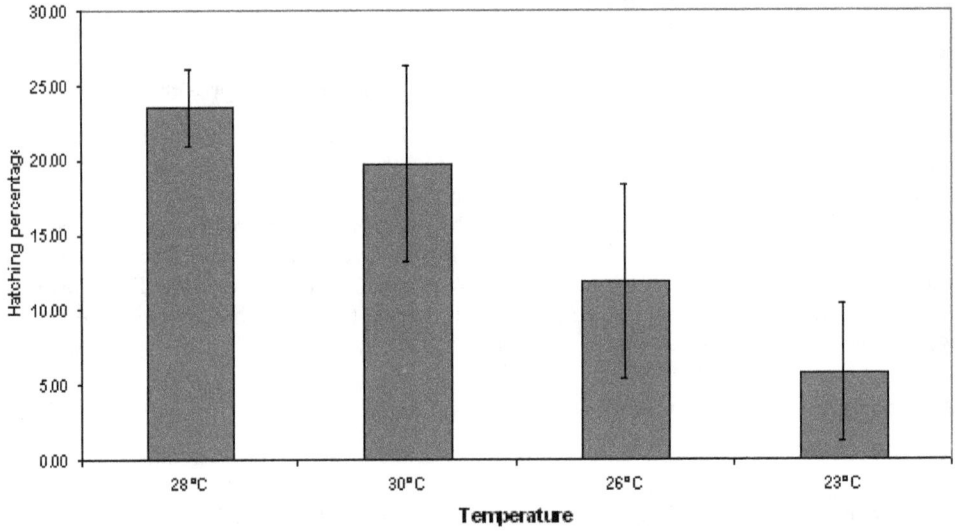

Fig. 3. Influence of temperature on percent hatch in snook larvae.

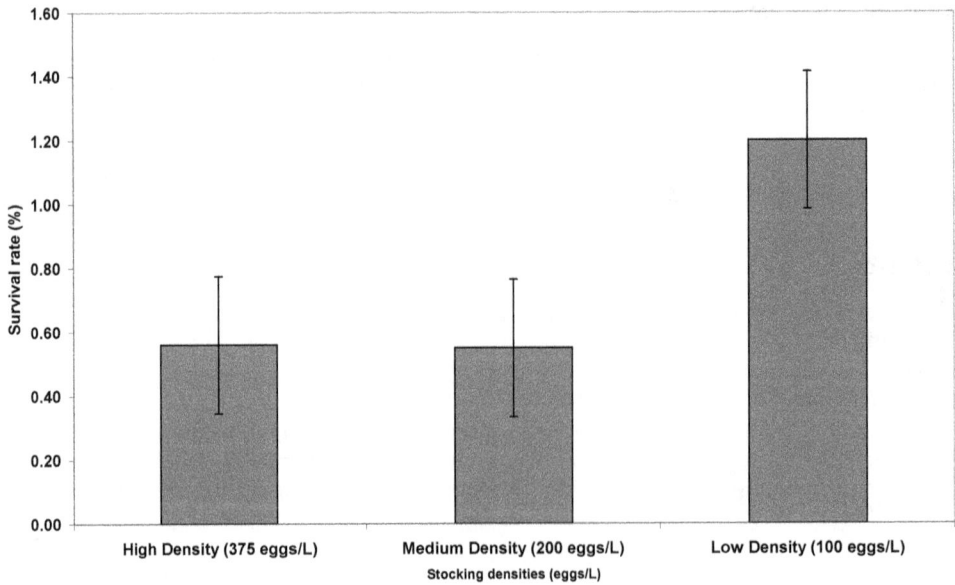

Fig. 4. Larval survival at three stocking densities at 14 DAH.

decreased to 19%by day 6. At 10 DAH, larval mean survival was only 5% from the initial stocking density. The slow flow (10 ml/min) treatment had a mean survival of 40% at 3 DAH, decreasing 3 days later to a 32% survival. Finally a mean survival of 21% was observed by 10 DAH (Figure 5). The third experiment had the lowest values in terms of survival. At high flow (30 ml/min), survival at 3 DAH was 16%, decreasing on Day 6 to a mean of 10% survival and by day 10 after hatch larval mean survival was 9% (Figure 5). No significant difference (p>0.05) was observed between the no flow and slow flow treatment at 3 DAH, but a significant difference (p< 0.05) in survival was found between all the treatments 6 and 10 DAH.

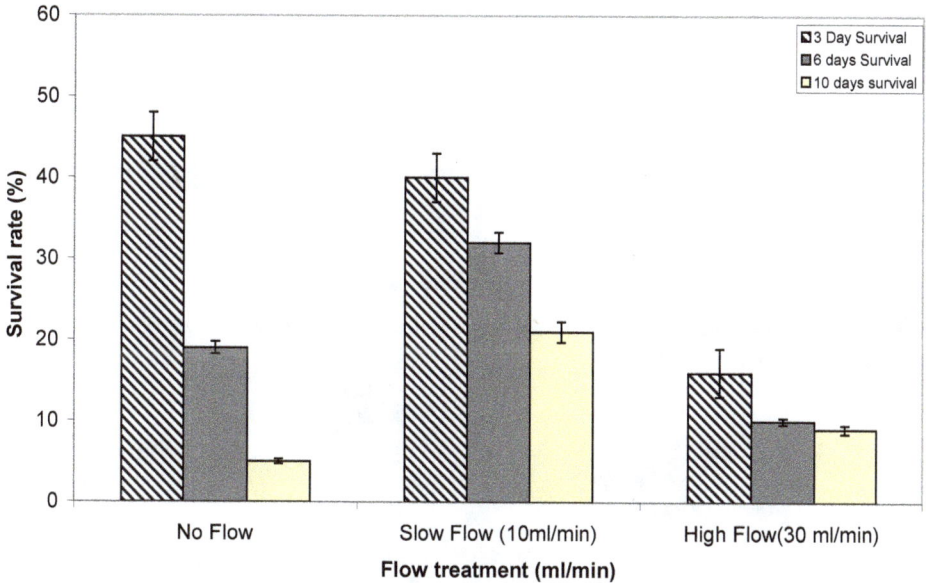

Fig. 5. Snook larval survival under three types of flow conditions.

3.4 Effect of background phytoplankton in the water (Green water technique) on larval survival and growth

Larvae were stocked in water with no algae and in water with algae (*N. occulata*) at a concentration of 1000/ml. Larvae tanks without algae are referred to as 'clear water' (Figure 6). The survival of larvae was significantly (p<0.05) influenced by the presence of *N. occulata* in the rearing water. The mean survival at 14 days post hatching in clear water was significantly lower at 0.17%, compared with 0.55% survival in green water tanks (Figure 6).

Length and myomere height in larvae from the clear water tanks had an average standard length (SL) of 3.20 mm and average myomere height of 0.70 mm (Figure 7); on the other hand, larvae in tanks subjected to green water technique had an average length of 3.34 mm and average myomere height of 0.73 mm. Although standard length and myomere height in the green water tanks were higher, no significant difference was found (p= 0.053).

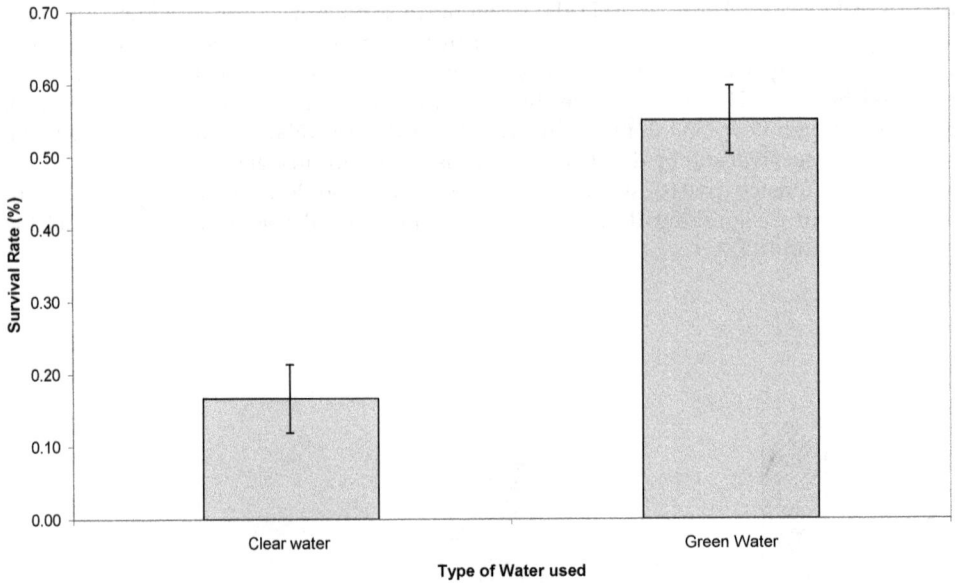

Fig. 6. 14 DAH snook larval survival from tanks with *N. oculata* (Green water) and without phytoplankton (clear water)

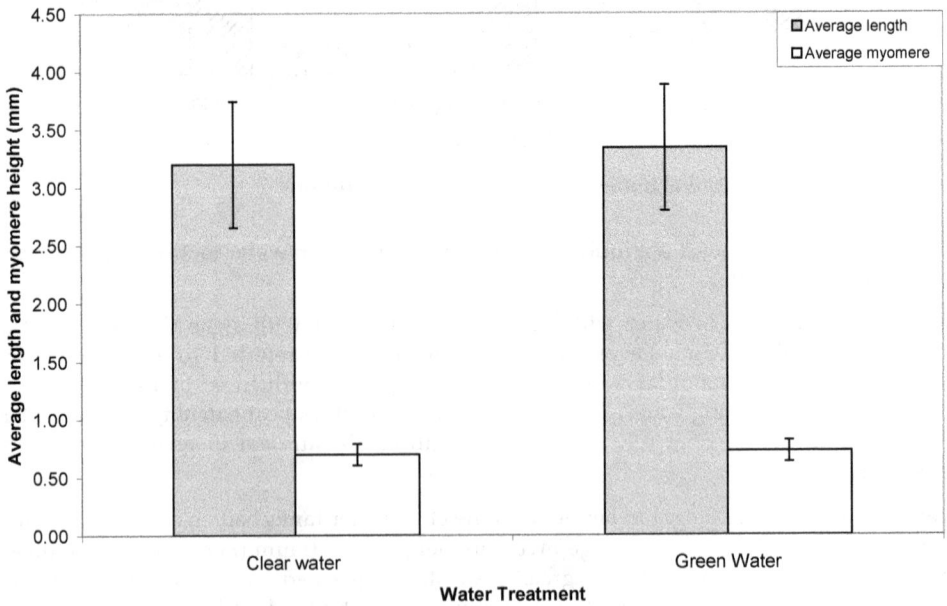

Fig. 7. Average snook larval length (SL) and myomere height from green water and clear water tank

3.5 Influence of rotifer density on larval survival

Three different rotifers densities were evaluated to determine the effect on growth and survival of common snook larvae (Figure 8). After 14 days, larvae fed 5 rotifer/ml diet had an average survival rate of 0.86% and only 13% of all tanks stocked had live larvae. All the tanks where common snook larvae were fed 15 rotifers/ml had a mean larvae survival of 2.5%, and 20% of tank of all tanks stocked had live larvae. In the third treatment with 30 rotifers/ml, 20% of the tanks had live larvae and a total average of 3.67% larvae survival. No significant differences was found in larval survival between tanks fed 15 and 30 rotifer/ml (p=0.053).

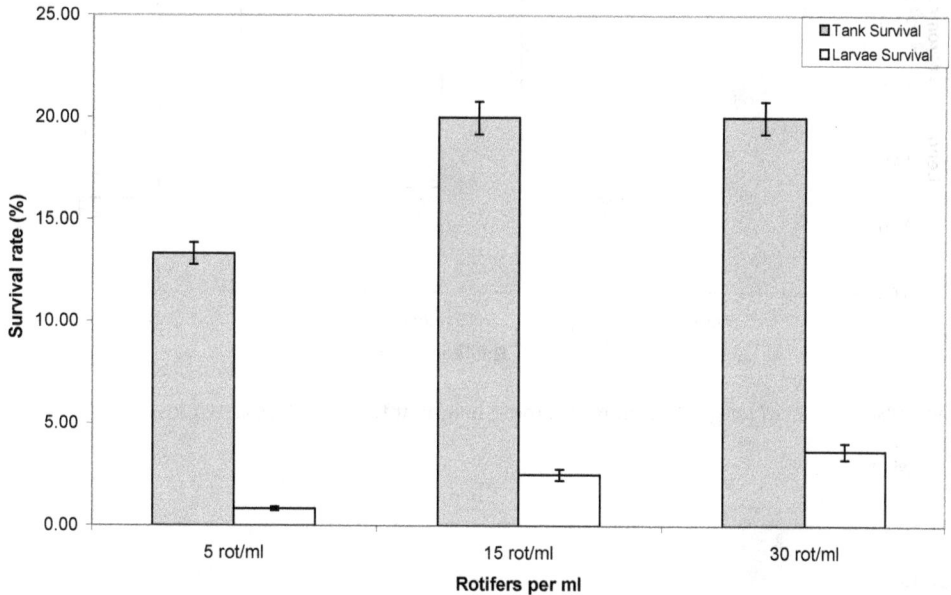

Fig. 8. Effect of 3 different rotifer densities on larval survival

Increasing rotifer concentrations from 5 to 30 rotifers/ml did not significantly (p>0.05) affect the larval length or myomere height (Figure 9). Larvae fed with 15 rotifers per ml had the highest mean length (SL) at 3.58 mm, followed by the other two treatments (3.48 and 3.49 mm in average respectively). Mean myomere height results were 0.74, 0.78 and 0.69 mm respectively.

3.6 Effect of alternative live food species on larval survival and growth

Four different diets were used in this experiment: 3 using copepods (*Acartia tonsa*) and SS type rotifers, one with 100% rotifers. The diet with 75% copepods and 25% rotifers (Figure 10) had the highest mean survival (1.44%), the second highest average larval survival occurred with the diet that had 100% rotifers (0.83%). The other two diets (50% rotifers and 50% copepods, and 25% copepods and 75% rotifers) had similar results with 0.61% mean survival. A significant difference was found between the 75/25 copepod/rotifer diet and the other three diets (p> 0.05). No significant difference was found between the 50/50 and the 25/75 diets (p<0.05).

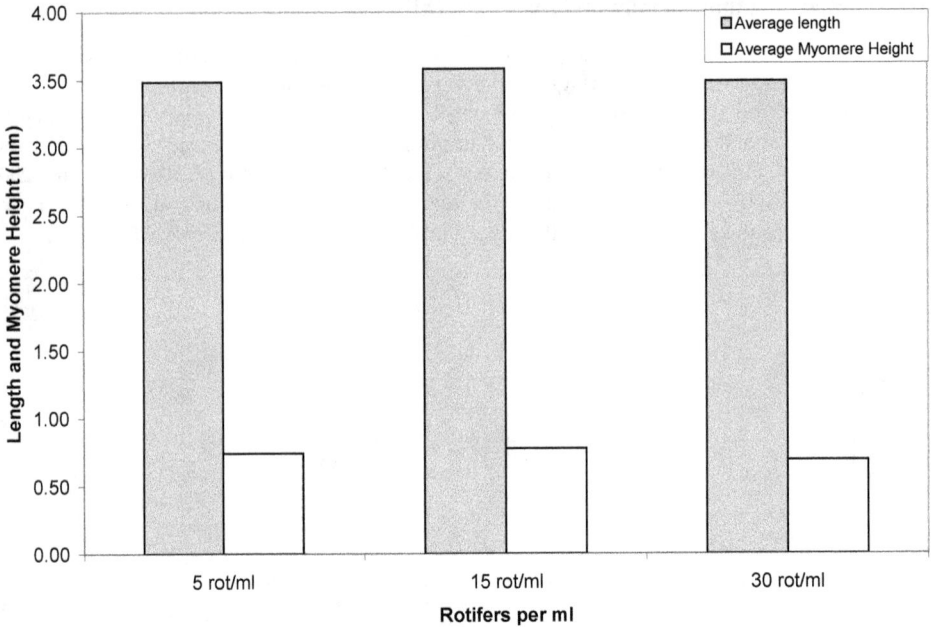

Fig. 9. Snook larval length (SL) and myomere height at three rotifers densities.

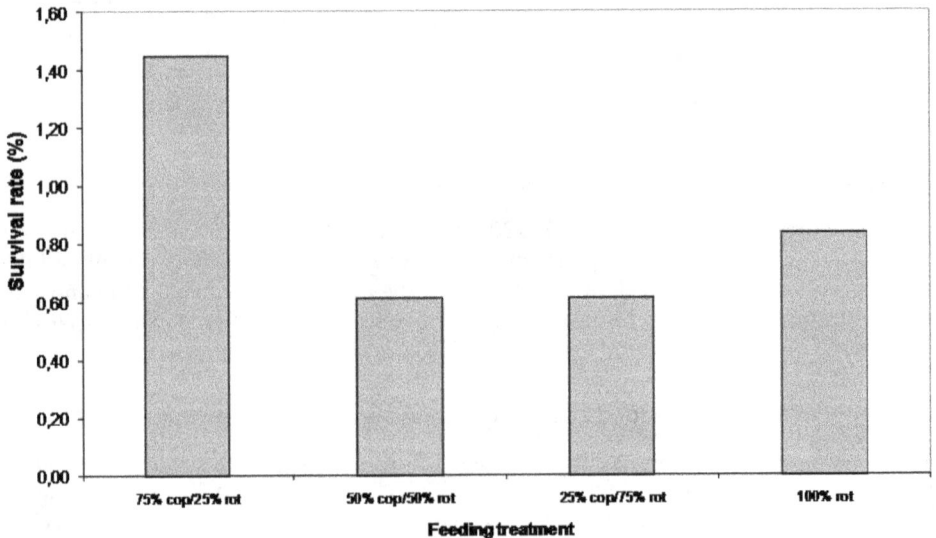

Fig. 10. Snook larval survival after 14 days associated with four diets that combined copepods and rotifers at different percentages. Feeding treatment 75/25 consisted of 75% copepods and 25% rotifers, 50/50 consisted of 50% copepods and 50% rotifers, and the 25/75 consisted of 25% copepods and a 75% rotifers.

The influence of live food combinations on larval growth is presented in Figure 11. Larvae at the 75/25 (copepod/rotifer) and the 50/50 tanks had the average highest length (4.70 mm), followed by the 25/75 (copepod/rotifer) with an average length (SL) of 4.32 mm, which was very similar to the 100% rotifer diet (average length (SL) of 4.28 mm). No significant difference was found between the first two treatments ($p > 0.05$) or between the other two (25/75 and 100 rot), but a significant difference was found between the first two and the last two ($p < 0.005$). Myomere height was similar on all the diets ranging from 1 mm to 1.04 mm.

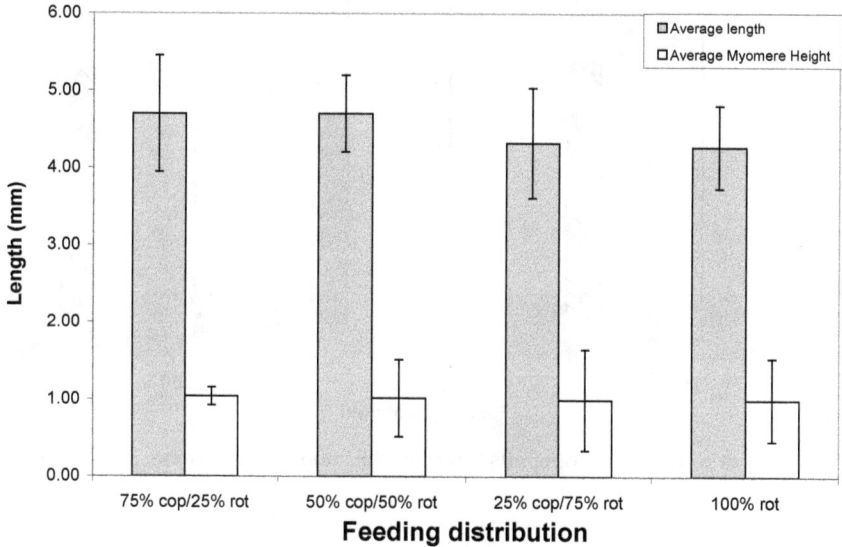

Fig. 11. Average length (SL) and myomere height from larvae exposed to four diets.

The stomach contents of 14 day old larvae were examined for food items (Figure 12). In the 75/25 treatment, 30.8% of the larvae had both rotifers and copepods in the guts, in the 50/50 diet, 30% of the larvae also had both rotifers and copepods in the guts, and in the 25/75 diet 18 % of the larvae had both rotifers and copepods in the guts. The rest of the larvae only had rotifers.

3.7 Acceptance of micro-diet feeding by larval snook

The stomach contents from larvae fed micro-diets were examined in this experiment. In tanks fed with rotifers, the percent of larvae with rotifers in the stomach after 7 days ranged from 10 to 37 (Figure 13). In tanks where 100 µm micro-diet food was offered, percent larvae with dry diet ranged from 6 to 28%. Finally, in tanks where the 150 µm micro-diet was offered the percent of larvae with dry food ranged from 7 to 27%.

4. Discussion

4.1 Importance of temperature during incubation

Fish are affected by many intrinsic and extrinsic factors. These factors can affect developmental controls resulting in phenotype alterations (Johnston et al.,1996; Adriaens &

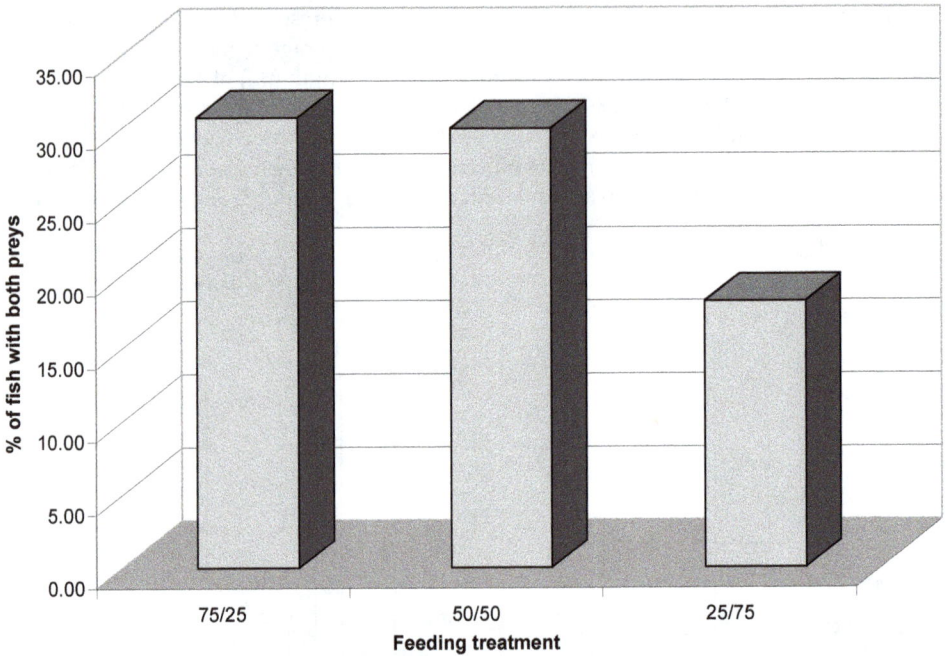

Fig. 12. Percent of rotifers and copepods found in the snook larvae stomach contents from four diets.

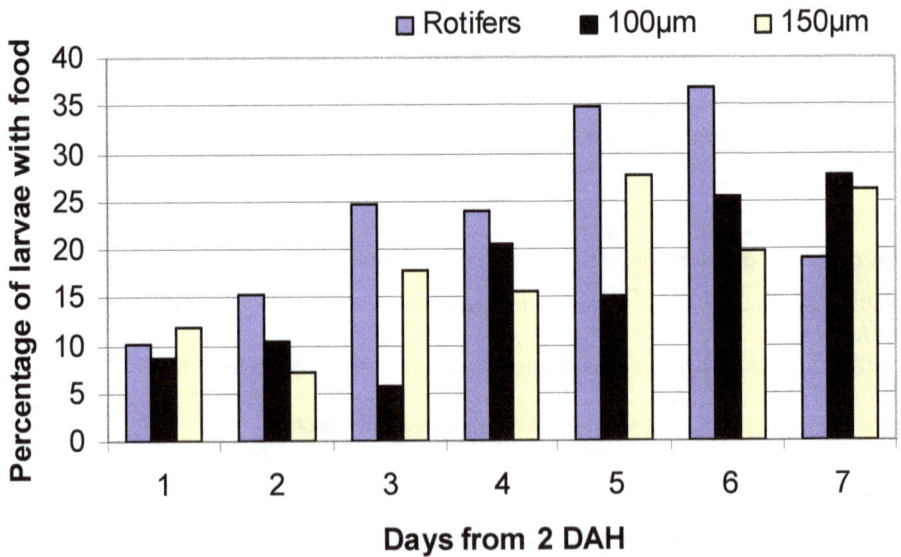

Fig. 13. Percent of larvae with food in the digestive system from the three diets (SS type rotifers, 100 and 150μm micro-diets).

Verraes, 2002). Among the abiotic factors, temperature has the most significant effect on development and growth (Blaxter, 1992; Kamler, 1992a; Hochachka & Somero, 2001), influencing developmental timing and formation and function of key tissues and structures (Kamler, 1992b; Fuiman et al., 1998; Koumoundouros et al., 1999) and the synchronization of these continuous developmental paths (Kovac, 2002). Temperature can also have a direct effect on physiology through its effects on enzyme reaction rates (Hochachka and Somero 2001).

Hatching rate of common snook varied with temperature in this study. Eggs incubated at 28°C showed the best hatching rates; however, no growth trials were conducted. These results confirm Rideout et al., (2004) findings of a direct influence of temperature during the incubation period on larval growth, which also has a direct effect on hatch rates (Pepin et al., 1997). It has also been argued that survival may be proportional to larval size, since larval mortality rates have been shown to be inversely proportional to growth rates (Pepin, 1991).

4.2 The effect of stocking density on larval survival

Egg stocking density is a key factor in larviculture; without an optimal stocking density, overall survival can be affected. The common snook stocking density experiments results showed the lower stocking density (100 eggs/l) to be the one with the higher survival. This finding agrees with Hernandez-Cruz et al., (1999), who obtained low survival in red porgy when eggs were stocked in high densities. On the other hand, Tagawa et al., (1997) found higher survival at higher rearing densities in Japanese flounder and this finding was confirmed in three other marine teleosts (Kaji et al., 1999). Tagawa et al., (2004). The higher survival at high densities was attributed to substances (proteins) secreted by larvae that were beneficial for their survival. Although more experiments need to be done on common snook density, the results clearly showed low survival at high densities, which might be due to the a high level of competition for prey or over-crowding the environment and reduced water quality conditions.

4.3 Influence of 'green water' on larval survival

Food availability is a key factor during first feeding and the consumption rate is dependent on availability. Additionally, the developmental stage of the individual affects the consumption rate (Houde and Schekter, 1980; Kentouri, 1985). Based on the previous findings, this study examined survival of larvae in green water and in clear systems. Final results showed a significant difference in survival between the two treatments, where larvae grown in a green water (N. occulata had ingested the algae and had higher survival than those reared in clear water. This result agrees with previously published data (Papandroulakis et al., 2002; Divanach et al., 1998; Oie et al., 1997; Holmejford et al., 1993). Also, it has been reported in cod (Van der Meeren, 1991) and halibut (Reitan, et al., 1991). This may support the idea that microalgae are used as a direct food source at the start of feeding confirming the important role that phytoplankton has during the early stages of several species. Another explanation for the increased survival in green water is the role of phytoplankton in stabilization and improvement of the rearing medium and its direct (Moffatt, 1981; Reitan et al., 1993; Van der Meeren, 1991) or indirect nutritional effect (Tamaru et al., 1993). Phytoplankton has been reported as being a protective agent, antagonistic towards pathogenic bacteria (Kennedy et al., 1998; Støttrup et al., 1995). Skjermo & Vadstein (1999) noticed that microflora in rearing tanks of Hippoglossus hippoglossus were more stable in the presence of phytoplankton, increasing the total bacteria population by about 45%. The same authors noticed also that bacteria in larval gut were similar to the ones in rearing water, being mostly species with low growth rate. In addition, Nicolas *et al.*, (1989) showed that stomach microflora affects survival during early stages of marine larvae. Other studies have hypothesized that green water produces a background effect that allows the fish to better locate its prey (Marliave, 1994). This effect has been documented to improved the larval rearing in a number of fish species, including dolphinfish *Coryphaena hippurus* (Ostrowski, 1989), yellow perch *Perca flavescens* (Hinshaw, 1985), walleye *Stizostedion vitreum* (Corraza & Nickum, 1981; Colesante, 1989), white bass *Morone chryops* (Denson & Smith, 1996), grouper *Epinephelus suillus* (Duray *et al.*, 1996), Dover sole *Solea solea* (Dendrinos *et al.* 1984), barramundi *Lates calcarifer* (Pearce 1991); and red porgy *Pagurs pagrus* (Rotllant *et al.* 2003).

4.4 Importance of rotifer density in common snook larviculture

Prey density is an important factor in successful larval rearing (Fushimi, 1983). Low density can cause larval starvation or nutritional deficiencies leading to high mortalities. High densities can deteriorate water quality and lead to system fouling, decreased oxygen levels, and increased ammonia levels. Finding the right prey density is crucial to avoid these problems and also to reduce costs associated with live food production.

During the rotifer density experiments, both 15 and 30 rotifers/ml treatment resulted in high survival. Although 30 rotifers/ml had the highest larval survival, the difference between the two treatments was not significant; therefore, the higher cost to produce 30 rot/ml treatment indicates that 15 rot/ml is the optimal density for the 2 and 6 L tank systems. Recent attempts to calculate optimal prey density for cod larvae (*Gadus morhua*) found that survival reaches a maximum level and then begins to decrease if prey densities are further increased (Puvanendran and Brown, 1999). The decrease in larval survival when higher prey densities are used may be a result of poor water quality due to the release of metabolites by the prey (Houde, 1975) or it may be related to a reduction in the ability of the larvae to capture prey; what Laurel et al. (2001) term a "confusion effect". Optimal rotifers densities differ between species, for instance the black sea bream needs 1-3 rotifers/ml (Kafuku & Ikenoue, 1983), and the red sea bream needs between 3-10 rotifers/ml (Fushimi, 1983). In the case of common snook, results showed that 5 rotifers/ml was inadequate to meet their food demand. Further experiments need to be carried out to find out the optimal density, this time focusing between 10 to 20 rotifers per ml.

4.5 Alternative prey for common snook larvae

The suitable size of prey for fish larvae varies with larval mouth size (Shirota, 1970), and fish larvae select larger prey size as they grow (Ivlev, 1961). Although many researchers have reported larval rearing trials with marine fishes, only a few studies have been conducted to compare the appropriate rotifer size among fish species and among different growth stages (Oozeki et al., 1992, in Hagiwara *et al.*, 2001). Four strains of rotifers and one copepod species were tested to find the optimal prey type for common snook larvae; prey that will suit the physical needs as the larvae develops and grows in size.

Previous work done on snook larval culture used L type rotifers and had little success. The experiments run with the L type rotifers in these experimental trials, showed extremely low survival and DHA analysis of larvae showed a steep decrease in DHA from 1 to 6 DAH. DHA values in 6 DAH larvae were below 1% of total lipid concentration (Yanes-Roca et al. 2009). In both experimental and production systems mass mortality regularly occurred between 5-6 DAH (Yanes-Roca et al. 2009) and 75-85% of all stocked tanks did not have live larvae after 6 DAH. These results along with the finding that only 5% of the larvae fed L type rotifers had food in their stomachs lead to the conclusion that the snook larvae were dying of starvation. This is likely due to the L type rotifer prey size, which was larger than the snook larvae mouth gape. Common snook larval rearing was successful when the larvae were fed copepods grown naturally in outdoor ponds (Lau & Shafland 1982), confirming the prey size hypothesis. Experimental trials comparing survival and growth with several strains of rotifers and copepods (*Acartia tonsa*) also showed an increase in snook larval survival. These findings agree with Doi *et al.*, (1997a,b) and Toledo *et al.*, (1997) found nauplii of copepods to be effective when fed to red spotted grouper, *Epinephelus coioides* and with Støttrup *et al.*, (1997) who found an increase in larval survival when rotifer feeding was supplemented with *Tisbe* sp copepod.

4.6 Acceptance of micro-diets by larval marine fish

Production of marine fish juveniles in commercial hatcheries still depends on the supply of live prey, such as rotifers and *Artemia*. Artificial micro-diet substitution for live prey is crucial to lower production costs and sustain production of high and constant quality juveniles (Cahu & Zambino-Infante, 2001). The use of micro-diets for common snook larvae during the pre-weaning period was also tested. Since the first rearing of plaice (*Pleuronectes platessa*) larvae to metamorphosis using an artificial diets (Adron *et al.*, 1974), many trials have been conducted, with different degrees of success, to utilize artificial diets in larval rearing of species, such as seabass *Dicentrarchus labrax* (Gatesoupe *et al.*, 1977; Cahu & Infante, 1994; Kolkovski *et al.*, 1997), sole *Solea vulgaris* (Gatesoupe *et al.*, 1977), Atlantic silverside *Menidia menidia* (Seidel *et al.*, 1980), red seabream *Chrysophrys major* and Ayu *Plecoglossus altivelis* (Kanazawa *et al.*, 1982). In all cases, poor results were obtained when live food was replaced completely by micro-diets. However, during the last decade, the pre-weaning period has been greatly reduced in many species, such as European sea bass *Dicentrarchus labrax* (Person Le Ruyet *et al.*, 1993, Zambonino-Infante *et al.* 1997). Cahu *et al.*, (1998) reported that 35% of European sea bass larvae were fed exclusively compound diet from mouth opening. In other marine species, some survival was obtained when fed compound diet from mouth opening, such as sea bream *Sparus aurata* (Fernandez-Diaz & Yufera, 1997) and red sea-bream *Pagrus major* (Takeuchi *et al.*,1998).

Common snook larvae were offered two sizes of micro-diets (100µm, 150µm) and rotifers for 7 days, from 2 to 9 DAH. Although there was higher percentage of larvae with food in their stomachs in tanks fed SS type rotifers, a significant number of larvae offered micro-diet had food in their stomachs. These results were observed in pike-perch (Ostaszewska *et al.*, 2005), where micro-diets were readily accepted, digested and absorbed as well as in the Japanese eel (Pedersen *et al.*, 2003), and the gilthead sea-bream (Salhi *et al.*, 1997). Earlier studies suggested that co-feeding with live food improved yellow perch growth and assimilation of artificial diets (Kolkovski *et al.*,1997), a method that could be applied to common snook, assuming that digestive enzymes of live food organisms supported digestive processes in fish larvae (Boulhic & Gabaudan 1992; Jones *et al.*, 1993). However, some publications have reported contradictory results (Cahu & Zambino-Infante, 1997; Kolkovski *et al.*, 1993).

Future research on micro-diets is needed to evaluate the effect on survival or growth. Results indicate that snook larvae seem to accept the artificial diet and based on the already mentioned literature findings on several marine larval species survival and growth, it appears that the common snook larvae pre-weaning period could be reduced.

5. Conclusions

Although common snook larval survival has improved, mortality is still high and more improvements in rearing techniques are needed to increase survival. Many factors could be responsible for the low survival rates observed, including the introduction of bacteria along with the live food (i.e., rotifers). Rotifers are major carriers of bacteria (Muroga and Yasunobu, 1987; Munro et al., 1993, 1994). Studies to evaluate the effect of bacteria on snook larval culture may help the overall survival. In these trials, two different experimental tank sizes were used (2 L and 6 L tanks) and survival was better in the larger tanks. More research is needed to determine the optimal tank size to examine growth and survival in larval snook. After four years of research on the snook larval rearing techniques, positive improvements have been made and the critical bottlenecks and potential solutions have been identified. Findings such as optimal rearing temperature (28°C), appropriate flow and water management (green technique) are basic for future research. Other critical results were: finding appropriate prey size (SS rotifers and copepods), optimal stocking and prey densities and the acceptance of microdiets prior to weaning.

6. Acknowledgements

This work was supported by grants from the Institute of Aquaculture at Stirling University, the Florida Fish and Wildlife

Conservation Commission, the National Oceanic and Atmospheric Administration funded research consortium, the Science Consortium for Ocean Replenishment (SCORE), and the Mote Scientific Foundation. Special thanks to the Center for Aquaculture Research and Development at Mote Marine Laboratory staff, especially to Nicole Rhody, Michael Nystrom and Dave Jenkins for their support and help through this four year research effort.

7. References

Adriaens, D. & Verraes, W. (2002). An empirical approach to study the relation between ontogeny, size and age using geometric morphologies. In: Aerts, P., D'Aout, K., Herrel, A. & VanDamme, R, (Ed.), *Functional and Ecological Vertebrate Morphology* , pp. 293–324. Maastricht: Shaker Publishing.]

Adron, J.W.; Blair, A. & Cowey, C.B. (1974). Rearing of plaice (*Pleuronectes platessa*) larvae to metamorphosis using an artificial diet. *Fishery Bulletin* 72, 353–357

Ager, L.A.; Hammond, D. E. & Ware. F. Artificial spawning of snook. (1978). Proceedings of the annual conference South-eastern Association Game Fish Commissioners 30, 158-166p.

Al-Abdul-Elah K.M. (1984). Procedures and problems of marine fish hatcheries with special reference to Kuwait. Master of Science Dissertation, University of Stirling.

Alderdice, D.F. & Velsen, F.P.J. (1978). Relationship between temperature and incubation time for eggs of chinook salmon (Oncorhynchus tshawytscha). Journal of the Fisheries Research Board of Canada 35 , 69–75

Batty, R.S. (1987). Effect of light intensity on activity and food-searching of larval herring (Clupea harengus): a laboratory study. Marine Biology 94, 323-327.

Baynes, S.M. & Howell, B.R. (1996). The influence of egg size and incubation temperature on the condition of the Solea solea (L.) larvae at hatching and first feeding. Journal of Experimental Marine Biology and Ecology 199, 59–77

Beacham, T.D. & Murray, C.B. (1990). Temperature, egg size, and development of embryos and alevins of five species of Pacific salmon: a comparative analysis. Transactions of the American Fisheries Society 119, 927–945

Blaxter, J.H.S. (1992). The effect of temperature on larval fishes. Netherlands Journal of Zoology 42, 336-357

Blaxter, J.H.S. (1986). Development of sense organs and behavior of teleost larvae with special reference to feeding and predator avoidance. Transactions of the American Fisheries Society 115, 98-114.

Boulhic, M & Gabaudan J. (1992). Histological study of the organogenesis of the digestive system and swim bladder of the Dover sole, (Solea solea) (Linnaeus1758). Aquaculture 102, 373-396.

Brannas, E. (1987). Influence of photoperiod and temperature on hatching and emergence of Baltic salmon (Salmo salar L.). Canadian Journal of Zoology 65, 1503–1508.

Brett, J.R. (1970) 3. Temperature. 3.3. Animals. 3.3.2. Fishes. In: Kinne O., (Ed.)   Marine Ecology , pp. 515-573 London: Wiley Interscience]

Bromley P.J & Howell B.R. (1983). Factors influencing the survival and growth of turbot larvae, Scophthalmus maximus L. during the change from live to compound feeds. Aquaculture 31, 31-40.

Bruger, G.E. & Haddad, K.D. (1986). Management of Tarpon, bonefish, and snook in Florida. Multi-jurisdcitional Management of Marine Fisheries: Marine Recreational Fisheries 11, 53-57.(Abstract)

Calzada, A.; Medina, A. & Gonzalez de Canales, M.L. (1998). Fine structure of the intestine development in cultured sea bream larvae. Journal of Fish Biology 53, 340-365.

Candreva, P.; Dhert, P.; Novelli, A. & Brissi, D. (1996). Potential gains through alimentation nutrition improvements in the hatchery. In: Chatain, B., Sargalia, M., Sweetman, J., Lavens, P. Eds. . 388 pp, 149–159. Oostende, Belgium.

Cahu. C. & Zambonino Infante J. (2001) Substitution of live food by formulated diets in marine fish larvae. Aquaculture 200, 161-180.

Cahu, C.L.; Zambonino Infante, J.L.; Escaffre, A.M.; Bergot, P. & Kaushik, S. (1998). Preliminary results on sea bass Dicentrarchus labrax larvae rearing with compound diet from first feeding. Comparison with carp Cyprinus carpio larvae. Aquaculture 169, 1-7

Cahu, C.L. & Infante, J.L.Z. (1994). Early weaning of sea bass (Dicentrarchus labrax) larvae with a compound diet: effect on digestive enzymes. Comparative Biochemistry and Physiology 109A, 213–222

Cerqueira, V.R.; Macchiavello, J.A.G. & Brugger, A.M. (1995). Producao de alevinos de robalo, Centropomus parallellus, (Poey 1890) atraves de larvicultura intensivaem laboratorio. 191-197.

Chapman, P.; Horel, G.; Fish. W., Jones. K. & Spicola. J. (1978). Artificial culture of snook, rookery Bay, 1977. Florida Game and Fresh Water Fish Commission.

Crisp, D.T. (1988) Prediction from temperature of eyeing, hatching and 'swim-up' times for salmonid embryos. *Freshwater Biology* 19, 41–48.

Colesante R.T. (1989). Improved survival of walleye fry during the first 30 days of intensive rearing on brine shrimp and zooplankton. *Progressive Fish-Culturist* 51, 109-111.

Corraza L. & Nickum J.G. (1981). Positive phototaxis during initial feeding stages of walleye larvae. Rapports et Proces-Verbaux Reunions du Conseil International pour l' Exploration de la Mer 178, 492-494p.

Dabrowski K; Lee K.-J & Rinchard J. (2003). The smallest vertebrate, teleost fish, can utilize synthetic dipeptide based diets. *Journal of Nutrition* 133, 4225-4229.

Dehasque, M.; De Wolf, T.; Candreva, P.; Coutteau, P. & Sorgeloos, P. (1998). Control of bacterial input through the live food in marine fish hatcheries. In: Grizel, H., Kestemont, P. , Aquaculture and Water: Fish Culture, Shellfish Culture and Water Usage. pp. 66-67. Oostende.: European. Aquaculture Society .

Denson M.R. & Smith T.I.J. (1996). Larval rearing and weaning techniques for white bass *Morone chrysops*. *Journal of World Aquaculture Society* 27, 194-201.

Dendrinos P.; Dewan S. & Thorpe J.P. (1984). Improvement in the feeding efficiency of larval, post larval and juvenile Dover sole (*Solea solea* L.) by the use of staining to improve the visibility of Artemia used as food. *Aquaculture* 38, 137-144.

Divanach, P.; Papandroulakis, N. & Kentouri, M. (1998). Bases de techniques de l'evage larvaire de la daurade *Sparus aurata* . 107–114p.

Doi, M.; Ohno, A.; Taki, Y.; Singhagraiwan, T. & Kohno, H. (1997a). Nauplii of the calanoid copepod, *Acartia injiensis* as an initial food organism for larval red snapper, *Lutjanus argentimaculatus*. *Suisan Zoshoku* 45, 31-40

Doi, M.; Toledo, J.D.; Golez, M.S.N.; Santos, M.D.L. & Ohno, A. (1997b). Preliminary investigation of feeding performance of larvae of early red-spotted grouper, *Epinephelus coioides*, reared with mixed zooplankton. *Hydrobiologia* 358, 259-263

Duray M.N; Estudillo C.B & Alpasan L.G. (1996) The effect of background color and rotifer density on rotifer intake, growth, and survival of the grouper (*Epinephelus suillus*) larvae. *Aquaculture* 146, 217-224.

Fernandez-Diaz, C. & Yufera, M. (1997). Detecting growth in gilthead seabream *Sparus aurata* L. larvae fed microcapsules. *Aquaculture* 153, 93–102.

Fore, P.L. & T.W. Schmidt (1973). Biology of juvenile and adult snook, *Centropomus undecimalis*, in the ten Thousand Islands, Florida. Atlanta, Georgia.

Fuiman, L.A.; Poling, K.R. & Higgs, D.M. (1998). Quantifying developmental progress for comparative studies of larval fishes. *Copeia* 199, 602–611.

Fushimi, T. (1983). III-5 ingestion by fish larave and juveniles. In: Koseishya-Koseikaku *The rotifers, Brachiomus plicatilis- Biology and mass culture*, pp. 69-93. Tokyo: Japanese Society of Science and Fisheries.

Fukusho, K. (1989). Biology and mass production of the rotifer, *Brachionus plicatilis*. *International Journal of Aquaculture and Fish Technology* 1, 292–299

Gatesoupe, F.J.; Girin, M. & Luquet, P. (1977). Recherche d'une alimentation artificielle adapte´e a` l'e´levage des stades larvaires des poissons. II-Application a` l'e´levage larvaire du bar et de la sole 4, 59–66p.

Hagiwara A.; Gallardo W.G.; Assavaaree M.; Kotani T & De Araujo A.B. (2001). Live food production in Japan: recent progress and future aspects. *Aquaculture* 200, 111-127.

Heath, M.R. (1992). Field investigation of the early life history stages of marine fish. *Advances in Marine Biology* 28, 2-174.

Heggberget, T.G. & Wallace, J.C. (1984) . Incubation of the eggs of Atlantic salmon, *Salmo salar*, at low temperatures. *Canadian Journal of Fisheries and Aquatic Sciences* 41 , 389–391.

Heming, T.A. (1982)　Effects of temperature on utilisation of yolk by chinook salmon (*Oncorhynchus tshawytscha*) eggs and alevins. *Canadian Journal of Fisheries and Aquatic Sciences* 39, 184–190

Heming, T.A. & Buddington, R.K. (1988) Yolk absorption in embryonic and larval fishes. In: S. Hoar and D.J. Randall, (Eds.) *In: Fish Physiology* , pp. 407–446　San Diego.: Academic Press]

Hernandez-Cruz C.M.; Salhi M.; Fernandez-Palacios H & Izquierdo M.S. (1994) Improvements in the culture of *Sparus aurata* L. larvae in relation to the use of antibiotics, phytoplankton and rearing system. *Aquaculture* 124 , 269-274.

Hernández Cruz, C.M.; M Salhi, M.; Bessonart, M.S.; Izquierdo, M.M.; González , H. & Fernández Palacios. (1999). Rearing techniques for red porgy (*Pagrus pagrus*) during larval development. Aquaculture 179:489-497.

Hinshaw J.M. (1985). Effects of illumination and prey contact on survival and growth of yellow perch, *Perca flavescens*. *Transactions of the American Philological Association* 114, 540-545.

Hirata, H. (1980). Culture methods of the marine rotifer, *Brachionus plicatilis*. *Mini Reviews and Data File of Fishreries Research* 1, 27-46.

Hochachka, P.W. & Somero G. N. (2001). *Biochemical Adaptation*. Oxford: Oxford University Press.

Holmejford, I.; Gulbrandsen, J.; Lein, I.; Refstie, T.; Leger, P.; Harboe, T.; Huse, I.; Sorgeloos, P.; Bolla, S., Olsen, I.; Reitan, K.I.; Vadstein, O.; Oie, G. & Danielsberg, A. (1993) An intensive approach to atlantic halibut fry production. *Journal of the World Aquaculture Society* 24 , 275-284

Houde, E.D. (1975). Effects of stocking density and food density on survival, growth and yield of laboratory reared larvae of sea bream *Archosargus rhomboidalis* (L.) (Sparidae). *Journal of Fish Biology* 7, 115- 127

Houde, E.D. & Schekter, R.C. (1980)　Feeding by marine fish larvae: developmental and functional responses. *Environ. Biol. Fishes* 5 4,

Howell B.R. (1979). Experiments on the rearing of larval turbot, *Scophthalmus maximus* L. *Aquaculture* 18, 215-225

Howell B.R. (1994). Fitness of hatchery-reared fish for survival in the sea. *Aquaculture and Fisheries management* 25, 3-17.

Howells, R.G.; Sonski, A.J.; Shafland, P.L. & Hilton B.D. (1990). Lower temperature tolerance of snook, *Centropomus undecimalis*. *Northeast Gulf Science* 11, 155-158.

Hunter, J.R. (1980). The feeding behavior and ecology of marine fish larvae. Fish Behaviour and its Use in the Capture and Culture of Fishes. International Center for Living Aquatic Resources. Management 512p.

Ina, K., Ryogi, Y. & Higashi, K. (1979). Color sensitivity of red sea bream, Pagrus major. *Bulletin of the Japanese Society of Sciencie Fisheries*. 45(1), 1-5

Ivlev, V.S. (1961). Experimental Ecology of Feeding of Fishes. New Haven, CT:　Yale University Press.

Jensen, A.J., Johnsen, B.O. & Saksgĺrd, L. (1989) . Temperature requirements in Atlantic salmon (*Salmosalar*), brown trout (*Salmo trutta*), and Arctic char (*Salvelinus alpinus*) from hatching to initial feeding compared with geographic distribution. *Canadian Journal of Fisheries and Aquatic Sciences* 46, 786–789

Johnston, I.A.; Vieira, V.L.A. & Hill, J. (1996). Temperature and ontogeny in ectotherms: muscle phenotype in fish. In: Johnston, I. A. & Bennett, A. F. *Phenotypic and Evolutionary Adaptations of Organisms to Temperature*, pp. 153–181. Cambridge: Cambridge University Press.]

Jones D.A.; Kamarudin M.S & LeVay L. (1993). The potential for replacement of live feeds in larval culture. *Journal of the World Aquaculture Society* 24, 199-210.

Jones A & Houde E. D., (1981). Mass rearing of fish fry for aquaculture. Realism in aquaculture: achievements, constrains and perspectives. 351-374p. European Aquaculture Society

Kafuku, T. & Ikenoue H. (1983). Modern methods of aquaculture in Japan. In: Anonymous *Developments in aquaculture and fisheries sciences*, pp. 216 Amsterdam: Tokyo and Elsevier]

Kane, T.R. (1988) Relationship of temperature and time of initial feeding of Atlantic salmon. *The Progressive Fish-Culturist* 50, 93–97

Kaji, T.; Tanaka, M. & Tagawa, M. (1999) Laboratory study of density-dependent survival after handling in yolksac larvae of tunas and a grouper. *Fishery Science* 65, 482– 483.

Kamler, E. (1992a). *Early Life History of Fish: An Energetics Approach*, London: Chapman & Hall.

Kamler, E. (1992b). Mixed feeding period. In: Anonymous *Early Life History of Fish*, pp. 176-181. London: Chapman & Hall

Kanazawa, A.; Teshima, S.; Inamori, S.; Sumida, S.; and Iwashita, T. (1982). Rearing of larval red sea bream and ayu with artificial diets. *Memoirs of the Faculty of Fisheries, Kagoshima University* 31, 185–192

Kanazawa, A. (1993). Essential phospholipids of fish and crustaceans. In: Kaushik, S.J., Luquet, P. *Fish nutrition in practice*, INRA edn. pp. 519-530. Paris, Les Colloques.

Kennedy, S.B..; Tucker, J.W..; Neidig, C.L..; Vermeer, G.K., Cooper, V.R.; Jarrell, J.L. & Sennet, D.G. (1998). Bacterial management strategies for stock enhancement of warm water marine fish: a case study with common snook *Centropomus undecimalis*. *Bulletin of Marine Science* 62, 573–58.

Kentouri, M. (1985). Comportement larvaire de 4 Sparides mediterraneens en elevage: *Sparus aurata, Diplodus sargus, Lithognathus mormyrus, Puntazzo puntazzo* Poissons teleosteens . Universite de Sciences et Techniques du Languedoc.

Klumpp, D.W. & Von Westernhagen, H. (1986) Nitrogen balance in marine fish larvae: Influence of developmental stage and prey density. *Marine Biology* 93, 189–200

Kolkovski S, Koven, W.M & Tandler A. (1997). The mode of action of Artemia in enhancing utilization of microdiet by gilthead seabream *Sparus aurata* larvae. *Aquaculture* 155, 193-205.

Koumoundouros, G.; Divanach, P. & Kentouri, M. (1999). Ontogeny and allometric plasticity of *Dentex dentex* (*Osteichthyes: Sparidae*) in rearing conditions. *Marine Biology* 135, 561–572

Kovac, V. (2002). Synchrony and heterochrony in ontogeny (of fish). *Journal of Theoretical Biology* 217, 499-507

Koven, W.M..; Tandler, A.; Kissil, G.W.; Sklan, D.; Friezlander, O. & Harel, M. (1990). The effect of dietary n-3 polyunsaturated fatty acids on growth, survival and swimbladder development in *Sparus aurata* larvae. *Aquaculture* 91 131-141.

Lau, S.R. & Shafland P. L. (1982). Larval development of snook *Centropomus undecimalis* (Pisces: *Centropomidae*). *Copeia* 618-627.

Laurence, G.C. (1978). Comparative growth, respiration and delayed feeding abilities of larval cod (*Gadus morhua*) and haddock (*Melanogrammus aeglefinus*) as influenced by temperature during laboratory studies. *Marine Biology* 50, 1– 7

Laurence, G.C. (1974). Growth and survival of haddock (*Melanogrammus aeglefinus*) larvae in relation to planktonic prey concentration. *Journal of the Fisheries Research Board of Canada* 31, 1415– 1419

Laurel, B.J.; Brown, J.A. & Anderson, R. (2001). Behaviour, growth and survival of redfish larvae in relation to prey availability. *Journal of Fish Biology* 59, 884– 901

Lazo J.P.; Dinis M.T.; Holt G.J.; Faulk C & Arnold C.R. (2000). Co-feeding microparticulate diets with algae: toward eliminating the need of zooplankton at first feeding in larval red drum (*Sciaenops ocellatus*). *Aquaculture* 188,

Liao, I.C.; Su, H.M. & Chang, E.Y. (2001). Techniques in finfish larviculture in Taiwan. *Aquaculture* 200, 1-31.

Lubzens, E. (1987). Raising rotifers for use in aquaculture. *Hydrobiologia* 147, 245–255.

May R.C. (1971). An annotated bibliography of attempts to rear the larvae of marine fishes in the laboratory.

Mani-Ponset, L.; Guyot, E.; Diaz, J.P. & Connes, R. (1996). Utilization of yolk reserves during post-embryonic development in three teleostean species: the sea bream, (*Sparus aurata*), the sea bass, (*Dicentrarchus labrax*) and the pike-perch (*Stizostedion lucioperca*). *Marine Biology* 126, 539-547.

Marliave J.B. (1994). Green water: optical rather than nutritional effect. AZA Conference Proceedings, 266-269p.

Marr, D.H.A. (1996) Influence of temperature on the efficiency of growth of salmonid embryos. *Nature* 212, 957–959

Matlock, G.C. & Osburn, H.R. (1987).Demise of the snook fishery in Texas, USA. *Northeast Gulf Science* 9, 53-58.(Abstract)

Moffatt, N.M. (1981). Survival & growth of northern anchovy larvae on low zooplankton densities as affected by the presence of a *Chlorella* bloom. In: Lasker, R., Sherman, K. *The Early Life History of Fish: Recent Studies*, pp. 475–480

Munk, P. & Kiørboe, T. (1985). Feeding behaviour and swimming activity of larval herring (*Clupea harengus*) in relation to density of copepod nauplii. *Marine Ecology Progress Series* 24, 15– 21

Munro, P.D., Barbour, A. & Birkbeck, T.H. (1994). Comparison of the gut bacterial flora of starting feeding larval turbot reared under different conditions. *Journal of Food Protection.* 40, 288

Munro, P.D.; Birkbeck, T.H. & Barbour, A. (1993). Bacterial flora of rotifers *Brachionus plicatilis*: evidence for a major location on the external surface and methods for reducing the rotifer bacterial load. In Reinertsen, H., Dahle, L.A., Jorgensen, L., Tvinnereim, K. *Fish Farming Technology*, pp. 93–100. Balkema, Rotterdam:

Muroga, K. & Yasunobu, H. (1987). Uptake of bacteria by rotifer. *Bulletin of the Japanese Society of Scientific. Fisheries* 53, 2091

Naas K.E, NÒss T & Harboe T. (1992). Enhanced first feeding of halibut larvae (*Hippoglossus hippoglossus* L.) in green water. *Aquaculture* 105, 143-156.

Nagata, W.D. & Hirata, H. (1986). Mariculture in Japan: past, present, and future prospective. *Mini Reviews and Data File of Fisheries Research* 4, 1–38.

Nellen W. (1981). Live animal food for larval rearing in aquaculture.- non-Artemia organisms Realism in Aquaculture: Achievements, Constrains, Perspectives 215-260. European Aquaculture Society.

Nicolas JL, Robic E & Ansquer D (1989). Bacterial flora associated with a tropical chain consisting of microalgae, rotifers and turbot larvae: influence of bacteria on larval survival. Aquaculture 83: 237–248

Oie, G., Makridis, P.; Reitan, K.I. & Olsen, Y. (1997). Protein and carbon utilization of rotifers *Brachionus plicatilis* in first feeding of turbot larvae *Scophthalmus maximus* L. *Aquaculture* 153, 103–122.

Oozeki, Y.; Hagiwara, A.; Eda, H. & Lee, C.S. (1992). Developmental changes of food selectivity of striped mullet *Mugil cephalus* during the Larval Stage. *Nippon Suisan Gakkaishi* 58, 1381

Ostaszewska T; Dabrowski, K.; Czuminska, K. & Olech W. (2005). Rearing pike-perch larvae using formulated diets- first success with starter feeds. *Aquaculture research* 36, 1167-1176.

Ostrowski A.C. (1989). Effect of rearing tank background color on early survival of dolphin larvae. *Progressive Fish- Culturist* 51, 161-63.

Papandroulakis N.; Divanach P & Kentouri M (2002). Enhanced biological performance of intensive sea bream *Sparus aurata* larviculture in the presence of phytoplankton with long photophase. *Aquaculture* 204, 45-63.

Pearce, D. (1991). Measuring Sustainable Development. Blueprint 2. Earthscan, London

Pedersen B.H.; Ueberschar B & Kurokawa T. (2003). Digestive response and rates of growth in pre-leptocephalus larvae of Japanese eel *Anguila japonica* reared on artificial diets. *Aquaculture* 215, 321-338.

Pedersen, B.H. & Hjelmeland, K. (1988). Fate of typsin and assimilation efficiency in larval herring *Clupea harengus* following digestion of copepods. *Marine Biology* 97, 467–476.

Pedersen, B.H.; Nilssen, E.M. & Hjelmeland, K. (1987). Variations in the content of trypsin and trypsinogen in larval herring *Clupea harengus* digesting copepod nauplii. *Marine Biology* 94, 171–18

Peña R.; Dumas S.; Saldivar-Lucio R.; Garcia G.; Trasvin a A & Hernandez-Ceballos D. (2004) The effects of light intensity on first feeding of the spotted sand bass *Paralabrax maculatofasciatus* Steindachner larvae. *Aquaculture Research* 35, 345-349.

Pepin, P. (1991). The effect of temperature and size on development and mortality rates of the pelagic early life history stages of marine fish. *Canadian Journal of Fisheries and Aquatic Sciences* 48, 503–518

Pepin, P.; Orr, D.C. & Anderson, J.T. (1997). Time to hatch and larval size in relation to temperature and egg size in Atlantic cod (*Gadus morhua*). *Canadian Journal of Fisheries and Aquatic Sciences* 54 (Suppl. 1), 2–10

Peterson, R.H.; Martin-Robichaud, D.J. & Berge, L. (1996) Influence of temperature and salinity on length and yolk utilization of striped bass larvae. *Aquaculture International* 89–103, 4.

Peterson, R.H. & Martin-Robichaud, D.J. (1995). Yolk utilization by Atlantic salmon (*Salmo salar* L.) alevins in response to temperature and substrate. *Aquacultural Engineering* 14, 85–99

Person Le Ruyet, J.; Alexandre, J.C.; Thebaud, L. & Mugnier, C. (1993). Marine fish larvae feeding: formulated diets or live preys?. *Journal of the World Aquaculture. Society.* 24, 211-224

Peterson, R.H.; Spinney, H.C.E. & Sreedharan, A. (1977) . Development of Atlantic salmon (*Salmo salar*) eggs and alevins under varied temperature regimes. *Journal of the Fisheries Research Board of Canada* 34, 31–43

Puvanendran, V. & Brown, J.A. (1999). Foraging, growth and survival of Atlantic cod larvae reared in different prey concentrations. *Aquaculture* 175, 77–92

Rainuzzo, J.R.; Reitan, K.I. & Olsen, Y. (1997). The significance of lipids at early stages of marine fish: a review. *Aquaculture* 155 , 103-115.

Rana, K, (1986). Parental influences on egg quality, fry production and fry performance in *Oreochromis niloticus* (Linnaeus) and *O. mossambicus* (Peters). PhD thesis, University of Stirling, Scotland.

Reitan K.I.; Bolla S,& Olsen Y. (1991). Ingestion and assimilation of microalgae in yolk sac larvae of halibut, *Hippoglossus hippoglossus* (L.). Larvi '91, Fish and Crustacean Larviculture Symposium 15, 332-334.p. European Aquaculture Society

Reitan, K.I.; Rainuzzo, J.R.; Oie, G. & Olsen, Y. (1993). Nutritional effects of algal addition in first-feeding of turbot *Scophthalmus maximus* L. larvae. *Aquaculture* 118, 257–275.

Rideout, R.M.; Trippel, E.A. & Litvak, M.K. (2004). Paternal effects on haddock early life history traits. *Journal of Fish Biology* 64, 695–701

Rombough, P.J. (1988). Respiratory gas exchange, aerobic metabolism, and effects of hypoxia during early life. In: Hoar S and D.J. Randall, (Ed.) *Fish Physiology*, pp. 59–61 San Diego: . Academic Press]

Rotllant J.; Tort L.; Montero D.; Pavlidis M.; Martinez M.; Wendelaar Bonga S.E & Balm P.H.M. (2003). Background colour influence on the stress response in cultured red porgy *Pagrus pagrus*. *Aquaculture* 223, 129-139.

Salhi M.; Izquierdo, M.S.; Hernandez de la Cruz C. M.; Socorro J & Fernandez-Palacios H. (1997). The improved incorporation of polyunsaturated fatty acids and changes in liver structure in larval gilthead seabream fed on microdiets. *Journal of Fish Biology* 51, 869–879

Scott A.P & Middleton C. (1979). Unicellular algae as a food for turbot (*Scophthalmus maximus*, L.) larvae the importance of dietary long-chain polyunsaturated fatty acids. *Aquaculture* 18, 227-240.

Seidel, C.R.; Schauer, P.S.; Katayama, T. & Simpson, K.L. (1980). Culture of Atlantic silverside fed on artificial diets and brine shrimp nauplii. *Bulletin of the Japanese Society of Scientific Fisheries* 46, 237–245

Shafland, P.L.& Koehl D.H. (1980). Laboratory rearing of the common snook. Proceedings of the annual Conference South-eastern Association of Fisheries and Wildlife Agencies 33 (425-431).

Skjermo, J. & Vadstein, O. (1999).Techniques for microbial control in the intensive rearing of marine larvae. *Aquaculture* 177, 333–343

Snell, T.W. (1991).Improving the design of mass culture systems for the rotifer, *Brachionus plicatilis*. In. Fulks, W, Main, K.L. Rotifer and Microalgae Culture Systems, 61-71.p.

Sorgeloos P.; Léger PH & Lavens P (1988). Improved larval rearing of European and Asian seabass, seabream, mahimahi, siganid and milkfish using enrichment diets for *Brachionus* and *Artemia*. *World Aquaculture* 19, 78-79.

Shirota, A. (1970). Studies on the mouth size of fish larvae. *Nippon Suisan Gakkaishi* 36, 353–368

Støttrup, J.G. (2000). The elusive copepods: their production and suitability in marine aquaculture. *Aquaculture Research*, 31, 703-711.

Støttrup J. G & Norsker N.H (1997). Production and use of copepods in marine fish larviculture. *Aquaculture* 155, 231-247.

Støttrup, J.G.; Gravningen, K. & Norsker, N.H. (1995). The role of different algae in the growth and survival of turbot larvae *Scophthalmus maximus* L. in intensive rearing systems. 201, 173–186p.

Tagawa M.; Kaji, T.; Kinoshita, M. & Tanaka, M. (2004). Effect of stocking density and addition of proteins on larval survival in Japanese flounder, *Paralichthys olivaceus*. *Aquaculture* 230, 517-525.

Tagawa M,; Kaji, T.; Kinoshita, M. & Tanaka, M. (1997). Effect of thyroid hormone supplementation to eggs on the larval survival of Japanese flounder. edn. p. 48p.

Takeuchi T.; Wang Q.; Furuita H.; Hirota T.; Ishida S & Hayasawa H. (2003). Development of microparticle diets for Japanese flounder *Paralichthys olivaceus* larvae. *Fisheries Science* 69, 547-554.

Takeuchi, T.; Ohkuma, N.; Ishida, S.; Ishizuka, W.; Tomita, M.; Hayasawa, H. & Miyakawa, H., (1998). Development of micro-particle diet for marine fish larvae 1st International Symposium on Nutrition and Feeding of Fish, Abstract only 193p.

Tamaru C.S.; Murashige R & Lee C.S. (1994). The paradox of using background phytoplankton during the larval culture of striped mullet, *Mugil cephalus* L. *Aquaculture* 119, 167-174.

Tamaru C.S.; Lee C, & Ako H. (1993). Improving the larval rearing of stripped mullet (*Mugil cephalus*) by manipulating quantity and quality of the rotifer. *Brachiounus plicatillis*. Proceedings of a US-Asia Workshop

Taylor R. G.; Whittington J. A.; Grier, H.J. & Crabtree R. E. (2000). Age, growth, maturation, and protandric sex reversal in common snook, *Centropomus undecimalis*, from the east and west coast of South Florida. *Fisheries Bulletin* 98, 612-624.(Abstract).

Toledo, J.D.; Golez, S.N.; Doi, M. & Ohno, A. (1997). Food selection of early grouper, *Epinephelus coioides*, larvae reared by the semi-intensive method. *Suisan Zoshoku* 45, 327-337

Tringali, M.D. & T. M. Bert (1996). The genetic stock structure of common snook (*Centropomus undecimalis*). *Canadian Journal of Fishery and Aquatic Science*. 53, 974-984.(Abstract)

Tsukamoto, K. (1993). Quality of fish for release. In: Kitajima, C. *Healthy Fry for Release, and Their Production Technique*, pp. 102-113 Koseisya Koseikaku, Tokyo.

Tucker, J.W.J.; Landau, M.P. & Falkner, B.E. (1985). Culinary value and composition of wild and captive common snook, *Centropomus undecimalis*, *Florida Scientist* 49, 196-200.(Abstract)

Van Der Meeren, T. (1991). Algae as first food for cod larvae *Gadus morhua* L. : filter feeding or ingestion by accident? *Journal of Fish Biology* 39, 225-237

Vasquez-Yeomans L.& Carrillo-Barrios-Gomez E & Sosa-Cordero E. (1990).The effect of the nanoflagellate *Tetraselmis suecica* on the growth and survival of grunion, *Leuresthes tenuis*, larvae. *Environmental Biology of Fisheries* 29, 193-200.

Watanabe, T. (1993). Importance of docosahexaenoic acid in marine larval fish. *Journal of World Aquaculture Society* 24, 495-501.

Webb J.F. (1999). Larvae in fish development and evolution. In: B.K. Hall & M.H. Wake. *The Origin and Evolution of Larval Forms* , pp. 109-158. San Diego, CA, USA: Academic Press]

Werner R.G & Blaxter J.H.S (1981) The effect of prey density on mortality, growth and food consumption in larval herring (*Clupea harengus* L.). *Rapports et Proces-verbaux du Conseil international pour l'Exploration de la Mer* 178, 405-408.

Wyatt, T. (1972). Some effects of food density on the growth and behaviour of plaice larvae. *Marine Biology* 14, 210-216

Yanes-Roca, C.; Rhody, N.; Nystrom, M.; Main, K. (2009). Effects of Fatty acid composition and spawning season patterns on egg quality and larval survival in common snook (*Centropomus undecimalis*). Aquaculture 287, pp 335-340

Yufera M.; Pascual E & Fernandez-Diaz C. (1999). A highly efficient microencapsulated food for rearing early larvae of marine fish. *Aquaculture* 177, 249-256.

Zambonino Infante, J.L.; Cahu, C.L. & Peres, A. (1997) Partial substitution of di- and tripeptides for native proteins in sea bass diet improves *Dicentrarchus labrax* larval development. *Journal of Nutrition* 127, 608-614.

11

Measurements Population Growth and Fecundity of Daphnia Magna to Different Levels of Nutrients Under Stress Conditions

Lucía E. Ocampo Q., Mónica Botero A. and Luis Fernando Restrepo

Antioquia University
Colombia

1. Introduction

In nature, zooplankton is the main nutritional source of poslarvae and young fish. The natural food offers essential nutrients to guarantee the survival and the growth of fish during their first development stages (Furuya et al. 1999). The description of feed value of living food, has been made by Watanabeetet al., (1998); Kraul, (2006). Living food has a vital job on seed production in fish farms. Without this living food, it is not possible to overcome an adequate survival rate, in species exclusively dependent (Kubitza, 1997; Lahnsteiner et al., 2009).

Micro crustaceans are highly important in aquaculture, mainly the freshwater genera *Moina* and *Daphnia spp*, these two are found in diverse natural environments (FAO, 1996).*Daphnia* genera includes *D. magna, D. pulex, D. longispina* among others. In crops of freshwater spices, poslarvae are fed with 2 or 3 organisms during the beginning of their first hexogen feeding, during their first 10 to 30 days (Lubzens & Zmora, (2003), as cited in Stottrup & Mc Evoy, 2003; Botero, 2004; Prieto, 2006). It is evident the importance of *D. magna* as live food. Authors such as Emmens, (1984), have been reported that *Daphnia spp.* is the best foodstuff for tropical fishes, frequently used food source in the freshwater larviculture (i.e. for different carp species) and in the ornamental fish industry (i.e. guppies, sword tails, black mollies and plattys etc.) (Delbare & Dhert, as cited in Lavens & Sorgeloos, 1996), native and foreign species, for example White Cachama *(Piaractusbrachypomus)*, Black Cachama, *(Colosoma macropomum)*, Bocachico *(Prochilodus magdalenae)*, Yamú *(Brycon amazonicus)*, Sabaleta *(Brycon henni)*, Dorada *(Brycon moreii)*, Striped Bagre *(Pseudoplatystoma fasciatum)*, Pacu *(Piaractus mesopotamicus)*, among others (Botero; 2004; Prieto, 2008).

It is known of certain difficulties on live food production, due to the fluctuations on the natural conditions of aquatic environments and the elevated infrastructure requirements, equipment, maintenance spending and working labor. It is not possible to produce a constant amount of live food in a regular basis (Kanasawa, 2000). On the other hand the laboratory research to develop culture techniques *of Daphnia magna* has been widely studied, this is because it is easy to cultivate and has a low cost in high densities. Additionally their maintenance in a small space makes them an economically viable alternative culture (Terra, et al., 2010).

The secondary production in lakes is supported by zooplankton, zoobenthos and fish; this means that this group is diverse from the taxonomic and functional point of view. On scale work of authors such as Stotz and Pérez (1992) and Andrade et al., (2009), emphasizes the necessity to recognize the production of a secondary source to determine variables such as maximum extraction. In fact secondary extraction is considered as one of the most important parameters to evaluate the population utilization sustainability (Andrade et al., 2009).This section presents production secondary variables, following the method designed by González, (1988) that is specific for the cladocerans, showing the results obtained at the laboratory scale.

This chapter presents an experimental evaluation of two *Daphnia magna* populations, the first population integrated by neonates and the other by adults in early reproductive stage under stress conditions. This stress condition on the test was made by using 3 cm^3 multi-cells, on each treatment, under controlled conditions of room temperature (21 – 25 °C), water temperature (22 – 23 °C) and pH (7.6). The diet used was *Saccharomyces cereviseae*, potatoes (*Solanum tuberosum*) and a fatty acid enriched environment n-6 (soy oatmeal). The diet and enrichment concentrations were 30 ppm and 15 ppm, factorial arrangement of 2^3, in concentrations of 15 and 30 ppm mixture of nutrients: yeast and potato and the same concentration for enrichment. Four replicas/treatments were made (32). The Feeding was on a daily basis for 20 days to determine the population performance effect. The productive variables were evaluated: maximum density (Dmáx) daily average density (Dmd), doubling time (Td),specific growth rate (k), performance (r) , numeric growth (PN), birth rate (b), (Edmodson equation), individuals average number ($\overline{N}$), biomass productivity (Pw), mortality rates (d), biomass (B), production rate (I de P) and final weight. Reproductive variables were: egg number/female (HPP), neonates number/female (NPP), egg maturity time (tm), first reproduction age (EPR), litter number (NC), reproduction frequency (FR), net reproduction rate (Ro) and generation time (Tc).

There were significant differences (p<0.05) on T2 from the population of adults, with concentrations of 15 ppm *S. cereviceae*, potato 15 ppm, soy oatmeal 30 ppm, with the highest specific growth, 0.50 ± 0.05 per day, less doubling time with 1.39 ± 0.14 days and the highest mortality rate with 0.49 ±0.07 per day. In the rest of the treatments there were no significant differences (p>0.05). There was evidence that the highest nutrient combinations strengthened the population growth in both adults and adults in juvenile reproductive stage. They reached in T6 with concentrations of 30 ppm *S. cereviceae*, potato 15 ppm, soy oatmeal 30 ppm, and 8.25 ± 1.70 and 15.0 ± 9.76 *Daphnias*/mL for each population respectively. Likewise in T6, was observed a higher value on egg number/female, 3.83 ± 0.82 and 3.55 ± 0.98, for each population respectively. *D. magna* presents a favorable adaptation under stress conditions, turning it into an excellent alternative for the living food for poslarvae production, with minimum infrastructure.

In order to use the *S. cereviseae* probiotic as an alternative of real control strategy, a meticulous evaluation must happen, evaluating their competing and functionality on the living food utilized with the different poslarvae species and their environments. It is imperative prerequisite to develop pathogenicity studies not only with the living food (*D. magna*) but also the selected probiotic from any commercial consideration; this means is necessary to explore this matter deeply (Austin & Brunt, 2009, as cited in Montet & Ray, 2009).

2. Cladocerans

Brachiopods are small crustacean with their legs flat as leafs. They can be found in freshwater habitats. Daphnia populations can be found in a range of water bodies, from huge lakes down to very small temporary pools, such as rock pools and vernal pools (seasonally flooded depressions). Often they are the dominant zooplankton and form, as such, an essential part of the food web in lakes and ponds. In many lakes, Daphnia are the predominant food for planktivorous fish, at least at times. As a consequence, the Daphnia species distribution and life history are closely linked with the occurrence of predators. Typically, Daphnia species are found in lakes with planktivorous fish they are smaller and more transparent than species found in fishless water bodies.

The cladocerans represent a key position in aquatic communities, not only as consumers herbivorous such as algae and bacteria but also as feedstuff for fish, birds and other aquatic predators (Dodson & Frey, 2001, as cited in Thorp & Covich, 2001; Brett et al., 2009, as cited in Arts et al., 2009). Taxonomically, the Branchiopods are grouped into six orders, 29 families, including the revisions suggested by Thorp & Covich (2001). Branchiopoda has the following orders and families. Order: Anomopoda, Ctenopoda, Onychopoda, and Family: Daphniidae, Moinidae, Bosminidae, Ilyocryptidae, Macrothricidae, Neothricidae, Acantholeberidae, Ophryoxidae, Chydoridae, Sididae, Holopediidae, Podonidae, Polyphemdida, Cercopagidae. Order: Haplopoda and Family: Leptodoridae. Order: Anostracai, and Family: Artemiidae, Branchinectidae, Branchipodidae, Chirocephalidae, Linderiellidae, Polyartemiidae, Streptocephalidae, Thamnocephalidaev. Order: Spinicaudat and Family: Cyclestheriidae, Cyzcidae, Leptestheriidae, Limnadiidae. Order: Laevicaudataand and Family: Lynceidae. Order: Notostraca and Family: Triopsidae (Dodson & Frey, 2001, as cited in Thorp &. Covich, 2001; Kobayashi et al., 2008, as cited in Suthers & Rissik, 2008).

2.1 Anatomy and physiology

The cladocerans do not have a segmented body, but they have a second segmented antenna. Most of them contain only one composed central eye during their adult stage and a clear transparent yellowish shell. Crustaceans are different from other arthropods because they have two pairs of the antennas. Cladocerans have a first pair of antennas (antennules, generally with one segment and other smaller antenna with chemical sense functions). The second pair of antennas is big and used to swim (Dodson &Frey, 2001, as cited in Thorp & Covich, 2001). Likewise other crustaceans, cladocerans mostly present in their heads: two pairs of antennas, one pair of jaws and two pair of jawbones. On the base of the head close to the shell they have a pair of short appendixes (antennules), normally this are shorter than the head and less visible, but sometimes can be longer (Moina) (Dodson &Frey, 2001, as cited in Thorp & Covich, 2001).

Behind their heads the bodies are composed by a thorax and an abdomen. The body finishes on a pair of claws (post-abdominal) that can show up from the shell. The chemistry of the shell is important, because cladocerans tend to have a hydrophobic exoskeleton, composed by chitin. The thorax and the abdomen displace in their shells when they are alive. The flat legs (called "phyllopod") have lines of mushrooms and spines that are used for feedstuff management, filtering, scraping and pumping out. The shell has a double wall, and between them there is a hemolymph flow, being part of the corporal cavity (Ebert, 2005).

The thoracic legs operate as electrostatic filters (not sifter), they collect algae and other particles that get attached to the flat surfaces and the mushroom combs. The intestine goes along with the mouth, in curves, continues over the body passing the thorax and the abdomen, and ends up in the anus close to the very end of the animal. It is divided in three regions. The previous intestine (water absorption using columnar cells) and the posterior intestine both aligned with the cuticle that wraps the exterior of the animal. The middle intestine (in the thorax) is aligned with the epithelium covered by microvillus, absorption site. On the head region of the Daphnia there is a pair of small bags (hepatic caecum) associated with the intestines. The heart is a muscular organ above the intestine and previous to the head. Cladocerans are between 0.5 mm and 6 mm long (Dodson & Frey, 2001, cited in Thorp &Covich, 2001; Ebert, 2005).

Males are distinguished from females by their smaller size, larger antennules, modified post-abdomen, and first legs, which are armed with a hook used in clasping. The genus Daphnia includes more than 100 known species of freshwater plankton organisms found around the world (figure 1) (Dodson & Frey, 2001, cited in Thorp &Covich, 2001).

3. Life cycle and development

Cladocerans have sexual and non-sexual reproduction, according to the environmental conditions; this is shown in figure 2. Most of the time the majority of the females has sexual reproduction (Dodson & Frey, 2001, cited in Thorp & Covich, 2001) and they develop eggs in a resting stage. Embryogenesis starts and ends in the incubation chamber, this is a space between the body and the shell. There are three types of eggs: - Diploid, they develop immediately in juveniles, - Resting eggs that come from haploid eggs, fertilized in embryo in early stages, the go on diapauses (resistant to heat, dryness and heating), - Pseudo-sexual eggs, this are diapausic embryos from non-sexual diploid eggs (Dodson & Frey, 2001, cited in Thorp & Covich, 2001; Ebert, 2005).

The non-sexual reproduction modality is an important characteristic in the implementation of controlled laboratory productions (by cyclical parthenogenesis). It is shown mainly that when there are satisfactory development conditions, producing female litters exclusively, that are able to succeed to the reproductive phase and to continue reproducing in a non-sexual way as long as the favorable conditions continue; feeding, low population density, and the main environmental factors and the water chemical quality maintain the adequate levels (Ebert, 2005).

Deteriorating environmental conditions stimulate the production of males; part of the brood will be established by males. When these males develop, they can give the possibility of the sexual reproduction fertilizing the females, and also a sexual resting eggs enclosed by anephippium, resistant structure, a resistant case made from the exoskeleton around the brood chamber. They contain one or two embryos (depending on the specie) on a sleeping stage, and they maintain the diapauses until the environmental conditions are suitable to restart the development of the embryo and then emerge juveniles that are able to start a new non-sexual reproductive cycle.

In the parthenogenesis reproductive system, the embryo development is direct and takes place on the female incubation chamber. From this chamber emerge juvenile organisms (neonates); they are freed to the environment when they complete its own development. The

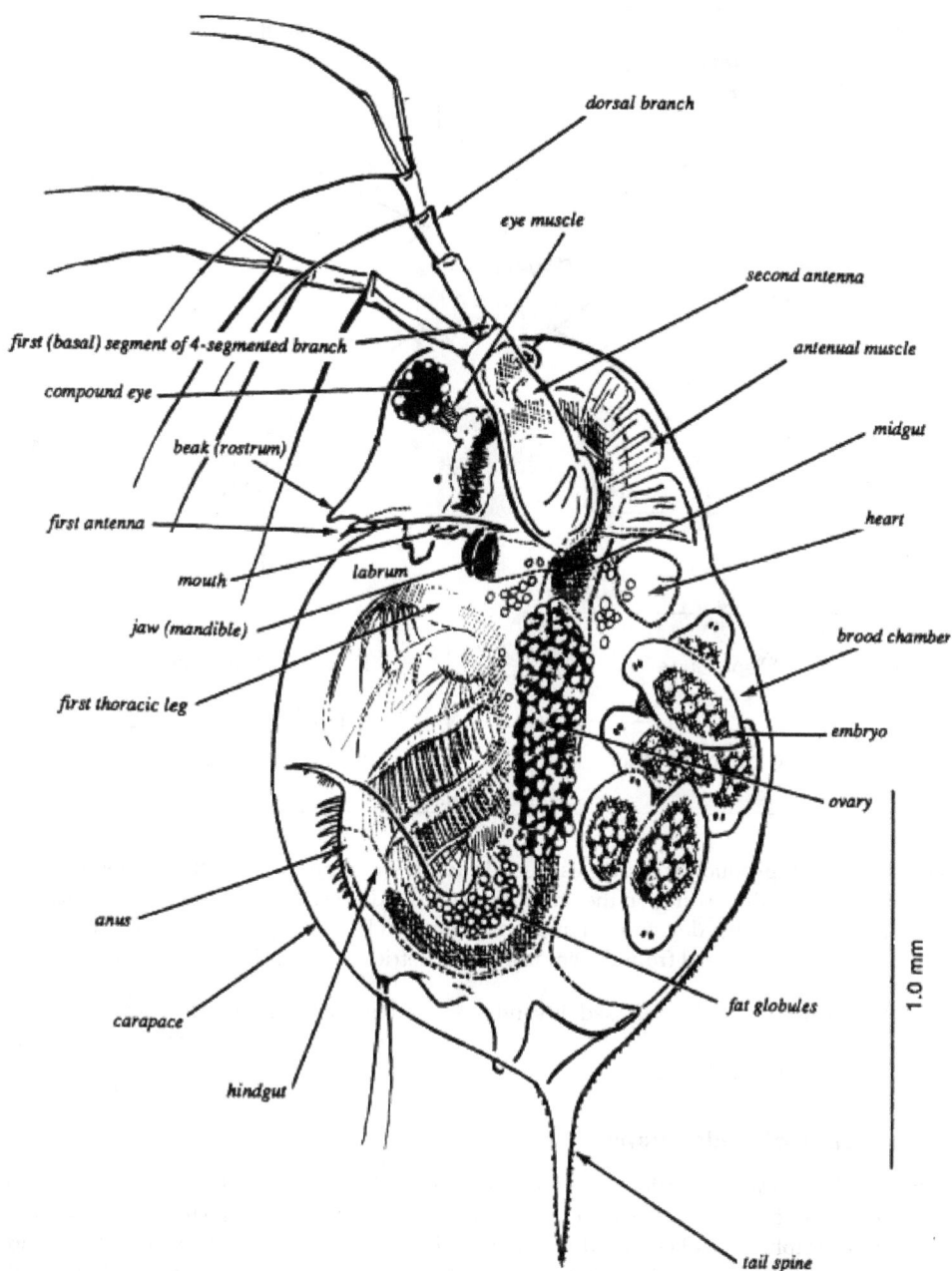

Fig. 1. Schematic drawing of the internal and external anatomy of retrieved *Daphnia*.
Retrieve from (Lavens & Sorgeloos, 1996)

```
┌─────────────────────────────┐        ┌─────────────────────────────┐
│          Natural            │        │          Artificial         │
│   Environmental Signals     │        │           Signals           │
│      •   Day length         │        │      •   Pesticides         │
│      •   Crowding           │        │  •  Industrial chemicals    │
└─────────────────────────────┘        └─────────────────────────────┘
```

Developmental Switch

Signals Absent

Signals Present

Reproduce Asexually

• Female offspring only

Reproduce Sexually

• Produce some males
• Produce some haploid eggs

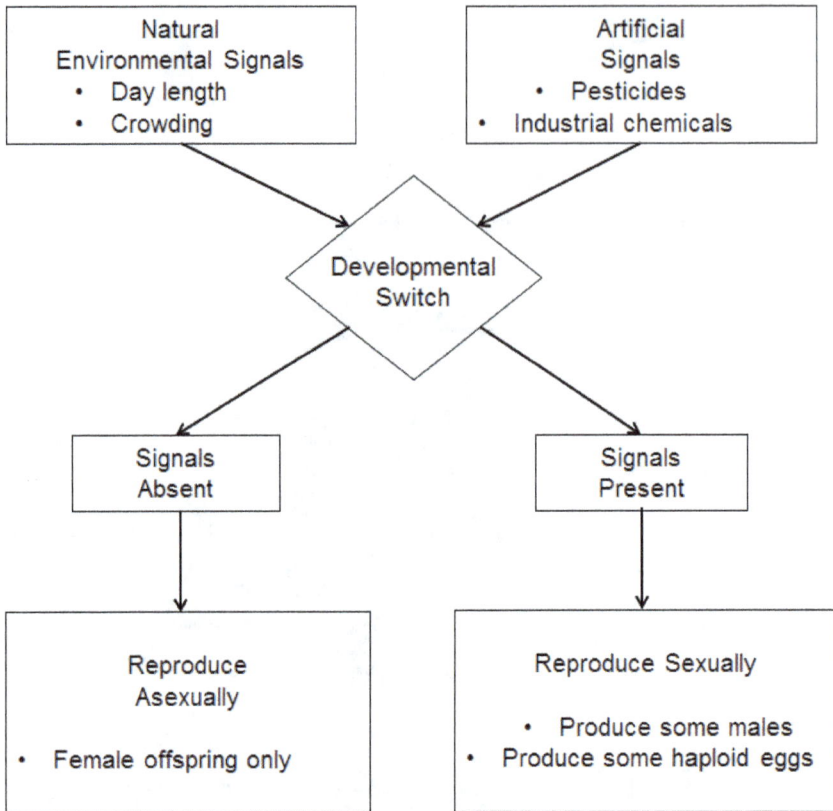

Fig. 2. Daphnia reproductive strategies. Female adults can reproduce on three different types of offspring, depending on the environmental conditions. Diploid eggs (emergent) reproduce non-sexually, develop in females and males. Haploid eggs are reproduced based on an exact chemical signal from the environment, Retrieve from (Thorp & Covich, 2001)

general appearance of the released juveniles is similar to the appearance of the adult, obviously with a minor size (Mitchell et al., 2004 as cited in American Public HealthAssociation [APHA], 2002).

4. Production of cladocerans

To keep a *Daphnia sp* production, it is required a basic laboratory infrastructure, expecting to have materials, equipment, reactive substances and organic strains such as algae and cladocerans. Daphnia has been used to evaluate chronicle and severe chemicals effects, and they are very sensitive to toxic substances. Breeding Daphnia allows to clone establishment with few genetic variation and reproductive results (American Public HealthAssociation, 2002).

4.1 Laboratory infrastructure

Laboratories should contain the minimal infrastructure necessary to develop controlled culture of micro-algae and to maintain cladocerans strains, which will be used for nutritional tests.

4.1.1 Materials

The containers used to grow the cladocerans reproducers should be made of the boron-silicate glass, adequate volume, depending on the production protocol that is used. They must be submitted to the appropriate cleaning routine and disinfection. To achieve these tests, enough and adequate glassware and minor equipment will be used to guarantee safe and repeatable proceedings, equipment like automatic calibrated pipettes, gagged glassware with the required capacity, stainless steel spatulas, etc.

4.1.2 Equipment

The essential equipment required is: analytical scale, optical microscope, stereomicroscope, autoclave (propane or electric), sterilized air or laminar flow hood for microbiological work, potentiometer, oximeter, salinometer, conductivimeter and spectrophotometer. It will be best to have a bioclimatic chamber for temperature control, lighting and photoperiods.

4.1.3 Reactives

Is necessary to have the indispensable reactives to prepare the dilution water (every time that cladoceran *Daphnia magna* production is made, it is essential to prepare the dilution water with the hardness specification required) (American Public Health Association, 2002), and culture medium, not only cladocerans but also for micro-algae. The compound should be analytical grade and free of contamination, as heavy metals. It is suggested the known commercial brands.

On each case it is suggested to prepare concentrated solutions (standard) and from this build up the culture medium and the dilution water. It must have a control over the proceedings to prepare the standard solutions and the culture medium and special attention on preserving substances that may be able to degrade or contaminate (vitamin solutions for example); in all cases it is necessary to apply the maintenance conditions to avoid the alteration on chemical concentration from the standard solutions.

5. Organisms for testing

Daphnia is widely distributed, this make it easy to obtain not only in laboratories but also in research institutions. 20 to 30 organisms are plenty to start the production. Additionally the reproduction of Daphnia allows establishing clones with low genetic variability and reproductive results (Martinez et al., 1998, as cited in Ramirez y Mendoza 2008). Some authors prefer other species of Daphnia, different from *Daphnia magna* because they have bigger neonates and some fish cannot use them for food (Prieto, 2001).

The Cladocerans species mentioned in this essay have been isolated from the Antioquia State (Ocampo et al., 2010). According to Martinez et al., (1998) the organisms used to do this evaluation must be obtained from laboratories with controlled strains. They can be obtained by purchasing them from distributors of the scientific material; this will guarantee specific and good quality biological material. It is possible to obtain them using recollection although in this case they will have to develop the harvest to an F2, also they must acquire an identity certificate issued by an accredited institution, in this institution they will have to deposit a fixed or live sample, this way it will get a catalog or collection code. It is not allowed to use organisms purchased as feedstuff on a pet store, because the origin, history and quality is unknown. Frequently this material does not guarantee reproducibility and this make the results less reliable (Martinez et al., 1998, as cited in Ramirez y Mendoza 2008, as cited in Ramirez y Mendoza 2008).

5.1 Water

Daphnia can be maintained in reservoirs of natural water; however it is suggested an environment of synthetic water and should be standard quality reconstituted in order to generate predictive results and allow an adequate reproduction and growth of the culture. Hardness water recommended is 160-180 mg CaCO3/L, dissolved in deionized water and aerated for few hours. pH closed to 8.0. Normally stays between 7.0 and 8.6, no monitoring or adjustment necessary. Requirements to prepare the hard water: 192 mg/L of NaHCO3, 120 mg/L of CaSO-4, 120 mg/L mgSO4, and 8.0 mg/L of KCl (American Public Health Association, 2002).

5.2 Food and fed

Daphnias usually are fed *in vitro* with different type of algae, generally *Ankistrodesmus falcatus, Selenastrun capricornutum,* and *Clorella.* A mixture is prepared in sterilized water, adding two drops of solution per each Daphnia adult culture. Another diet to use is adding 6.3 g of trout pellet, 6.2 g of yeast and 0.5 g of alfalfa. Mix for 5 minutes with sterilized water, gauge to 500mL, decant for an hour and get rid of the surplus. Freeze 50 to 100 mL portions and do not save it for more than eight days. Use It to feed only 0.5 mL per each 1000 mL of culture medium, with a 3 times a week frequency; in an aerated environment and replaced every week (American Public Health Association, 2002). The yeast (*Saccharomyces cereviseae*) is a probiotic. FAO (2001) declare that probiotics are "living microorganisms, when managed in adequate amounts, present to the host health benefits" (Austin & Brunt, 2009, as cited in Montet & Ray, 2009).

5.3 Temperature

Temperature changes can induce death or induce ephippium production or sexual eggs. The ideal temperature is between 20°C and 25°C.

5.4 Light

Light intensity variations between 538 and 1076 lux and the prevalence of the light during the day/night cycles; do not affect the reproduction and growth of Daphnia cultures significantly. Try to provide a minimum of 12 hours of light / 12 hours of darkness.

5.5 Containers

Glass or plastic containers should be used, in order to observe easily the cultures, they should have a capacity of 3 L, with 2.75 L of medium, and 30 Daphnias, will be able to produce approximately 300 neonates and young Daphnias per week. It is necessary to clean weekly, removing the food left and the death Daphnias. Wash the containers monthly with detergent and replace the medium completely.

5.6 Air supply

Daphnia can survive with oxygen levels up to 3 mg/L, but to grow requires 6 mg/L. To aerate the harvest, it is possible to use motors to aerate aquariums.

5.7 Maintenance of cultures

The media should be replaced weekly; it is not necessary if massive media (100 L) is used. Approximately 30 Daphnias per culture should be selected every week; to retrieve them a glass or plastic Pasteur pipette must be used. This activity is suggested to do at the same time when the medium is changed, this system will avoid cross-linking (American Public Health Association, 2002).

5.8 Selection of organisms for test

The examination should initiate with the second or third offspring, selecting neonates with ≤ 24 birth hours. Before the initiation any test, 30 to 40 young Daphnias should be isolated on a 400 mL beaker with 300 mL of algae mix (*Ankistrodesmus falcatus*, *Selenastrun capricornutum*, and *Clorella*). Do not use ephippiums for Daphnia cultures, because its presence is an indicator of unfavorable conditions (American Public Health Association, 2002).

6. Methodology

Potatoes (*Solanum tuberosum L*) and yeast (*Saccharomyces cereviceae*) contain more than 60% of protein, this consumption by the post- larva is consistent with the protein requirement, and also this source contains immune system stimulant compounds such as β-glucan, nucleic acids like manna oligosaccharides that increase the immune response (Champagne et al., 2009). When yeast gets attached to the intestine of the larvae, on their 27th day after birth, it produces a higher secretion of amylase and this stimulates the enzymes located on the membrane cells in the brush border shape (Austin & Brunt, 2009, as cited in Montet & Ray, 2009).

Potatoes (*Solanum tuberosum L*) have been used as a nutrient because they play an energetic roll due to their high levels of starch (60% – 80%), of the dry matter is starch. In addition, the potato is low in fat and rich in several micronutrients, especially vitamin C. It is also a good source of vitamins B1, B3, B6, folate, pantothenic acid, riboflavin and minerals, such as potassium, phosphorus and magnesium, Potatoes also contain dietary antioxidants, which may play a part in preventing diseases related to ageing (Murniece et al., 2011; FAO, 2008a).

The enrichment with flour provides protein (50%) and high level of lecithin to the living food. Few authors have shown the incidence in growth and intestine morphology improvement including soy oatmeal (Murray et al., 2010). Despite its limitations, the incidence of scoliosis and mouth twisting in the larvae is reduced with the addition of lecithin Salze et al., 2010). Gutiérrez-Espinoza and Vásquez-Torres (2008), confirm that soy in different presentations has a high digestibility index in White Cachama juveniles (*Piaractus brachypomus*) and can be used without restrictions on this specie (Guillaume et al., 2004; Sealey et al., 2009).

The mutual action of these three nutrients over the productive and reproductive parameters of Daphnia magna can be demonstrated with this research, thorough its exposition to eight diets. These diets include yeast, potato and soy oatmeal, with 15 and 30 ppm concentrations of nutrient and enrichment compound, determining productive and reproductive parameters, creation of life history and comparing populations under stress conditions.

Daphnia magna was used, with two populations: one with adults and the other with young breeders (32 specimens / population). *Daphnia magna* was seeded individually (1 Daphnia/tank) in multicells with 24 tanks, each one with 3 mL capacity, covered with parafilm. The compartments were fully filled with the culture medium according to the treatment. The daily exchange for a culture medium was 50%. Concentrations of 30 mg/L and 15 mg/L were prepared for the culture medium for each of the nutrient with discolored water. The treatments used were the following:

Treatment 1: 15 ppm *S. cereviceae*, 15 ppm potato, 15 ppm soy oatmeal flour.

Treatment 2: 15 ppm *S. cereviceae*, 15 ppm potato, 30 ppm soy oatmeal flour.

Treatment 3: 15 ppm *S. cereviceae*, 30 ppm potato, 15 ppm soy oatmeal flour.

Treatment 4: 15 ppm *S. cereviceae*, 30 ppm potato, 30 ppm soy oatmeal flour.

Treatment 5: 30 ppm *S. cereviceae*, 15 ppm potato, 15 ppm soy oatmeal flour.

Treatment 6: 30 ppm *S. cereviceae*, 15 ppm potato, 30 ppm soy oatmeal flour.

Treatment 7: 30 ppm *S. cereviceae*, 30 ppm potato, 15 ppm soy oatmeal flour.

Treatment 8: 30 ppm *S. cereviceae*, 30 ppm potato, 30 ppm soy oatmeal flour.

The whole production was preserved in a laboratory with the following environmental conditions: 1.200 masl (meters above the sea level), 22 ± 3 °C room temperature, 12 light hours, 12 dark hours, 7.0 ± 0.6 pH water, during 20 days. Eight treatments were used with four replica/treatment.

These are the productive and reproductive parameters of Daphnia magna:

6.1 Growth and corporal weight: Productive variables

Maximum density (Dmax):

$$\text{Dmáx} = \frac{\text{Final population}}{\text{Volume}} \equiv \left[\frac{\text{Individuals}}{\text{Volume}} \right] \tag{1}$$

Daily average density (Dmd):

$$\text{Dmd} = \frac{\text{Final population}}{T} \equiv \left[\frac{\text{Individuals}}{\text{Time}} \right] \tag{2}$$

Doubling Time (Td):

$$Td = \frac{0.693}{k} \equiv [\text{días}] \tag{3}$$

Specific growth rate (k):

$$k = \frac{Ln(\text{ final population}) - Ln(\text{PoblInitial population inicial})}{\text{TiempoTime}} \equiv \left[\frac{1}{\text{day}}\right] \tag{4}$$

Performance (r):

$$r = \frac{(\text{Final populationl}) - (\text{Initial population})}{\text{Time}} \equiv \left[\text{individuals} * \frac{1}{\text{Time}}\right] \tag{5}$$

Numeric growth (PN):

$$P_N = b * \bar{N} \equiv \left[\text{\# individuals} * \frac{1}{\text{Time}}\right] \tag{6}$$

N= individual media number
b= birth rate population, in a period of time.

Birth rate (b) (Edmodson equation):

$$b = ln(E/D + 1) \equiv [\text{\# neonates}] \tag{7}$$

E= egg number /female
D= egg development time (days).

Another way to calculate the birth rate is:

$$k = b + d \tag{8}$$

therefore:

$$b = k - d \tag{9}$$

Individuals average number ($\bar{N}$):

$$\bar{N} = \frac{(No + N_t)}{2} \tag{10}$$

No=individuals initial number
Nt=individual number after a period of time t

Biomass productivity (Pw):

$$P_W = b * \bar{N} * \bar{W} \equiv \left[\frac{\text{Biomass}}{\text{Volume}} * \frac{1}{\text{Time}}\right] \tag{11}$$

W=Final weight at the end of the period.
N= Individuals media number
b= birth rate population, in a period of time.

Mortality rate (d):

$$d=\frac{Ln(\text{Final pop. of neonates})-Ln(\text{Initial pop. of neonates})}{T}\equiv\left[\frac{1}{day}\right]\tag{12}$$

Biomass (B):

$$B=\frac{(\text{Average number})*(\text{Final weigth})}{T}\equiv\left[\frac{\text{Biomass}}{\text{Volume}}\right]\tag{13}$$

Production rate (I de P):

$$I\ de\ P=\frac{P}{B}\tag{14}$$

Final weight (pf) was determined as dry weight. Every Daphnia was dried up with tissue paper (analytical scale, 0.0001g. precision), expressed in mg/L.

6.2 Reproductive variables

Egg number/female (HPP). The egg number/female was determined in the microscope (10x and 40x).

Neonate number/female (NPP). This count was made directly.

The measurement of the rest of the variables was taken directly from the data base, such as:

Egg development time (tm), number of days passed between the appearance of the egg in the incubation chamber and the presence of the neonate.

Offspring number (NC)

First reproduction age (EPR), in days.

Production frecuency (FR), in hours.

The born individuals were moved to another tank with the same treatment. Neonate mortality was evaluated here.

6.3 Life history parameters

Daily measurements were made in 32 young breeders and in 32 adults, keeping a follow up during the 21 days (duration of the research). The frequency of the observation was 12 hours to differentiate seeded population and produced population.

The reproduction net rate (Ro) and the generation time (Tc), were calculated according to the Lotta equation (1913).

$$\sum_{X=0}^{n} l_x m_x (\exp^{-rx}) = 1 \tag{15}$$

lx= survival in a specific time starting in the birth period.
mx= fertility in a specific period
x=period (from 0 to 21 days)
r= intrinsic grow rate

Net reproduction rate (Ro):

$$R_o = \sum_{X=0}^{n} l_x m_x \tag{16}$$

Generation time (Tc):

$$T = {}^{1}\!/_{R_o} \sum_{X=0}^{n} l_x m_x X \tag{17}$$

Where Ro= net reproductive rate and Tc= generation time.

6.4 Statistical methodology

An experimental classification design was used, fully randomized, fixed effect, symmetric, balanced, 23 factorial arrangement, with 4 replicas, n=32 to each population. The results were compared through ANOVA and Tukey test with α=0.05. An exploratory descriptive analysis, one-dimensional for each condition type, was established. Data transformation to the square root function was used on egg number/female and neonates number variables. For survival and mortality percentages, arcsine transformation was applied. SAS version 9.1 Statistical package was used.

$$T = {}^{1}\!/_{R_o} \sum_{X=0}^{n} l_x m_x X$$

$$Y_{ijsk} = \mu + L_i + P_j + E_s + LP_{ij} + LE_{is} + PE_{js} + LPE_{ijs} + \varepsilon_{k(ijs)} \tag{18}$$

Where Y_{ijsk} : represented de number variable of *Daphnia magna*.

μ :   Experiment average effect

L_i :   Yeast effect

P_j :   Potato dosage effect

E_s :   Enrichment effect

$\varepsilon_{k(ijs)}$:   Experimental error

Additionally, a comparison test with adult population and young breeders was made. To do this, the Mann Whitney test was applied, for the productive variables: daily average density (Dmd), doubling time (Td), specific growth rate (k), performance (r) and biomass productivity (Pw) and for reproductive parameters: egg number/female (HPP), neonates number/female (NPP), litter number (NC), net reproduction rate (Ro) and generation time (Tc). A correlation analysis was completed between the variables. For reproduction net rate and generation time, one way analysis of the variance and the Tukey test took place, in order to compare the life history parameters between the different diets.

7. Results

In the adult population were found significant differences (p<0.05) on the average treatment in the daily growth rate, duplication time and mortality. The daily growth rate from T1 to T8 presented the following values: 0.15 ± 0.05^b, 0.50 ± 0.05^a, $0.33 \pm 0.12^{a,b}$, $0.23 \pm 0.03^{a,b}$, $0.31 \pm 0.08^{a,b}$, $0.33 \pm 0.18^{a,b}$, $0.40 \pm 0.20^{a,b}$ and 0.19 ± 0.11^a cladocerans per day, respectively. The major growth rate occurred on T2 with 0.50 ± 0.05^a cladocerans per day, presenting a significant differences p (<0.05) in the treatment T1, that showed a value of 0.15 ± 0.05^b cladocerans per day compared with T2 with values of 0.50 ± 0.05^a cladocerans per day, as can be seen on table 1.

The duplication time of T1 to T8 had values of 4.96 ± 1.52^a, 1.39 ± 0.14^b, $2.31 \pm 0.90^{a,b}$, $2.99 \pm 0.48^{a,b}$, $2.27 \pm 0.53^{a,b}$, $2.80 \pm 1.83^{a,b}$, $2.10 \pm 1.06^{a,b}$ and $4.40 \pm 2.43^{a,b}$ days respectively. This parameter presented significant differences p (<0.05) during the treatment T1, where a duplication time of 4.96 ± 1.52^a was observed, in relation to T2 that presented a minor duplication time with 1.39 ± 0.14^b days, as can be seen on table 1.

The mortality rate from T1 to T8 presented the following values: 0.15 ± 0.05^b, 0.49 ± 0.0^a, $0.32 \pm 0.12^{a,b}$, $0.21 \pm 0.04^{a,b}$, $0.31 \pm 0.08^{a,b}$, $0.31 \pm 0.17^{a,b}$, $0.40 \pm 0.20^{a,b}$, $0.21 \pm 0.10^{a,b}$, respectively. The highest mortality rate was presented in T2 with 0.49 ± 0.07^a cladocerans per day, presenting significant difference p (<0.05) betweenT1 with 0.15 ± 0.05^b cladocerans per day with regard to T2 with 0.49 ± 0.07^a cladocerans per day (table 2).

Even though there were no significant differences (p>0.05) between the treatment media in the rest of the evaluated parameters, not only in the adult population group, but also the young breeding group, it is important to notice in a general matter, that it was a better population performance in T2 and T6 (tables, 3, 4, 5 and 6). In T2 where growth rate was the highest with 0.37 ± 0.13^a cladocerans per day, the minor duplication time with 2.02 ± 0.59^a days, and the mayor performance with 1.23 ± 0.46^a cladocerans per period. Likewise T6 presented the highest maximum density and the mayor egg number per female with 15.0 ± 9.76^a and 8.25 ± 1.70^a cladocerans, and 5.0 ± 0.0^a and 3.83 ± 0.82^a eggs per female, respectively. This information can be seen on table 4.

7.1 Population comparisons

The daily average density (Dmd) of *D. magna* in the adult population was 4.0 individuals per cell and in the population of young breeders were 2.0 individuals. The maximum Doubling Time (Td) in adults was 2.43 days and in the population of young breeders was 12 days. The maximum Performance (r) was, in adults 2.7% and in young breeders 0.58%. The

Treatments	Adult Populations					Young Breeder populations				
	Dmd (Clad/vol.) x±DE	Dmáx (Clad/vol.*d) x±DE	k (Clad/day) x±DE	Td (days) x±DE	r (Clad/vol.*d) x±DE	Dmd (Clad/vol.) x±DE	Dmáx (Cla/vol*d) x±DE	k (clad/day) x±DE	Td (days) x±DE	r (Clad/vol.*d) x±DE
1	4.7±4.78a	0.42±0.47a	0.18±0.08 a	4.34±2.05 a	0.66±0.47 a	4.0 ±2.94 a	0.48 ±0.31a	0.15±0.05 b	4.96±1.52 a	0.35±0.27 a
2	6.25±3.86a	1.12±0.72a	0.37±0.13 a	2.02±0.59 a	1.23±0.46 a	4.5±4.35 a	1.66±0.57 a	0.50±0.05 a	1.39±0.14 b	1.56±0.33 a
3	2.75±3.5 a	0.29±0.15 a	0.14±0.0 a	5.0±0.0 a	0.47±0.0 a	10.75 ±3.2 a	1.86 ±0.75 a	0.33±0.12 a, b	2.31±0.90 a, b	1.42 ±0.70 a
4	4.25±2.75 a	0.47±0.40 a	0.16±0.13 a	7.65 ±7.32 a	0.43±0.38 a	8.0±4.58 a	0.98±0.20 a	0.23±0.03 a, b	2.99 ±0.48 a, b	0.71 ±0.32 a
5	4.75±4.92a	0.60±0.63 a	0.20±0.12 a	5.1±4.29 a	0.63±0.65 a	10.25±2.69 a	1.47±0.59 a	0.31±0.08 a, b	2.27 ±0.53 a, b	1.26 ±0.61 a
6	8.25±1.70 a	0.82±0.45 a	0.22±0.15 a	3.94±1.61 a	0.71±0.36 a	15.0±9.76 a	1.91±1.48 a	0.33±0.18 a, b	2.80±1.83 a, b	1.76 ±1.46 a
7	7.75±5.37 a	0.67±0.38 a	0.20±0.08 a	3.66±1.19 a	0.75±0.10 a	9.75±7.22 a	1.84±0.91 a	0.40±0.20 a, b	2.10 ±1.06 a, b	1.39 ±0.99 a
8	7.75±2.87 a	0.50±0.18 a	0.12 ±0.02 a	5.72±1.31 a	0.41±0.19 a	10.66±7.50 a	1.07±0.82 a	0.19±0.11 a, b	4.40 ±2.43 a, b	0.91 ±0.89 a

Dmáx: maximum density; Dmd: Daily media density; k: intrinsic growth rate; Td: Duplicating time; r: Performance. DE= Standard deviation. Different letters indicate statistical difference ($p<0.05$) between columns, per population.

Table 1. Average ± DE, for population parameters during the evaluated period in two populations of *Daphnia magna,* according to treatment

Treatments	Adult Populations					Young Breeder populations				
	Pw (mg/mL*day) x ± DE	B (mg/mL) x ± DE	P/B (I de Prod./day) x ± DE	b (Clad/day) x ± DE	d (Clad/day) x ± DE	Pw (mg/mL*day) x ± DE	B (mg/mL) x ± DE	P/B (I of Prod/day) x ± DE	b (Clad/day) x ± DE	d (clad/day) x ± DE
1	·	·	·	0.57 ± 0.27 a	0.39 ± 0.19 a	1.25 ± 1.36 a	3.5 ± 2.18 a	0.30 ± 0.15 b	0.34 ± 0.15 a	0.15 ± 0.05 b
2	3.06±0.0 a	5.9 ± 0.0 a	0.52 ± 0.0 a	0.64 ± 0.10 a	0.27 ± 0.13 a	2.55 ±1.17 a	5.22 ± 2.21 a	0.48± 0.02 a	0.48± 0.02 a	0.49±0.07 a
3	2.07±0.0 a	7.47±0.0 a	0.28 ± 0.0 a	0.28 ± 0.0 a	0.14 ± 0.0 a	2.81 ± 0.0 a	5.32 ± 0.0 a	0.53 ± 0.00 a	0.38 ± 0.24 a	0.32 ± 0.12 a,b
4	6.48±0.0 a	7.3 ± 0.0 a	0.89 ± 0.0 a	0.53 ± 0.34 a	0.37 ± 0.21 a	4.49 ± 3.71 a	7.87 ± 6.02 a	0.54 ± 0.07 a	0.58 ± 0.06 a	0.21 ± 0.04 a,b
5	12.15± .0 a	12.8 ± 0.0 a	0.95 ± 0.0 a	0.55 ± 0.35 a	0.35 ± 0.28 a	4.07 ± 2.05 a	7.69 ± 3.58 a	0.53 ± 0.04 a	0.54 ± 0.04 a	0.31 ± 0.08 a,b
6	5.44±0.77 a	7.15 ± 1.08 a	0.77 ± 0.17 a	0.74 ± 0.16 a	0.52 ± 0.03 a	7.63 ± 8.69 a	12.18 ± 8.30 a	0.51 ± 0.23 a	0.52 ± 0.22 a	0.31 ± 0.17 a
7	7.68 ± 6.59 a	9.31 ± 5.52 a	0.74 ± 0.26 a	0.65 ± 0.24 a	0.44 ± 0.27 a	5.29 ± 4.69 a	8.14± 6.15 a	0.56 ± 0.19 a	0.57 ± 0.18 a	0.40 ± 0.20 a,b
8	6.27 ± 4.61 a	9.35 ± 3.52 a	0.58 ± 0.28 a	0.58 ± 0.28 a	0.45 ± 0.26 a	5.48 ± 6.03 a	8.95 ± 6.75 a	0.45 ± 0.32 a	0.48 ± 0.30 a	0.21 ± 0.10 a,b

Pw: Biomass productivity; B: Biomass; PB: Productivity index, b: birth rate and d: Mortality rate. DE= Standard deviation. Different letters indicate statistical difference (p<0.05) between columns, per population.

Table 2. Average ± DE, for population parameters during the evaluated period in two populations of *Daphnia magna,* according to treatment

Treatments	Adult Populations				Young Breeder Populations			
	NPP (Neona/Fem.) x±DE	NCAM Birth/perd. x±DE	EPR (days) x±DE	FR (hours) x±DE	NPP (Neona/fem.) x±DE	NCAM birth/perd. x±DE	EPR (days) x±DE	FR (hours) x±DE
1	7.50±3.53 [a]	2.50±0.70 [a]	13.0±2.82 a	48.0±33.94 [a]	4.0 ±2.64 [a]	1.33 ±0.57 [a]	5.0±0.0 [a]	7.50±3.53 [a]
2	7.0±2.0 [a]	2.0±0.0 [a]	14.33±0.57 [a]	76.0±18.33 [a]	7.0±2.82 [a]	1.50 ±1.73 [a]	2.5±2.12 [a]	7.0±2.0 [a]
3	7.0 ±0.0 [a]	3.0±0.0 [a]	13.0±00 [a]	60.0 ±00 [a]	9.75 ±3.20 [a]	4.25 ±1.25 [a]	2.75±0.50 [a]	7.0 ±0.0 [a]
4	4.33±2.08 [a]	1.66 ±0.57 [a]	14.0 ±1.0 [a]	42.0 ±8.48 [a]	8.50 ±4.79 [a]	2.75 ±1.25 [a]	4.0 ±2.58 [a]	4.33 ±2.08 [a]
5	5.0±5.19 [a]	2.0 ±1.0 [a]	13.0±1.0 [a]	84.0±31.17 [a]	9.25±2.62 [a]	2.5 ±0.57 [a]	4.75 ±2.62 [a]	5.0±5.19 [a]
6	7.25±1.70 [a]	3.0±0.81 [a]	13.75±1.5 [a]	45.0±24.7 [a]	14.0±9.76 [a]	3.50 ±2.08 [a]	4.33±2.51 a	7.25±1.70 [a]
7	9.0±3.60 [a]	2.66±0.57 [a]	12.0±1.0 [a]	92.0 ±45.4 [a]	8.75±7.22 [a]	2.25 ±1.50 [a]	3.50 ±1.00 a	9.0±3.60 [a]
8	6.75±2.87 [a]	2.0±0.81 [a]	14.25±1.28 [a]	36.0 ±12.0 [a]	11.0 ±6.68 [a]	2.75 ±1.25 [a]	4.33±2.88 a	6.75 ±2.87 [a]

NPP: Neonates per female, NCP: Offspring number, EPR: First reproduction age y FR: Reproduction frequency. DE= Standard deviation. Different letters indicate statistical difference (p<0.05) between columns, per population.

Table 3. Average ± DE, for population parameters during the evaluated period in two populations of *Daphnia magna*, according to treatment

Treatments	Female Adult Population		Young Breeder Population	
	HPP (N/female/day) x ± DE	D (days) x ± DE	HPP (N/female/day) x ± DE	D (days) x ± DE
1	2.36 ± 0.89 a	3.75 ± 1.0 a	1.8 ± 0.50 a	5.0±2.82 a
2	2.26 ± 1.57 a	3.33 ± 0.28 a	1.79 ± 0.34 a	2.5±0.70 a
3	1.72 ± 0.32 a	4.33 ± 0.0 a	2.24 ± 1.04 a	2.85 ± 1.03 a
4	2.53 ± 0.61 a	4.33 ± 2.51 a	3.46 ± 1.07 a	2.77 ± 1.18 a
5	2.7 ± 1.0 a	4.39 ± 1.66 a	3.05 ± 0.97 a	3.41 ± 0.68 a
6	3.83 ± 0.82 a	3.54 ± 0.76 a	3.55 ± 0.98 a	3.36 ± 2.76 a
7	3.23 ± 1.08 a	4.16 ± 1.25 a	2.81 ± 1.19 a	2.69 ± 1.33 a
8	3.63 ± 1.17 a	5.20 ± 2.07 a	3.19 ± 1.60 a	2.49 ± 1.37 a

HPP: Egg number, D: Development time. DE= Standard deviation. Different letters indicate statistical difference (p<0.05) between columns, per population.

Table 4. Average ± DE, for population parameters during the evaluated period in two populations of *Daphnia magna*, according to treatment

Treatments	Female Adult Population			Young Breeder Population		
	Pf (mg/mL) x ± DE	PN (Ind*mL) x ± DE	N average (Cladoc.) x ± DE	Pf (mg/mL) x ± DE	PN (Ind*mL) x ± DE	N average (Cladoc.) x ± DE
1	1.77± 0.26 a	0.70 ± 0.70 a	2.0 ± 1.32 a	2.03 ± 0.0 a	2.67 ± 2.17 a	3.0 ± 2.5 a
2	1.57 ± 0.30 a	3.59 ± 1.28 a	3.75 ± 1.76 a	1.97± 0.0 a	2.64 ± 1.03 a	4.0 ± 1.0 a
3	1.28 ± 0.49 a	3.75 ± 2.03 a	5.37 ± 1.60 a	2.05 ± 0.25 a	1.11 ± 0.0 a	4.0 ± 0.0 a
4	1.70 ± 0.42 a	1.76 ± 0.96 a	4.37 ± 2.52 a	2.03 ± 0.56 a	1.46 ± 1.22 a	2.33 ± 1.15 a
5	1.90± 0.48 a	3.22 ± 1.67 a	4.87 ± 1.25 a	2.01 ± 0.16 a	2.24 ± 2.98 a	3.0 ± 2.59 a
6	1.69 ± 0.21 a	5.13 ± 5.42 a	7.37± 4.85 a	1.89 ± 0.23 a	2.96 ± 0.17 a	4.12 ± 0.85 a
7	1.91 ± 0.14 a	3.49 ± 2.98 a	4.50 ± 3.67 a	1.98 ± 0.16 a	3.44 ± 2.31 a	5.0 ± 1.80 a
8	1.78 ± 0.08 a	3.16 ± 2.57 a	5.87 ± 3.56 a	2.66 ± 0.49 a	2.35 ± 1.82 a	3.62 ± 1.43 a

Pf: Final weight, PN: Numeric average, N prom: Number average. DE= Standard deviation. Different letters indicate statistical difference (p<0.05) between columns, per population.

Table 5. Average ± DE, for population parameters during the evaluated period in two populations of *Daphnia magna*, according to treatment

Treatments	Female Adult Population		Young Breeder Population	
	Reproductive net rate 1/day x ± DE	Regenerating time Days x ± DE	Reproductive net rate 1/day x ± DE	Regenerating time Days x ± DE
1	6.33 ± 0.00 [a]	7.00 ± 0.00 [a]	7.5±3.53[a]	15.75 ± 7.42 [a]
2	7.00 ± 2.82 [a]	8.00 ± 2.82 [a]	6.74 ± 2.08 [a]	16.48 ± 4.67 [a]
3	9.63 ± 3.21 [a]	11.00 ± 3.46 [a]	7.00 ± 0.0 [a]	15.00 ± 0.00 [a]
4	8.50 ± 4.79 [a]	13.50 ± 0.70 [a]	4.33 ± 2.08 [a]	14.46 ± 6.31 [a]
5	9.08 ± 2.32 [a]	9.00 ± 1.00 [a]	5.16 ± 5.05 [a]	13.93 ± 1.88 [a]
6	13.55 ± 9.30 [a]	14.00 ± 11.26 [a]	7.5 ± 1.91 [a]	15.56 ± 5.13 [a]
7	10.00 ± 7.50 [a]	10.33 ± 8.62 [a]	8.65 ± 4.40 [a]	6.06 ± 10.47 [a]
8	11.18 ± 6.65 [a]	11.75 ± 7.58 [a]	6.95 ± 2.53 [a]	12.08 ± 7.48 [a]

Ro: Reproductive net rate; Tc: Regenerating time. DE= Standard deviation. Different letters indicate statistical difference ($p<0.05$) between columns, per population

Table 6. Reproductive net rate, regenerating time in two populations of *Daphnia magna*, according to treatment

Specific growth rate (k) was 0.65 cladocerans per day in adults and 0.52 in young breeders. The maximum Biomass (B) was 24.73 mg/cell in adults and 13.22 mg/cell in the population of young breeders.

Adult *D.magna* presented a maximum value of 27 neonates/ adult female per each volume unit during the evaluation period, different from the population of young breeders when their fecundity levels reached 12 neonates/female.

The offspring maximum number was 6 offspring/period for the adult population, and 4 offspring/period for the population of young breeders. The maximum age for the first reproduction was 7 days in adults, meanwhile in the population of young breeders was 16 days. The maximum level of reproduction frequency in the adult population was 216 hours and for the population of young breeders was 120 hours.

In the following figures 3 and 4, a μ adjustment or specific growth rate founded experimentally can be observed. The regression coefficient for the treatments on the adult population oscillated between r (0.8966 – 0.9364). The dots in both graphics correspond to the experimental data and the curves correspond to de adjustments. In the graphics the differences in the growth rate for both populations can be observed as well. Note that the adult population it is found in a reproductive process and growth in the beginning of the experiment, meanwhile the population of neonates must conquer its maturity to initiate reproduction and growth.

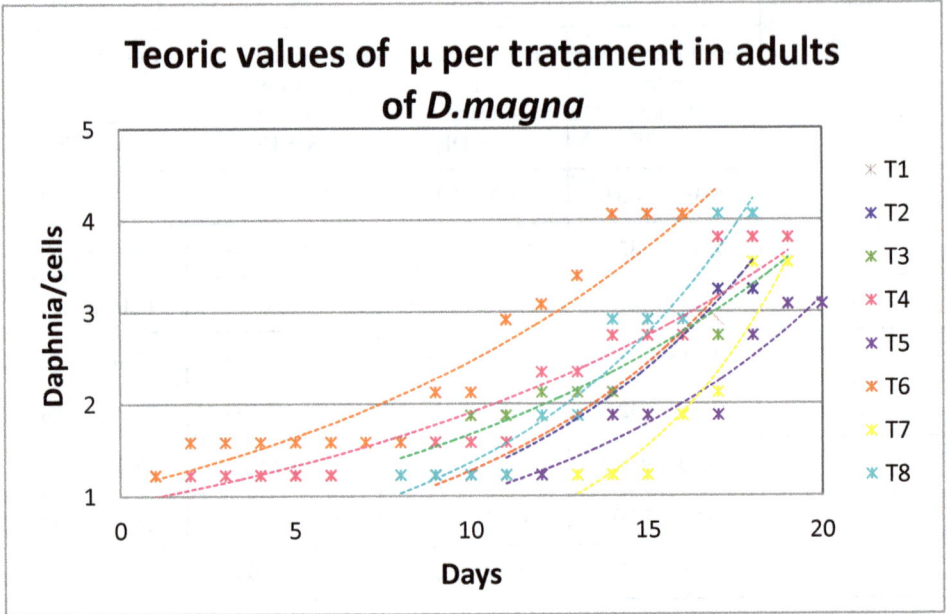

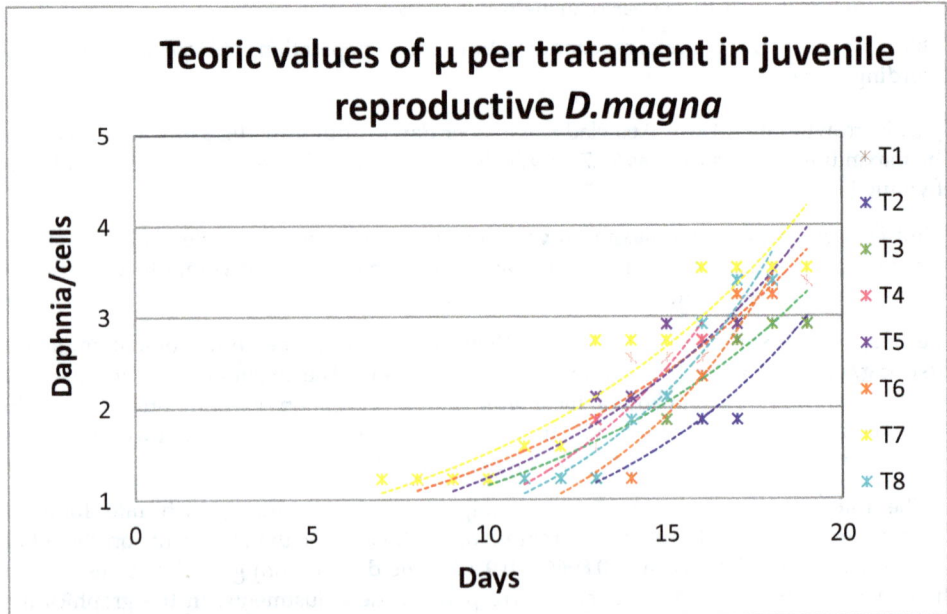

Figs. 3. and 4. Population growth of *D. magna*. The trajectory shows an exponential growth with μ values common for cladocerans. Correspond to nutritional situation, specific conditions of each treatment, depreciable mortality and water temperature closed to 20°C.

7.2 Interactions

Even though there was not statistic relation (p>0.05) between the parameters of maximum density, daily media density and final weight on the adult population, there was a nutrient significant correlation (p<0.05): The yeast nutrient was highly significant (p<0.01) on the 30 ppm concentration, on the maximum density with 11.46 ± 6.79 cladocerans per volume, while the treatments with a concentration of 15 ppm of yeast, had 6.7 ±4.43 cladocerans per volume, on a concentration of 30 ppm. Similarly happened with the final weight, that presented values of 1.63 ± 0.35 mg in 15 ppm and 1.82 ± 0.26 in 30 ppm, presenting a mayor value on final weight on the diet with 30 ppm yeast concentration.

A highly significant relation was found (<0.01) between the three nutrients. With yeast and potato, with concentrations of 30 ppm and 15 ppm respectively, the highest value on the variable maximum density was found, with 12.62 ± 7.0 cladocerans per volume. The daily media density presented significance (p<0.05) in the relation between the potato and the enrichment, presenting the highest value for 1.85 ± 0.7 cladocerans per volume per day, in the concentrations of 30 ppm and 15 ppm respectively. The daily media density does not seem influenced by yeast addition.

The mortality rate presented significant differences in relation with yeast and enrichment nutrients. The mayor mortality was shown in the treatment with yeast 30 ppm and enrichment of 15 ppm with 0.31 ± 0.12 cladocerans per day. Also it was a highly significant difference (p<0.01) with 0.37 ± 0.16 using potato and enrichment nutrients in the same concentration of 15 ppm. It seems that enrichment concentrations have an impact in the mortality of adult populations of cladocerans.

About the relation of nutrients found in the growth rate, duplication time and performance, two of the three nutrients presented a highly significant relation, in the adult population. The growth rate presented its highest value with 0.38 ± 0.16 per day, in concentrations of 15 ppm and 30 ppm of potato and enrichment respectively, followed by the concentrations of 30 ppm and 15 ppm of yeast and enrichment respectively, with a maximum value of 0.36 ± 0.15 per day.

In the variable duplicative time highly significant differences were observed p (<0.01) reputedly in two nutrients, as follows: In yeast and enrichment nutrients, with 30 ppm and 15 ppm concentrations respectively, the minor duplicative time was observed with 2.18±0.7 days. Additionally, with potato and enrichment nutrient, with concentrations of 30 ppm and 15 ppm respectively, a duplicative time of 2.21± 0.9 days was presented.

8. Discussion

The present study is a experimental job that specifies the effect variability of two nutrients and enrichment media (*S. cereviseae*, potato and soy oatmeal flour) in two concentration level used (15 ppm and 30 ppm) over the population dynamics of *D. magna*.

It is known that the relative fecundity on cladocerans populations is hard to determine. Even though it is known that conditions and the food amount have an influence on this variable. Martínez-Jerónimo et al. (2008), fed *D. magna* with *Ankistrodesmus falcatus and*

Scenedesmus incrassatulus, in 6, 12 and 18 mgL concentrations and 19°C of temperature, and they observed with a mayor food concentration there was less survival and the egg number per female was on a production interval of 9 and 23, the mayor values where observed in less food concentration, agreeing with mayor fecundity. In the essay, the mayor number of egg per female in adult population and the young breeders was 5.0 ± 0.0 [a] y 3.83 ± 0.82[a] eggs per female respectively, in treatment T6, that correspond to high concentrations of food. Similarly, the highest maximum density was observed with 15.0 ± 9.76[a] and 8.25 ± 1.70[a] cladocerans, and a minor population survival, under special stress conditions and plenty of food, with a room temperature of 21-25°C, showing a similar tendency on the obtained results by the previously mentioned authors in relation with the amount of food, population reproduction and survival.

A work from Hülsmann (2001), with *Daphnia galatea* showed an obvious dependency between fertility and food concentration. He found as well, a relation between the offspring maximum and the size of the body (Hülsmann, 2001). Other authors have demonstrated a clear relation between the food concentration and the number of eggs per female (Müller-Navarra &Lampert, 1996).

Data about productivity index in Daphnia magna cultures only is reported by the authors Jana & Pal (1983), whom used farmyard leftovers, caw manure and Mahua (*Madhuca indica*) substratum. Their values agree with the values originated in this research. It is important to indicate that the study of Jana & Pal (1983) was completed under normal conditions.

Sevrin-Reyssac (1993) found a production interval of *D. magna* between 200-400 g/m³/week, feeding with micro-algae in a media of pork manure, in 2 m³ cells during summer time (18-25 °C). During winter time they only reached 30 g/m³/week, even though they were fed with high micro-algae concentration. In contrast the minimum value found during this research was 4.816 g/m³/week, in a population of adult *D. magna*, under room stress conditions, feeding with *S, cereviseae*, potato and enrichment media of soy oatmeal flour with 21-25°C room temperature.

In cladocerans, food activity depends on the temperature of the concentration of the food; in concentrations of food over the threshold concentration of incorporation, the nutrition rate increases, and at the end there is a high quantity of energy available for growth and reproduction, according to Heugens (2006). Therefore a better performance of the reproductive and productive variables is expected from the populations where there are nutritional conditions of abundance and appropriate temperatures (close to 20°C) (Heugens et al; 2006).

The previous information was confirmed in this present work, where an excellent performance was presented on the growth rate, number of neonates per female, number of eggs per female, final weight, net reproduction rates obtained on T6, T7 and T8 Treatments. These treatments presented the highest concentrations of food, improving this way the averages of these variables not only for adult population but also for the young breeder population.

These facts are suggesting that the use of probiotics such as *Saccharomyces cereviceae*, soy oatmeal flour and potato for concentrations of 15 ppm and 30 ppm, as food for *Daphnia magna*, possibly improve the biochemical composition quality as life food, since the present work reflected a better performance of this productive and reproductive variables in both populations, although they were under stress conditions. Additionally, the nutritional value of Daphnia it is not the optimal for fish post-larvae, because Daphnia does not fulfill the nutritional requirements of all the fish poslarvae (Watanabe et al., 1983). Because of this situation in the nutritional fish larvae studies made previously, is important to determine the biochemical composition of the organisms used as life food.

The duplicative time of the culture, allow us to predict the abundance of cladocerans that are able in any moment of the growth curve and it is an intrinsic characteristic of each culture under those growth conditions (Cerna et al; 2009). This concept is similar to the generational time and can be estimated using the life tables as generational time or to start from the exponential equation of growth as the duplicative time. These estimations are different, and this is because the life tables were the generation times are estimated, they do not consider the alive and death percentage on its construction and the estimation that is made for the duplicative time on the exponential equation (Werdin y Ferrero; 2008; Cerna et al; 2009).

In this essay experimental design interactions were found, with a significant relation p (<0.05) in two of the three nutrients used in the adult population; this mean, the enrichment nutrient was present on the interactions that influenced on the productive variables (daily media density, growth rate, duplicative time and performance).With the nutrients potato and yeast did not happen the same, their reflected interaction with the enrichment media, separately. This allows as confirming that one of the two nutrients yeast or potato was added unnecessarily. Excepting the variables maximum density and daily media density; where both nutrients influence notoriously. Particularly on the variables of maximum density and final weight, and which got influenced by the yeast significantly. It is possible to confirm that the use of probiotics as *S. cereviseae* improves the productivity of *Daphnia magna*.

9. Conclusions

From the diet provided to the young breeder population and the adult population, the productive variables that showed better performance were T2 and T6.

Both populations presented a poor performance on the diets with T1, were the diet supply had the minor nutrient concentration (15 ppm).

In reference to the abundance of cladocerans, Treatments T6, T7 and T8 presented a notorious performance; these treatments had the biggest nutrient concentration (30 ppm).

In the young breeder population, a better homogeneity in data from all the treatments was observed. Including the data front e treatment T6 from both populations. This suggests that to evaluate nutrients, use a neonate population with less than 24 hours.

The present work permitted compile the existing information about maintenance of zooplankton cultures, common calculations used to study the secondary production of zooplankton and the assessment of the productive and reproductive parameters on the dynamic population of *Daphnia magna* together with the reproduction net rate and generation time, proper parameters of biological harvests with continue growth like Daphnia magna.

Although the collection of common productive variables cited in books was made, it is sufficient to evaluate some of them, for example: maximum density, growth rate and duplication time. Regarding the reproductive variables, the number of eggs per female and the age on the first offspring, reflected this performance. Also the productivity index P/B is important to calculate, because this index reflects the turnover rate of the harvest, and allow it to predict the productivity during a period of time.

The maintenance of each individual isolated avoid intra-specific and inter-specific competency for food, guaranteeing the wellness of the organisms, the best use of metabolic energy and the inadequate concentration of metabolic waste.

From the previous work we can say that yeast is a key nutrient in 30 ppm concentrations over the maximum density and the final weight, not considering the addition of enriched media, this last one did not present a significant relation (p<0.05). Additionally, the parameters growth rate, duplication time and performance, are highly affected by the enrichment nutrient. This let us make sure that one of this ingredients yeast or potato was added unnecessary.

It is important to design a diet with the enrichment media used on this research (soy oatmeal), keep in mind not selecting nutrients such as potato and yeast at the same time, because they present the same effect in the valuation on the adult population.

Finally *Daphnia magna* can be considered a specie with reproductive potential, because of it easy management and resistant to manipulation, even under stress conditions, this can be important in life food production in small infrastructures.

A study of pathogenicity should be made on species that uses probiotics, as well as life food.

10. Acknowledgment

The authors express their gratitude to the professor Martha Olivera for her constant support on the consecution of the equipment needed to develop this project, as well as to the zootechnician Luis Fernando Zuluaga that assisted with the translation of the document.

11. References

Abdel-Tawwab Mohsen, Ahmad Mohammad, Khattab Yassir, Shalaby Adel. (2010). Effect of dietary protein level, initial body weight, and their interaction on the growth, feed utilization, and physiological alterations of Nile tilapia, Oreochromis niloticus (L). *Aquaculture*, 298, (Octubre 2009), 267–274, ISBN 0044-8486

American Public Health Association (APHA), American Water Works Association (AWWA) and Water Environment Federation (WEF). (2002). *Standard Methods for the*

Examination of Water and Wastewater. (Vigésima edición). American Public Health Association., Washington D.C.

Andrade Claudia, Montiel Américo Quiroga Eduardo. (2009). Estimation of Secondary Production and Productivity From an Intertidal Population Of Trophon Geversianus, (Laredo Bay, Straits Of Magellan). *Anales Instituto Patagonia,* 37, 1, (Junio 2009), pp. 73-84, 0718-686X

Austin B. & Brunt J.W. (2009). The Use of Probiotics in Aquaculture, In: *Aquaculture Microbiology and Biotechnology.* Montet, D., Ray, R. C., pp. 185-208. Science Publishers, ISBN 978-1-57808-574-3, India

Brett Michael T., Müller-Navarra Dörthe C., and Persson Jonas. (2009). Crustacean Zooplankton Fatty Acid Composition, In *Lipids in Aquatic Ecosystems.* Arts, M.T., M.T. Brett and M. Kainz, pp. 115-146, Springer, ISBN: 978-0-387-88607-7, USA

Botero M. (2004). Selecciones: Comportamiento de los peces en la búsqueda y la captura del alimento. Rev Colomb Cienc Pecu; 17, (Octubre 2003), 63-75, ISBN 0120-0690

Cerna E, Badii MH, Ochoa Y, Aguirre LA, Landeros J. (2009). Tabla de vida de Oligonychus punicae Hirst (Acari: Tetranychidae) en hojas de aguacate (Persea americana Mill) variedad hass, fuerte y criollo. 25(2):133-140. *Universidad y Ciencia, Trópico Húmedo.* Vol 25, núm 2, pp. 133-140, (Noviembre) 2009, ISSN: 0186-2979

Champagne, C.P., Green-Johnson, J.M., Raymond, Y., Barrette, J., and Buckley, N.D. (2009). Selection of probiotic bacteria for the fermentation of a soy beverage in combination with Streptococcus thermophilus. *Food Research International,* (2008), 42, 5-6, pp. 612-621. ISSN: 0963-9969

Covich Alan P. & Thorp, James H. (2001). Introduction to the Subphylum Crustacea, In: *Ecology and Classification of North American Freshwater Invertebrates.* Thorp, James & Covich Alan P. pp. 777-780, Academic Press., ISSN 0-12-690647-5, USA.

Dodson Stanley I. & Frey David G. (2001). Cladocera And Other Branchiopoda, In: *Ecology and Classification of North American Freshwater Invertebrates.* Thorp, James & Covich Alan P. pp. 849-862, Academic Press., ISSN 0-12-690647-5, USA

Ebert D. (2005). *Ecology, Epidemiology and Evolution of Parasitism in Daphnia.* Thomas Zumbrunn, Universität Basel, Retrieve from http://www.ncbi.nlm.nih.gov/entrez/query.fcgi?db=Books

Emmens, C. W. (1984). *How to Keep and Breed Tropical Fish.* Neptune City, NJ : T.F.H. Publications, Inc., (3 ed), ISBN 0876664990, Australia.

FAO, (2008a). International Year of the Potato 2008. New Light on a Hidden Treasure. In: *Food and Agriculture Organization of the United Nations, Rome.* Enero 2011< http://www.potato2008.org/es/lapapa/IYP-6es.pdf>

Furuya Valéria Rosseto Barriviera, Hayashi Carmino, Furuya Wilson Massamitu, Soares Claudemir Martins e Galdioli Eliana Maria. (1999). Influência de plâncton, dieta artificial e sua combinação, sobre ocrescimento e sobrevivência de larvas de curimbatá (Prochilodus lineatus). *Acta Scientiarum* 21(3), (Agosto 1999), 699-703, ISBN 1415-6814

González de Infante, A. (Eva Chesneau). (1988). *El plancton de las aguas continentales,* Secretaría General de la Organización de Estados Americanos. The General

Secretariat of the Organization of American States. ISBN 0-8270-2736-2, Washington, D. C.

Guillaume, .J, Kaushik, S., Bergot, P. and Metaille, R. (2002). *Nutrición y alimentación de peces y crustáceos*, Mundi-Prensa, ISBN: 84-8476-150-9, Madrid

Gutierrez-Espinoza M.C. & Vasquez-Torres, W. (2008) Análisis de *digestibilidad* de Glicine max L, soya en juveniles de Cachama blanca *Piaractus brichipomus* Cuvier 818. 12, 2 (Noviembre 2008). Pp. 141-148, ISSN: 0121-3709

Kanazawa A. (2000). Nutrición de larvas de peces, *Proceedings of Avances en Nutrición Acuícola IV. Memorias del IV Simposium Internacional de Nutrición Acuícola*, ISBN 970-694-51-0. La Paz, B.C.S., México, Noviembre-1998

Kraul S. (2006). Live food for marine fish larvae. *Proceedings of Memorías del Octavo Simposium Internacional de Nutrición Acuícola*, ISBN 970-694-331-5, Mazatlán, Sinaloa, México Noviembre 15-17 2006

Kubitza, F. (Acqua Supre Com.). (1999). *Nutricao e Alimentacao Dos Peixes Cultivados*. Jundiaí : F. Kubitza, ISBN 8590101762. Campo Grande, Rio do Soul

Heugens, E., Tokkie, L., Kraak, M., Hendriks, J., Straalen, N. M. and Admiral, W. (2006). Population growth of *daphnia magna* under multiple stress conditions: joint effects of temperature, food, and cadmium. *Environmental Toxicology and Chemistry*, Vol. 25, No. 5, pp. 1399-1407, ISSN: 1552-8618

Hülsmann, S. (2001). Reproductive potential of *Daphnia galeata*in relation to food conditions: implications of a changing size-structure of the population. *Hydrobiologia*, 442, pp. 241–252, ISSN: 1573-5117

Jana, B. B. & Pal, G. P. (1985). Some life history parameters and prodution of *Daphnia carinata* (King) growth in different culturing media. *Water Res.*, vol 19, pp. 863-867, ISSN: 0043-1354

Lahnsteiner, F., Kletzl, M., Weismann, T. (2009). The risk of parasite transfer to juvenile fishes by live copepod food with the example Triaenophoruscrassus and Triaenophorusnodulosus. *Aquaculture* 295 (June 2009), 120–125, ISSN 0044-8486

Lavens, P., & Sorgeloos, P. (1996). *Manual on the production and use of live food for aquaculture* FAO (Ed.) *Fisheries Technical Paper* Retrieved from ftp://ftp.fao.org/docrep/fao/003/w3732e/w3732e00.pdf

Lubzens Esther & Zmora Odi. (2003). Production and Nutritional Value of Rotifers, In: *Live Feeds in Marine Aquaculture*, Støttrup Josianne G. & McEvoy Lesley A, (31-42), Blackwell Science Ltd, Retrieved from www.blackwell-science.com

Martínez-Jeronimo F. (2008). Ensayo de toxicidad aguda con cladoceros de la familia Daphnidae, In: *Ensayos toxicologicos para la evaluacion de sustancias quimicas en agua y suelo La experiencia en Mexico.* Ramirez P. y Mendoza Cantu A. pp. 99-115, Secretaría de Medio Ambiente y Recursos Naturales (Semarnat). ISBN: 978-968-817-882-9, Mexico

Mitchell, S. E, Read, A. F. and Tom, J. L. (2004). The effect of a pathogen epidemic on the genetic structure and reproductive strategy of the crustacean Daphnia magna. *Ecology Letters*, 7, 848–858, ISSN: 1461-0248

Müller-Navarra, D. C. & Lampert, W. (1996): Seasonal patterns of food limitation in *Daphnia galeata*: separating food quantity and food quality effects. *J. Plankton Res.*, vol 18, pp. 1137-1157, ISSN: 1464-3774

Murray, H., Lall, S., Rajaselvam, R., Boutilier, L., Blanchard, B., Flight, R., Colombo, S., Mohindra, V., Douglas, S. (2010). A nutrigenomic analysis of intestinal response to partial soybean meal replacement in diets for juvenile Atlantic halibut, Hippoglossushippoglossus, L. *Aquaculture*, 298, (Noviembre 2009), pp. 282-293, ISSN: 0044-8486

Murniece Irisa, Karklina Daina, Galoburda Ruta, Santare Dace, Skrabule Ilze, Costa Helena S. (2011). Nutritional composition of freshly harvested and stored Latvian potato (SolaNo. tuberosum L.) varieties depending on traditional cooking methods. *Journal of Food Composition and Analysis*, 24, (septiembre 2010), pp. 699 - 710, ISSN: 0889-1575

Ocampo Lucía E, Botero Mónica C, Restrepo Luis F. (2010). Evaluación del crecimiento de un cultivo de Daphnia magna alimentado con Saccharomyces cereviseae y un enriquecimiento con avena soya. *Revista Colombiana de Ciencias Pecuarias*. Vol 23, 1 (Noviembre 2009), 78-85, ISBN 0120-0690

Prieto, M. G. (2001). Aspectos reproductivos del cladócero *moinodaphnia sp.* en condiciones de laboratorio. *Revista MVZ-Córdoba*, vol 6, núm 2, (2001), 102-110, ISSN: 1909-0544

Prieto G, Martha & Atencio G, Victor. (2008). Zooplankton In Larviculture Of Neotropical Fishes. *Rev.MVZ Cordoba*, vol.13, 2, (mayo-agos. 2008), pp.1415-1415, ISSN: 1909-0544

Pérez E.P. & W.B. Stotz. (1992). Comparaciones múltiples de parámetros gravimétricos entre poblaciones submareales de *Concholepas concholepas* (Bruguiére, 1789) en el Norte de Chile. *Revista de Biología Marina, Valparaíso*, vol 27, núm2, (Noviembre 1992) p. 175-186

Salze, Guillaume, McLean, Ewen, Battle, P. Rush, Schwarz, Michael H. Craig, Steven R. (2010). Use of soy protein concentrate and novel ingredients in the total elimination of fish meal and fish oil in diets for juvenile cobia, Rachycentron canadum. *Aquaculture*, (Noviembre 2009), 298, pp. 294-299, ISSN 0044-8486

Sevrin-Reyssac, J. (1993). Performances and constraints of intensive rearing of the cladocerans Daphnia magna Straus, utilization of produced biomass. 1a European Crustacean, *Proceedings of the first European Crustacean Conference, (1992)*, vol 64, No. 3, pp. 357-360, Paris (France), 31 Aug-4 Sep, 1992

Sealey, W., Barrows, F., Smit,h C., Overturf, K. & La Patra, S. (2009). Soybean meal level and probiotics in first feeding fry diets alter the ability of rainbow trout Oncorhynchusmykiss to utilize high levels of soybean meal during grow-out. *Aquaculture*, 293 (Abril 2009), pp. 195-203, 0044-8486

Terra Nara Regina, Feiden Ilda Rosa, Lucheta Fabiane, Pereira Gonçalves Silvana and Schons Juliana Gularte. (2010). Bioassay using *Daphnia magna* Straus, 1820 to evaluate the sediment of Caí River (Rio Grande do Sul, Brazil). *Acta Limnologica Brasiliensia*, 2010, vol. 22, no. 4, p. 442-454, 51-55, 2179-975X

Watanabe, T., Kitajima, C., Fujita, S. (1983). Nutritional value of live organisms used in Japan for mass propagation of fish: a review. *Aquaculture* 34, (October 1982), pp. 115-143, ISSN: 0044-8486

12

Advances in Domestication and Culture Techniques for Crayfish *Procambarus acanthophorus*

Martha P. Hernández-Vergara[1] and Carlos I. Pérez-Rostro[2]
[1]Native Crustacean Aquaculture Laboratory,
Graduate Study and Research Division,
Instituto Tecnológico de Boca del Río,
Boca del Río, Veracruz
[2]Genetic Improvement and Production Laboratory,
Graduate Study and Research Division,
Instituto Tecnológico de Boca del Río,
Boca del Río, Veracruz
México

1. Introduction

1.1 Ecological importance

The crayfish are a group of crustaceans that habit in different environments in the world, both in lotic systems as lentic, in addition to caverns, which makes them cosmopolitan organisms with a wide range of tolerance to environmental conditions. Over 600 crayfish species are known to exist in the worldwide, with at least 100 species in Australia and about 300 in the Americas (Holdich, 1993), mostly (85%) in North and Central America (Rojas, 1998). Species in Mexico include one in the *Orconectes* genus, 10 in the *Cambarellus* genus and 44 in the *Procambarus* genus, the latter also distributed in Belize, Honduras and the United States (Villalobos, 1948; Hobbs, 1984; Rojas, 1998; López, 2006). The genus *Procambarus* habits in temporary water bodies, during the dry season can be seen in small holes in the soil, similar to the anteaters, which conduct to tunnels and chambers with sufficient moisture for the crayfish to survive to the drying (López, 2008). The crayfish have been adapted in various ways, according to environmental conditions that occur in the places they want to colonize. The first adaptation is their ability to spend a lot of time, even months, faced with the lack of water and breathe atmospheric oxygen (Huner, 1995) in some cavemen environments it has been recorded that these organisms exhibit a diminution in their effective breathing rate as a response to the decrease in the concentration of oxygen, and undersupply of food (Mejía, 2010). Despite their abundance, less than a dozen crayfish species are cultivated worldwide and only two species constitute sizable commercial fisheries (Huner, 1994). Crayfish have a high potential for use in aquaculture systems because they are at the bottom of the trophic chain, feeding largely on carrion and detritus, for that they are therefore considered fundamental for maintaining ecological balance in natural ecosystems (Rojas, 1998), and is possible their maintained in control conditions. The global diversity of crayfish allows to establish a productive activity associated with diverse environments, even in reduced environments with eutrophication, sulfate-reducing bacteria and that there is a proliferation of algae (Sánchez et al., 2009).

1.2 Aquaculture importance

Crayfish like other decapods crustaceans, have some biological characteristics that make them potentially important species for aquaculture, among which include: adaptation to conditions of captivity and handling; accept artificial feeds of different origins (shrimp, fish aquatic plants, vegetables), can even be fed diets with vegetable protein (75%); have a relatively short life cycle (two years or less). First studies indicates that these organisms can breed in captivity at early age (about four months), and first spawning had high survival rates (> 75%), with reproduction all year, and some females had more than one spawning per year.

The physiological characteristics of crayfish allow them to adapt to extreme climatic variations, diversifying their potential habitats, ensuring reproduction and contributing to progeny survival under adverse conditions. This occurs under natural or artificial conditions, making them promising organisms for use in aquaculture systems (Gutiérrez-Yurrita, 1994; Rodríguez-Almaraz & Mendoza-Alfaro, 1999). Of the crayfish species which can be cultivated in subtropical environments, *Procambarus acanthophorus* stands out for its biological attributes, such as high number of progeny per spawn, resistance to a wide range of environmental and water quality conditions, and successful performance in captivity (Arrignon, 1985; Cervantes, 2008; Cervantes-Santiago et al., 2010a).

1.3 Advances in laboratory research for aquaculture facilities

1.3.1 Environmental requirements of the species

Advances in laboratory research indicate that crayfish *P. acanthophorus* has a high potential for being used in aquaculture, for that reason different trials were done to determinate the best biotechnology for semi intensive and intensive culture conditions in monoculture and polyculture facilities.

Studies under laboratory conditions had showed that crayfish *P. acanthophorus* (Villalobos, 1948; 1993), can be quickly adapted to conditions of captivity, despite coming from natural environments. In the study the organisms were maintained at an average temperature of 26 ± 2°C and photoperiods between 12 and 14 h light: dark, oxygen concentrations from 0.5 to 5 mgL-1, indicating that they may like other crustaceans endure low contents O_2. The pH tolerance of the species ranges from 6 to 9, the ammonium concentration is about 0.5 mlL-1, and hardness greater than 200 mlL-1 as calcium carbonates.

1.3.2 Feeding

Even though it is known that crayfish accept balanced food for aquaculture species as shrimp. Laboratory experiment with 20 formulated diets containing different protein (200, 250, 300, 350 and 400 gkg-1) and lipid (60, 80, 100 and 120 gkg-1) levels (Table 1) on growth and survival in juvenile crayfish (*P. acanthophorus*) during 12-week nutritional trial, indicate

Ingredients (g·kg⁻¹)	200/60	200/80	200/100	200/120	250/60	250/80	250/100	250/120	300/60	300/80	300/100	300/120	350/60	350/80	350/100	350/120	400/60	400/80	400/100	400/120
Fish meal[1]	77.1	77.1	77.1	77.1	96.3	96.3	96.3	96.3	115.6	115.6	115.6	115.6	134.8	134.8	134.8	134.8	154.1	154.1	154.1	154.1
Soya meal[2]	327.9	327.9	327.9	327.9	409.8	409.8	409.8	409.8	491.8	491.8	491.8	491.8	573.8	573.8	573.8	573.8	667	681	695.1	702.6
Wheat bran[3]	58.1	58.1	58.1	58.1	72.6	72.6	72.6	72.6	87.2	87.2	87.2	87.2	101.7	101.7	101.7	101.7	88.3	53.5	18.6	0.0
Fish oil[4]	22.3	32.3	42.3	52.3	20.4	30.4	40.4	50.4	18.4	28.4	38.4	48.4	16.5	26.5	36.5	46.5	14.6	24.6	34.6	44.6
Corn oil[5]	20.8	30.8	40.8	50.8	18.6	28.6	38.6	48.6	16.3	26.3	36.3	46.3	14	24	34	44	12.4	23.3	34.2	44.7
Corn starch	431.3	411.3	391.3	371.3	319.8	299.8	279.8	259.8	208.3	188.3	168.3	148.3	96.7	76.7	56.7	36.7	1.1	1.0	0.9	0.0
Vitamin and mineral premix[6]	30.0	30.0	30.0	30.0	30.0	30.0	30.0	30.0	30.0	30.0	30.0	30.0	30.0	30.0	30.0	30.0	30.0	30.0	30.0	30.0
Carboxy-methylcellulose	15.0	15.0	15.0	15.0	15.0	15.0	15.0	15.0	15.0	15.0	15.0	15.0	15.0	15.0	15.0	15.0	15.0	15.0	15.0	15.0
Soya lecithin	5.0	5.0	5.0	5.0	5.0	5.0	5.0	5.0	5.0	5.0	5.0	5.0	5.0	5.0	5.0	5.0	5.0	5.0	5.0	5.0
Pigment	10.0	10.0	10.0	10.0	10.0	10.0	10.0	10.0	10.0	10.0	10.0	10.0	10.0	10.0	10.0	10.0	10.0	10.0	10.0	10.0
Vitamin C[7]	2.5	2.5	2.5	2.5	2.5	2.5	2.5	2.5	2.5	2.5	2.5	2.5	2.5	2.5	2.5	2.5	2.5	2.5	2.5	2.5
Proximate composition (g·kg⁻¹ dry base)																				
Moisture	63.1	54.9	63.6	59.4	66.2	61.3	59.4	52.3	92.5	82.4	82.9	79.1	62.7	57.4	75.5	53.7	66.5	75.2	67.4	56.1
Crude protein	217.6	219.6	220.0	227.7	276.9	271.9	271.2	268.2	317.9	316.9	315.4	319.4	377.2	373.0	376.5	381.5	434.4	424.5	425.9	420.9
Ether extract	68.3	95	116.9	134.8	67.5	91.1	114.7	136	62.6	81.4	98.4	112.8	63.4	84.1	110.7	135.1	62.3	86.4	104.9	126.1
Ash	42.3	46.0	44.8	45.9	56.4	55.3	54.8	54.6	65.5	63.9	64.5	64.6	74.1	77.3	74.5	77.3	83.5	83.5	82.1	99.4
Crude fiber	10.7	11.9	11.0	12.3	10.8	16.1	20.4	20.9	25.8	22.9	21.0	20.1	29.3	31.6	34.1	40.8	34.8	36.5	30.4	28.2
NFE	598.0	572.6	543.7	519.9	522.2	504.3	479.5	468.0	435.7	432.5	417.8	404.0	393.3	376.6	328.7	311.6	318.5	293.9	289.3	269.3
Energy (MJ·100g⁻¹)	18.92	18.83	20.01	17.24	18.42	18.67	18.75	17.24	18.29	18.25	18.46	19.00	18.54	18.92	19.17	18.83	18.08	18.04	17.58	18.46

Table 1. Formulation and proximate composition of experimental diets containing different protein/lipid ratios fed *P. acanthophorus*.

that the protein requirement for young *P. acanthophorus* is in a range between 210 and 280 gkg^{-1}, without observing a specific requirement of lipids, the results also suggest that in culture, it is possible to use foods with a maximum of 279 gkg^{-1} of protein and 60 gkg^{-1} lipid for better growth in crayfish, which can use up to 75% protein of vegetal source and only 25% from animal source, with growth performance and uptake efficient (Cervantes, 2006) (Table 1).

The study results indicate that nutritional diets can be used with protein content between 211 and 232 gkg^{-1} to feed growing crayfish in order to minimize feed costs, indicating that these organisms consume protein from vegetal source and assimilated efficiently (1.09:1 FCR) regardless of sex. This was verified by the assessment of carcass composition of crayfish fed with 20 experimental diets and by sex, where females were found to store more lipids (%) and body caloric energy (MJ/100 g) without significant differences in relation with the values reported for males. Other important information obtained during the nutritional study was the detection of ovigerous females in treatments 200/120, 250/60 and 400/120, but could not be attributed to the generation of egg protein or lipids tested, which suggests that in general all the diets allowed the bodies to cover their energy requirement for basic functions, but also could reproduce and promote sexual maturation.

The crayfish survival ranged from 66% to 86%, without differences between treatments. Because of the lowest survival recorded in the juvenile fed with the 250/12 gkg^{-1} diet, although this treatment yielded the most efficient parameters (WG, SGR and DWG), a correlation analysis was performed between weight gain and survival to determine a possible influence of mortality on growth. The weight gain survival ratio was not significant ($r^2 > 0.037$; $P > 0.1365$), indicating that crayfish growth was only affected for the experimental diets.

1.3.3 Life cycle and reproduction

A vital aspect to consider when determining an organism culture potential, is its reproductive capacity under controlled conditions, which in turn depends on its ability to adapt to the culture system, feed and water quality. Factors reported to significantly affect crayfish reproductive capacity include water temperature, photoperiod, and sex ratio (Yeh & Rouse, 1995; Carmona-Osalde et al., 2002; 2004a, 2004b). The results in laboratory conditions demonstrate that the crayfish *P. acanthophorus* is a candidate for aquaculture production in a closed cycle since it effectively reproduces in captivity. During the time that organisms captured from the wild remained in captivity, it was observed mating and reproduction, from which ovigerous females were obtained, indicating that breeding in captivity could be obtained. Under the study conditions, P. *acanthophorus* exhibited the peak of the reproductive activity during November and December, when average water temperature was 25 °C. The lowest reproductive activity occurred in February and March, when wide variations in water temperature (25 ± 5°C) may have affected organism metabolism and consequently their reproductive cycle (Figure 1).

In a similar researches, Rodríguez-Serna et al. (2000) reported that reproduction in *P. llamasi* occurs year around, although, in contrast to *P. acanthophorus*, this species has three spawning peaks between November and June. Its maximum activity is in May and June and its minimum in August and October, when temperatures above 26.7 °C negatively affect its reproductive efficiency. Temperature is clearly a limiting factor for reproduction in

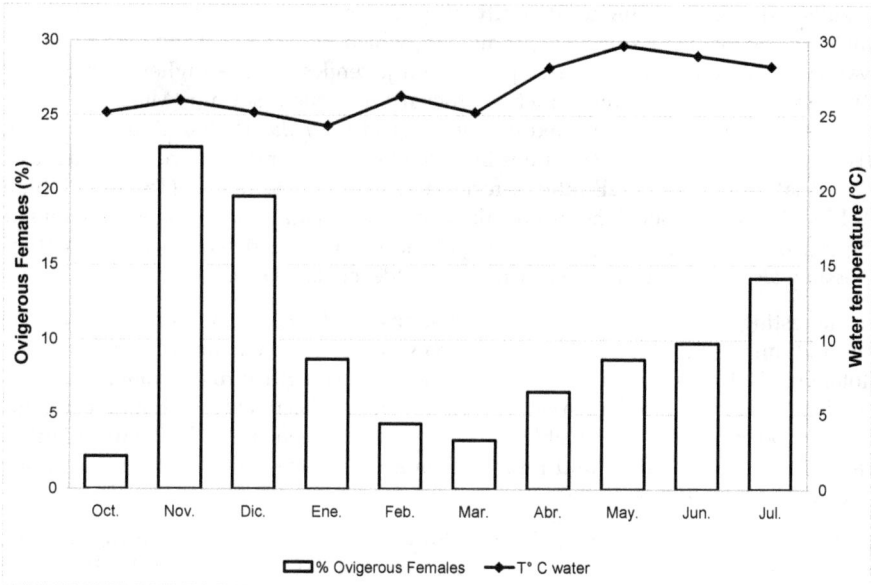

Fig. 1. Proportion of ovigerous females versus water temperature (°C) during a ten-month period.

P. llamasi, with optimum spawning at 21 °C, although breeder length and degree of female sexual maturity can also affect spawning (Carmona-Osalde et al., 2004a).

In the study, a total of 92 ovigerous females were recorded from the 192 placed in the reproduction tanks during the study. ANOVA showed no significant variation for the number of ovigerous females at the three sex ratios used in the treatments, with 30 ovigerous females at 1:1, 40 at 1:3 and 22 at 1:5 male: female ratios, respectively (Figure 2).

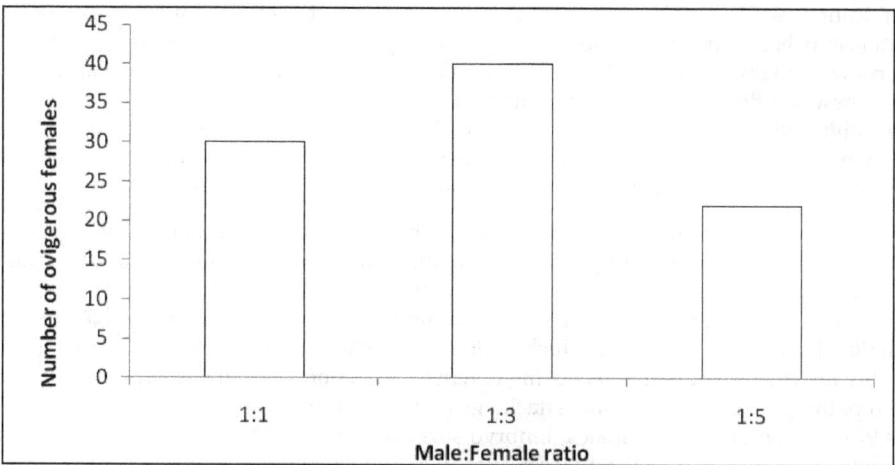

Fig. 2. Number of ovigerous females per sex ratio treatment

The results indicate that a higher quantity of gravid females can be obtained if sex ratio of 1:3 (male: female) is using during reproduction, which indicates that the ratio male: female is a variable that can affect the mass production juveniles. This is higher than the 1M: 1F ratio reported for optimum reproductive efficiency in peneid shrimp (Martínez, 1999), but lower than the 1M: 5F ratio reported for *Cherax quadricarinatus* (Yeh & Rouse, 1995). In the crayfish *Astacus astacus*, a 1M: 3F ratio is inefficient since all females do not reproduce at this ratio, for that reason a 1M: 2F ratio is recommended for best results (Taugbol & Skurdal, 1990). In *P. llamasi*, higher densities resulted in higher mating and ovigerous female rates during reproduction, meaning density significantly influenced female sexual maturation and possibly also male maturation (Carmona-Osalde, et al., 2004a).

Other interesting result is that mating was first observed when females reached an average length of 30 mm, although the first viable spawns did not occur until females reached 37 mm total length. This is therefore considered the female length at first sexual maturity for *P. acanthophorus*. This is similar to *P. llamasi* females, which exhibit initial reproductive behavior at 30 mm but have viable spawns only between 40 and 60 mm length. This indicates that crayfish mature and reproduce at an early age, an advantage for production under controlled conditions.

Aspects of reproductive biology such as fertility vary by species, ranging from as little as five eggs in *Astacus pachypus* up to 960 in *Cherax destructor* (Lee & Wickins, 1992). Egg counts for *P. acanthophorus* females in the present study ranged from 77 to 467 per female, with an average of 240.9. The ranges are similar with reported average fertility ranges for other crayfish species: 100-700 eggs in *P. clarkii* (Lee & Wickins, 1992; McClain & Romaire, 2007); 200-700 eggs in *P. llamasi* (Rodríguez-Serna et al., 2000); 300-400 eggs in *P. zonangulus* (Reynolds, 2002); and 323 eggs in *P. leniusculus* (Celada et al., 2005).

Egg count per female in crayfish depends on organism age and length. For instance, egg production in *A. leptodactylus* varies according to female length, with organisms measuring 47 to 76 mm producing between 200 and 400 eggs (average = 305.9), and 72 mm long females producing a average maximum of 588 eggs (Köksal, 1988; Mustafa et al., 2004). In the present study, total female length in *P. acanthophorus* had a linear, positive and significant ($p \leq 0.01$) relationship ($r^2 = 0.654$) with egg count. This is similar to the positive relationship between cephalothorax length and egg count reported for *Austropotamobius pallipes* where females with a 25 mm minimum carapace length produced a maximum of 80 eggs (Brewis & Bowler, 1985). The number of eggs per spawn can also depend on egg size. The crab *Cambaroides japonicus* produces only 22 to 75 eggs per spawn but these are 2.13 to 2.50 mm in diameter; this is significantly larger than mean egg size in other crustaceans and may contribute to this species high survival rate (Nakata & Goshima, 2004).

Independent of egg count, not all eggs hatch. This can be caused by biotic and/or abiotic factors that lead to losses during embryo growth. Temperature was the main factor causing egg loss in the present study since at higher temperatures fungi began to grow on the egg masses, affecting water quality in the hatchery units. Rodríguez-Serna et al. (2000) reported a similar incident in *P. llamasi* which seriously affected egg survival. Water quality and stability are clearly key elements during crayfish incubation since this is apparently a phase when pathogenic microorganisms attack eggs. Maximum egg viability in the present study was 97.4% in 56 mm long females. Embryo survival varied according to female length but did not exhibit a clear pattern; it was highest (42%) in 56-60 mm females and almost absent in 66-70 mm females. This decrease may be the result of female age since egg quality, and therefore viability, generally decreases as age increases, although this cannot be emphatically stated in the present case because the breeders were collected from the wild and their ages were therefore not exactly known. Two suggestions arise from the above results for commercial production of *P. acanthophorus*. First, breeders should be between 41 and 60 mm in length to ensure the highest possible egg and viable progeny counts. Second, adequate female nutritional condition and genetic quality need to be ensured since these are expressed in progeny quality and survival, perhaps by maintaining well-fed breeder stocks and employing constant selection to improve genetic quality.

1.3.4 Physicochemical parameters

Water chemical and quality parameters during the crayfish reproduction trial were within tolerance ranges for organisms of the same sex (Malone & Burden, 1988; McClain & Romaire, 2007; Cervantes-Santiago et al., 2010b): temperature, 23.8±2.2 °C; dissolved oxygen, 5.7±0.18 mg L-1; total hardness, 110 mg L-1CaCO3; pH, 8.67±0.13; ammonium, 0.18±0.10 mg L-1; N-nitrite, 0.25±0.20 mg L-1; and N-nitrate, 32.5±20.6 mg L-1.

1.3.5 Fertility

The ratio between total female length and egg counts ($r^2=0.6541$) was positive, linear and significant ($p \leq 0.01$), defined by the equation y= 8.4126X - 216.4313. Average ovigerous female length was 54.4 mm (max = 71 mm; min = 37 mm) and average egg count per female was 240.9 (±S.D. 93.08) (max = 467; min = 77) (Figure 3).

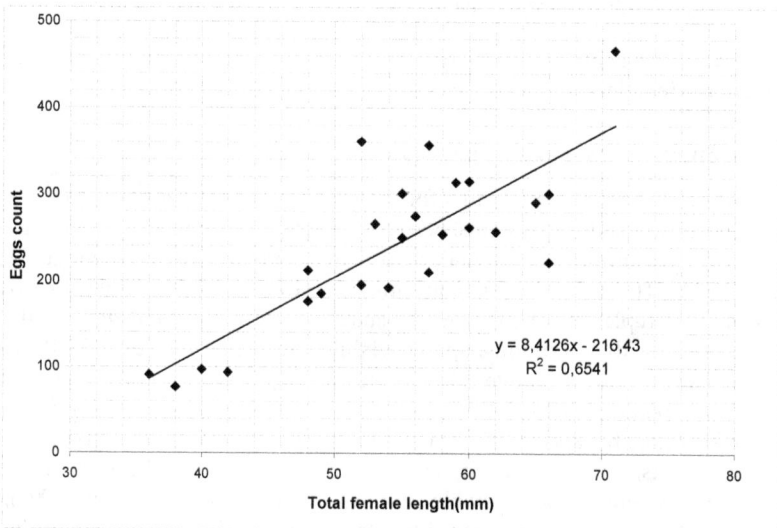

Fig. 3. Ratio of egg count to total female length in *Procambarus acanthophorus* during study.

1.3.6 Egg viability

Average egg viability was 29.1% (± S.D. 31.7; n=66), with a maximum of 97.4% and a minimum of zero. Overall, females between 41 and 60 mm had the highest egg viability

(97.4%). Those within the 46-50 mm size had an average viability greater than 40%, while those in the 66-70 mm had the lowest (2.9%) (Figure 4).

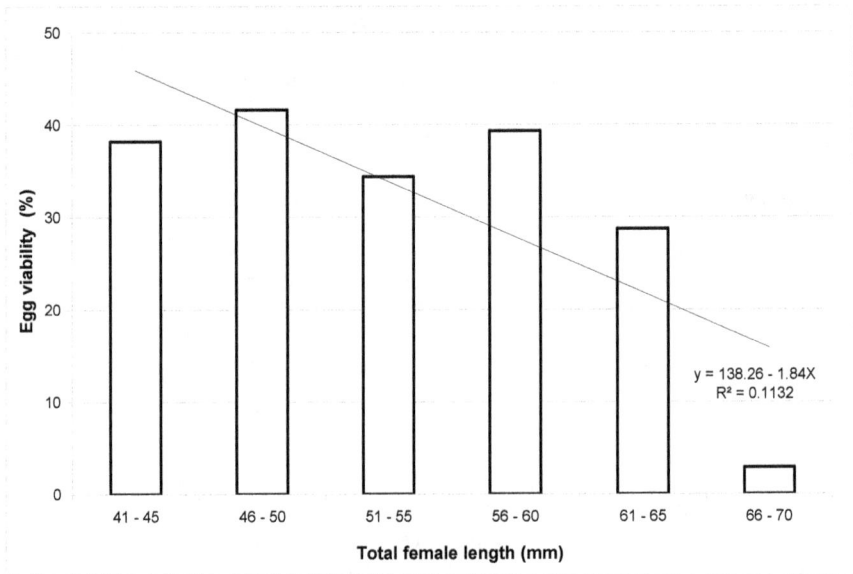

Fig. 4. Ratio of egg viability to total female length in *Procambarus acanthophorus* during study.

Water temperature probably affected egg viability since when it surpassed 30 °C, the eggs detached from the female's abdomen and were soon infected by fungi.

1.3.7 Female length at first sexual maturity

The present reproductive biology results for the crayfish *Procambarus acanthophorus* show it to be apt for use in aquaculture systems. Based on specimens grown in captivity, it is known that it reproduces year around and females are sexually mature at 37 mm length (13 weeks) (Cervantes, 2008). This allows for constant reproduction and implementation of a continuous breeder replacement program to ensure high quality and genetically variable spawns. Under the study conditions, spawn viability was variable (29.1- 97.4%), although this could probably be kept above 44% by controlling environmental parameters (e.g. keeping water temperature below 30 °C) and thus providing a regular supply of juveniles for culture. In addition, the species did not exhibit aggressiveness or territoriality during the study. These results, in conjunction with previous studies indicating these species preference for feeds containing vegetable protein sources (Cervantes et al., 2007), confirm that the crayfish *P. acanthophorus* is a candidate for culture under controlled conditions, be it as a preservation strategy for commercial purposes.

1.3.8 Embryology development

Incubation of crayfish *P. acanthophorus* eggs has a duration period of three to four weeks, depending on temperature and embryonic development inside the egg. At the time of hatching, the larvae present physical characteristics and eating behavior similar to an adult (Cervantes, 2006; 2008). A detailed description of embryonic development until juvenile stage, suggests that the development lasts 21 to 27 days on average, where it can identify nine embryonic stages, four post-embryonic and one juvenile. Also was observed that embryonic development, presents 11 color changes of the eggs, which are however asynchronous the same ovigerous mass, so it is considered that the coloring of eggs during embryonic development is not a clear indicator the stage of development.

Under the laboratory conditions, fourteen embryonic development stages were identified for *P. acanthophorus* using the structure descriptions and nomenclature of Anderson (1982). The females produced fertile eggs which exhibited nine embryonic stages, with an average total elapsed time of 15±3 days. After embryonic stages, crayfish had four post-embryonic stages and a final juvenile stage, which lasted an average of 10 days. Total elapsed time of development from fertilization through juvenile stage was on average 25 ± 3 days. Egg diameter and later embryo and juvenile length increased constantly until reaching a final length 600 times larger than initial diameter of a recently fertilized egg) (Figures 5 to 18).

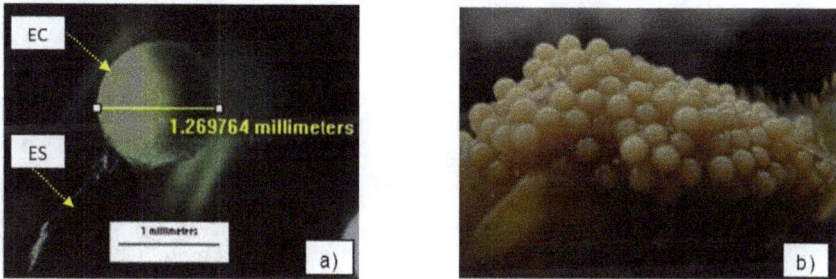

Fig. 5. Stage 1 (days 1-2; 0-11% development). - a) The recently fertilized eggs have a spherical shape and large quantities of mucus are present evidence of recent spawning. b) Egg mass has uniform light beige color which corresponds to the vitellus. By day 2 the vitellus has divided and small drops appear, probably the beginning of scission. It is observed the egg capsule (EC) and egg stolon (ES).

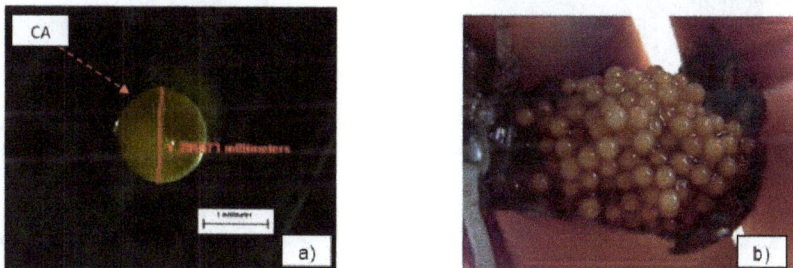

Fig. 6. Stage 2 (days 3-4; 11-22%). - a) Cellular division continues and the vitellus is completely divided into small drops. A region of greater cell accumulation (CA) is observed which corresponds to the zone where the blastopore will form. b) The egg mass begins to change color (yellow-olive green).

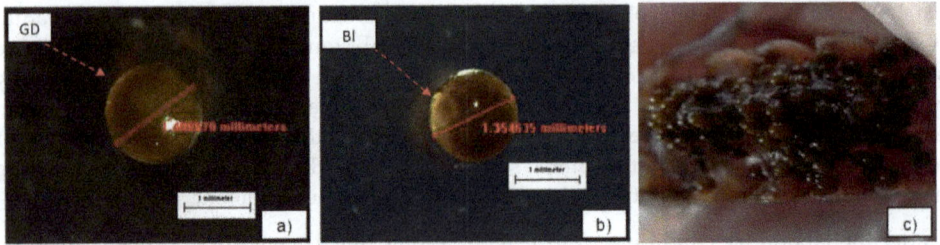

Fig. 7. Stage 3 (days 5-6; 22-33%). - a) Cell division continues and a group of cells begins to form on the egg ventral surface, corresponding to the germinal disk (GD). Starting on day 6, the cell layer begins to expand and form a depression, corresponding to the gastrula. b) The gastrula's ventral plate sinks in to form a groove, which is how the blastopore (Bl) appears; the forward portion of the caudal papilla starts to develop. The blastopore then closes and the rear portion of the caudal papilla appears. d) The egg mass changes in color from dark beige to a translucent olive green.

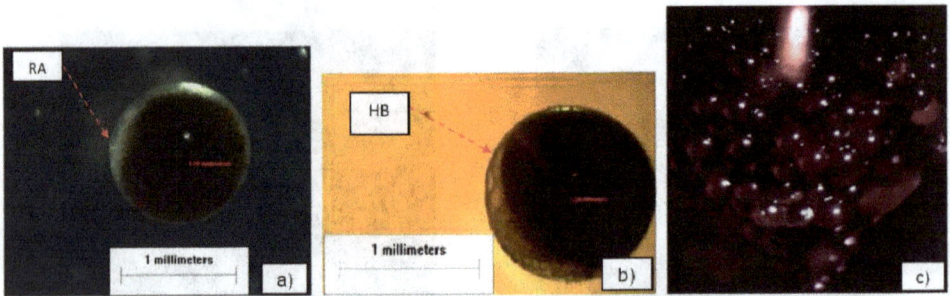

Fig. 8. Stage 4 (days 7-8; 33-44%). - a) An embryo with rudimentary anterior appendages (RA) is evident. Outlines of the frontal lobules and antennules, antennae and mandible can be distinguished. b) On day 8 the heart beat (HB) and embryo contractions can be. c) The egg mass slowly changes in color from olive green to brown.

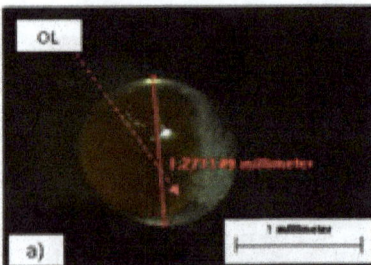

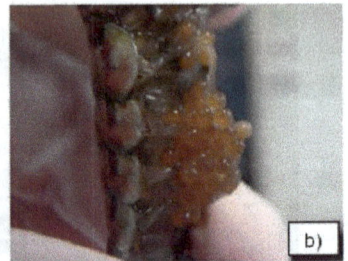

Fig. 9. Stage 5 (days 9-10; 44-55%).- a) Primordial eyes or ocular lobules (OL) appear as two elevations in front of the body, a transversal groove appears in the vitellus, crossing the middle of the egg, and a long, thin, anterior-curved caudal papilla is present. The embryo contracts more frequently and the vitellus is clearly visible. b) The egg mass has a translucent, light yellow color and the ocular spots appear as black dots.

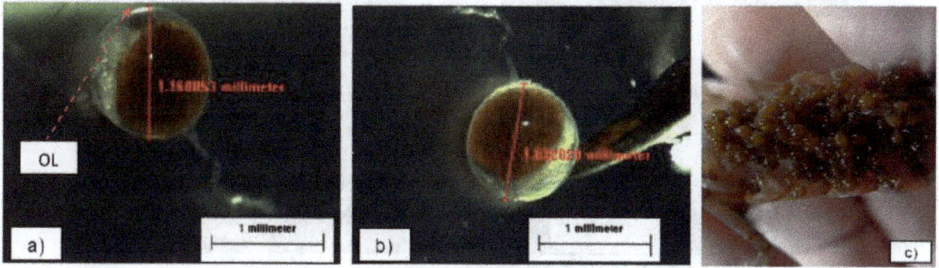

Fig. 10. Stage 6 (days 11-12; 55-66%). - a) OL are well-defined on the anterior body, while the abdominal somites and periopods remain rudimentary, although the chelae are visible. b) The caudal papilla is folded after and covered by the periopods, almost reaching the head. c) The egg mass is heterogeneously colored, with tones varying from light brown, to olive green and khaki.

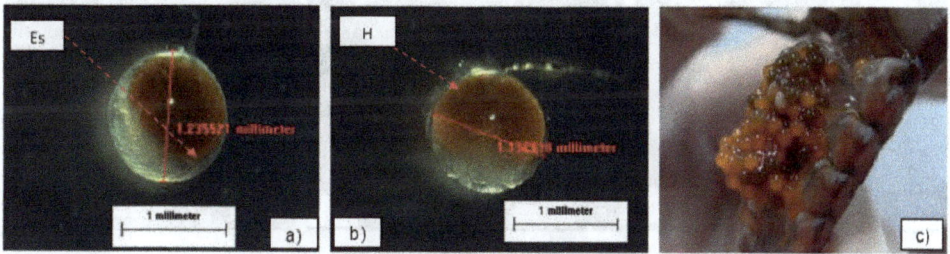

Fig. 11. Stage 7 (days 13-14; 66-77%). - a) Eyespots (Es) are visible, and deep grooves cross through the vitellus along the dorsal medial line. b) The heart (H) can be seen to beat strongly and regularly, Embryo interior is clear and more complex, the periopods are elongated and thin, and a small rostrum appears between the eyes. c) Egg mass coloring is heterogeneous, varying from olive green to bright orange.

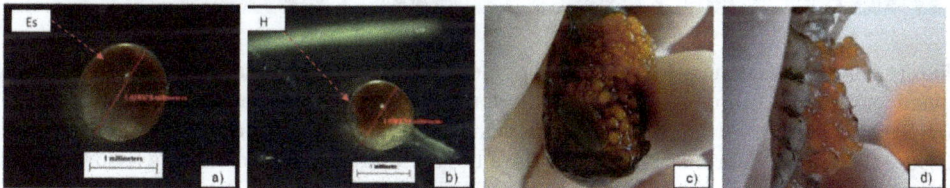

Fig. 12. Stage 8 (days 15-17; 77-88%). - a) The embryo occupies approximately three quarters of the egg ventral surface. b) The thoracic appendages are more developed and the chelae are totally formed. The eyes are sessile and elongate. c) The egg mass has taken on a translucent bright orange color. d) The embryo and eyes are clearly visible.

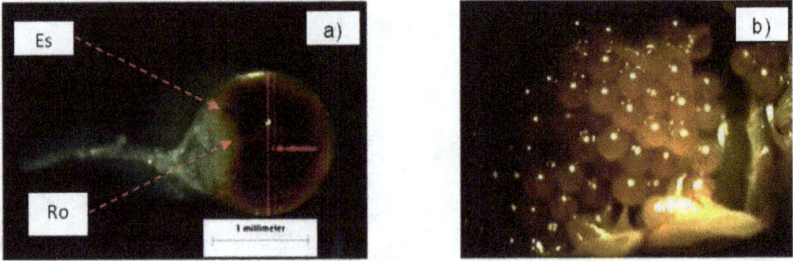

Fig. 13. Stage 9 (day18; 88-99%). - a) Shortly before hatching, the embryo appears compressed inside the egg such that the appendages seem flat and overlapped; there is no space remaining in the chorion. The chelae have grown in front of the eye base, the rostrum (Ro) is visible between the eyes and a groove sagittally crosses half the embryo. b) Egg mass color is bright yellow, rudimentary appendages are visible on the translucent embryos and the eyes are clearly identifiable.

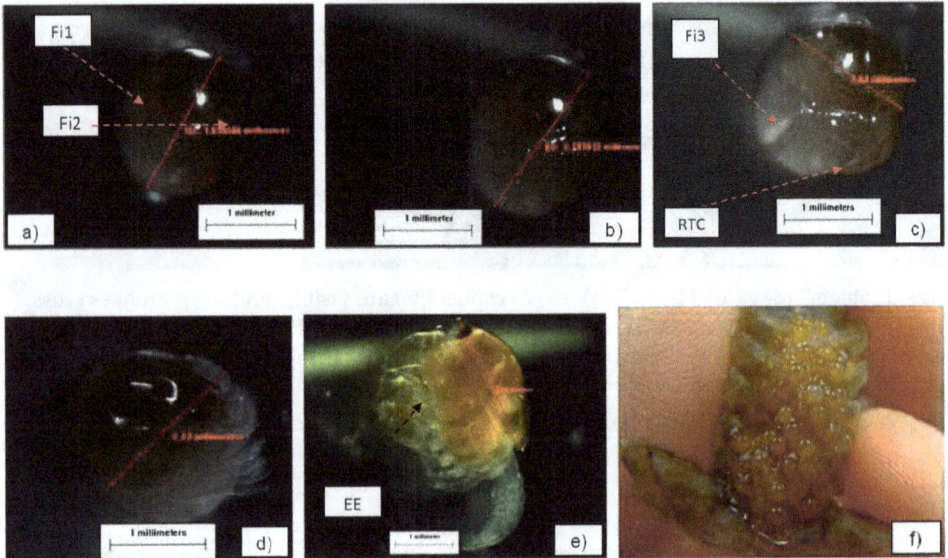

Fig. 14. Stage 10. Hatching (day 18; Development 100%). - The hatching process lasts an average of 10 to 15 min, from when the chorion breaks to when the embryo is completely free. a, b) It begins with the first fissure (Fi1) in the surface of the chorion barely visible, fissure (Fi2) of the chorion covers the folds of the eyes and maxillae; c) notable fissure (Fi3) accompanied by total rupture of the chorion (TRC). d) Breaking the chorion, the periopods are the first to get out. e) Expulsion of the embryo (EE) to the outside of the chorion. f) External appearance of hatchings, births shown asynchronism. This is considered a critical phase in embryo survival since exiting the chorion requires considerable energy expenditure.

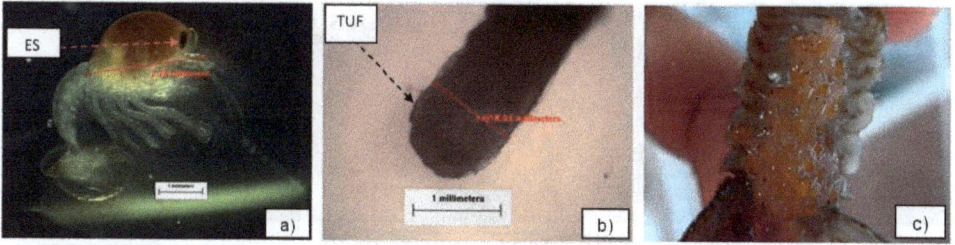

Fig. 15. Post-embryonic stage I (days 18-19). - a) The cephalothorax is formed by a yellow, elongated dorsal hump containing the remaining vitellus, which supplies nutrients to the organism during the following post-embryonic stages; the rest of the body is translucent. The eyes are round and sessile (ES) and contain dark pigment on about one quarter of the overall surface. The antennae and antennules are caudally curved and have sensory villi. b) The partially developed telsons and uropods are fused together in the membranous ligaments that attach the organisms to the chorion interior. c) The hatched organisms remain attached to the mother's pleura.

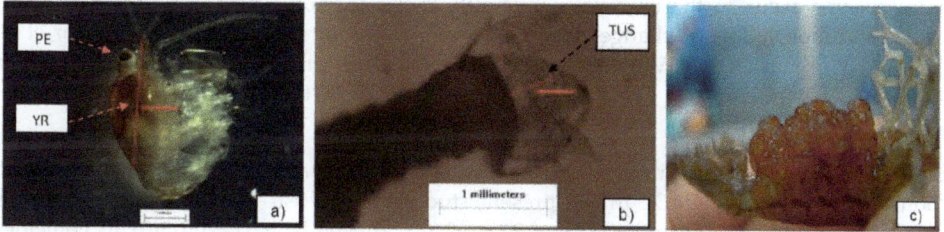

Fig. 16. Post-embryonic stage II (days 20-21).- a) The eyes are pedunculate (PE) with dark pigment, the dorsal hump which corresponds to the yolk remaining (YR) that nourishes the organism is smaller than in the previous stage (almost half its original size) and the cephalothorax has almost reached it final anatomy. Red dots begin to cover the entire body, the beginning of chromatophore pigmentation. b) The telson and uropods (TUS) appear to be separate with bristles at the ends. c) The organisms remain attached to the mother's pleura.

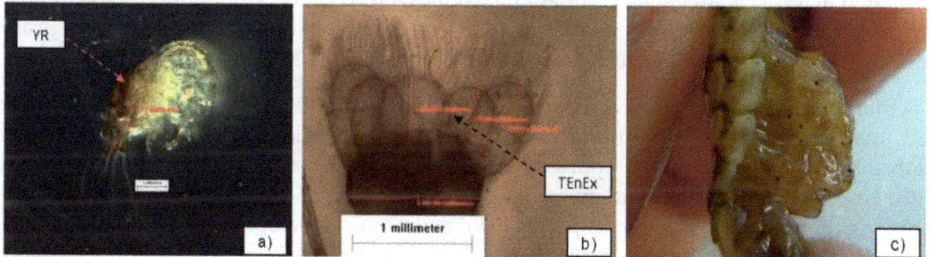

Fig. 17. Post-embryonic stage III (days 22-23). - a) The number and size of chromatophores increases over the entire body, but peduncles eyes not have dark pigment. Yolk Reserves (YR) are still present in the vitellus but almost exhausted, no exogenous feeding activity is observed. b) Telson and uropods are larger-well defined, this last are divided in endopodites and exopodites (TEnEx) which are still immobile and short but well-defined. c) Independent locomotion does not yet occur and the organisms remain attached to the mother's periopods.

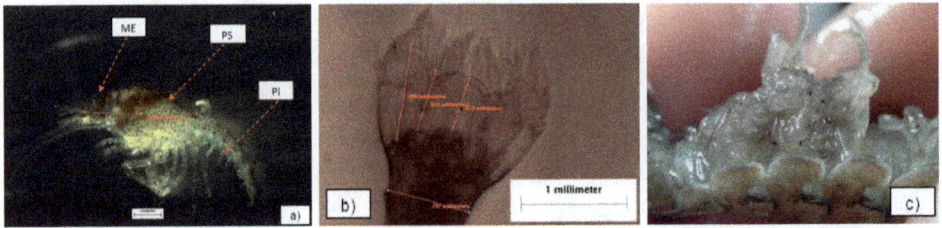

Fig. 18. Post-embryonic stage IV (days 24-25).- a) Fully development eyes or mature eyes (ME), pigmentation spots (PS) dispersed throughout the body, pleopods (Pl) are visible; the vitellus is exhausted but exogenous feeding has not begun. b) Telson elongated and uropods compound (protop, endopod, exopod) with bristles and sensorial filaments at the ends; but independent movement does not yet occur. Three pairs of pleopods (Pl) are visible on the abdominal somites, and c) the organisms remain attached to the mother's pleopods.

Descriptions were based on the external morphological changes observed in the embryos during ontogeny because the most important ontogonic events during development in *P. acanthophorus* occur in the embryos while still in the chorion. These define the different stages and morphological changes in ways similar to those reported by Montemayor et al. (2010) in *P. regiomontanus* (Villalobos, 1954) (Cambarideae family); Sandeman & Sandeman (1991) in *C. destructor*, and García-Guerrero et al. (2003) in *C. quadricarinatus* (both of the Parastacidae family).

Embryonic development in some crustaceans is highly dependent on water temperature (Bottrell, 1975; Herzig, 1983). In the present study, the crayfish *P. acanthophorus* embryos developed at an average temperature of 23.8 ± 2.2 °C during January-March. This coincides with Cervantes (2008), who reported that this species can reproduce year round under laboratory conditions as long as average temperature is kept at 25 °C. Auvergne (1982) stated that optimum temperature for each life stage in crustaceans is species dependent; for instance, *Astacus astacus* (Linnaeus, 1758) has a range of between 18 to 20 °C whereas *Procambarus clarkii* (Girard, 1852) requires a range of 22 to 26 °C. In further examples, Sandeman & Sandeman (1991) reported satisfactory development in *C. destructor* eggs incubated at 19 °C; García-Guerrero et al. (2003) described successful embryo development in *C. quadricarinatus* at 26°C; and García-Guerrero & Hendrickx (2009) reported proper development in fertilized *Macrobranchium americanum* eggs at 24 °C.

As embryo development progressed, egg length increased; initial egg diameter was 1.3 mm and total length of juvenile organisms was 6±1 mm. This development trajectory differs from the 16 stages (22 days) reported for *P. regiomontanus*, with eight embryonic stages, eight post-embryonic stages and an average juvenile size of 2 cm (Montemayor et al., 2010). However, both these *Procambarus* species have embryonic development periods near the 30-day average for cambarid crustaceans. *Procambarus clarkii* completes its embryonic development in an average of three to four weeks (McClain & Romaire, 2007), and *P. llamasi* completes it in 27 to 30 days (Rodríguez-Serna et al., 2000). These contrast with the longer development periods of other species. *C. quadricarinatus* has a 42-day development period with ten embryonic stages, two post-embryonic stages and a juvenile stage (García-Guerrero et al., 2003), while C. destructor has a 40-day development period with an unknown number of embryonic stages, at least two post-embryonic stages and a juvenile stage (Sandeman &

Sandeman, 1991). In stark contrast to all the above species, *Austropotamobius pallipes* (Lereboullet, 1858), a cold water species, requires over seven months to complete embryo development (Holdich & Lowery, 1988).

During the embryonic development trial, selected live eggs were kept in Petri dishes to evaluate survival. Artificially incubated fertile eggs were found to remain viable after development stage eight. This means that egg lots could be artificially incubated to reduce development period and synchronize times in mass production settings and/or if only small lots of reproductive-age females are available. The present study is the first report of embryonic development and artificial incubation of eggs in *P. acanthophorus*. It constitutes a significant contribution to the biology of this and other decapods crustaceans with potential for use in alternative, sustainable aquaculture systems.

1.3.9 Culture

1.3.9.1 Optimum density for growth

Advances in knowledge for the commercial culture of crayfish have shown that in this specie territorial habits are not present, allowing stocking densities as high as 100 orgm^{-2} without affecting the growth density and survival, which is considered an advantage in production. In addition, laboratory studies indicate that it is possible to keep under recirculation systems the biculture of crayfish and tilapia at densities between 30-50 orgm^{-2}, where the main species is tilapia, and crayfish support sustainable use of water.

1.3.9.1.1 Response parameters

A experimental trial was done when different densities of culture where evaluated, survival (%), showed no significant differences between treatments and remained in a range between 67 and 85%, with the increase of survival were crayfish were kept at densities of 50 and 90 orgm^{-2}, while the lowest occurred where the crayfish were kept at a density of 60 orgm^{-2}, without a positive relation between survival and density (Table 3).

Treat.	SUP. (%)	IW (g)	FW (g)	IWG (gday^{-1})	WG (%)	SGR (%day^{-1})	IFC (gday^{-1})	FCR
T1	85.2±17	1.0±0.3	5.6±1	0.1±0	450±58.9	1.0±0.1	0.2±0	3.1±0.1
T2	67.5±16	1.0±0.3	5.6±0.6	0.1±0	507±207	1.1±0.2	0.2±0	3.4±0.7
T3	72.9±40	1.0±0.1	5.6±0.6	0.1±0	474±33.2	1.1±0	0.2±0	3.3±0.2
T4	80.4±10	1.0±0.3	5.3±1.2	0.1±0	463±166	1.0±0.2	0.2±0	3.5±0.8
T5	84.4±14	1.1±0.2	5.3±0.7	0.1±0	412±85.8	1.0±0.1	0.2±0	3.5±0.4
T6	75.0±30	0.9±0.2	5.3±0.6	0.05±0	504±62.1	1.1±0.1	0.2±0	3.5±0.1

T1=50, T2=60, T3=70, T4=80, T5=90, T6=100 (org/m2)

Treat.= Treatments (org/m^2); SUP= Survival; IW= Initial Weight; FW= Final Weight; IWG= Individual weight gain; SGR= Specific growth rate; IFC= Individual food consumed and FCR= Feed Conversion rate

Table 3. Response parameters ± s.d of the crayfish was culture at different densities.

With respect to growth, the final weight (FW g), weight gain (WG%) and specific growth rate (SGR%/day) for all treatments was similar. The organisms in treatment 1, had a greater weight gain, during the first 30 days of culture maintained a growth rate similar to that of crayfish of the other treatments, then increased its rate of growth, although not differ significantly from other organisms under study, it was considered that the density did not significantly affect growth or survival of the crayfish. The juvenile kept it in treatment 5 and 6 had the lowest growth without significant differences between treatments (95% confidence).

1.3.9.2 Biculture

Actually research on the feasibility of polyculture systems including fish and crustaceans, has produced inconsistent conclusions, for that reason is difficult to determinate the potential of polyculture in recirculation system for sustainable aquaculture. In a study with red claw crayfish *C. quadricarinatus* and tilapia *Oreochromis niloticus* in a polyculture system in earthen ponds, Brummett & Alon (1994) reported positive results for the crayfish growth and survival; whereas using the same species combination Rouse & Kahn (1998) reported that competition for feed and space between species negatively affected survival in *C. quadricarinatus*. An alternative to the above system is the use of recirculation systems in crustacean/fish polyculture because water quality and feed supply can be controlled, and shelters can be provided for crustaceans to prevent territorial competition, allowing in both species a survive and grow properly (Karplus et al., 2001). Polyculture is particularly appealing since it makes extremely efficient use of resources and can increase production. It can be quite viable for producers as long as appropriate species are identified in terms of biology and market demand. Using a polyculture with common carp, grass carp, silver carp, tilapia, mullet and Malaysian prawn, Cohen et al. (1983) reported efficient water use and an increase in production from 3.5 to 11 tons/ha/year. In a polyculture system for carp (*Cyprinus carpio*) and crayfish (*Cambarellus montezumae*) growth in artificial ponds in which feeding was focused mainly on the carp, Auró et al. (2000) reported that these species could coexist and use the food in the system, making it a viable system.

Competition is increasingly high for good quality water sources for productive activities such as aquaculture. There is also a need to optimize space during production and to promote productive and sustainable alternative activities in rural areas. Under these conditions, the most adequate option for polyculture systems is to use recirculation systems, ensuring that the species in the system do not compete for resources and have established market niches (Kazmierczak & Caffey, 1995). With the objective to determine the feasibility of a crayfish *P. acanthophorus*/ tilapia *O. niloticus* polyculture and monoculture using a water recirculation system, as aquaculture sustainable alternative, a experimental research in outdoor facilities was conducted. Six plastic tanks (3 m diameter x 1.2 m depth) in a recirculation system with a biological and sand filter were used. During 90-day experimental period, three treatments were evaluated with two replicates per treatment in a completely random design. T1: crayfish monoculture (1.02±0.2 g); T2 polyculture: crayfish (1.04±0.2 g) and tilapia (2.99±0.1 g); T3: tilapia monoculture (3.45±0.6 g).

Survival in the T2 crayfish was significantly lower (34.7%) compared to that in the T1 crayfish (72%). In contrast, the tilapia in both T2 and T3 had similar survival (>95%) and growth rates (83-86 g) with no apparent effect from the presence of the crayfish (Table 4).

Treatment		S (%)	IW (g)	FW (g)	IWG (g)	WG (%)	SGR (%)
T1	Crayfish	72±12.7[a]	1.02±0.2[a]	4.8±0.4[a]	0.042±0[a]	373.5±55[a]	68.4±3.3[a]
T2	Crayfish	34.7±20.5[b]	1.04±0.1[a]	3.9±0.3[b]	0.033±0[b]	282.6±73[b]	59.9±3.3[b]
	Tilapia	98.7±0[a]	2.99±0.1[a]	88.5±9.9[a]	0.95±0.1[a]	2861.2±26[a]	194.2±5.3[a]
T3	Tilapia	93.7±9.9[a]	3.45±0.6[a]	84.3±10.7[a]	0.89±0.1[a]	2344.9±71[a]	192.1±5.6[a]

[1]Values in the same column with the same superscript are not statistically different (p>0.05)

S% = survival rate; IW = initial weight; FW = final weight; IWG = individual weight gain; WG% = percentage weight gain; SGR = specific growth rate.

Table 4. Growth and efficiency parameters in crayfish and tilapia monocultures and crayfish/tilapia polyculture in a water recirculation system[1].

The lower crayfish survival rate in T2 had no apparent effect on growth, as might be expected due to the density effect, since IWG and WF in T2 (0.033±0g; 3.9±0.3g, respectively) were significantly lower than in T1 (0.042±0g; 4.8±0.4g, respectively). Tilapia growth in T2 and T3 followed a steeply-sloped exponential curve whereas crayfish growth was constant but with a lesser slope, reflecting their lower growth rate (Figure 19).

Water quality parameter values during the trial were within the ranges tolerated by tilapia fingerlings and crayfish culture. Dissolved oxygen (DO) concentration was 3.8 mgL[-1] throughout the experimental period, lower than the 5 mg/L recommended for optimum growth in P. acanthophorus (Cervantes-Santiago et al., 2007). This level coincides with the >3 mgL[-1] DO level recommended for P. clarkii (Huner, 1994), and suggests that P. acanthophorus can adapt to environmental variations during cultivation. Like crayfish, tilapia can also tolerate low DO levels (<2 mgL[-1]), although levels greater than 3 mgL[-1] are recommended for good growth (El-Sayed & Abdel-Fattah, 2006). This DO level also favors proper functioning of the biological filter in the recirculation system and prevents the death of nitrifying bacteria (Yousef et al., 2003). Increased temperature during the experimental period improved growth in the tilapia and crayfish, although efficient growth in P. acanthophorus is reported to occur at temperatures <28°C. Optimum growth in tilapia occurs at 28°C, even though the species can tolerate a range of 15-35°C. Apparently, environmental parameters are no impediment to polyculture of Nile tilapia and P. acanthophorus. Higher temperatures (23-33°C) have also been reported to increase growth in a polyculture of C. quadricarinatus and tilapia (Rouse & Kahn, 1998). In addition, the temperature tolerance exhibited by P. acanthophorus coincides with overall temperature tolerance (10-38 °C) among Procambarus genus crayfish (Holdich, 2002). Of course, individual species have specific optimum temperature ranges for growth; for instance, P. clarkii prefer temperatures from 22-30°C (Holdich, 2002) with optimal levels around to 20-25°C (Huner & Gaude, 2001). This tolerance for a wide range of environmental temperatures highlights the potential for crayfish cultivation in commercial systems.

Water pH levels were adequate for proper growth in both tilapia and crayfish, although both can tolerate pH from 3.5 to 12, another advantage for polyculture of these species (Huner, 1994; El-Sayed & Abdel-Fattah, 2006). Ammonium, N-nitrite and N-nitrate (mgL[-1]) values were below sublethal and lethal levels for the two cultured species (Huner, 1994;

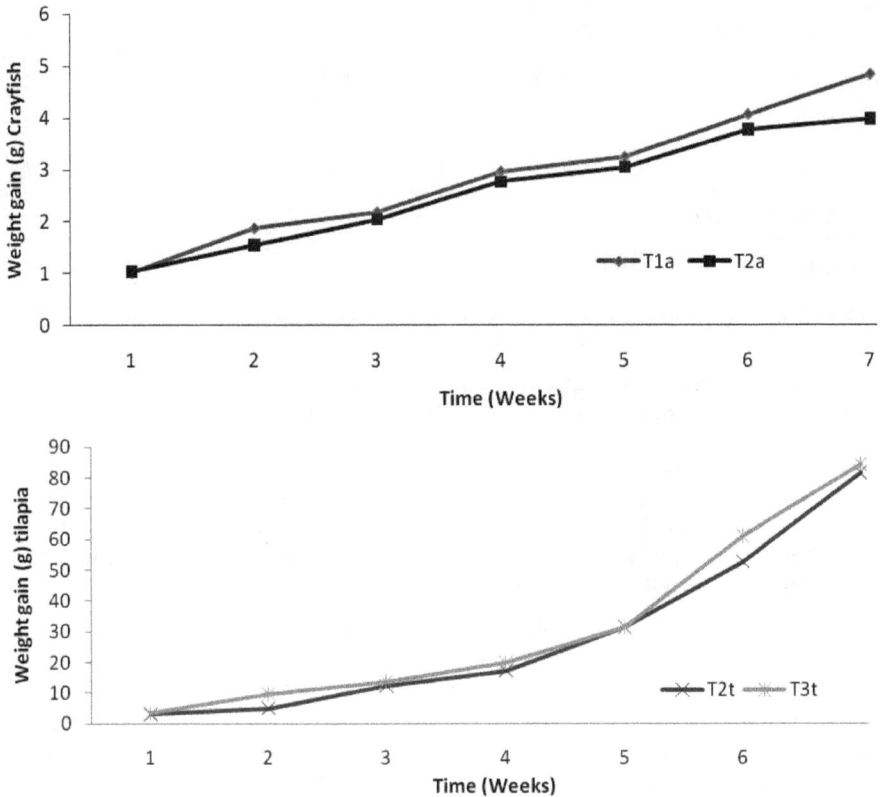

T1: Crayfish, T2: Crayfish/Tilapia, T3: Tilapia

Fig. 19. Weight gain in crayfish and tilapia monocultures and crayfish/tilapia polyculture in a water recirculation system.

El-Sayed & Abdel-Fattah, 2006; Cervantes, 2008), indicating proper biological filter functioning.

This is the first report of a crayfish *P. acanthophorus* / Nile tilapia *O. niloticus* polyculture. The feasibility of this species pair in polyculture as a productive alternative when a recirculation system to increase water efficiency use was evaluated, because both tilapia and crayfish are readily available, and the *Procambarus* genus is diverse around the world, similar than tilapia. The polyculture system was successful in that both species developed properly during the experimental period, although survival among the *P. acanthophorus* (37.4%) was significantly lower than in the crayfish monoculture (72%), and crayfish IWG and FW were lower in the polyculture than in the monoculture. Presence of the tilapia in the same tanks was apparently the main potential cause of this overall lower growth performance in the polyculture since all other variables were within proper ranges for crayfish. There may have been interspecies competition for space and/or feed, or predation by the tilapia of the crayfish during molting partially due to a substantial size difference between species. The latter possibility was not evaluated in the present study since the same initial stocking sizes were used in all treatments: average initial stocking sizes were 1.03±0.77 cm (crayfish) and 3.15±1.04 cm (tilapia), giving a clear growth advantage to the tilapia. The tilapia did not necessarily need to be antagonistic for them to prey on the much smaller crayfish in T2. Different types and quantities of shelters could help to protect the crayfish from the larger tilapia, and manipulating initial stocking size of the different species might help to reduce predation.

Auró et al. (2000) highlighted the feasibility of crayfish/fish polyculture in a study using carp *Cyprinus carpio* and crayfish *Cambarellus montezumae* in artificial ponds. The species coexisted and had enough food resources at densities of up to 50 Org m^{-3} for both species under good water quality conditions. The crayfish *P. acanthophorus* has excellent potential for use in aquaculture systems, because it can be culture to high densities, tolerates handling, adapts to variable environmental conditions and accepts different kinds of artificial diets in captivity, although it does not reach sizes as large as other crustaceans, such as *C. quadricarinatus* (Cervantes, 2008; Cruz-Ordoñez, 2009). Its use in sustainable rural aquaculture production systems is promising under monoculture, and possibly under polyculture conditions after further research into optimum initial stocking size, densities and crayfish shelter type and quantity. Greater production of alternative protein sources using aquaculture in rural areas is an important step towards increasing food availability and diversity.

In addition to the above, the biculture system may be associated to aquaponic production of herbs (cilantro, basil and aquaponic green fodder) with satisfactory results.

1.3.9.3 Genetic improvement

An experiment was designed to estimate genetic variability of growth as heritability (h^2) of the weight at different ages. For which was captured 2135 organisms (4.1g ± 1.79) from its natural medium (G0) of which 10% heavier were selected (i = 1,755) for each sex: 140 females (5.62g ± 1.97) and 48 males (6.02g ± 1.9) to form the progenitors of the line selection (LS), while the control line (LC) consisted of organisms take it at random. These organisms were maintained for reproduction in two rectangular fiberglass tanks of 2.4 m^2 and 0.15 m deep (one tank per line), and a density of 49 orgm^{-2}, and relation females: males (3:1). Biweekly organisms were reviewed to identify gravid females From these organisms 30 full-sib families per line (LC and LS) were obtained (F1), and grown individually for five months, tracking them individually in a recirculation system consisting of 60 rectangular plastic tubs (54 cm x 37 cm x 22 cm) distributed in three levels with mechanical and biological filtration under laboratory conditions and fed 2 times day with balanced feed for shrimp (35% protein). Once the organisms in lines arrived at three months old, and due to differential mortality in the families and in order to reduce the effect of the environment, the density was standardized to 11 organisms per family (55 orgm^{-2}) to continue growth in the same. Heritability for growth in broad sense (h^2) and in each age, for both, the control (LC) and selection line (LS) was estimated from variance components (ANOVA method REML) using a full-sib design from the formulas described by Roff (1997). Growth between the lines in F1 was also compared for each age.

Table 4 shows estimates of heritability from full-sib design for F1. The value of h^2 estimates for LC was 0.48 initially and decreased until the fourth month of age, not to suffer variations due to the rearrangement of the population in the third month compared to LS which began

Age (month)	Heritabilty estimate (h²) ± E.S.	
	LC	LS
1	0.48 ± 0.11	1.10 ± 0.14
2	0.08 ± 0.04	0.61 ± 0.12
3	0.23 ± 0.07	0.27 ± 0.08
4	0.20 ± 0.10	0.58 ± 0.14
5	0.27 ± 0.11	0.34 ± 0.12

Table 4. Estimates of heritability (h^2) ± E.S in F1 at the different ages (months), in selection line (LS) and control line (LC)

with a value of 1.1 reducing the third month (0.27) have to rise again in the fourth month (0.58). Due to the differential survival and in order to reduce the effect of common environment which is reported in previous studies with crustaceans (Benzie et al., 1997; Hetzel et al., 2000; Pérez-Rostro & Ibarra, 2003), the density of families in the third month were standardized, leaving 11 organisms per family in F1.

Estimates of heritability (combined males and females) in F1 at the end of the trial was similar for both lines (0.27 ± 0.11 and 0.34 ± 0.12, LC and LS, respectively), coinciding with results obtained by Pérez-Rostro & Ibarra (2003) in white shrimp *Litopenaeus vannamei*, where values of 0.20 were obtained at 17 weeks of culture as well and those of Cameron et al. (2004) in the red claw crayfish *C. quadricarinatus*, where a decrease in the value of heritability of 0.38 for the first year to 0.13 after 4 years.

When growth gain between the lines was compared, it was seen that LS was significantly heavier (9.6%) than the control line (Figure 20).

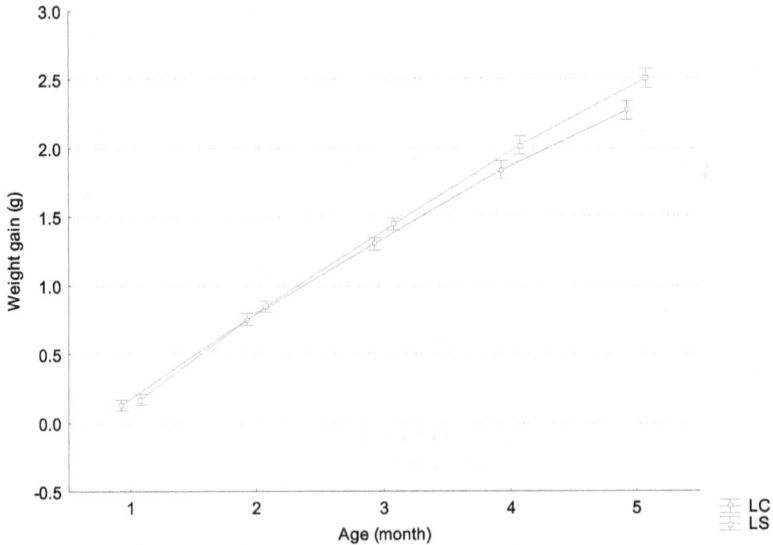

Fig. 20. Weight gain between selection line (LS) and control line (LC) during the five months of culture.

The values of heritability indicate that the species has a positive response to selection, so they can continue to implement a screening program to improve the growth of the species and thus promote commercial cultivation.

2. References

Anderson, D. (1982). Embryology. In: D. Bliss and L. Abele (eds.), Embryology and Genetics. (Academic Press, New York). *The Biology of Crustacea* 1:1-42.

Arrignon, J. (1985). *Cría del cangrejo de Río*. Editorial Acribia. Zaragoza, España. 222 Pp.

Auró, A.A.; Fragoso, C.; Ocampo, C.M.; Sumano, L.H. & Osorio, D.S. (2000). Evaluación del crecimiento de carpas (*Cyprinus carpio* var. *rubrofuscus*) y acociles (*Cambarellus moctezumae*) en bicultivo, alimentados con ensilado de cerdaza empastillado, en un embalse artificial y en tanques de fibra de vidrio.
http://albeitar.portalveterinaria.com. México, 418/634 pp 5.

Auvergne, A. (1982). *El cangrejo de río. Cría y explotación*. Editorial Mundi-Prensa. Spain. 119 pp.

Benzie, J.A.H; Kenway, M. & Trott, L. (1997). Estimates for the heritability of size in juvenile *Penaeus monodon* prawns from half-sib matings. *Aquaculture* 152: 49-53.

Bottrell, H.H. (1975). Generation time, length of life, instar duration and frequency of moulting, and their relationship to temperature in eight species of Cladocera from the River Thames. Reading. *Oecology* (Berlin), 19:129-140.

Brewis, J.M.; Bowler K. (1985). A study of reproductive females of the freshwater crayfish *Austropotamobius pallipes*. *Hydrobiology Journal* 121:145–149.

Brummett, R.E. & Alon, N.C. (1994). Polyculture of nile tilapia (*Oreochromis-niloticus*) and australian red claw crayfish (*Cherax-quadricarinatus*) in earthen ponds. *Aquaculture* 122: 47-54

Cameron, P.M.; Jones, C.M. & Shanks, A.S. (2004). Selection for increased weight at 9 months in redclaw crayfish (*Cherax quadricarinatus*). *Aquaculture* 237: 131-140.

Carmona-Osalde, C.; Rodríguez-Serna, M.; Olvera-Novoa, M.A.; Gutiérrez-Yurrita, P.J. (2004a). Gonadal development, spawning, growth and survival of the crayfish *Procambarus llamasi* at three different water temperatures. *Aquaculture* 232: 305–316

Carmona-Osalde, C.; Rodríguez-Serna, M.; Olvera-Novoa, M.A.; Gutiérrez-Yurrita, P.J. (2004b). Effect of density and sex ratio on gonad development and spawning in the crayfish *Procambarus llamasi*. *Aquaculture* 236: 331–339.

Carmona-Osalde, C.; Rodríguez-Serna, M.; Olvera-Novoa, M.A. (2002). The influence of the absence of light on the onset of first maturity and egg laying in the crayfish *Procambarus (Austrocambarus) llamasi* (Villalobos, 1955). *Aquaculture* 212: 289–298

Celada, J.D.; Antolín, J.I.; Carral, J.M.; Sáez-Royuela, M.; Rodríguez, R. (2005). Successful sex ratio of 1M: 4F in the astacid crayfish *Pacifastacus lenisculus* Dana under captive breeding conditions. *Aquaculture* 244: 89–95.

Cervantes, S.E. (2006). *Efecto de la sustitución de harina de pescado por harina de soya, en dietas para el crecimiento y supervivencia del acocil Procambarus (Astrocambarus) acanthophorus*, Villalobos, 1948, bajo condiciones de laboratorio. Tesis profesional de licenciatura. Instituto Tecnológico de Boca del Río. Boca del Río, Veracruz México. 55 pp.

Cervantes, S.E. (2008). *Relación proteína-lípidos y fuentes alternas de proteína en dietas para el aprovechamiento sustentable del acocil Procambarus (Austrocambarus) acanthophorus*.

Thesis for degree of Master in Sciences in Aquaculture. Instituto Tecnológico de Boca del Río, Veracruz, México. 93 pp.

Cervantes-Santiago, E.; Hernández-Vergara, M.P.; Pérez-Rostro, C.I. (2007). Sustitución de harina de pescado por harina de soya en dietas para crecimiento y supervivencia del acocil *Procambarus (Austrocambarus) acanthophorus*, en condiciones de laboratorio. Memorias en extenso del *Simposium Internacional Aquamar 2007*, y VI Feria Internacional de Acuacultura y Pesca. Boca del Rio, Veracruz, México. Impulso Tecnológico, Edición especial, pp. 107–114. ISBN:1405-0323.

Cervantes-Santiago, E.; Hernández-Vergara, M.P.; Pérez-Rostro, C.I.; Olvera-Novoa, M.A. (2010a). Protein/lipids ratios for efficient growth and survival in the crayfish *Procambarus (austrocambarus) acanthophorus* under controlled conditions. *Aquaculture Research* 41(9): 62-71.

Cervantes-Santiago, E.; Hernández-Vergara, M.P.; Pérez-Rostro, C.I.; Olvera-Novoa, M.A. (2010b). Reproductive performance of the crayfish *Procambarus (austrocambarus) acanthophorus* Villalobos 1948 under controlled conditions. *Aquaculture* 308 (1-2): 66-70.

Cohen D., Z. Ra'anan & Barnes A. (1983). Production of the freshwater prawn *Macrobrachiurn rosenbergii* in Israel. Integration into fish polyculture systems. *Aquaculture* 3(1): 67-76

Cruz-Ordoñez, S.B. (2009). *Cultivo piloto del acocil Procambarus (Austrocambarus) acanthophorus en bicultivo con tilapia Oreochromis niloticus en un sistema de recirculación.* Thesis for degree of Master in Sciences: Aquaculture. Instituto Tecnológico de Boca del Río, Veracruz México. 78 pp.

El-Sayed & Abdel-Fattah, M. (2006). *Tilapia culture.* CABI Publishing. Oceanography Department, Faculty of Science, Alexandria University. Alexandria, Egypt. pp. 139-145.

García-Guerrero, M.U. & Hendrickx, M.E. (2009). External description of the embryonic development of the prawn, *Macrobrachium americanum* Bate, 1868 (decapoda, palaemonidae) based on the staging method. *Crustaceana* 82 (11): 1413-1422.

García-Guerrero, M.U.; Hendrickx, M.E. & Villareal, C.H. (2003). Description of the embryonic development of *Cherax quadricarinatus* Von Martens, 1868 (Decapoda Parastacidae) based on the staging method. *Crustaceana* 76 (3): 269-280.

Gutiérrez-Yurrita, P.J. (1994). *Estudios fisioecológicos sobre algunos aspectos del metabolismo energético de Procambarus bouvieri y Procambarus digueti* (Crustacea: Decapoda: Cambaridae). Master in Science Thesis. Faculty of Sciences, Universidad Nacional Autónoma de México. México. 180 pp.

Herzig, A. (1983). The ecological significance of the relationship between temperature and duration of embryonic development in planktonic freshwater copepods. *Hydrobiología* 100: 65-91.

Hetzel, J.S.D.; Crocos, J.P.; Davis, P.G.; Moore, S.S.; Preston, C.N. (2000). Response to selection and heritability for growth in the Kuruma prawn, *Penaeus japonicus*. *Aquaculture* 181: 215-223.

Hobbs, H.H. Jr. (1984). On the distribution of the crayfish genus *Procambarus* (Decapoda: Cambaridae). *Journal of Crustacean Biology*, 4: 12-24.

Holdich, D.M. & Lowery, R.S. (1988). *Freshwater crayfish, biology, management and exploitation.* Timber Press. 480 pp.

Holdich, D.M. (1993). A review of astaciculture freshwater crayfish farming *Department of Life Science, Univm~iiyoJ Noiiingham, Nottingham NG7 ZRn, U.K. Aquat. Living Kesour* 6:307-317.

Holdich, D.M. (2002). *Biology of freshwater crayfish.* Blackwell Science. EUA pp. 702

Huner, J.V. (1994). *Freshwater Crayfish Aquaculture in North America, Europe, and Australia.* Food Products Press. USA. pp. 311

Huner, J.V. (1995). Ecological observations of red swamp crayfish *Procambarus clarkii* (Girard, 1852) and White crayfish *Procambarus zanangulus*, Hobbs & Hobbs 1990 as regards their cultivation in eathen ponds. *Freshwater Crayfish*, 10: 456-468 p.

Huner, J.V. & Gaudé III., A.P. (2001). Hatcheries for red swamp crawfish. *Crawfish Tales* 8 (1): 23-25.

Karplus, I.; Harpaz, S.; Hulata, G.; Segev, R. & Barki, A. (2001). Culture of the Australian red-claw crayfish (*Cherax quadricarinatus*) in Israel - IV. Crayfish incorporation into intensive tilapia production units. *Journal of Aquaculture-Bamidgeh* 53 (1): 23-33.

Kazmierczak, Jr. & Caffey, R.F. (1995). Management ability and the economics of recirculating aquaculture production systems. *Marine Resources Economies.* 10, 187–209

Köksal, G. (1988). *Astacus leptodactylus* in Europe. In:Holdich, D.M., Lowery, J.S. (Eds.), Freshwater Crayfish: Biology, Management and Exploitation. Croom Helm, London, pp. 365–400.

Lee, D.O'C. & Wickins, J.F. (1992). *Crustacean Farming.* Blackwell Scient, Oxford. 392 pp.

López, M.M. (2006). Diversificación de *Procambarus (Villalobosus)* Hobbs, 1972. (Crustacea: Decapoda: Cambaridae). Tesis doctoral. Posgrado en Ciencias del Mar y Limnologia. UNAM. 225-228 p

Lopéz, M.M. (2008). *Sistemática de los cangrejos de río de México.* In: Álvarez F., Rodríguez-Almaraz G. A. (1ª Ed.) Crustáceos de México. Estado Actual de su Conocimiento. Universidad Autónoma de Nuevo León. PromeP, Mexico:115-165

Malone, R.F.; Burden, D.G. (1988). Design of recirculating soft crawfish shedding systems. Louisiana Sea Grant College Program. 73 pp.

Martínez, C.L.R. (1999). Cultivo de camarones peneidos, principios y prácticas. AGT Editor, S.A., México. 282 pp.

McClain, W.R. & Romaire, R.P. (2007). Procambarid crawfish life history and biology. *Southern Regional Aquaculture Center.* Publication No. 2403. EUA. 6 pp.

Mejía, O.L.M. (2010). Pautas de comportamiento alimentario y adaptaciones progresivas en los apéndices auxiliares en *Agostocaris bozanici* y *Barbouria yanezi* (Crustacea: Decapoda: Caridea: Agostocarididae, Hyppolitidae). *Revista mexicana de biodiversidad.* S193-S201. 81 p

Montemayor, L.J.; Mendoza, A.R.; Aguilera C. & Rodríguez A.G. (2010). Influencia de la alimentación, sobre la reproducción y crecimiento del cangrejo de rio regio *Procambarus regiomontanus*, especie en peligro de extinción. July- September 2010. *Ciencia* UANL 13(3) 276-286.

Mustafa, H.M.; Barım, Ö.; Türkgülü, I.; Harlıoglu, A.G. (2004). Potential fecundity of an introduced population, Keban Dam Lake, Elazıg˘,Turkey, of freshwater crayfish, *Astacus leptodactylus leptodactylus* (Esch, 1852). *Aquaculture* 230: 189–195

Nakata, K. & Goshima, S. (2004). Fecundity of the Japanese crayfish, *Cambaroides japonicus*: ovary formation, egg number and egg size. *Aquaculture* 242: 335–343

Pérez-Rostro, C.I. & Ibarra A.M. (2003). Quantitative genetic parameter estimates for size and growth rate traits in Pacific White shrimp, *Penaeus vannamei* (Boone 1931) when reared indoors. Aquaculture Research 34: 1-11.

Reynolds, J.D. (2002). *Growth and reproduction*. In: Holdich, D.M. (Ed.), Biology of Freshwater Crayfish. Blackwell, USA, pp. 152–184.

Rodríguez-Almaraz, G.A. & Mendoza-Alfaro, R. (1999). *Crustáceos nativos de agua dulce*: Conocimiento y Utilización. 3ª Reunión de Redes de Acuacultura, Cuernavaca, Morelos. Pp. 181-191.

Rodríguez-Serna, M.; Carmona-Osalde, C.; Olvera-Novoa, M.A.; Arredondo-Figueroa, J.L. (2000). Fecundity, egg development and growth of juvenile crayfish *Procambarus (Austrocambarus) llamasi* (Villalobos 1955) under laboratory conditions. *Aquaculture Research* 31: 173–179

Roff, D.A. (1997). Evolutionary quantitative genetics. Editorial Ghapman y Hall. Printed in USA. 41-42 p.

Rojas, P.Y.R. (1998). Revisión taxonómica de ocho especies del genero *Procambarus* (Crustacea: Decapada: Cambaridae) del centro de Veracruz, México. Tesis de licenciatura, Facultad de Ciencias, UNAM. México. 157 PP.

Rouse, D.B. & Kahn, B.M. (1998). Production of Australian Red Claw *Cherax quadricarinatus* in Polyculture with Nile Tilapia *Oreochromis niloticus*. *Journal of the World Aquaculture Society* 29(3): 340-344

Sánchez, M.P.E.; Romero, J.J.; Negrete, R.P. (2009). Aprovechamiento de los ambientes reducidos para la producción de organismos acuáticos susceptibles a cultivo, para el consumo humano. *Vet. Mex.*, 40 (1): 55-67 p.

Sandeman, R. & Sandeman, D. (1991). Stages in development of the embryo of the freshwater crayfish *Cherax destructor*. *Roux Archives of Developmental Biology* 200: 27-37.

Taugbol, T.; Skurdal, J. (1990). Effect of density on brood size in noble crayfish *Astacus astacus* L., subjected to indoor rearing conditions. *Aquaculture Research* 21: 17–23

Villalobos, F.A. (1948). Estudios de los cambaridos mexicanos VII. Descripción de una Nueva Especie del Género *Procambarus*. *Procambarus acanthophorus* n. sp. Anales del Instituto de Biología de la Universidad Nacional Autónoma de México 19: 175-182.

Villalobos-Hiriart, J.L.; Cantú Díaz-Barriga, A. y Lira-Fernández, E. (1993). Los crustáceos de agua dulce de México. *Revista de la Sociedad Mexicana de Historia Natural* 44: 267-290

Yeh, H.S. & Rouse, D.B. (1995). Effects of water temperature, density, and sex ratio on the spawning rate of red claw crayfish *Cherax quadricarinatus* (von Martens). *Journal of the World Aquaculture Society* 26: 160–164.

Yousef, S.; Al-Hafedh, A.A. & Alfaque, M.A. (2003). Performance of plastic biofilter media with different configuration in a water recirculation system for the culture of Nile tilapia (*Oreochromis niloticus*). *Aquacultural Engineering* 29: 139–154.

13

An Updating of Withebait Farming (*Galaxias maculatus*) in Chile

Alfonso Mardones[1] and Patricio De los Ríos-Escalante[2]
[1]*Escuela de Acuicultura, Facultad de Recursos Naturales,*
Universidad Católica de Temuco, Casilla 15-D, Temuco
[2]*Escuela de Ciencias Ambientales, Facultad de Recursos Naturales,*
Universidad Católica de Temuco, Casilla 15-D, Temuco
Chile

1. Introduction

The withebait *Galaxias maculatus* is an endemic species from Argentina and Chile (Cussac et al., 2004), and it is important in ecological webs in Argentinean and Chilean inland waters (Soto & Zúñiga, 1991; Soto et al., 1994; Modenutti et al., 1998). This species has a maximum size of 17 cm and its migratory juvenile stage is transparent or "crystalline" (Mardones et al., 2008). In Chile the juvenile stage is known as "puye" and the pigmented adult stage as "ao". In Argentina in the larval stage is known as "puyen", whereas in New Zealand is known as "withebait", whereas the adults are known as "inanga" (McDowall, 1971).

This species inhabits rivers and littoral zones of lakes in Chile between Huasco river to Tierra del Fuego Island (28-53° S), G. *maculatus* predates mainly microinvertebrates, its reproductive period is in southern spring and summer, and it is endangered due predation by exotic species (salmonids) and pollution of their habitats (Vila et al., 2006). Also this species has exposed to fisheries activities that in consequence generated a capture decreasing during the last decades, that would generate a collapse of fisheries, one tone of crystalline puye juveniles corresponds to approximately 3 million fish of 0.3 g individual body weight (Mardones et al., 2008). In this scenario there is legal management from Fisheries Subsecretary (SubPesca-Chile) and there are studies with the aim of G. *maculatus* farming that is developed by the Universidad Católica de Temuco.

2. Technical considerations for the establishment of whitebait cultivation

Galaxias maculatus develops part of its life cycles in freshwater, specifically in rivers and littoral zones of lakes as well as in estuarine environments (Vila et al., 2006). The first stages of its life cycle are carried out in freshwater and it comprises (a) conditioning of crystalline juveniles to obtain marine reproducers; (b) spawning; and (c) fertilization. The second stage is realized under brackish water conditions and comprises (a) incubation and (b) larval culture (Mitchell, 1989; Vega et al., 1996; Valdebenito & Vega, 2003; Mardones et al., 2008; Hicks et al., 2010).

The most important elements and details to consider for G. *maculatus* culture are: (a) selection of the species, (b) selection of the location and the water source for cultivation, (c) appropriate cultivation systems and (d) selection and training of personnel (Mardones et al., 2008).

For reproduction it is necessary a broodstock proportions of one male: three females, and the average production of one female is approximately 600 fertile eggs (Valdebenito & Vega, 2003), and it is possible close the reproductive cycle in a period of 390 days (Tables 1 and 2; Mardones et al., 2008), and it is possible reach crystalline juveniles with commercial weight and size in 180 days after fertilization (Mitchell, 1989; Mardones et al., 2008).

ID	Stages	Day
1	Fertilization	1
2	Incubation & hatch	30
3	First feeding	40
4	Crystalline larvae	140
5	Maturation	180
6	Total duration	390

Table 1. Time of duration of the stages of the life cycle of *Galaxias maculatus* (Cf: Mardones et al., 2008)

Operation	% Survival	% Mortality	Number of inviduals
Number of males			103237
Number of females			309710
Fertilization*			185825893
Incubation**	50	50	148660714
Hatching	80	20	74330357
First nourishment	80	20	59464286
Postlarva	70	30	47571429
Crystalline larvae***			33300000

* Fertilization: total number of eggs obtained from the spawning, 80 % being adequate for incubation.
** Incubation: number of incubated eggs.
*** One kilogram of product = 3330 crystalline larvae approximately.

Table 2. Survival performance required to culture 10 tonnes of *Galaxias maculatus* (Cf: Mardones et al., 2008)

2.1 The selection of sites

This procedure is based in two aspects: the first is the determination the availability of optimal water quality and the topography of the terrain. Desirable locations have the following features: (a) close to sea level, (b) availability of water with different salinities including brackish water from 5-11 ppt and (c) coastal gravel or sand material, which will allow the installation of stakes in the sea or the sinking of deep sub-superficial wells for marine water (30 m) and deep wells for freshwater (100 m)(Mardones et al., 2008).

2.2 Overview of the production process

The production process begins with the getting of wild crystalline juveniles at the time of their migratory returns in September-October, that corresponding to southern spring (Allibone & Townsend, 1997; Valdebenito & Vega, 2003; Boy et al., 2006). They are transported to the culture facilities, where they are held up to six months to reach sexual maturity and as first step, the live specimens are placed in quarantine and given a prophylactic treatments to eliminate bacterial and parasite diseases brought from wild environment. Once mature broodstock are stripped by abdominal pressure and the eggs are fertilized and incubated over a period of 25-30 days (Valdebenito & Vega, 2003).

Once incubation occurs, the small larvae remain in the first feeding stage for 38-40 days, during which time they are fed ad libitum. Subsequently the larvae continue to a fattening stage where they remain for 140 days until they reach the weight and size of the commercial crystalline juveniles to be harvested and processed. A fraction of animals of this group is retained to be conditioned as reproducers for the next cycle. The total duration of the first production cycle late approximately 480 days, it begins from the purchase of the natural fry through conditioning as breeders until harvest and processing of 10 tonnes of crystalline juveniles (Valdebenito & Vega, 2003).

2.3 Purchase and transport of wild juveniles

Captures of wild juveniles are not constant as they are very dependent on the climate and other factors, fluctuating between 0.5 to 20 tonnes/season, this period requires approximately 90 days, and once, approximately 8 kg of juveniles are captured daily and the juveniles are placed in two tanks of 8 m³ each one, until sufficient amounts are reached to realize the transport to the culture centre, where they will be conditioned (Valdebenito & Vega, 2003). The purchase price at this stage is approximately USD 20/kg (Mardones et al., 2008). The carrying density is 12 kg*m⁻³, and it is used tanks capacity of 3 m³ the mortality ranges from 0 to 5 % during transportation but is less when adults are captured (Valdebenito & Vega, 2003). The juveniles easily adapt to the captivity, and the total mortality from the arrival of the fish at the culture centre until pre-spawning is 60 %. The total capture of wild juveniles required to ensure sufficient reproducers in the first year considers: (a) each juvenile weighs approximately 0.3 g; (b) mortality due to transport is 0 %; (c) mortality up to pre-spawning is 60 %; (d) 30 % of the animals will not reproduce; and (e) the sex ratio at purchase is 1:1 as it is only possible to determine the sex once the reproducers are mature (Mardones et al., 2008). The table 3 shows the number and total biomass of juveniles that need to be purchased.

2.4 Broodstock conditioning

Once they arrived the juveniles are held in twelve 2 m³ tanks for approximately 270 days until they become in optimal reproductive conditions (Valdebenito & Vega, 2003). In according to literature (Valdebenito & Vega, 2003; Mardones et al., 2008) the water quality parameters at this stage are 10-15° C, 0 ppm salinity, a density of 7 kg/m³ and a daily exchange rate of 2. There are two main periods: (1) quarantine period and (2) period of sexual maturity.

Items	Ammount
Number of reproducers required for spawning	412946
Number of exemplary in a reproductive state	619420
Virginal reproducer (30 %)	884885
Mortality until pre-spawning (60 %)	2212213
Mortality (0 %)	2212213
Number of wild alevins to buy	2212213
Kilograms alevins	664

Table 3. Number of crystalline juvenile required. Cf: Mardones et al., (2008)

A single quarantine it is necessary when wild fish enter the culture center. This process consists of periodic application of prophylactic treatments from the day following entry to the hatchery until 14 days in order to eliminate potential ectoparasites and external bacteria. The prophylactic treatments are calculated based on saline and oxytetracycline baths and oxytetracycline oral administration (Table 4; Valdebenito & Vega, 2003; Mardones et al., 2008).

Treatment	Concentration	Application
Saline bath	6 g L⁻¹	Two times per week until 40 day, bath of 30 min.
Oxytetracycline – bath	35 mg L⁻¹	Day 2, bath of 30 min.
Oxytetracycline – oral	75 mg kg⁻¹	Day 3-14 incorporated into feed.

Table 4. Types of prophylactic treatments applied and their specifications. Cf: Mardones et al., (2008)

The juveniles are fed with beef liver in the first week; 75 % liver and 25 % oil-free salmonid starter diet in the second week; 50 % liver and 50 % starter in the third week; 25 % liver and 75 % starter in the fourth week, and 100 % starter in the fifth week. Once quarantine period finished, the juveniles are conditioned for 140 days to become reproducers and during this period they are fed a pelletized freshwater salmon diet that provides good growth and disease resistance (Table 5). The diet ration is at 2 % body weight daily in 10 daily rations by automatic feeders, depending on the biomass and the growth of the fish (Table 6; Valdebenito & Vega, 2003; Mardones et al., 2008).

The broodstock are reared in 8 m³ fibreglass tanks 3 m in diameter. They remain in these tanks during vitellogenesis and they are transferred avoiding excess manipulation, only during the spawning period for separation according to their maturity level (Valdebenito & Vega, 2003). The broodstock are spawned three times during their lifetime and so a program of annual replacement and genetic management is required in order to maintain the reproductive cycle. The broodstock are replaced from juvenile produced, as well as from those fish that are not sexually mature from the previous year (Mardones et al., 2008).

Sex can be differentiated by colour and abdomen size. The state of maturity, once reached, is classified by visual check of the external morphology of the specimens and abdominal pressure on a monthy, biweekly basis according to the descriptions of (Valdebenito & Vega, 2003). This maturity stage is determined weekly for the stages mature 2 and hatched 2 with

Diet	Total length (mm)	% Proteins	% Lipids	% Carbohydrates	Crude energy MJ kg^{-1}
Starter 00	0.3 – 0.5	57.9	30.2	11.9	22.5
Starter	0.5 – 1.0	55.6	33.4	11.0	22.9
Alevins crumble 1	1.0 – 1.5	53.5	36.4	10.1	23.4
Alevins crumble 2	1.5 – 2.0	51.4	39.2	9.4	23.8

Table 5. Ongrowing diet sequence. Cf: Mardones et al., (2008)

Stage (day)	Unitary weight	Number of individuals	Biomass (kg)	Feed day (kg)	Feed total (kg)
Crystalline 1 – 60 days	0.3	2212213	664	13	796
Young 61 – 120 days	1	2212213	2212	44	2655
Pre-spawning reproduction 121-180 days	3	884885	2655	53	3186

Table 6. Characteristics of the fish and feed supplement quantities. Cf: Mardones et al., (2008)

the purpose of reducing the stress level of reproducers. During the post-spawning period for females and males are transferred to unoccupied tanks and a series of additional tanks are required for the juveniles obtained to replenish the broodstock (Mardones et al., 2008).

3. Spawning and fertilization

The maturity of the broodstock is not instant or simultaneous and the spawning period is extended over whole year, with a peak between May and September (Alliborne & Townsend, 1997; Valdebenito & Vega, 2003). The main period covers approximately 90 days, with a peak interval of 30 days (Valdebenito & Vega, 2003; Dantagnan et al., 2007). The spawning of captive individuals can be synchronized through management of the photo-thermal period manipulation, variations of the temperature or hormonal management. It must be remembered that, of the total, of the total reproducers surviving until the end of the conditioning stage 30% of these will be virginal and will be used in the next spawning.

After sex determination, males and females are separated into different tanks. Once they reach maturity stage 3, they are selected for spawning using abdominal massage (Mardones et al., 2008). The spawners are anaesthetized using MS222 (Boy et al., 2006) at a concentration of 0.3 mL*L^{-1}, and through slight abdominal pressure the gametes of females and males are released. Approximately 100 females can be spawned per hour. The mature reproducers have an average weight of 3 g, and once spawned they lose 29% of their initial weight (Valdebenito & Vega, 2003).

fertilization is achieved by the first extracting the semen from the males into a petri dish, and the obtained eggs are deposited in mono-layers in the incubation trays, which consist of a net of 0.2 mm mesh (Valdebenito & Vega, 2003). The eggs are fertilized using sperm diluted with water and left to rest for between 15 and 30 min, during which time they adhere to the base of the incubation trays, one male is used for every three females (Valdebenito & Vega, 2003; Dantagnan et al., 2007; Mardones et al., 2008). After this time, the eggs are washed repeatedly with clean water to eliminate the excess semen and contaminating particles. It is known that each female produces an average of 600 eggs and that approximately 80% of the eggs are fertilized, and for incubation those lots with more than 60 % embryos are used, the remainder being eliminated (Valdebenito & Vega, 2003; ., 2008).

4. Incubation and hatching

This stage extends from 45-120 days from the first incubated eggs to the last hatching, once the hydration process has occurred and the eggs are fertilized, the tare placed in an incubation room for approximately 25-30 days until the hatching of the larvae, with has a controlled environment and high humidity (Barile, 2003; Mardones et al, 2008). The trays are placed in modules of 10 U in hanging structure, and during this period (Barile, 2003; Mardones et al., 2008), they are protected from light specially sunlight because the natural ultraviolet radiation damages the DNA in the eggs (Battini et al., 2000). The optimal parameters for incubation have been found to be 10-12°C, 90% humidity and 10-16 g*L⁻¹ salinity. In addition to providing adequate environment, the eggs require to be sprayed with water at salinity of 10-16 g*L⁻¹ (Barile, 2003; Mardones et al., 2008). The number of daily sprays and the number of incubation trays to be used are mainly dependent on the number of females that are spawned daily and the duration of this stage, the optimal number of pulses is three, if 3200 females are spawned every day (Barile, 2003; Mardones et al., 2008). Nevertheless, the females do not mature in a synchronized form and so the total incubation period may extend for up 90 days from the first hatching of larvae, and typically, 96000 females can be spawned over 30 days (Barile, 2003; Valdebenito & Vega, 2003; Dantagnan et al., 2007;Mardones et al., 2008).

For the production of 10 tonnes of juveniles, a total of 185826000 eggs are required, using 3200 incubation trays, where they will remain until hatching, the hatching under natural conditions is a completely asynchronous process, covering a long period until are released (Dantagnan et al., 2007; Mardones et al., 2008). In captivity conditions, the hatching can be synchronized through temperature management (Barile, 2003; Mardones et al., 2008). Once the eggs are about to hatch, they are transferred to the hatching systems, which consist of small tanks with a capacity of 10 trays each, finally these are closed systems with salinity from 10 to 16 g*L⁻¹ and controlled temperatures varying between 12 - 15°C (Mardones et al., 2008),

5. Larviculture

In according to literature (Borquez et al., 2003; Dantagnan et al., 2007) larviculture extends for 270 days, from the time the first larvae are obtained until the harvest of the last juveniles, this stage has two significant period: (a) first feeding and (b) culture of postlarvae. One day after hatching, the small larvae are transported to fattening tanks, where they will remain for 180 days, the first 38-40 days corresponds to the stage of the first feeding and the remaining days to the culture of postlarvae, after which the crystalline juvenile are ready (Bórquez et al., 2003; Dantagnan et al., 2007; Mardones et al., 2008).

The larvae absorb the yolk sack between 5 and 8 dph (days post-hatching), if the larvae do not receive food immediately after this, high mortality takes place due starvation, the highest mortality rate of about 20% being observed between 12 and 13 dph (Bórquez et al., 2003; Dantagnan et al., 2007; Mardones et al., 2008). The optimal conditions for larval development are $13 \pm 1°C$, 16 g*L^{-1} salinity, a density of 90 larvae * L^{-1} and a rate of exchange of 1 * day^{-1} (Barile et al., 2003). Larval densities are reduced from 90 larvae * L^{-1} at 1-20 dph to 20 larvae * L^{-1} at 20-40 dph. In according to literature (Bórquez et al., 2003; Dantagnan et al., 2007; Mardones et al., 2008) it is possible found high survival due low essential fatty acids (EPA and DHA) at salinities between 10-15 ppt, whereas an inverse situation was observed with freshwater (Dantagnan, 2003). At hatching, the larvae are able to accept small particles and they are fed daily, by hand, to satiation on rotifers enriched with fatty acids and *Artemia* nauplii. The feed sequence and quantities are specified in table 7 (Mardones et al., 2008).

Period (day)	Nourishment feed	Daily in rotifers (kg; 1*10^6 rotifers = 0.15 g)	Daily ration (kg; 3 * 10^9 nauplius = 0.6 g)	Total amount (kg)
0 – 20	40 rotifers * ml^{-1}	4	0	79286
21 – 30	(40 rotifers + 4 nauplius) * ml^{-1}	18	0.002	178417
31 – 40	(30 rotifers + 10 nauplius) * ml^{-1}	13	0.006	133854
Total		35	0.008	391556

Table 7. Diet for the stage of first feeding. Cf: Mardones et al., (2008)

6. Postlarvae culture

After first feeding, the larvae pass on to the postlarval culture stage, the obtained larvae are fed only with inert diets for a period about 140 days, the optimum conditions in this phase are 13° C, 15-16 g*L^{-1}salinity, a density of 10 kg * m^{-3}, an exchange ratio 0.33 of the tank capacity daily, for the postlarvae culture, a total of 33300000 individuals are used with an equivalent biomass of 10000 kg (Bórquez et al., 2003; Dantagnan et al., 2007; Mardones et al., 2008). It is necessary a density of 10 kg * m^{-3}, 1000 m^3 of culture capacity, that representing 125 tanks of 8 m^3 in use for 300 days, based on the quantitative data presented here, at the end of the larval culture phase 10 tonnes of junveniles will be produced (Mardones et al., 2008).

7. Harvest and commercialization

At the end of the larval culture period, the commercial product will be obtained as eel like that is called, crystalline juvenile with no, or only slight, coloration a body length of 4-6 cm and a body weight of 0.3 g, before harvesting, the fish are starved for 2 days and are placed live in tanks of clean water for depuration (Mardones, 2003; Dantagnan et al., 2007; Mardones et al., 2008). These specimens are slaughtered by asphyxia, washed in a solution of 5 g*kg^{-1} chlorine and citric acid or ascorbic acid at 5 g*kg^{-1}, ready for subsequent transportation to the packaging plants (Mardones, 2003; Mardones et al., 2008). Currently , whitebait is sold fresh as well as frozen in Chile and New Zealand, at a price that fluctuates between 10 and 45 USD*kg^{-1}. It is known that, in Chile the product is sold fresh on the beach at price of 10 USD*kg^{-1} is required (Mardones, 2003; Mardones et al, 2008).

8. Discussion

Puye or whitebait culture has considerable aquaculture potential and the production technology and package outlined here already runs successfully (Dantagnan et al., 2007; Mardones et al., 2008). Nonetheless, there are several ways in which the system could be further optimized, the good quality domesticated broodstock is a key stage that would benefit from further work. Reproducers with a higher fecundity and better survival rates, including greather resistance to endo-parasitic infection, would be beneficial. These domestic broodstock could then more reliably support production of eggs and larvae at a lower cost.

Domestication and improvement of management and culture technologies must focus on the reduction in mortality that takes place up to the spawning stages in the broodstock as well as in the incubation and postlarval stages. For broodstock and juveniles produced entirely in captivity, it may not be necessary to use of intensive prophylaxis, diminishing fish manipulation, stress and mortality, that is a frequent procedure in aquaculture.

Development of technologies to increase the densities during the broodstock conditioning stage and to coordinate the spawning under controlled conditions would be useful. It is also necessary to increase the larviculture densities, improve the enrichment of live food and further automate all culture processs, including massive incubation and larval feeding.

While the results to date are very positive, the cost of production is high, bearing in mind that, at least a price relatively similar to that obtained by fresh product on the beach (USD 10) must be obtained. In the financial analyses conducted to date, the costs of training have not been considered, but for full development but for full development culture industry, the early training of staff is absolutely necessary.

9. References

Allibone, R.M., & Townsend, C.R. (1997) Reproductive biology, species status and taxonomic relationships of four recently discovered galaxiid fishes in a New Zealand river. *Journal of Fish Biology,* 51: 1247-1261.

Barile, J., (2003). Antecedentes generales del puye. In: *Antecedentes para el cultivo del puye Galaxias maculatus (Pisces: Galaxiidae,)* J. Barile (Ed). Universidad Católica de Temuco, 11-24 pp.

Battini, M. Rocco V., Lozada M., Tartarotti B. & Zagarese H.E. (2000). Effects of ultraviolet radiation on the eggs of landlocked *Galaxias maculatus* (Galaxiidae, Pisces) in Northwestern Patagonia. *Freshwater Biology* 44, 547-552.

Bórquez, A., Salgado I., Quevedo J., & Dantagnan, P. (2003). Producción de larvas cristalinas. *Antecedentes para el cultivo del puye Galaxias maculatus (Pisces: Galaxiidae,)* J. Barile (Ed). Universidad Católica de Temuco, 81-128 pp.

Boy, C.C., Morriconi E. & Calvo, J. (2006). Reproduction in puyen, *Galaxias maculatus* (Pisces: Galaxiidae), in the southernmost extreme of distribution. *Journal of Applied Ichthyology* 23, 1-8

Cussac, V., Ortubay S., Iglesias G., Milano D., Lattuca M., Barriga J., Battini M. & Gross, M. (2004). The distribution of South American galaxiid fishes: the role of biological traits and post glacial history. *Journal of Biogeography* 31, 103-122.

Dantagnan, P (2003). Requerimientos de ácidos esenciales en larvas de puye (*Galaxias maculatus*, Jenyns, 1842): efecto de la salinidad. Tesis Doctoral, Universidad de las Palmas de Gran Canaria, Departamento de Biología, 204 pp.

Dantagnan, P., Valdebenito I., Bórquez A., Quintana N., Rodríguez A. & Ortega A. (2007). Estado actual del cultivo larvario del puye (*Galaxias maculatus*). In: *Producción de larvas de peces.* Dantagnan, P., Bórquez A., Valdebenito I. & Hernández A. (Eds). Editorial Universidad Católica de Temuco, Temuco, Chile. 29-42 pp.

Hicks, A., Barbee N.C., Swearer S.E., & Downes B.J. (2010) Estuarine geomorphology and low salinity requirement for fertilization influence spawning site location in the diadromus fish, *Galaxias maculatus*. *Marine and Freshwater Research*, 61, 1252-1258.

Mardones, A., Vega R. & Encina, F. (2008). Cultivation of whitebait (*Galaxias maculatus*) in Chile. *Aquaculture Research* 39, 731-737.

Mardones, A., (2003). Pesquería, procesamiento y mercado. *Antecedentes para el cultivo del puye Galaxias maculatus (Pisces: Galaxiidae,)* J. Barile (Ed). Universidad Católica de Temuco, 128-144 pp.

McDowall, R.M., (1971) The galaxiid fishes of South America. *Zoological Journal of Linnean Society* 50, 33-73.

Mitchell, C.H. (1989). Laboratory culture of *Galaxias maculatus* and potential applications. *New Zealand Journal of Marine and Freshwater Research* 23, 325-336.

Modenutti, B.E., Balseiro E.G., Queimaliños C.P., Añón Suárez D.A., Dieguez M.C. & Albariño R.J., (1998). Structure and dynamics of food webs in Andean lakes. *Lakes, Reservoir Research and Management*, 3, 179-189.

Soto D., Campos H., Steffen W., Parra O. & Zúñiga L., (1994). The Torres del Paine lake district (Chilean Patagonia): a case of potentially N-limited lakes and ponds. *Archiv für Hydrobiologie*, 99, 181-197.

Soto, D. & Zúñiga L.R., (1991). Zooplankton assemblages of Chilean temperate lakes: a comparison with North American counterparts. *Revista Chilena de Historia Natural*, 64, 569-581.

Valdebenito, I., & Vega R. (2003). Reproducción y producción de ovas. *Antecedentes para el cultivo del puye Galaxias maculatus (Pisces: Galaxiidae,)* J. Barile (Ed).. Universidad Católica de Temuco, 25-56 pp.

Vila, I., Pardo R., Dyer B. & Habit, E. 2006. Peces limnicos: diversidad, origen y estado de conservación. In: Vila I, A Veloso, R Schlatter & C Ramírez (Eds). Macrófitas y vertebrados de los sistemas límnicos de Chile. Editorial Universitaria, Santiago de Chile, 73-102 p.

Vega, R., Valdebenito I., Barile J., Dantagnan P., Bórquez A. & Mardones, A. (1996). Estado actual de la investigación y desarrollo de la tecnología para el cultivo comercial del puye *Galaxias maculatus* En: Silva, A. & G. Merino (Eds). Actas IX Congreso Latinoamericano de Acuicultura y 2° Simposio Avances y Perspectivas de la Acuicultura en Chile. Asociación Latinoamericana de Acuicultura, Coquimbo, Chile, 81-84 p.

14

In Vitro Culture of Freshwater Pearl Mussel from Glochidia to Adult

Satit Kovitvadhi[1] and Uthaiwan Kovitvadhi[2,*]
*[1]Department of Agriculture, Faculty of Science
and Technology, Bansomdejchaopraya
Rajabhat University, Bangkok
[2]Department of Zoology, Faculty of Science,
Kasetsart University, Bangkok
Thailand*

1. Introduction

The culture of freshwater pearl mussel is divided into three steps, i.e. glochidia, juveniles and adults. Juvenile has been successfully cultured in the laboratory by attaching glochidia to fish until they could transform into the early juvenile stage (Fukuhara et al., 1990; Panha, 1992; Buddensiek, 1995; Uthaiwan et al., 2003; Hanlon & Neves, 2006). Furthermore, sterilized artificial media could be utilized for the culture of glochidia (to bypass the parasitic stage); the progress of this technique can be followed in a succession of reports by Isom & Hudson (1982, 1984a,b), Keller & Zam (1990), Uthaiwan et al. (2001, 2002), Kovitvadhi et al. (2006, 2007, 2008, 2009), Areekijeree et al. (2006), Lima et al. (2006), Supannapong et al. (2008), Srakaew et al. (2010) and Chumnanpuen et al. (2011). Moreover, some species of freshwater mussel glochidia cultured in artificial media could develop to adulthood as well as inducing gonadal development to sexual maturity and the marsupia could develop, namely *Hyriopsis (Limnoscapha) myersiana* (Kovitvadhi et al., 2006, 2008), *Chamberlainia hainesiana* (Kovitvadhi et al., submitted) and *Hyriopsis (Hyriopsis) bialatus* (Kovitvadhi & Kovitvadhi., in preparation).

Therefore, in this chapter author will explain each step from preparation and culture of glochidia in artificial media as well as all techniques of rearing early juvenile through the adult. In addition, water qualities and food suitable for rearing from early juvenile until adult will be described including morphological development under light microscope and scanning electron microscope.

2. Culture of glochidia in artificial media

Glochidia were cultured in artificial media according to Kovitvadhi, (2000); Uthaiwan et al., (2002) and Kovitvadhi et al., (2006) until they transformed to 0-day-old juveniles. The details in each step of glochidia culture in artificial media were as follows:

---

* Corresponding Author

2.1 Composition of artificial media

The composition of artificial media for culture of glochidia is shown in Table 1. Artificial media were based on those improved by Keller & Zam (1990), which consisted of a modification from the formulae of Isom & Hudson (1982). The main differences concerned the composition of the commercial media M199. While the protein source was exclusively horse serum in Keller & Zam (1990) and common carp, *Cyprinus carprio* fish plasma was used as an alternative support to the medium for cultured glochidia of *H. (L.) myersiana* (Uthaiwan et al., 2002; Kovitvadhi et al., 2006, 2007, 2008, 2009), *Hyriopsis (Hyriopsis) bialatus* (Areekijeree et al., 2006; Supannapong et al., 2008; Srakaew et al., 2010; Chumnanpuen et al., 2011; Kovitvadhi & Kovitvadhi., preparation), *Chamberlainia hainesiana* (Kovitvadhi et al., submitted) and *Anodonta cygnea* (Lima et al., 2006). Glochidia could transform into juvenile in the media containing common carp fish plasma as protein source. These glochidia were completely transformed within 8-11 days with a survival rate up to 93% except 34% of *A. cygnea*. All surviving larvae transformed into the juvenile stage except *A. cygnea* <34%. For these reasons, composition of artificial medium (Table 1) was according to Uthaiwan et al. (2002), based on Isom & Hudson (1982) and Keller & Zam (1990), was proposed for culture of glochidia.

Composition of media	The ratio of artificial medium
M199	2
Common carp fish plasma	1
Antibiotics and antimycotic	0.5

Table 1. Composition of artificial medium for culture of glochidia

2.2 Preparation of glochidia media

2.2.1 M199 preparation

Dissolved one packet of M199 powder (Gibco, No. 6231100-035) in 1 liter volume of sterile distilled water and added 2 g of $NaHCO_3$. Thereafter, M199 was filtered through 0.45 and 0.20 µm filter paper, respectively and kept at 4 °C.

2.2.2 Fish plasma preparation

Common carp was anesthetized with 50 mg/l of quinaldine. Fish blood was collected from the caudal vein, in the tail area, using a syringe needle no. 18 which was coated with sodium heparin at 1000 unit/ml. The blood sample was placed into sterile plastic test tube and centrifuged at 1000 and 3000 rpm for 10 min each. Plasma portion (clear yellow in colour) was separated and placed into the new test tube and centrifuged at 3000 rpm for 10 min. Then, plasma was separated and filtered through 0.45 and 0.20 µm filter paper, respectively and kept at -10 to -20 °C.

2.2.3 Antibiotics and antimycotic preparation

The composition of antibiotics and antimycotic chemicals (Isom & Hudson, 1982) is shown in Table 2.

Compound	Concentration (μg / ml)
Antibiotics	
Carbenicillin	100
Gentamicin sulfate	100
Rifampin	100
Antimycotic	
Amphotericin B	5

Table 2. Combination of antibiotics and antimycotic for culture of glochidia.

2.3 Glochidia media preparation

Medium199 (see Section 2.2.1), fish plasma (see Section 2.2.2) and antibiotics/antimycotic (see Section 2.2.3) were mixed in the ratio 2:1:0.5, respectively (Table 1). The artificial media were divided into sterile plastic test tubes and kept at -10 to -20 °C for stocking culture media.

2.4 Glochidia preparation

Adult freshwater mussel were collected from the natural habitat. They were sexually identified by microscopic observation of sperms and eggs in fluid sucked from the gonads by use of a sterile syringe. Fifteen female and fifteen male adult mussels were cultured together in a cylinder net cage (diameter 50 cm × height 50 cm) in an earthen pond for the production of mature glochidia. They were allowed to feed freely on natural food. After 1-2 weeks, all females were observed marsupial colour by tongs to open the shell slightly, which marsupial colour indicates the development of larval stage. In the immature stage, glochidia was yellow in colour, while partially brown colour was at the beginning of maturity. Only completely brown marsupia of gravid mussel was selected in order to examine the strong and suitable glochidia for culturing in artificial media. Thereafter, the outer shell of gravid mussel with completely brown marsupia was washed with tap water and then sterile tap water. The glochidia were sucked by using a sterilized 1 ml syringe and discharged into depression or well slide with sterilized distilled water. Then, the glochidia were observed under a light microscope (×400). If their shells periodically closed, they were sucked according to abovementioned. Later, they were cleaned to eradicate tissue residues, mucus and glochidia shell fragments by spraying sterilized distilled water onto them. Complete cleaning and stronger glochidia were used for culture. Glochidia from gravid mussels should be cultured in artificial media within 5 h after harvesting.

2.5 Glochidia culture

Approximately 5000–6000 glochidia were transferred into a culture dish (90×15 mm) containing 10 ml of artificial medium (Table 1). The culture dishes were placed in a low-temperature incubator at 25 °C with 5% CO_2. The culture medium was removed and replaced with fresh medium in the middle of cultured period. Finally, 4 ml of sterilized distilled water was added to the culture dish to stimulate the transformation when the mantle (Fig. 1A) was observed before 1 day of transformation from glochidia into juveniles. Juvenile transformation was observed under a light microscope (×400) for the movement of juvenile foot (Fig. 1B) and also juvenile movement as an indicator of the glochidia transformation success into juvenile stage.

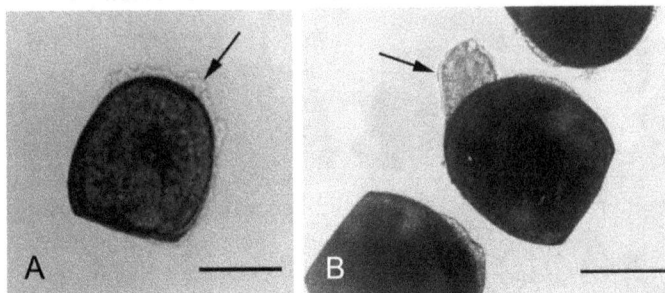

Fig. 1. Light microscope of glochidial development to early juvenile. A; Glochidial development with the mantle edge bordering shell outside (arrow). B; Early juvenile with a foot (arrow). Bar = 100 μm (Uthaiwan et.al., 2001).

3. Selection of phytoplankton food species for juveniles

Phytoplankton has proven to be a vital source of nutrient for several species of freshwater mussel juveniles (Hudson & Isom, 1984; Gatenby et al., 1996; Gatenby et al., 1997; ÓBeirn et al., 1998; Uthaiwan et al., 2001). Moreover, Kovitvadhi et al. (2000) reported that phytoplankton contributed to 99% of the gastrointestinal tract content of the adult freshwater pearl mussel, *H. (L.) myersiana* in natural habitat. Consequently, phytoplankton from the gastrointestinal tract of adult were cultured and selected for juvenile feeding. Collecting mussels of different sizes from natural habitat, which phytoplankton existed in the gastrointestinal tract then it was cutting and sucking to culture in sterilized water with f/2 media (Guillard & Ryther, 1962). Then, they were cultured under light not less than 10,000 Lux for 12 h as well as in mixed about 3% carbon dioxide. From there, phytoplankton to be cultured were in the process of sub-culture and purified every 2-10 days by streak plating technique (Hoshaw & Rosowski, 1973). Streak plates were placed under light until the single colonies of phytoplankton appeared which might last for 10-30 days. Thereafter, those phytoplankton were kept in slant tube. Whenever phytoplankton were required for feeding juveniles, those phytoplankton were multiplied in 1 liter bottle by using the same separated formulae. Kovitvadhi et al. (2006) as abovementioned found ten species of phytoplankton; *Ankistrodesmus gracilis*, *Chlamydomonas* sp., *Chlorella* sp.1, *Chlorella* sp.2, *Kirchneriella incurvata*, *Monoraphidium* sp., *Navicula* sp., *Scenedesmus* sp., *Stichococcus* sp. and *Coccomyxa* sp. Thereafter, all phytoplankton were selected for suitable cultured juvenile which should be considered based on 6 criteria according to Areekijseree et al. (2006); Kovitvadhi et al. (2006); Supannapong et al., (2008): (1) size (2) capability of filter into gastrointestinal tract by observing under microscope within 30 min and 1.30 h after giving phytoplankton (3) the movement of cilia at gill, mantle and foot by observing under microscope (4) color changes in gastrointestinal tract which should be from green to yellow or brown and shape from normal to debris (5) carbohydrate and protein contents of phytoplankton (6) efficiency of digesting carbohydrate and protein of phytoplankton by using crude enzyme extracts from juveniles (*in vitro* digestibility). The major phytoplankton were suitable for culturing juvenile of freshwater mussel, namely *Chlorella* sp. 2 and *K. incurvata* (Table 3).

Phytoplankton	Size[1] (μm)	Filtration time (h)[1]		Cilia activity[1]		Phytoplankton change[1]	Carbohydrate[2] (%)	Protein[2] (%)	Digestibility[2]	
		0.30	1.30	Gill	Foot				Carbohydrate	Protein
Ankistrodesmus gracilis	3×20	-	-	-	-	-	43.50	13.67	122.22	9.93
Chlamydomonas sp.	8×9	+	-	-	-	+	25.67	20.00	158.74	11.50
Chlorella sp.1	3	+	+	-	+	+	45.80	22.50	130.74	11.25
Chlorella sp.2	4	+	+	-	+	+	37.32	25.45	231.77	15.95
Kirchneriella incurvata	4	+	+	+	+	+	21.64	12.27	248.82	16.67
Monoraphidium sp.	12×20	-	+	-	-	+	26.29	17.50	155.08	11.23
Navicula sp.	4×16	+	+	+	+	+	24.2	8.31	-	-
Scenedesmus sp.	16×20	-	+	-	+	-	37.47	16.00	158.73	12.67
Stichococcum sp.	3-4	+	+	-	-	-	82.31	16.33	131.95	11.46
Coccomyxa sp.	3×5	+	+	+	+	+	21.89	25.83	168.47	9.50

Table 3. Characteristic of phytoplankton isolation from gastrointestinal tract of adult freshwater mussel in natural habitat and *in vitro* digestibility for carbohydrate (μg maltose mg plankton[-1]) and protein (μg DL-alanine equivalent mg plankton[-1]) of ten algal species at seven days old, using crude enzyme extracts from 30-day-old juveniles. [1] Kovitvadhi et al. (2006), [2] Supannapong et al. (2008).

4. Juvenile culture

After glochidia transformed to juvenile in artificial medium, water of recirculating aquaculture system was added in the ratio of medium to water equaled 3:1, 1:1, 1:3 for 30 min, added *K. incurvata and Chlorella* sp. in ratio 1:1 at density of 1×10^5 cells per ml in ratio medium to water equaled 1:3. From there, cultured juvenile was transferred to culture units. Culture of juvenile stage could be divided into 3 stages, namely first stage beginning from 0 days old juvenile until two shell mussel completely closed. Kovitvadhi et al. (2008) cultured juvenile 0-120 days old of freshwater mussel, *H. (L.) myersiana* by recirculating aquaculture system (Fig. 2). This system comprised of three filter cabinets made of 6 mm thick acrylic (particulate filter cabinet, macrophytes filter cabinet and biological filter cabinet), one water resting cabinet (Fig. 2E) and nine plastic culture units (Fig. 2F). The size of particulate filter cabinet (Length×Width×Height×Water level = 46×35×51×42 cm) was divided into two equal parts, of which the first part was filled with a 30 cm thick nylon filter (Fig. 2B). Water flowed through this filter and via the second to the macrophytes filter cabinet. The size of macrophytes filter cabinet (Length×Width×Height×Water level = 80×40×51×42 cm) was divided into four equal units. Each unit contained 57 ambulia plants, *Limnophila heterophylla* (Raxb.) Bentham; these, 228 plants in total, were introduced when they were 6 cm in height and had an average weight of 2.69±0.13 g (Fig. 2C). The plants were removed and replaced when their tips reached the water surface. The upper parts of the cabinets were equipped with three fluorescent lamps (each 20 W) 25 cm above the water surface (light intensity at water surface equal to 5320 lux; 24 h) (Fig. 2G). The water then flowed into the biological filter cabinet (60×34×51×42 cm) filled with bioball to full capacity (Fig. 2D) and then to the resting cabinet (46×41×51×42 cm). In the resting cabinet there were two water pumps: The first returned water to the particulate filter cabinet at the rate of 1 l per minute continuously and the second pumped water at 20 ml per minute to nine plastic culture units (each 84×14×15×7 cm). This pump was stopped for 1 h after phytoplankton was introduced into the culture unit. The bottom of the culture unit was filled with sand (<120 μm) at 0.27 g/cm². The inside of the culture unit was divided into two section, as described previously, but of different sized (section 1-66.1×14×15×7 cm; section 2-17.9×14×15×7 cm). The first section also consisted of five acrylic sheets jutting from the walls on alternate sides. Juveniles were fed *Chlorella* sp. and *K. incurvata*. Each species of algae was collected from the 100 l by being pumped through 0.3 μm ceramic filters and then separated from the water by centrifuging at 8000 ×g. The sediments of the two algal species were mixed at a ratio of 1:1 wet weight and kept in a freezer. When required, the mixture was brought to room temperature then sucked by Pasteur pipette into the all plastic culture unit to an algae density of 1×10^5 cells per ml. Algae were supplied twice a day (06.00 h and 18.00 h), and the frozen stock was usually used within 7 days of collection.

Second stage, thirty-five juveniles (120 days old) were transferred to culture units (20×12×72 cm). The culture units had all four vertical sides lined with nylon net (0.42 mm mesh size) and each had a plastic lid with holes to cover the top. The lower part of the culture unit consisted of a section 2 cm in height, which fitted snugly into the culture unit, from which it could be removed (Fig. 3). This lower part contained 400 g of sand (<425 μm in size). The juveniles were placed directly on the sand. The culture unit was then hung in the earthen pond; the base of the culture unit was adjusted to a position approximately 50 cm below the water surface. The juveniles fed by filtering phytoplankton from the water in the earthen pond. All mussels from the culture unit were rinsed every 10 days.

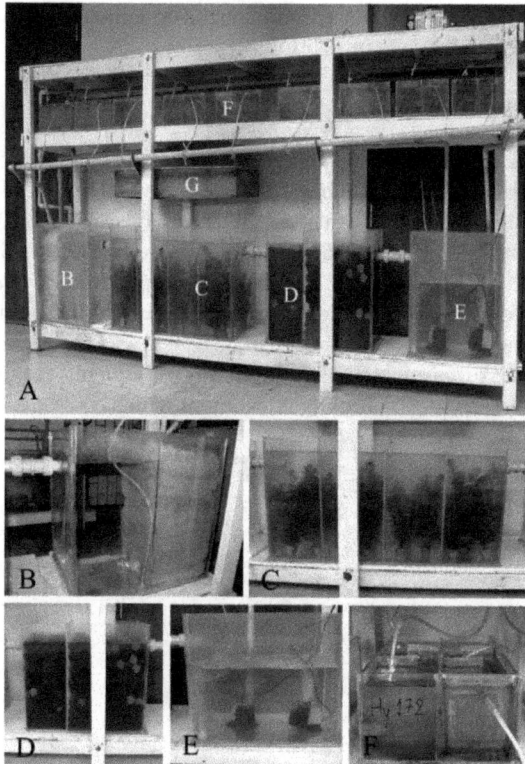

Fig. 2. Photographs of the recirculating used to rear freshwater pearl mussel juveniles. A; Recirculating aquaculture system, B; Particulate filter cabinet, C; Macroplants filter cabinet, D; Biological filter cabinet, E; Water resting cabinet, F; Plastic culture unit, G; Fluorescent box.

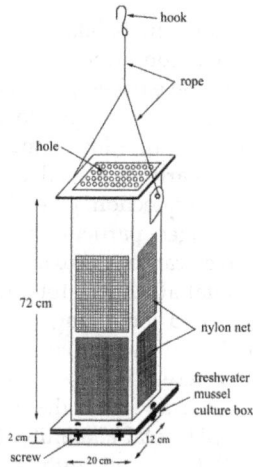

Fig. 3. Rearing container of juveniles (Kovitvadhi et al., 2006).

Third stage, juveniles (180 days old) were transferred to culture into natural habitat or the earthen pond by a cylinder net cage until adult.

5. Adult culture

At present, adult of some species freshwater pearl mussels had been successfully cultured in an earthen pond and natural habitat. They had high survival and could produce glochidia stage such as H. (L.) myersiana (Uthaiwan et al., 2002; Kovitvadhi et al., 2006, 2007, 2008, 2009), H. (H.) bialatus (Areekijeree et al., 2006; Supannapong et al., 2008; Srakaew et al., 2010; Chumnanpuen et al., 2011; Kovitvadhi & Kovitvadhi., preparation), C. hainesiana (Kovitvadhi et al., submitted). Fifteen female and fifteen male adult mussels, were cultured together in a cylinder net cage (diameter 50 cm × height 50 cm). Then, it was hung under the raft (Fig. 4) at 1.5-2 m deep from water surface which phytoplankton were plenty at this level. The cage was shaken every week for protecting biofouling attachment which could mass mortality.

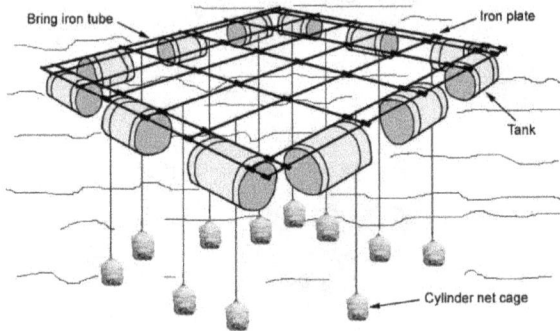

Fig. 4. Raft for adult freshwater pearl mussel culture. (Kovitvadhi, 2008).

6. Water analysis

Prior to culturing mussel, water quality in habitat, was studied in water improvement suitable to growth and survival which conformed to Kovitvadhi et al. (2006, 2008) that juvenile cultured in the laboratory and the earthen pond nearby natural habitat (Kovitvadhi et al., 1998) (Table 4). Water quality parameters should be analyzed for freshwater pearl mussel culture: water temperature, pH, tubidity, conductivity, dissolved oxygen, total alkalinity, free carbon dioxide, total hardness, total ammonia nitrogen, nitrite, nitrate, phosphorus, silica and calcium. In this connection, juvenile stage had been more sensitive to environmental changes than another stages, particularly water qualities suitable and rather stable; water temperature, pH, free carbon dioxide, dissolved oxygen, nitrate and phosphorus and decreasing values; total alkalinity, total hardness, total ammonia nitrogen, silica, and calcium except nitrite that had increasing value (Fig. 5). When averaged water quality value was calculated to relationship with averaged survival value and shell length with equation: $Y = b_0 + b_1 X + b_2 X^2 + b_3 X^3$ where Y is the survival or shell length, X is age (days), and b_0, b_1, b_2 and b_3 are parameters. It was found that survival of 0-120-day-old juveniles would have direct relationship with pH, total alkalinity, total hardness, silica and calcium with highly significant difference (P<0.01) and with reverse relationship to free carbon dioxide and nitrite (Table 5) (Kovitvadhi et al., 2008).

Water quality	Culturing of juvenile in laboratory[1]		Culture of mussel in the earthen pond[1] 120 -360 days old (mean±SD)	Culture of mussel[1] 0 -360 days old (mean±SD)	Mussel Habitat[2] (min. – max.)
	0 - 60 day old (mean±SD)	60 - 120 day old (mean±SD)			
Water temp. (°C)	25±0.74	28±0.54	26.5±1.1	24.5-28.5	23.8-31.6
pH	7.03±0.02	7.51±0.04	7.65±0.46	6.85-8.08	6.92-8.14
Dissolved oxygen (ppmO$_2$)	8.1±0.07	7.5±0.04	5.0±0.6	4.2-8.2	2.5-9.0
Total alkalinity (ppmCaCO$_3$)	52±1.41	52.75±0.35	83.7±20.3	50-114	62.5-115.0
Free carbondioxide (ppmCO$_2$)	10±0.04	4.25±0.35	3.95±2.7	0-10.2	0-6.0
Total hardness (ppmCaCO$_3$)	154±2.83	123±9.90	196.8±12.6	121-222	90-133
Ammonia nitrogen (ppmNH$_3$-N)	0.42±0.02	0.28±0.01	0.44±0.19	0.20-0.82	0.22-0.88
Calcium (ppmCaCO$_3$)	139±9.9	89±4.24	101.1±5.1	86-142	65-105
Phosphorus (ppmP)	0.12±0.06	0.19±0.07	0.17±0.1	0.01-0.45	0.08-0.88
Silica (ppmSiO$_2$)	4.85±0.6	4.05±0.5	5.75±1.6	3-8	0.2-5.5

[1]Kovitvadhi et al. (2006), [2]Kovitvadhi et al. (1998).

Table 4. Water quality during culturing of 0-360 day-old juveniles and the adult mussel habitat of *H. (L.) myersiana* in the Mae Klong River, Kanchanaburi Province.

Parameter	Survival	Shell length	Shell height
Water temperature	-0.093[ns]	0.075[ns]	0.107[ns]
pH	0.716**	-0.597*	-0.590*
Dissolved oxygen	-0.118[ns]	-0.055[ns]	-0.035[ns]
Total alkalinity	0.841**	-0.849**	-0.827**
Free carbon dioxide	-0.634*	0.481[ns]	0.476[ns]
Total hardness	0.769**	-0.764**	-0.751**
Total ammonia nitrogen	-0.152[ns]	-0.051[ns]	-0.061[ns]
Nitrite	-0.716**	0.709**	0.688**
Nitrate	0.203[ns]	-0.200[ns]	-0.218[ns]
Phosphorus	0.003[ns]	-0.085[ns]	-0.091[ns]
Silica	0.914**	-0.913**	-0.091**
Calcium	0.817**	-0.751**	-0.761**

Table 5. Coefficient of correlation between average survival rate and water quality; average growth rate and water quality of juvenile *H. (L.) myersiana* cultured in recirculating aquaculture system every 10 days. (Kovitvadhi et al., 2008) (* = $P<0.05$, ** = $P<0.01$, ns = not significant difference, $P>0.05$).

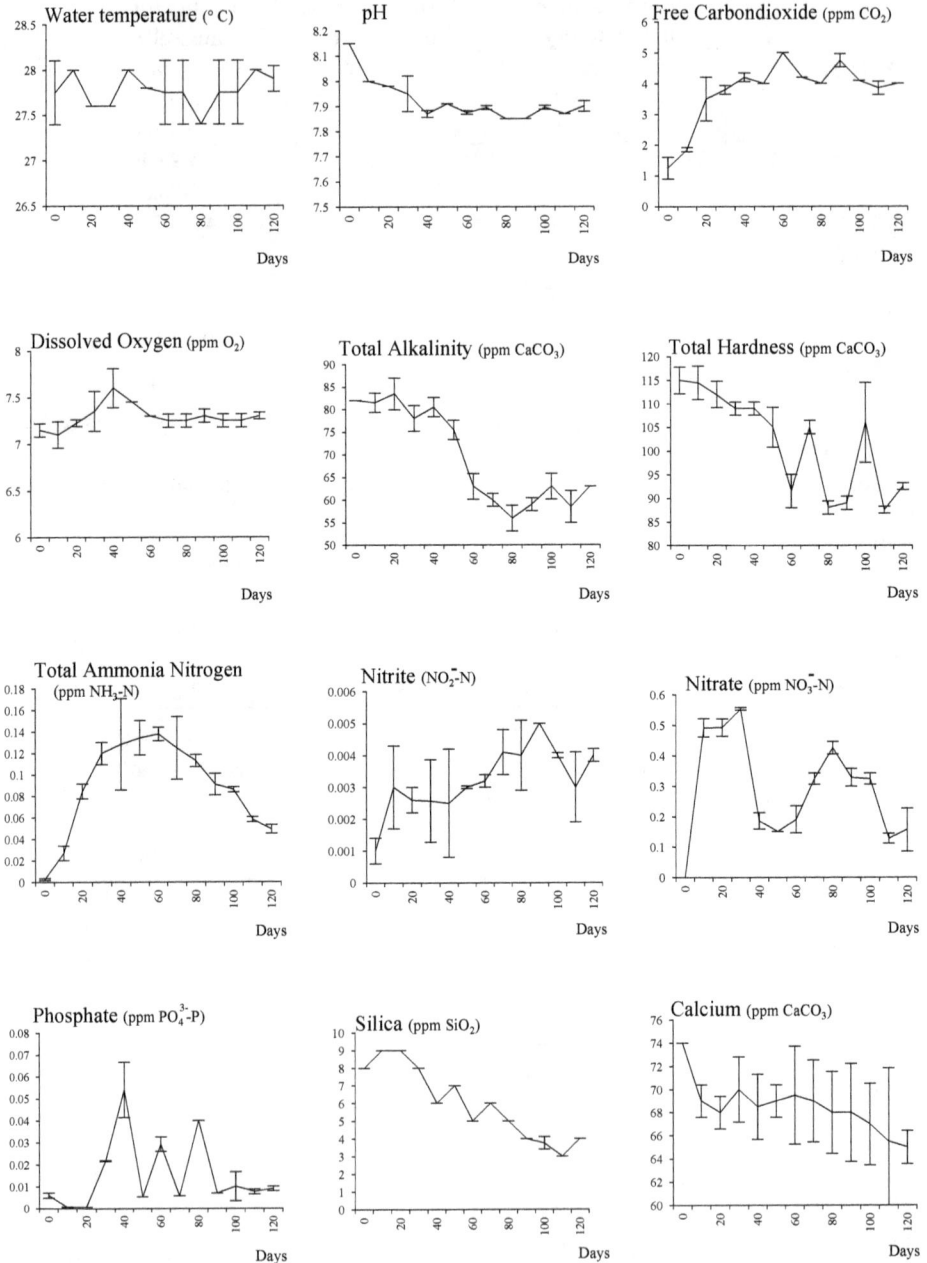

Fig. 5. Water quality during culture for 0-120 days of *Hyriopsis* (*Limnoscapha*) *myersiana* juveniles in recirculating aquaculture system (Kovitvadhi et al., 2008).

7. Phytoplankton communities

Phytoplankton was found in the gastrointestinal tract, which it was a source of nutrient for several species of freshwater mussel (Hudson and Isom, 1984; Gatenby et al., 1996; Gatenby et al., 1997; ÓBeirn et al., 1998; Kovitvadhi et al., 2000, 2001). This finding is consistent with gut content analyses from other bivalve species (Gale & Lowe, 1971; Huca et al., 1983; Paterson, 1986; Parker et al., 1998). From abovementioned data, it is confirmed that freshwater pearl mussels will filter phytoplankton as the main food. Therefore, culture of freshwater pearl mussels from juvenile to adult, it is necessary to have available phytoplankton species used as food for juvenile and in suitable amount throughout culturing period both in the laboratory and natural resource since freshwater mussel has to filter phytoplankton all the time. Thus, density of phytoplankton used for feeding juvenile or adult is not necessary in surplus but the food must be available throughout the culturing period which will result in increasing survival and growth.

8. Morphological development of freshwater mussel

Since the glochidia freshwater pearl mussel of H. (L.) myersiana was cultured in the artificial media that could develop to the adult (Kovitvadhi et al., 2006). Therefore, Kovitvadhi et al. (2007) could study for the morphological development of the juvenile through the adult H. (L.) myersiana. The mussels were collected in sequential developmental stages between 0 and 360 days old. Morphological development was observed by light microscope and SEM. SEM observations were prepared in fixative solution containing 10% neutral buffered formalin for 24 h and stored in 5% neutral buffered formalin for further process. The samples for SEM were thoroughly washed under running water for 30 min and then dehydrated in a graded series of ethanol and dried to critical point. Thereafter, they were mounted on SEM specimen stubs with conductive silver paint and coated with gold and observed with a Jeol Model JSM-5410LV scanning electron microscope operated at 25 KV. All samples before fixation, they were anesthetized in 2% chloral hydrate to observe the internal regions.

The morphological development of H. (L.) myersiana juveniles in culture (0-360 days old) is shown in Fig. 6. The early juvenile of H. (L.) myersiana at 0 days old after transformation has semi-oval, equivalve shells with an equilateral valve, presenting the same size and shape as the glochidium (Fig. 6A). Anterior shell growth was clearly seen in the first day of juvenile development (Fig. 6B), while posterior shell growth followed afterwards (Fig. 6C). The shells of 0-40-day-old juveniles were thin and transparent as seen under light microscope (Fig. 6D). The inner organs (i.e., stomach, intestine, gills, heart, foot, mantle, and cilia at the gills, mantle and foot) were clearly observed through the shell in this period (Fig. 6E). The shell, however, became thicker during the developmental process and covered all the inner organs (Figs. 6E-L). The first anterior and posterior wings appear in 50-day-old juveniles (Fig. 6E), with the posterior wing becoming dominant relative to the anterior from the 140-day-old stage (Fig. 6H). The mantle lobes of 0-50-day-old juveniles are joined dorsally and are free ventrally (Fig. 6F). The incurrent siphon and excurrent siphon appear after 50 days (Figs. 6G-6L). The complete adult morphology is apparent from 160 days old (Fig. 6I). Males and females reproductive organs are sexually mature about 270-360 days old mussel which depend on environment (Kovitvadhi et al., 2006; Srakaew et al., 2010) (Figs. 6K-6L).

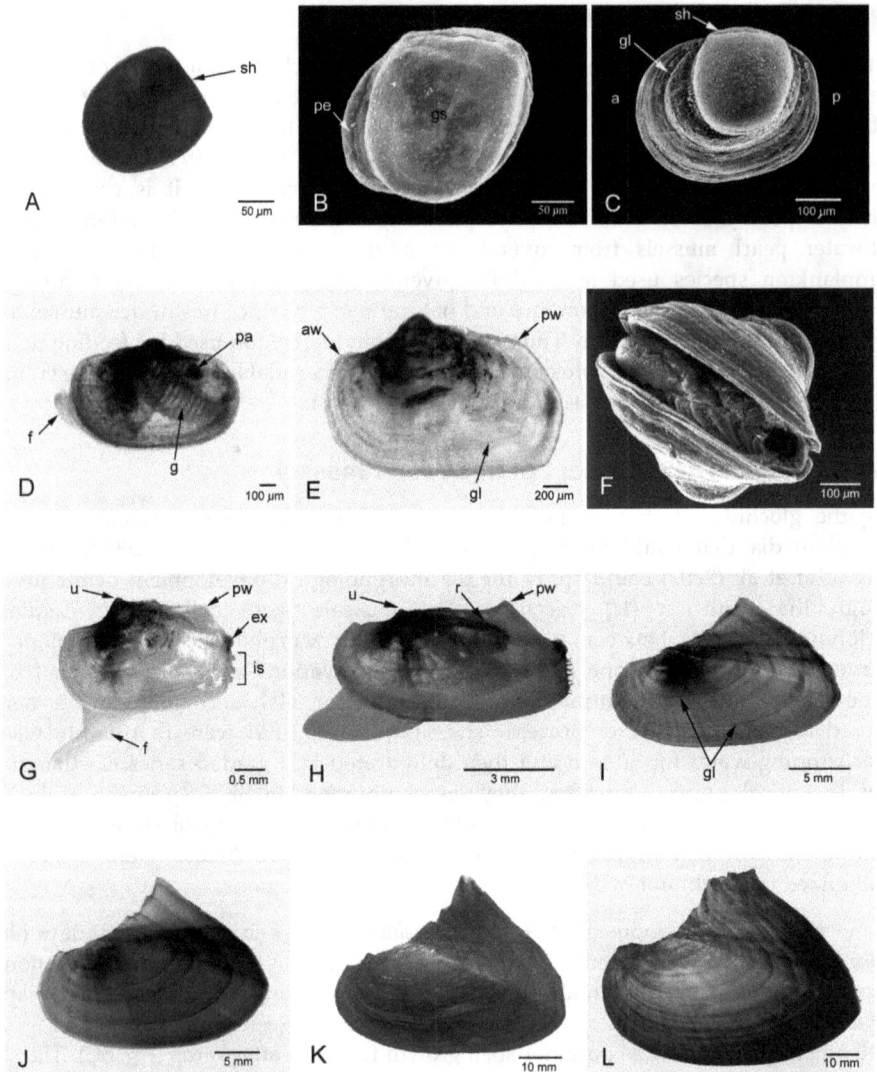

A; Light microscopy of early juvenile (after transformation, 0-day-old) shell, note shell hinge (sh). B; SEM micrograph of 1-day-old juvenile, appearance in anterior region of new soft periostracum (pe), note glochidium shell (gs). C; Juvenile 10 days old, anterior (a) region appears before and grows more than the posterior (p), note growth line (gl). D; Light microscopy of development of shell, 40-day-old juvenile, note foot (f), gill (g), posterior adductor muscle (pa). E; Light microscopy of development of shell, 50-day-old juvenile, note anterior wing (aw); posterior wing (pw). F; SEM of ventral side of 50-day-old juvenile. G; Light microscopy of external morphology of 90-day-old juvenile, note excurrent siphon (es), incurrent siphon (is), umbo (u). H; Light microscopy of external morphology of 140-day-old juvenile, note rectum (r). I-L; External morphology of shell, 160, 180, 270 and 360 days old, respectively.

Fig. 6. Morphological development of 0–360-day juveniles of *Hyriopsis* (*Limnoscapha*) *myersiana*. (Kovitvadhi et al., 2006; 2007).

9. Summary

Culture of freshwater pearl mussel is divided into the three consecutive steps: (1) culture of glochidia larvae in artificial media, (2) rearing juveniles and (3) rearing adult. The results of several studies indicate that glochidia in some species of freshwater mussel could be cultured in artificial media containing mixtures of M199, common carp plasma, antibiotics and antimycotic, and could have fully developed adult and gametogenesis was complete. The important factors in juvenile culture included culturing systems, water quality, substrate (sand) and food. The laboratory-scale recirculating aquaculture system, which water quality change was rather stable and sand could attached materials for food such as organic matter or microorganisms. Furthermore, the juveniles can burrow into the sand as they do in nature, and this helps them to prevent the attachment to the shell of feces and pseudofeces with many protozoa and later flatworms and eventual death of the juveniles. However, the size of sand should appropriate for each size of the juvenile. Phytoplankton was a vital source of nutrients, which has suitable size and shape to move into the mouth of the juveniles so that juveniles can digest them. Prior to transfer to outdoor, mussel organs have to fully developed for ingesting food, particularly gills, the incurrent and excurrent siphon, and their shells must close completely. Moreover, the water quality and food were also important factors to growth and survival.

10. Acknowledgments

We thank the Department of Aquaculture, Faculty of Fisheries, Kasetsart University, for providing a pond for culturing the mussels. We are very grateful to Director of the Kanchanaburi Inland Fisheries Development Center, Department of Fisheries and Mrs. Oodeum Meejui, whose supply of freshwater mussel *Chamberlainia hainesiana* was greatly appreciated.

11. References

Areekijseree, M., Engkagul, A., Kovitvadhi, S., Kovitvadhi, U., Thongpan, A. & Rungruangsak Torrissen, K. (2006). Development of digestive enzymes and *in vitro* digestibility of different species of phytoplankton for culture of early juveniles of the freshwater pearl mussel, *Hyriopsis (Hyriopsis) bialatus* Simpson 1900. *Invertebrate Reproduction and Development* 49, pp. 255-262, ISSN 2157-0272

Buddensiek, V. (1995) The culture of juvenile freshwater pearl mussels *Margaritifera margaritifera* L. in cages: a contribution to conservation programmes and the knowledge of habitat requirements. *Biological Conservation* 74, pp. 33-40, ISSN 0006-3207

Chumnanpuen, P., Kovitvadhi, U., Chatchavalvanich, K., Thongpan, A. & Kovitvadhi, S. (2011) Morphological development of glochidia in artificial media through early juvenile of freshwater pearl mussel, *Hyriopsis (Hyriopsis) bialatus*. *Invertebrate Reproduction and Development* 55, pp. 40-52, ISSN 2157-0272

Fukuhara, S., Nakai, I. & Nagata, Y. (1990) Development of larvae of *Anodonta woodiana* (Bivalvia) parasitized on host fish. *Venus* 49, pp. 54-61, ISSN 0042-3580

Gale, W.F. & Lowe, R.L. (1971) Phytoplankton ingestion by the fingernail clam, *Sphaerium transersum* (Say), in pool 19, Mississippi River. *Ecology* 52, pp. 507-512, ISSN 0012-9658

Gatenby, C.M., Neves, R.J. & Parker, B.C. (1996) Influence of sediment and algal food on cultured juvenile freshwater mussels. *Journal of the North American Benthological Society* 15, pp. 597–609, ISSN 0001-4966

Gatenby, C.M., Parker, B.C. & Neves, R.J. (1997) Growth and survival of juvenile rainbow mussels *Villosa iris* (Lea, 1829) (Bivalvia: Unionidae), reared on algal diets and sediment. *American Malacological Society* 14, pp. 57–66, ISSN 1000-3207

Guillard, R.R.L. & Ryther, J.H. (1962) Studies of marine planktonic diatoms. I *Cyclotella nana* Hustedt and *Detonula confervacea* (Cleve). Gran. *Canadian Journal of Microbiology* 8, pp. 229-239, ISSN 0008-4166

Hanlon, S.D. & Neves, R.J. (2006) Seasonal growth and mortality of juveniles of *Lampsilis fasciola* (Bivalvia: Unionidae) released to a fish hatchery raceway. *American Malacological Bulletin* 21, pp. 45–49, ISSN 0740-2783

Hoshaw, R.W. & Rosowski, J.R. (1973) *Methods for microscopic algae*. In: Stein, J.R. (ed.), Handbook of Phycological Methods, Culture Methods and Growth Measurements. Cambridge University Press, USA. ISBN 0-521-20049-0

Huca, G.A., Brennerand, R.R. & Niveiro, M.H. (1983) A study of the biology of *Diplodon delodontus* (Lamarck, 1819). *Veliger* 25, pp. 51–58, ISSN 0042-3211

Hudson, R.G. & Isom, B.G. (1984) Rearing juveniles of the freshwater mussels (Unionidae) in a laboratory setting. *Nautilus* 98, pp. 129–135, ISSN 0028-1344

Isom, B.G. & Hudson, R.G. (1982) *In vitro* culture of parasitic freshwater mussel glochidia. *The Nautilus* 96, pp. 147-151, ISSN 0028-1344

Isom, B.G. & Hudson, R.G. (1984a) Freshwater mussels and their fish hosts; physiological aspects. *Journal for Parasitology* 70, pp. 318-319, ISSN 0022-3395

Isom, B.G. & Hudson R.G. (1984b) Culture of freshwater mussel glochidia in an artificial habitat utilizing complex liquid growth media. U.S. *Patent* 4449480

Keller, A.E. & Zam, S.G. (1990) Simplification of *in vitro* culture techniques for Freshwater mussels. *Environment Toxicology and Chemistry* 9, pp. 1291-1296, ISSN 0730-7268

Kovitvadhi, S. (2008) *In vitro culture of freshwater mussel juvenile Hyriopsis(Limnoscapha) myersiana (Lea, 1856)*. Ph.D. Thesis. University of Porto, Portugal.

Kovitvadhi S., Kovitvadhi U. & Meejui O. 2011.Growth and survival of the freshwater pearl mussel *Chamberlainia hainesiana* (Lea, 1856) from juvenile to adult reared in different densities and locations. *Aquaculture* (submitted)

Kovitvadhi, S., Kovitvadhi, U., Sawangwong, P. & Machado, J. (2007) Morphological development of the juvenile through to the adult in the freshwater pearl mussel, *Hyriopsis (Limnoscapha) myersiana*, under artificial culture. *Invertebrate Reproduction and Development* 50, pp. 207–218. ISSN 2157-0272

Kovitvadhi, S., Kovitvadhi, U., Sawangwong, P. & Machado, J. (2008) A laboratory-scale recirculating aquaculture system for juveniles of freshwater pearl mussel *Hyriopsis (Limnoscapha) myersiana* (Lea, 1856). *Aquaculture* 49, pp. 255-262, ISSN 0044-8486

Kovitvadhi, S., Kovitvadhi, U., Sawangwong, P., Thongpan, A. & Machado, J. (2006) Optimization of diet and culture environment for larvae and juvenile freshwater pearl mussels, *Hyriopsis* (*Limnoscapha*) *myersiana* (Lea, 1856). *Invertebrate Reproduction and Development* 49, pp. 61–70, ISSN 2157-0272

Kovitvadhi, S., Kovitvadhi, U., Sawangwong, P., Trisaranuwatana, P. & Machado, J. (2009) Morphometric relationship of weight and length of cultured freshwater pearl mussel, *Hyriopsis (Limnoscapha) myersiana* (Lea, 1856) under laboratory conditions and the earthen pond phases. *Aquaculture International* 17, pp. 57-67, ISSN 0967-6120

Kovitvadhi, U. (2000) *Culture of glochidia of freshwater pearl mussel Hyriopsis myersiana (Lea, 1856) in artificial media*. Ph.D. Thesis. University of Porto, Portugal.

Kovitvadhi, U., Chaopaknam, B., Nagachinta, A., Jongrungwit, K. & Kulayanamit S. (1998) Ecology of freshwater pearl mussel, *Hyriopsis(Limnoscapha) myersiana* (Lea, 1856) in the Maeklong River, Kanchanaburi Province. *Kasetsart Journal : Natural Science* 32, pp. 1-12, ISSN 0075-5192

Kovitvadhi, U., Nagachinta, A. & Aungsirirut, K. (2000) Species composition and abundance of plankton in the gut contents of freshwater pearl mussel, *Hyriopsis* (*Limnoscapha*) *myersiana*. *Journal of Medical and Applied Malacology* 10, 203–209, ISSN 1053-6388

Lima, P., Kovitvadhi, U., Kovitvadhi, S. & Machado, J. (2006) *In vitro* culture of glochidia from the freshwater mussel *Anodonta cygnea*. *Invertebrate Biolology* 125, 34-44, ISSN 1077-8306

ÓBeirn, F.X., Neves, R.J. & Steg, M.B. (1998) Survival and growth of juvenile freshwater mussels (Unionidae) in a recirculating aquaculture system. *American Malacological Bulletin* 14, pp. 165–171 ISSN 1000-3207

Panha, S. (1992) Infection experiment of the glochidium of a freshwater pearl mussel *Hyriopsis myersiana* (Lea, 1856). *Venus* 51, pp. 303–314, ISSN 0042-3580

Parker, B.C., Patterson, M.A. & Neves, R.J. (1998) Feeding interactions between native freshwater mussels (Bivalvia: Unionidae) and zebra mussels (*Dreissena polymorpha*) in the Ohio River. *American Malacological Bulletin* 14, pp. 173–179, ISSN 1000-3207

Paterson, C.G. (1986) Particle-size selectivity. I The freshwater bivalve *Elliptio complanata* (Lightfoot). *Veliger*, 29, 235–237, ISSN 0042-3211

Srakaew, N., Chatchavalvanich, K., Kovitvadhi, S., Kovitvadhi, U. & Thongpan, A. (2010) Histological structure of developing gonads in the freshwater pearl mussel, *Hyriopsis* (*Limnoscapha*) *myersiana*. *Invertebrate Reproduction and Development* 54, pp. 203-211, ISSN 2157-0272

Supannapong, P., Pimsalee, T., A-komol, T., Engkagul, A., Kovitvadhi, U., Kovitvadhi, S., & Rungruangsak Torrissen, K. (2008) Digestive enzymes and *in vitro* digestibility of different species of phytoplankton for culture of the freshwater pearl mussel, *Hyriopsis (Hyriopsis) bialatus*. *Aquaculture International* 16, pp. 437-453, ISSN 0967-6120

Uthaiwan, K., Noparatnaraporn, N. & Machado, J. (2001) Culture of glochidia of the freshwater pearl mussel *Hyriopsis myersiana* (Lea, 1856) in artificial media. *Aquaculture* 195, pp. 61-69, ISSN 0044-8486

Uthaiwan, K., Pakkong, P., Noparatnaraporn, N., Vilarinho, L. & Machado, J. (2002) Study of a suitable fish plasma for *in vitro* culture of glochidia *Hyriopsis myersiana* (Lea, 1856). *Aquaculture* 209, pp. 197-208, ISSN0044-8486

Uthaiwan, K., Pakkong, P., Noparatnaraporn, N., Vilarinho, L. & Machado, J. (2003) Studies on the plasma composition of fish hosts of the freshwater mussel, *Hyriopsis myersiana*, with implications for improvement of the medium for culture of glochidia. *Invertebrate Reproduction and Development* 44, pp. 53-61, ISSN 2157-0272

15

Potency of Barnacle in Aquaculture Industry

Daniel A. López[1,2], Boris A. López[1],
Christopher K. Pham[3] and Eduardo J. Isidro[3]
[1]*Department of Aquaculture and Aquatic Resources,*
Universidad de Los Lagos, Osorno
[2]*Advanced Research Center, Universidad de*
Playa Ancha, Valparaíso
[3]*Department of Oceanography and Fisheries,*
Universidade dos Açores, Horta,
[1,2]*Chile*
[3]*Portugal*

1. Introduction

1.1 Biological characteristics

Barnacles belong to the Cirripedia group (*cirri*: cirri, *pedia*: feet) and are, for the most part, sessile crustaceans that live permanently adhered to a substrate that can be inorganic (rock), organic (coral, molluscs, turtles, whales) and even artificial (plastic, wood) (Southward, 1987). They mainly inhabit marine environments; although some species can resist low salinities in estuarine zones (Arenas, 1971), and are distributed bathymetrically from the high intertidal zone to depths of over 1,000 m (Anderson, 1994). They are gregarious, forming dense "patches" of individuals that can completely cover substrates.

The Thoracica superorder is the principal group that assembles the majority of Cirripedia species described. In contrast to other crustaceans, they are characterized by lacking development of the abdomen. This group is divided into three suborders: Lepadomorpha (Fig. 1A), Balanomorpha (Fig. 1B), Verrucomorpha. In general, they present a fragile, chitinous exoskeleton, as a result of which, the majority of species also possess a set of calcareous plates (mural plates and opercular plates) that surround the individual, providing support and defence, as well as, in the case of the opercular valves, associated to feeding, respiration and moulting processes in the specimen. Furthermore, some species have a calcareous base (Southward, 1987).

The thoracic barnacles are filtering organisms with a series of modified articulated appendages, such as a cirral fan, that enables them to capture particles suspended in the water column, feeding, principally, on microscopic algae and the larvae of other invertebrates (Anderson, 1994). These species are simultaneous hermaphrodites, undertaking crossed fecundation or copulation between adjacent individuals by means of elongation of a penis. The ovary is located in the basal part of the organism. After fecundation, the egg mass becomes compact, forming the so called "ovigerous lamellae", within which embryonic and early larval development of specimens occurs. Barnacles incubate eggs within the body cavity up to the larval nauplius I stage, which is released into the water column, initiating a planktonic, free-swimming life that can last two to four weeks. This larval stage, and principally their transport, is influenced by local oceanographic factors, as well as meso and macro-scale processes, such as currents, winds and upwelling (Roughgarden et al., 1988; Gaines and Bertness, 1992; 1993; Pineda, 2000). Larvae grow in size during this stage, increasing the number of appendages and advancing through successive stages (nauplius larvae II to nauplius VI), until converting into the competent larvae denominated *cyprid*. The *cyprid* presents a bivalve form, is lecitotrophic and possesses a sensorial system that enables it to search for an adequate substrate, where texture, colour, quantity of light and biofilm are variables that determine attachment. Immediately after *cyprid* larval settlement, they undergo metamorphosis, adopting the form of an adult specimen, where factors such as depredation, competition and physical disturbances (Gaines & Roughgarden, 1985; Minchinton and Scheibling, 1991, Thomason et al., 1998) play an important role in the early survival of individuals. Under laboratory conditions, it has been described how aspects, such as type and concentration of food, temperature and larval density are key factors influencing duration and survival during larval development (Qiu and Qian, 1999; Mishra et al., 2001; Thiyagarajan et al., 2003a; 2003b). In addition, with respect to induction of larval settlement, it has been observed that a great variety of chemical signals trigger this response, associated, principally, with three types of source: presence of conspecific indivuduals, presence of prey species and microbial films (Rodríguez et al., 1992).

The lepadomorph barnacles ("goose-neck" or "stalked barnacles") are characterized by their fleshy stalk used to adhere to the substrate, as well as calcareous plates, found only in the apical zone of the animal, denominated capitulum. These species mainly inhabit the intertidal zone and are considered neustonic species, adhering to floating structures (Fig.1A). On the contrary, the balanomorph barnacles ("acorn barnacles" or "non-stalked barnacles") do not possess a stalk or peduncle, and adhere directly to the substrate by means of a base that can be membraneous or calcareous. They possess greater development of the calcareous plates, with between 4 and 8 overlapping plates that form a supporting

Fig. 1. General view of the barnacles. A. Lepadomorpha. B. Balanomorpha.

structure. Furthermore, two pairs of plates are located in the upper zone, forming the opercule (terga and scuta) (Fig. 1B). These species are habitual members of the shallow intertidal and subtidal communities of the coastal area. Finally, the verrucomorph barnacles ("wart barnacles") are small organisms, without a peduncle, characterized by their asymmetric opercular plates, and are more frequent in deep-waters, where they appear as epizoos of other invertebrates (Anderson, 1994).

1.2 Commercially important barnacle species worldwide

In spite of the wide variety of barnacle species on a global scale, only a few are commercially important, due to the small size of the majority of these organisms (Table 1). Of the lepadomorph barnacles, *Pollicipes pollicipes* (Leach, 1817), the "goose neck barnacle", is the best known culinary resource. It is extracted along the coasts of Galicia (Spain) by fishermen´s organizations, known as unions ("perceberos"), and can reach a height of 10 to 12 cm. Approximately 300-500 ton/year are captured, and demand is high (Molares &

	Scientific name	Common name	Geographic distribution	Production type
"Goose barnacles"	*Pollicipes pollicipes*	"Spanish goose barnacle"	Spain, Portugal, France, Morroco	Fisheries by fishermen´s unions ("perceberos")
	Pollicipes polymerus	"Canadian goose barnacle" "Leaf barnacle"	Canada, USA	Local fisheries
	Pollicipes elegans	"Pacific goose barnacle"	Mexico, Ecuador, Peru	Local fisheries
	Capitulum mitella	"kamenote" or "Japanese goose barnacle"	South of Japan	Artisanal fisheries
"Acorn barnacles"	*Austromegabalanus psittacus*	"picoroco" "giant barnacle"	Perú, Chile, Argentina	Artisanal fisheries/ semi industrial aquaculture
	Balanus nubilus	"giant acorn barnacle"	Alaska, Canada, USA.	Local fisheries
	Balanus rostratus	"mine fujit subo"	Russia, Northern Japan	Local fisheries.
	Megabalanus rosa	"aka fujit subo" "rose barnacle"	Japan	Local fisheries.
	Tetraclita japonica	"kuro fujit subo" "black barnacle"	Japan	Local fisheries.
	Tetraclita kuroshioensis	"hat fujit-subo"	Japan, Indonesia	Coastal fishermen
	Megabalanus azoricus	"craca"	Azores Islands (Portugal)	Local fisheries/ experimental aquaculture
	Megabalanus tintinnabulum	"claca"	Entire tropical zone of the Atlantic and Indian Oceans. Canary Islands (Spain)	Local fisheries

Table 1. Commercially important barnacle species and species used for human consumption worldwide. Scientific and common names, geographic distribution and production types. Source: Modified from López et al., (2010).

Freire, 2003). Overexploitation of natural banks of this resource has prompted regulation of extraction activity. Other species of the same genus, such as *P. polymerus* (Sowerby, 1833) and *P. elegans* (Lesson, 1830), are commercially significant on a local scale in Canada and Perú, respectively, and, secondarily, are used to supply the Iberian market (Bald et al., 2006; Jacinto et al., 2010). On the other hand, the species *Capitulum mitella* (Linnaeus, 1758) or "stalked barnacle", distributed along the southern coast of Japan, is smaller (< 5 cm height) and is only consumed on the local market.

Of the balanomorph barnacles (Table 1), the most commercially important species is *Austromegabalanus psittacus* (Molina, 1782) "picoroco" or "giant barnacle" a large species that can reach a height of 30 cm. Distribution ranges from southern Perú to the austral zone of southern Chile and the southern coast of Argentina (Pilsbry, 1916; Young, 2000, López et al., 2007a). It is exploited on a small scale by artisanal fisheries and catches are concentrated in southern Chile (42°S), with landings that fluctuate between 200-600 ton/year (Fig. 2). Average commercial size is 2.2 cm carino rostral length. Another giant species is *Balanus nubilus* (Darwin, 1854), distributed along the Pacific coast, from Alaska to North America, and consumed by indigenous coastal communities (Morris et al., 1980). *Megabalanus azoricus* (Pilsbry, 1916) "craca", is a subtidal species, limited to the Azores Islands archipelago (Portugal), in the Atlantic Ocean (Southward, 1998; Regala, 1999; Santos et al., 2005). It is exploited locally, and specimens over 1 cm carino rostral length are consumed fresh (Lotaçor, 2006; Pham et al., 2008). The analysis of the enterprises LOTAÇOR S.A data, from 1980 to 2010 shows that reported landings are modest, ranging from 1 to 3.3 ton/year on the 80´s, stabilized bellow 1ton/year in the 90´s, have raised considerably between 1999 to 2003, when a historic maximum of 7ton/year was auctioned. Since 2003 the landings have been dropping to about 3.7ton/year in 2010. From 2005 to 2010, the average first selling price has been around 3.90€/kg, with a minimum of 0.20€/kg and a maximum of 20€/kg. The real catches and economic value of the resource may be substantially higher than the reported. The species is highly appreciated locally, and traditionally caught along the shores, by the people and for their own consumption. Dionisio et al., (2009), based on LOTAÇOR data, described some socio-economic aspects of this fishery and estimate a *per-capita* consumption between 88 and 241g/year. The *Megabalanus tintinnabulum* (Linnaeus, 1758) "claca" species is extracted in the Madeira archipelago (Portugal), and the Canaries (Spain). It is a cosmopolitan species whose distribution covers the entire tropical zone of the Pacific, Atlantic and Indian oceans (Young, 1998).

In the Japanese market, balanomorph barnacles are referred to as "fujit-subo". In northern Japan, mainly in the Aomori prefecture, the *Balanus rostratus* (Hoek, 1833) "mine fujit-subo" is commercialized, and can reach a height of 5 cm, with landings of around 10 ton/year. Other smaller species of balanomorph barnacles, *Megabalanus rosa* (Pilsbry, 1916) "aka fujit-subo" and *Tetraclita japonica* (Pilsbry, 1916); "kuro fujit-subo", are also extracted locally in the central-southern zone of Japan. Similarly, in Indonesia the *Tetraclita kuroshioensis* Chan, Tsang & Chu, 2007, "hat fujit-subo", is consumed by coastal communities.

2. State of barnacle culture on a world scale

In spite of the commercial importance of various species of barnacles, development of activities associated with cultures is limited to a few species (López et al., 2010). Among the lepadomorphs, previous information on *Pollicipes pollicipes* is available related to obtaining

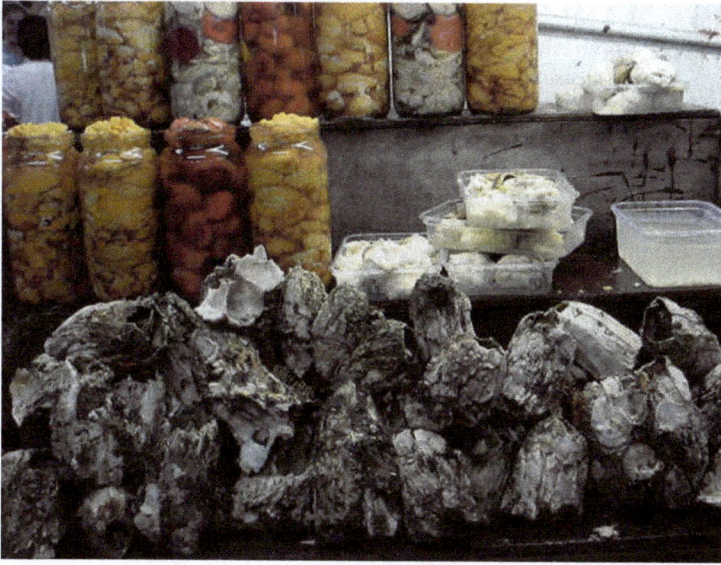

Fig. 2. "Giant barnacle" in a local market in Puerto Montt city, southern Chile.

juveniles in artificial floating substrates (Goldberg, 1984). Results have been scarce, mainly because, in this species, larval settlement is determined by the presence of conspecific adults (Molares et al., 1994; Cruz et al., 2010). Among balanomorph barnacles, similar experiments have been undertaken on *Megabalanus azoricus* in the Azores islands, where technologies have been designed for installing collectors in the water column. The results indicate that larval settlement occurs, although it is still not possible to ensure that levels are sufficient for industrial scale cultures (Pham et al., 2008; Pham & Silva, 2010; López et al., 2010). Thus, progress has been made, on an experimental level, with these two species, in the area of spat collection from the wild. The only species where studies have advanced from experimental to semi-industrial cultures, is the "giant barnacle","picoroco", *Austromegabalanus psittacus*, on the Chilean coast. In this species, it has been shown that levels of spat collection from the wild will permit the development of commercial cultures. Similarly, the growth rates enable specimens of a commercial size to be obtained over a short period of time (López, 2008; López et al., 2010). Furthermore, low cost production technologies have been developed to obtain spat and to enhance growth. The yield and economic feasibility of various products have also been evaluated (Bedecarratz et al., 2011).

In *Megabalanus azoricus*, culture activities are incipient. This species is an intertidal barnacle found strictly in areas characterized by strong water flows around the Azorean coastline. It is a highly appreciated shellfish, whose natural populations are subject to intense exploitation (Santos et al., 1995). As a result, the regional authorities have identified the need to conduct trials with a view to developing aquaculture technologies for this species (Pham et al., 2008). Initial results for this species are encouraging (Pham et al., 2011), but much more research is required.

One aspect critical to barnacle species culture is present levels of knowledge with regard to: the particular biological characteristics of life cycle; the dispersion and behaviour of competent larvae, as well as ecological and functional aspects of juveniles during growth. From experiences undertaken to date, the main difficulties affecting culture are the following: (a) juveniles cannot be obtained from the wild, or can only be obtained in very limited quantities. It may also be possible to procure juveniles through larval cultures in hatcheries; however this increases costs considerably. The limited amount of spat obtained from the wild can be associated with various aspects, such as: - low supply of competent *cyprid* larvae; - substrates that are inadequate for the exploratory and attaching behaviour of the competent *cyprid* larvae; - lack of synchronization between period of maximum competent larvae quantities and the installation of artificial substrates (spat collectors) in the water column; - high spatial and temporal variation in spat collection, associated with climatic, oceanographic and topographic factors (Goldberg, 1984; Lagos et al., 2008; Andrade et al., 2011). (b) high levels of unpredictability in competent larvae supplies, that operate on a macro and meso-scale in the ocean. Consideration must be given to the fact that barnacles can form metapopulations and larval dispersion can be very wide (Lagos et al., 2005); (c) problems during growth related to the presence of predators or species that compete for the substrate or food (Pham & Silva, 2010); (d) density-dependent effects associated with mass recruitment (Hills & Thomason, 2003); (e) heavy weight of the shell, that occurs mainly in balanomorphs species. This also generates low harvest yields.

On the other hand, the main advantages associated with the biological characteristics of the barnacle species are: (a) possibility of obtaining spat from the wild; (b) gregarious larval settlement, which, if suitable substrates are available, can ensure an adequate supply of spat; (c) internal embryonic development in the case of balanomorphs barnacles, that limits early mortality and d) filter feeding, that permits suspended cultures without provision of exogenous feed.

Future projections for barnacle culture are aided by the wide-ranging and diversified literature available on larval development in various species; specifically, on aspects such as: duration, effects of environmental factors and feeding, as well as ecological, physiological and behavioural factors associated with larval settlement and recruitment (Walker, 1995; Jenkins et al., 2000; Dionisio et al., 2007; Tremblay et al., 2007; Pineda et al., 2009). Although this information is mainly related to species that are not commercially important, it can be used to optimize barnacle culture.

3. Spat collection

3.1 Barnacle life cycle

Knowledge about the barnacle life cycle is essential in order to design culture strategies. Identification of the embryonic and larval development particularities - when, where and in what magnitude they occur – will permit the elaboration of production technologies that are suitably adapted to these characteristics.

The life cycle of barnacles that are cultured, or in the process of being cultured, such as the "giant barnacle" and the "craca", presents characteristics that facilitate semi-intensive cultures. The fact that fecundation is internal and that the embryonic stages are incubated, determines that mortality during the early stages of development is lower than in many invertebrate species, such as mussels and clams, where fecundation is external and the entire ontogenetic development occurs in the water column. Nevertheless, during planktotrophic larval development, mortality occurs according to the duration of this phase. Similarly, larvae experience wide dispersion that determines high variability in the quantity of competent larvae, both spatially and temporally. This aspect is important when determining the location of the culture centre, in terms of provision of spat from the wild. On the other hand, the high selectivity of the competent larvae, conditions the choice of artificial substrates,that must satisfy the biological and physical-chemical requirements necessary for larval settlement. The *cyprid* larva possesses a complex sensorial apparatus associated with the need to evaluate the substrate where larval settlement will occur. This stage is crucial to larval effectiveness and determines the subsequent viability of specimens.

3.2 Barnacle spat collection from the wild. Culture technologies: The cases of *Austromegabalanus psittacus* (Chile) and *Megabalanus azoricus* (Portugal)

Technologies for obtaining spat from the wild have only been developed for two species of balanomorphs. In the case of the "giant barnacle", *Austromegabalanus psittacus*, it is possible to obtain quantities of spat that can sustain the development of commercial cultures.

The "giant barnacle" is bathymetrically distributed in the shallow subtidal zone (5-7 m depth), although, occasionally, it can also be found in the lower intertidal zone. Furthermore, it is an encrusting species, with a considerable presence associated with the fouling of ships, culture systems and pier support structures (Brattström, 1990; López et al., 2007a). In cultures, it has been possible to take advantage of its high spatial variability and the wide range of substrate-types on which larval settlement develops. Different types of collectors have been used as spat catchment from the wild, including artificial substrates that differ in texture, size, form, colour and buoyancy. In this species, materials tested included 900 cm^2 rectangular plates of expanded polystyrene, polythene and synthetic felts ("bidin"), treated with tar. In addition, polythene tubular substrates measuring 70 to 100 cm (equivalent to 2,200 to 3,000 cm^2 useable attachment surface) were tested (Fig. 3). Each collector system is composed of three substrate units, joined by polypropylene cables that are placed vertically in the water column from surface levels down to 8 m depth, suspended from floating systems (long-lines or culture rafts) in shallow, wave protected or semi-protected bays .

According to results, artificial spat collectors with the following characteristics ensure high spat collection levels: they should be innocuous (do not affect settled organisms), with rough textures, dark colour and low buoyancy. Spat preference for these characteristics is explained due to the sensorial characteristics of the competent larvae at the moment of attachment to the substrate. These possess negative phototaxism (preference for low luminosity), positive geotaxism (preference towards the bottom) and rugofilic behaviour (preference for rough substrates).

Spat density levels have varied according to type of collectors, obtaining values ranging from 0.01 individuals/cm^2 up to 0.2 individuals/cm^2, with substrate coverage of close to 100%.

In the Azores, spat collection and growth experiments in *Megabalanus azoricus* were initially conducted on PVC tubes with recruitment densities reaching up to 800 barnacles/ m^2*month

Fig. 3. "Giant barnacle", *Austromegabalanus psittaccus*, spat collection from the wild, southern, Chile.

during late spring/early summer (Pham et al., 2011). Recent experiments showed that the choice of materials and their orientation are important factors to be considered for optimizing recruitment values and future spat collector designs (Pham et al., unpublished data). Densities on material placed horizontally, whatever its category, are higher than when material is placed vertically. This difference is particularly pronounced for the most efficient material tested, PVC plates. PVC plates should be considered for the design of spat collection structures.

The private sector has tested a pilot structure for the production of barnacles (Pham & Silva, 2010). The structure tested was designed for both phases of the culture cycle: (a) spat collection and (b) ongrowing. The system was capable of holding 12 PVC tubes within a single structuring frame and was placed at 8 m depths over a period of 1.5 years. Results were satisfactory, as more than 9,000 barnacles of commercial size were produced with low mortality rates. Work is being currently being conducted to improve the design and efficiency of the system, based on recent results.

3.3 Spatial and temporal variability

Spat collection from the wild can vary in spatial on different scales (between latitudes, between locations , between depths, between types of substrate) and temporal terms (inter-annual, seasonal, monthly), given that natural populations of barnacles function as metapopulations, that is, there is a continuous exchange of larvae between geographically adjacent populations by means of passive transport by currents. Thus, abundance of individuals at a given site does not depend solely on the stock of breeders from that location, but also on that of more distant zones. Local oceanographic factors exert an influence, as do variables operating on a meso and macro-scale. In the case of the "giant barnacle", spat collection has been evaluated using artificial collectors in different sites along the Chilean coast, and at different depths and periods of the year. There is a temporal pattern in larval settlement, given that spat capture in artificial collectors is greater in the spring-summer period, principally between October and January, observing a second settlement period in the autumn, during the months of March and April. This is associated with the fact that gonadal maturity reaches its highest values when water temperature and food supply increases. Latitudinal differences have been evaluated in three sites along the Chilean coastline: North Totoralillo (29o29'S; 71o42'W), Tongoy Bay (30o36'S; 71o37'W) and Metri Bay (41o36'S; 72o42'W); the former two are located in the northern zone of Chile and the third, in the south of the country. All three sites are characterized as being shallow, wave-protected or semi-protected bays. There are marked climatological and oceanographic differences along the length of the Chilean coast, principally in temperature, which is higher in the northern zone. After 9 months, density of specimens recruited (number of specimens/cm^2) in the collectors differed between sites, with significantly higher densities in the locations of North Totoralillo than in Tongoy Bay and Metri Bay. Thus variability in spat collection does not have a latitudinal pattern, but rather varies from location to location. In addition, differences have been reported between types of collectors, spat collection being greater in the tubular plastic collectors than in the plates, which could be associated with the different surface/volume relationship.

Bathymetric variability also exists between collectors located on the surface and those located at greater depths. Cultures are more effective at greater depths, although no variations where recorded as at depths of over 4 m; this could be associated with the positive geotactic and negative phototrophic behaviour of the competent larva.

In some locations pronounced inter-annual variations in spat collection of spat from the wild have been verified. Oceanographic aperiodic events, such as ENSO (El Niño Southern Oscillation) and salinity variation on a local scale, could account for these results.

Spatial-temporal variability in spat collection from the wild limits the development of industrial cultures. As a result, it may be advisable to develop mass cultures of the "giant barnacle" in locations that are particularly apt for spat collection, and to establish other sites for the growth process. This practice has been used for many years in mussel cultures (*Mytilus chilensis*) in southern Chile.

Recent records indicate that the probability of local "giant barnacle" presence is associated with variables that are related to productivity (e.g. nitrate concentration, phosphate). This may be related with processes of coastal upwelling,that modify the physical-chemical regime of the coastal ocean. Furthermore, chlorophyll concentration and its variability are important variables that account for the presence of juveniles and adults, probably related to the trophic relationships of these filtering organisms.

In *M. azoricus*, results collectec monthly from settlement panels located in two sites off Faial Island, over a period of 2.5 years, revealed that recruitment takes place all-year round, with a peak in the summer, from early June until the end of September, coinciding with an increase in water temperature. This same study showed that newly settled barnacles (less than 15 day old) have a base diameter ranging between 0.79 and 2.80 mm and after two months can reach up to 12 mm. This large size and rapid initial growth increases the competitive capacity of *M. azoricus* recruits in relation to other barnacle species settling on the plates (*Chtamalus montagui* and *Balanus trigonus*).

3.4 Optimization models for spat collection from the wild

Dynamic models have been applied to "giant barnacle" spat collection from artificial substrates in the wild (Andrade et al., 2011). These models enable us to establish the variables involved in the spat collection process, the relationships between them and their relative effect on spat harvest. The dynamic hypothesis proposes that the number of "giant barnacle" spat obtained from the wild is influenced by competent larval abundance (*cyprid* density) over time, substrate availability and mortality after larval settlement. The number of factors affecting their spatial-temporal variability, illustrates the complexity of the spat collection process and the difficulties that exist with respect to predicting results. Simulation tests to establish consistency of the model were undertaken, using empirical background information obtained from semi-industrial cultures in southern Chile. In these cultures, average spat density in artificial collectors fluctuated between 0.1 and 3 spat/cm^2, while early mortality was below 90%. If artificial collectors of 10,000 cm^2, suspended from 100 m long lines are used, between 8,000 and 9,000 "giant barnacle" spat will be required to produce a harvest of 1 gross ton. Synchronization between *cyprid* abundance levels and installation of artificial substrates in the water is critical to achieve maximum collector efficiency. A period of less than 1 week out of synchronization, produces significant losses (60-70%) in spat production. When deployment of collectors and maximum quantity of competent larvae are synchronized, an increase of almost double the number of spat can be obtained. Quantity of competent larvae, as well as substrate quantity and quality, are key factors to achieving adequate spat provision from the wild, enabling the development of commercial cultures.

3.5 Spat's production in hatcheries

To date, barnacle larval cultures have not been undertaken on a commercial scale; however considerable experience has been gained with regard to the larval development of different species, studying, not only the effect of environmental and feeding factors on survival and duration of the entire development process, but also the variables that influence larval settlement. Larval development, and the factors that influence it, has been studied in various species of lepadomorphs and balanomorphs (Miller et al., 1989; Molares et al., 1994; Yan & Chan, 2001; Dionisio et al., 2007; Li et al., 2011).

Spat production in hatcheries possesses various advantages over obtaining spat by collection from the wild. Production can be programmed over time, and is not dependent on the spatial-temporal variability of the larval pool in the environment. The energetic quality of the *cyprid* larvae can also be improved, that will be reflected in the percentage of larval settlement, size and growth rate of juveniles (Thiyagarajan et al., 2002; Thiyagarajan et al., 2003b; Desai & Anil, 2004). Post-settlement mortality can be, for the most part, reduced or completely eliminated.

Experiences related to the production of balanomorph barnacle larvae in the laboratory, date back considerably (Knight-Jones, 1953; Rittschof et al., 1984; Brown and Roughgarden,

1985). *Cyprid* have been obtained in various species and larval settlement trials undertaken (Yan, 2003; Desai and Anil, 2004; Anil et al., 2010; Li et al., 2011).

In species such as *Balanus amphitrite*, trials have been carried out with different diets and temperatures, establishing development periods that can fluctuate between 4 and 18 days (Qiu and Quian, 1999), which indicates the influence of these factors on culture success. Microalgae densities of the genera *Chaetoceros sp*, *Skeletonema* and *Isochrysis*, vary from $1x10^5$ to $10x10^5$ cells/ml. In this species, survival has exceeded 90% (Qiu and Quian, 1999; Mishra et al., 2001). Percentage of larval settlement can vary from 10% to 80%, depending on the culture conditions, the substrate type and the use of inductors (Rittschoff et al., 1984; Mishra and Kitamura, 2000; Mishra et al., 2001; Thiyagarajan et al., 2003b).

With regard to larval settlement induction, a great variety of chemical signals have been described as fundamental to trigger the larval settlement response. Natural inductors have been characterized, based on studies of larval settlement response as a consequence of natural substrates and have been associated, mainly, with three types of sources: conspecific individuals, prey species and microbial films (Rodríguez et al., 1992). Different kinds of microbial biofilms have been described as important inductors of larval settlement in benthonic marine invertebrates. Induction has been observed both by diatom and cyanobacteria biofilms (Akashige et al., 1982), as well as by bacterial biofilms (Kirchman et al., 1982; González et al., 1987). Induction by bacteria has been widely studied; the positive effect of these types of biofilm has been observed in barnacles (Maki et al., 1988, Maki et al., 1990). The settlement response associated with these biofilms would be generated, apparently, by the presence of polysaccharides or extracellular glycoproteins attached to the bacterial wall (Kirchman et al., 1982, Hadfield, 1986), or, alternatively, soluble compounds released by the biofilms (Bonar et al., 1990). An increase in the settlement inductive potential of older bacterial biofilms was a recurring factor observed (Maki et al., 1989, Pearce and Scheibling, 1990, Mataxas and Saunders, 2009). Similarly, the contrary effect has been described (Maki et al., 1989).

In synthesis, sufficient knowledge is available on which to base the design of barnacle hatcheries, although the transition still has to be made from the experimental to commercial level. No information is available with regard to the costs that this would imply. As commercial cultures are implemented, the development of spat production in hatcheries will become a necessity, in order to improve predictability and quality in spat production.

4. Growth

4.1 Production technologies

Technological designs must ensure specimens have the best access to food and water flow during the growth phase in suspended barnacle cultures. Similarly, space availability must be sufficient to maintain animals up to the harvest stage. As is the case in spat collection systems from the wild, materials used must be innocuous. Weight must also be compatible with suspension from rafts or long-lines, ensuring easy manipulation and costs that are compatible with a commercial activity.

There are two types of production technologies for the growth phase of the "giant barnacle": "Cultch" systems - corresponding to technologies that maintain the specimens over the period from spat collection, during growth, through to the harvest stage. Substrates used have included plates and 900 cm^2 to 2,200 cm^2 polythene tubes, suspended vertically in groups of three units. One advantage is that harvesting of spat is not necessary and this system can be more effective, because specimens are less exposed to mechanical manipulation, and, therefore, suffer fewer mortalities. In the "cultchless" system, two different production technologies are used, one for obtaining spat from the wild and another for growth. Spat harvest occurs approximately four months post-larval settlement. Subsequently, individuals are transferred to a growth system where they remain until the harvest stage. The surfaces used as artificial collectors must permit the innocuous extraction of juveniles. Since specimens are cemented to the collector substrate, a surface must be used that avoids damaging the calcareous structures, particularly of the basal zone, when the specimens are detached. Expanded polystyrene and felts ("bidin") covered with tar have been used as substrates, with an area of 900 cm^2 to 1,600 cm^2. For subsequent growth, trays can be used or specimens can be attached with adhesive to a new surface. The best results have been obtained with the "cultch" systems. Both the "cultch" and "cultchless" systems are located on floating structures (long-lines or rafts) suspended from surface levels to a depth of 8 m, in shallow wave-protected or semi-protected bays. Maintenance and cleaning of the systems is necessary during the growth phase eliminating, periodically, the accumulation of sediment and fouling (mainly mytilids, ascidians and other species of barnacles), in order to avoid mortalities by suffocation and overgrowth by other species.

4.2 Spatial and temporal variability: Results in the "giant barnacle"

In balanomorph barnacles, a density-independent estimator of growth is the carino-rostral length or opercular length, corresponding to the maximum rectilinear distance between the carinal plate and the rostral plate (López et al., 2007b). Other growth estimators, such as mural plate height or basal diameter, are inadequate growth indicators, because they are modified by growth density.

Average commercial size of the "giant barnacle" in Chile is 3.5 cm ± 0.6 carino-rostral length (approximately 10 cm height). Specimens of around 2 cm carino-rostral length can be found in local markets,. Nevertheless, specimens from natural banks can reach sizes in excess of 6 cm carino-rostral length and more than 30 cm height. Shell height depends on the degree of aggregation during growth, due to the development of a modified base generated in the hummocks (López et al., 2007b).

Growth rate varies according to depth. Twenty months after larval settlement, average carino-rostral length of "giant barnacle" specimens growing at depths of between 4 and 6 m, was 3.4 ± 0.17 cm, while the average size of individuals growing at depths of between 1 and 2 m, was 2.92 ± 0.14 cm (López et al., 2008a).

If minimum commercial size on the Chilean market is considered (2.1 cm carino-rostral length), the culture period, from larval settlement up to harvest, can be 7-8 months. Nevertheless, recent studies in the northern zone of Chile show a relative increase in growth in comparison to the results obtained in the southern zone, probably associated with the differences in water temperature. In the north, specimens of a commercial size (> 2 cm carino-rostral length) have been obtained over a period of only 4-6 months (Fig. 4).

Fig. 4. "Giant barnacle" culture growth system in northern Chile.

Instantaneous growth rates decrease with age, fluctuating from 0.03 cm/day carino-rostral length at the beginning of growth, to 0.0005 cm/day carino-rostral length after 20 months.

For the Chilean market, "giant barnacle" harvest should occur between 18 – 24 months, depending on the season the collectors are placed in the water. Prior to harvesting, the majority of specimens must have reached over 3 cm carino-rostral length. Nevertheless, the culture period can vary according to the requirements of each market. If, eventually, the "giant barnacle" were to be used as a simil of the "fujit subo" on the Japanese market, the culture period would be substantially less.

Productivity in "giant barnacle" culture systems is extremely high, taking production by extractive fisheries as a point of reference. National average gross production of this resource by small-scale local fisheries is around 200 ton/year; this could be obtained with only 20 to 30 long lines. A double, 100 m floating long-line can produce between 7 and 10 gross ton of total biomass in one season. This can signify approximately 2 ton wet weight of soft parts. Each tubular system of approximately 1 m in length, can produce, on average, 15 ± 8 kg of biomass, equivalent to an average of 100 specimens of commercial size, with a yield corresponding to 20-30 g per individual, depending on the maturity of the female gonad. The results obtained in "giant barnacle" cultures on the Chilean coast indicate that there are spatial differences in growth, both on a latitudinal and local level, as well as bathymetrically and according to the type of culture system. Inter-annual and seasonal temporal differences also exist. Thus, production planning is still subject to a high degree of unpredictability. On the other hand, the heavy weight of the shell, that can fluctuate between 70 to 85% of total weight, together with high variability in the mortality rate during the growth stage, are also limiting factors for cultures. Nevertheless, in view of its rapid growth and productivity, as well as the simple and economic production technologies used during the fattening phase, culture of the "giant barnacle", and other balanomorphs species, constitutes an interesting alternative for diversification of Aquaculture.

4.3 Factors influencing growth

Experiments conducted under controlled conditions have examined the role of environmental variables in the spatial-temporal variability of "giant barnacle" growth (López et al., 2008a). It has been reported that photoperiod and temperature can synergistically affect the metabolism, influencing molt frequency and growth in some barnacle species (Crisp & Patel, 1960; El-Komi & Kajihara, 1991). These variables change seasonally within the geographic distribution range of the "giant barnacle". Temperature and photoperiod affect the instantaneous growth rate, which is greater at 10°C than at 16°C and at photoperiods of 8h light :16 h darkness, than at 16 h light:8 h darkness. Specimens in the natural environment presented higher instantaneous growth rates at depths of between 4 and 6 m than between 1 and 2 m depth, under conditions of greater darkness and lower temperature, in accordance with the results obtained in the laboratory. In other species of barnacles, increased growth has also been reported under darkness conditions (Costlow & Bookhout, 1953; 1956). Futhermore, as occurs in other species of barnacles, growth and molting frequency are not correlated (Costlow & Bookhout, 1956). In the "giant barnacle" molt frequency was greater at 16°C than at 10°C, with no evidence of variation due to photoperiod. This can be associated with a decreasing metabolism as a result of lower temperatures, or to effects at the neuroendocrine system level. However, the inter-molt periods were longer at 10°C than at 16°C and during photoperiods 8: 16 than 16:8 (López et al., 2008a). If this information is applied to suspended "giant barnacle" cultures, the fattening phase should be undertaken at depths of over 4 m and at low temperatures.

5. Economic aspects

Of the economically significant species of barnacle, only the "goose barnacles" have an established international market; nevertheless, none of them are cultured on a commercial scale. Species of "acorn barnacles" are consumed on very restricted local markets, although progress has been made in the culture of two species, the "picoroco" in Chile and the"craca" in Azores Islands. Thus, in spite of the interesting projections associated with barnacle culture, commercialization remains limited.

5.1 Product, yield and market

"Goose barnacles", together with most of the "acorn barnacles", are consumed fresh. Only the "giant barnacle "*Austromegabalanus psittacus* and the "mine fujit subo" *Balanus rostratus,* are available in various types of elaborated product. The "giant barnacle" is consumed fresh, frozen and canned, principally on the local Chilean market (Fig. 5). The "mine fujit subo" is consumed fresh, as tempura and as soup flavouring on the Japanese market.

The yield of cultured "giant barnacle" has been determined in a processing plant as 15-20% of total harvest weight. Loss through processing is approximately 10% (Bedecarratz et al., 2011). At harvest, specimens reach an average total weight of 150 g, most of which is attributable to the calcareous shell, consisting of six mural plates and two opercular plates. Two, double 100 m long-lines can produce an average yield of approximately 420 gross ton/year. This doubles the average annual production of artisanal fisheries activities over the last few years in Chile.

Fig. 5. Canned products of "giant barnacle".

With regard to markets and prices, the "goose barnacles" can reach prices of between €15-25/Kg on the Iberian market. Of the "acorn barnacle" species, the "giant barnacle" is commercialized on the local Chilean market at prices that range between US$1.5/Kg and 20/Kg. The "mine fujit subo" reaches between US$12-15/Kg on the Japanese market and the "craca", between €2.5-5/Kg on the local market of the Azores Islands, Portugal (López et al., 2010). Market studies of the "giant barnacle" in Japan indicate that some products can reach up to US$25/Kg. " Giant barnacle" cultures on the Chilean coast have generated an interest in its commercialization in the USA and on the Iberian market.

5.2 Economic feasibility of "giant barnacle" culture. Financial indicators

"Giant barnacle" culture is technically and economically feasible based on spat obtained from the wild (Bedecarratz et al., 2011). Nevertheless, the commercial success of these cultures depends on three factors: (a) appropriate selection of culture locations, according to food availability, as well as environmental, oceanographic and climatic conditions. These variables determine spat supply and growth rate, that in turn define the period from spat collection to harvest; (b) effective culture management in order to reduce mortality; (c) ensuring the best prices on external markets by implementing an adequate commercial marketing strategy and exporting the most profitable products.

The financial indicators obtained for frozen and canned products are adequate. The net present value (NPV) is positive for both products; the internal rate of return (IRR) varies between 31-61% and the discounted payback period (DPBP) varies between 3 and 5 years.

The economic profitability index (EPI) fluctuates between 1.7 and 2.9 and confirms that this aquaculture activity is economically attractive, given that current value of expected cash inflows (at 19.6% discount rate) exceeds initial investments (Bedecarratz et al., 2011).

Greatest cash outlay corresponds to production technology (material and labour costs associated with the construction and installation of production technologies). Sensitivity analysis indicates that positive NPC values are maintained, in spite of significant changes in each critical variable, such as: spat density, gross weight, mortality up to harvest, processing and packing costs, FOB sale prices and exchange rate. For elasticity analysis, the most relevant variables are: gross weight at harvest and sale price FOB, while the variable with lowest impact on NPV is processing and packing costs (Bedecarratz et al., 2011).

6. Conclusions and projections

6.1 Advantages of barnacle culture

"Giant barnacle" culture benefits from advantages over other crustacean delicate meats, such as lobsters or prawns, summarized in López et al., (2010):

a. They are omnivorous filter feeders that consume a wide size-range of phytoplankton and zooplankton. Thus, they have a low trophic level and cultures do not require provision of exogenous food, as a result of which costs and environmental impact are moderate.

b. Although they are gregarious, they do not create culture problems associated with the territorial behaviour of other species. "Giant barnacles" living in groups do not generate density-dependent effects during growth or reproduction, because of their ability to modify their base, thus limiting demand for areas of adhesion to the substrate (López et al., 2007b).

c. Spat can be obtained from the wild in sufficient quantities to permit the development of commercial cultures. This means production costs are low. Spat production in hatcheries ensures reliable supplies, but clearly increases costs (López et al., 2005; 2010; Andrade et al., 2011).

d. They have high growth rates, thus specimens can reach a commercial size in periods that, depending on the requirements of the market, fluctuate between 6 and 18 months (López et al., 2010).

e. They possess early sexual maturity and high fecundity.

f. They have high resistance to environmental variables, particularly hypoxia and to salinity variations, factors that are usually subject to frequent changes, affecting the survival of organisms maintained in suspended cultures (López et al., 2003).

g. Culture technologies are simple and economical, and potential for contamination and mass environmental effects, is low. They occupy limited space.

h. Although real demand has not been completely defined, commercialization prices on international markets are high.

6.2 Effect of giant barnacle culture on aquaculture diversification in Chile

Potential for Aquaculture diversification in Chile is high, due to the presence of a large number of economically significant native species of fish, invertebrates and algae. Many of them are exported at high prices; the majority in limited volumes. Fisheries and aquaculture exports in Chile reach around US$4 billion and are destined to more than 130 countries. Nevertheless, limitations, generated by the overexploitation of natural populations, exist. Culture options could become a reality if the level of biological knowledge and trial experiences aimed at developing adequate production techniques, increase; to date, experimental cultures have been undertaken using around a dozen species (López et al., 2008b). The development of industrial cultures of native species, permits greater predictability in production volumes as well as improved product quality. At present Aquaculture production exceeds 700,000 ton/year. A total of seventeen species are cultured - four algae (11.5% of production), five species of fish (62.6%) and eight species of molluscs (25.9% of production). Around 63% correspond to introduced species, principally salmonids and, to a lesser extent, abalones and the Japanese Oyster. Twenty-five percent includes various species of native molluscs, principally mytilids, the Chilean oyster and the northern scallop (Sernapesca, 2009). However, they are cultured with introduced technologies, developed for similar species in other countries. Only around 14% correspond to two species of native algae, cultured using national technologies.

Traditional fisheries activities extract 130 species of commercial interest, of which 68 are fish, 29 molluscs, 19 crustaceans, 11 algae and 3 "other" groups. Annual landings reach approximately 2,000,000 ton/year. Seventy-three percent of landings correspond to fish, 19% algae, 5.2% molluscs and 0.7% crustaceans (Sernapesca, 2009). Data reveal interesting and varied alternatives for aquaculture diversification, based on the incorporation of mass cultures of native species. Among these, is the "giant barnacle", about which considerable knowledge has been accumulated regarding biological aspects associated with aquaculture, such as: growth (López et al., 2008a); density-dependent effects (López et al., 2007b); reproduction and life cycle (López & Toledo, 1979); effects of environmental factors, such as salinity and hypoxia (Vial et al., 1999; López & López, 2005; López et al., 2003; Simpfendörfer et al., 2005; 2006); and behaviour (López & López, 2005; López et al., 2008a). Furthermore, the development of production technologies for obtaining spat from the wild and for growth, have permitted culture of the "giant barnacle" on a semi-industrial scale (López, 2008; López et al., 2010).

Although cost and experience of the workforce are important factors affecting the economic feasibility of cultures, present conditions are favourable, due to the increased availability of manual labour with aquaculture experience in southern Chile. This is a consequence of reduced demand for workers in the salmonid production sector, which has been negatively affected by the presence of viral disease (Mardones et al., 2009; Vike et al., 2009). This same causal factor has increased the availability of maritime concessions, processing plants, transport systems and other facilities for new aquacultures activities.

7. Acknowledgements

Funds were provided by the Chilean Government (FONDEF Projects D03I1116 and D07I1042), the Autonomous Regional Government of the Azores, Portugal (Projects EPA-I PRODESA 2004.91.001646.0 and PEDAPCA) and the Universidad de Los Lagos, Osorno, Chile, aimed at developing barnacle culture. This financial support is gratefully acknowledged, as is the collaboration of the private companies JAL Fisheries SA (Chile), Cultivos Cholche (Chile), Aquanorte (Chile) Plastimar Ltda (Chile) Octopus Mar SA (Chile), and Sea Expert SA (Azores Islands). In particular, the assistance of José A. López, Justo García, Hugo Ulloa, Pedro González, Luis Pereira, and Henrique Ramos was especially significant. Similarly, the authors appreciate the support of the Universidade dos Açores and the collaboration of the Chilean researchers María Luisa González, Sergio Arriagada,

Óscar Mora, Verónica Riquelme, Paula Bedecarratz, Mauricio Pineda, José Uribe and Lorenzo Andrade as well as the students Victor Oyarzún, Erwin Aros, Elías Romeri, Haroldo Aburto, Diego Barriga and Alexis Catalán. The contribution of the researchers from the Azores Islands, Mirko Di Girolamo, Rui Guedes and Emanuel Silva, is much appreciated. Our thanks also go to Pedro Ségure of FONDEF-CHILE. Finally, the cooperation of Sandra Mancilla, in the administrative work, Susan Angus, in the translation of the manuscript, and the Editor comments, are much appreciated.

8. References

Anderson, D.T. (1994). *Barnacles: structure, function, development and evolution.* Chapman and Hall, London. U.K.

Andrade, L.I., López, D.A. & López, B. A. (2011). Dynamic models applied to giant barnacle culture. *Aquaculture International* (in press).

Akashige, S.T., Kano-noy, H.S. & Nomura, T. (1981). Effects of gamma-aminobutyric acid and certain neurotransmitters on the settlement of the larvae of *Haliotis discus hannai* Ino (Gastropoda). *Bulletin Tohoku Regional Fishery Research Laboratory* 43:37-45.

Anil, A.C., Khandeparker, L., Desai, D.V., Baragi, L.V. & Gaonkar, C.A. (2010). Larval development, sensory mechanisms and physiological adaptations in acorn barnacles with special reference to *Balanus amphitrite. Journal of Experimental Marine Biology and Ecology* 392 (1-2): 89-98.

Arenas, J. (1971). Distribución de *Elminius kingii* Gray (Cirripedia) en el estuario del río Valdivia. *Beiträge Zur Neotropical Fauna* 6 (3): 199-206.

Bald, J., Borja, A. & Muxika, I. (2006). A system dynamics model for the management of the gooseneck barnacle *(Pollicipes pollicipes)* in the marine reserve of Gaztelugatxe (Northern Spain). *Ecological Modelling* 194 (1-3): 306-315.

Bedecarratz, P.C., López, D.A., López, B.A. & Mora, O.A. (2011). Economic feasibility of aquaculture of the giant barnacle, *Austromegabalanus psittacus* in southern Chile. *Journal of Shellfish Research* 30 (1): 147-157.

Bonar, D., Coon, M., Walch, M., Weiner, R. & Fitt, W. (1990). Control of oyster settlement and metamorphosis by endogenous and exogenous chemical cues. *Bulletin of Marine Sciences* 46: 484-498.

Brattström, H. (1990). Intertidal ecology of the northern part of the Chilean Archipelago. *Sarsia* 75: 105-160.

Brown, S. & Roughgarden, J. (1985). Growth, morphology, and laboratory culture of *Balanus glandula* larvae (Cirripedia:Thoracica). *Journal of Crustacean Biology* 5: 574-590.

Costlow, J.D. & Bookhout, C. G. (1953). Molting and growth in *Balanus improvisus. The Biological Bulletin* 105: 420-433.

Costlow, J.D. & Bookhout, C. G. (1956). Molting and shell growth in *Balanus amphitrite niveus. The Biological Bulletin* 110 (2): 107-116.

Crisp, D.J. & Patel, B. S. (1960). The moulting cycle in *Balanus balanoides* (L.). *The Biological Bulletin* 118: 31-47.

Cruz, T., Castro, J.J. & Hawkins, S.J. (2010). Recruitment, growth and population size structure of *Pollicipes pollicipes* in SW Portugal. *Journal of Experimental Marine Biology and Ecology* 392 (1-2): 200-209.

Desai, D.V. & Anil, A.C. (2004). The impact of food type, temperature and starvation on larval development of *Balanus amphitrite* Darwin (Cirripedia: Thoracica). *Journal of Experimental Marine Biology and Ecology* 306; 113-137.

Dionisio, M.A., Rodrigues, A. & Costa, A. (2007). Reproductive biology of *Megabalanus azoricus*, (Pilsbry) the Azorean barnacle. *Invertebrate Reproduction & Development* 50(3): 155-162.

Dionisio, M.A., Rodrigues, A., Pires, P. & Costa A. (2009). Bases para a gestão e conservação de *Megabalanus azoricus*. *1° Congresso de Desenvolvimento Regional de Cabo Verde / 15° Congresso da APDR*. Cidade da Praia, Cabo Verde, Junho de 2009: 319-337.

El-Komi, M.M. & Kajihara, T. (1991). Breeding and molting of barnacles under rearing conditions. *Marine Biology* 108: 83-89.

Gaines, S. & Roughgarden, J. (1985). Larval settlement rate: a leading determinant of structure in an ecological community of the marine intertidal zone. *Proceedings of the National Academy of Sciences (USA)* 82: 3703–3711.

Gaines, S.D. & Bertness, M.D. (1992). Dispersal of juveniles and variable recruitment in sessile marine species. *Nature* 360: 579-580.

Gaines, S.D. & Bertness, M. (1993). The dynamics of juvenile dispersal - why field ecologists must integrate. *Ecology* 74: 2430-2435.

Goldberg, H. (1984). Posibilidades de cultivo del percebe, *Pollicipes cornucopia* Leach, en sistemas flotantes. *Informe Técnico. Instituto Español de Oceanografia* 19: 1-13.

González, L., Castilla, J. & Guisado, C. (1987). Effect of larval diet and rearing temperature on metamorphosis and juvenile survival of the sea urchin *Loxechinus albus* (Molina, 1782) (Echinodermata: Echinoidea). *Journal of Shellfish Research* 6:109-115.

Hadfield, M . (1986). Settlement and recruitment of marine invertebrates: a perspective and some proposals. *Bulletin of Marine Sciences* 39: 418-425.

Hills, J. M. & Thomason, J. C. (2003). The 'ghost of settlement past' determines mortality and fecundity in the barnacle, *Semibalanus balanoides*. *Oikos* 101: 529-583.

Jacinto, D,. Cruz, T., Silva T. & Castro, J.J. (2010). Stalked barnacle (*Pollicipes pollicipes*) harvesting in the Berlengas Nature Reserve, Portugal: temporal variation and validation of logbook data. *ICES Journal of Marine Science*. 67: 19-25.

Jenkins, S.R., Aberg, P., Cerving, G., Coleman, R.A., Delany, J., Della, S.P., Hawkins, S.J., Lacroix, E., Myers, A.A., Lindegarth, M., Power, A.M., Robert, M.F. & Hartnoll, R.G. (2000). Spatial and temporal variation in settlement and recruitment of the intertidal barnacle *Semibalanus balanoides* (L.) (Crustacea : Cirripedia) over a European scale. *Journal of Experimental Marine Biology and Ecology* 243: 209-225.

Kirchman, D., Graham, S., Reish, D. & Mitchell, R. (1982). Bacteria induce settlement and metamorphosis of *Janua* (*Dexiospira*) *brasiliensis* Grube (Polychaeta: Spirorbidae). *Journal of Experimental Marine Biology and Ecology* 56: 153-163.

Knight-Jones, E.W. (1953). Laboratory experiments on gregariousness during setting in *Balanus balanoides* and other barnacles. *Journal of Experimental Biology* 30: 584– 598.

Lagos, N.A., Navarrete, S.A., Veliz, F., Masurero, A. & Castilla, J.C. (2005). Meso-scale spatial variation in settlement and recruitment of intertidal barnacles along the coast of central Chile. *Marine Ecology Progress Series* 290: 165-178.

Lagos, N.A., Castilla, J.C. & Broitman, B. (2008). Spatial environmental correlates of intertidal recruitment: a test using barnacles in northern Chile. *Ecological Monographs* 78 (2): 245-261.

Li, H.X., Miao, S.Y., Yan, Y., Yu, X.J. & Zhang, L.P. (2011).Larval Development the Barnacle, *Microeuraphia withersi* (Cirripedia,Thoracica,Chthamalidae) reared in the laboratory source. *Crustaceana* 84 (2): 129-152.

Lotaçor (2006). *Pescado comercializado nas Lotas de região autónoma dos Açores*. Reviewed April 12. Available from:
http://www.lotacor.pt/estatisticas/free/publicacoes_antigas/2006.pdf.

López, B.A., & López, D.A.(2005). Moulting frequency and behavioural responses to salinity and diesel oil in *Austromegabalanus psittacus* (Molina) (Cirripedia: Balanidae). *Marine and Freshwater Behaviour and Physiology* 38 (4): 249-258.

López, D. (2008) Giant barnacle "picoroco" culture in Chile, *in* C. K. Pham, R. M. Higgins, M. De Girolamo & E. Isidro (Eds.), *Acta of the International Workshop: Developing a Sustainable Aquaculture Industry in the Azores. Arquipelago. Life and Marine Sciences.* Supplement 7: 56-57.

López, D. & Toledo, G. (1979). Estudio descriptivo comparado de estados larvales tempranos y de cypris de balanomorfos chilenos. *Acta Zoologica Lilloana* 35: 547- 561.

López, D.A., Castro, J.M., González, M.L. & Simpfendörfer, R.W. (2003). Physiological responses to hypoxia and anoxia in the giant barnacle, *Austromegabalanus psittacus* (Molina 1782). *Crustaceana* 76: 533-545.

López, D.A., López, B.A., González, M.L. & Arriagada, S.E. (2005). Aquaculture diversification in Chile: potential culture of giant barnacles. *Global Aquaculture Advocate* 8 (2): 73-74.

López, D.A., López, B.A. & González, M.L. (2007a). Bibliographic index on aquatic biodiversity of Chile: Crustacea, Cirripedia, Thoracica. *Ciencia y Tecnología del Mar* 30 (1):161-165.

López, D.A., López, B.A., Burgos, C., Arriagada, S.E. & González, M.L. (2007b). Consequences of base modification in hummocks of the barnacle Austromegabalanus psittacus. *New Zealand Marine and Freshwater Research* 41 (3): 291-298.

López, D.A., Espinoza, E.A., López, B.A & Santibañez, A.F. (2008a). Growth and moulting behaviour in the giant barnacle *Austromegabalanus psittacus* (Molina, 1782). *Revista de Biología Marina y Oceanografía* 43 (3): 607-613.

López, D. A, López, B.A. & González, M.L. (2008b). Shellfish culture in Chile. *International Journal of Environment and Pollution* 33 (4): 401-431.

López, D.A., López, B.A., Pham, C.K., Isidro, E.J. & De Girolamo, M. (2010). Barnacle culture: background, potential and challenges. *Aquaculture Research* 41: e367-e375.

Maki, J.S., Rittschof, D., Costlow, J.D. & Mitchell, R. (1988). Inhibition of attachment of larval barnacles, *Balanus amphitrite*, by bacterial surface films. *Marine Biology* 97: 199–206.

Maki, J.S., Rittschof, D., Schmidt, A.R., Snyder, A. & Mitchell, R. (1989). Factors controlling attachment of bryozoan larvae: a comparison of bacterial films and unfilmed surfaces. *The Biological Bulletin* 177: 295-302.

Maki, J.D., Rittschof, O., Samuelsoon, U., Szewzky, A., Yule, B., Kjelleberg, S., Costlow, J.D. & Mitchell, R. (1990). Effect of marine bacteria and their exopolymers on the attachment of barnacle cypris larvae. *Bulletin of Marine Science* 46 (2):499-511.

Mardones, F., Pérez, A. & Carpenter, T. (2009). Epidemiologic investigation of the re-emergence of infection salmon anemia virus in Chile. *Diseases of Aquatic Organisms* 84 (2): 105-114.

Metaxas, A. & Saunders, M. (2009). Quantifying the bio-components in biophysical models of larval transport in marine benthic invertebrates: Advances and Pitfalls. *The Biological Bulletin* 216:257-272.

Miller, K.M., Blower, S.M., Hedgecock, D. & Roughgarden, J. (1989).Comparison of Larvaland adult stages of *Chthamalus dalli* and *Chthamalus fissus* (Cirripedia, Thoracica). *Journal of Crustacean Biology* 9: 2(242-256).

Minchinton, T.E. & Scheibling, R.E. (1991). The influence of larval supply and settlement on the population structure of barnacles. *Ecology* 72: 1867-1879.

Mishra, J. K. & Kitamura, H. (2000). The effect of mono-amino acids on larval settlement of the barnacle, *Balanus amphitrite* Darwin. *Biofouling* 14 (4): 299-303.

Mishra, J.K., Kitamura, H. & Tomoda, F.I. (2001). Laboratory culture of *Balanus trigonus* larvae by the method of *Balanus amphitrite* for establishing a multi-species settlement assay. *Sessile Organisms* 18 (1):1-6.

Molares, J. & Freire, J. (2003). Development and perspectives for community-based management of the goose barnacle *(Pollicipes pollicipes)* fisheries in Galicia (NW Spain). *Fisheries Research* 65 (1-3): 485-492.

Molares, J., Tilves, F., & Pascual, C. (1994). Larval development of the pedunculate barnacle *Pollicipes cornucopia* (Cirripedia: Scalpellomorpha) reared in the laboratory. *Marine Biology* 120: 261-264.

Morris, R.H., Abbott, D.P. & Haderlie, E.C. (1980). Intertidal invertebrates of California. Stanford University Press, Palo Alto, California, USA.

Pearce, C.M. & Scheibiling, R.E. (1990). Induction of settlement and metamorphosis in the sand dollar *Echinarachnius parma*: evidence for an adult-associated factor. *Marine Biology* 107: 363-369.

Pham, C.K. & Silva, E. (2010). New species in european shellfish Aquaculture: barnacle farming in the Azores. *Aquaculture Europe* 35 (2): 27-29.

Pham, C.K., Higgins, R.M., De Girolamo, M. & Isidro, M. (2008). Acta of the International Workshop: Developing a Sustainable Aquaculture Industry in the Azores. Arquipelago. Life and Marine Sciences Supplement 7: 1- 81.

Pham, C.K., De Girolamo, M. & Isidro, E.J. (2011). Recruitment and growth of *Megalabanus azoricus* (Pilsbry, 1916) on artificial substrates: first steps towards commercial culture in the Azores. *Arquipelago. Life and Marine Sciences* 28: 47-56.

Pineda, J. (2000). Linking larval settlement to larval transport: assumptions, potentials, and pitfalls. *Oceanography of the Eastern Pacific* 1: 84-105.

Pineda, J., Reyns, N.B., Starczak, V.R. (2009). Complexity and simplification in understanding recruitment in benthic populations. *Population Ecology* 51: 17–32.

Pilsbry, H. (1916). The sessile barnacles (cirripedia) contained in the collection of the U.S. National Museum including a monograph of the American species. *Bulletin of United States Natural Museum* 93: 1-366.

Qiu, J.W. & Qian, P.Y. (1999) .Effects of salinity and temperature on the life-history of *Balanus amphitrite*: effects of past experience. *Marine Ecology Progress Series* 188: 123-132.

Regala, J.T. (1999). Contribuição para o estudo da biologia da craca dos Açores, Megabalanus azoricus (Pilsbry 1916). Relatório de Estágio do Curso de Licenciatura em Biologia Marinha. DOP, Universidade do Algarve, Faro. Portugal

Rittschof, D., Branscomb ,E.S. & Costlow, J.D. (1984). Settlement and behavior in relation to flow and surface in larval barnacles, *Balanus amphitrite* Darwin. *Journal of Experimental Marine Biology and Ecology* 82: 131- 146.

Rodríguez, S.R., Ojeda, F.P. & Inestrosa, N.C. (1992). Inductores químicos del asentamiento de invertebrados marinos bentónicos: importancia y necesidad de su estudio en Chile. *Revista Chilena de Historia Natural* 65: 297-310.

Roughgarden, J., Gaines, S. & Possingham, H. (1988). Recruitment dynamics in complex life cycles. *Science* 241: 1460–1466.

Santos, R.S., Hawkins, S.J., Monteiro, L.R., Alves, M. & Isidro, E.J. (1995).) Marine research resources and conservation in Azores. *Aquatic Conservation Marine and Freshwater Ecosystems* 5: 311–354.

Sernapesca. (2009). *Anuario Estadístico de Pesca. Servicio Nacional de Pesca. Subsecretaria de Pesca, Ministerio de Economía, Fomento y Turismo. Gobierno de Chile.* Reviewed June 2011. Available from http://www.sernapesca.cl/index.php?option=com_remository&Itemid=54&func= select&id=2.

Simpfendörfer,R.W., Oelkers, K.B., Nash,D. & López, D.A. (2005). Kinetic properties of the muscular pyruvate kinase from the giant marine barnacle, *Austromegabalanus psittacus* (Molina , 1782) (Cirripedia, Balanomorpha). *Crustaceana* 78 (1): 1203-1218.

Simpfendörfer, R.W., Oelkers, K.B & López, D.A. (2006). Phosphofructokinase from muscle of the marine giant barnacle *Austromegabalanus psittacus*: Kinetic characterization and effect of in vitro phosphorylation. *Comparative Biochemistry and Physiology C-Toxicology & Pharmacology* 142 (3-4): 382-389.

Southward, A. J. (Ed.). (1987). *Crustacean Issues 5. Barnacle Biology.* A. A. Balkema. Rotterdam, Netherlands.

Southward, A. J. (1998). New observations on barnacles (Crustacea: Cirripedia) of the Azores region. *Arquipelago. Life and Marine Sciences* 16a: 11-27.

Thiyagarajan, V., Nair, K.V., Subramoniam, T. & Venugopalan, V.P. (2002). Larval settlement behaviour of the barnacle *Balanus reticulatus* in the laboratory. *Journal of the Marine Biological Association of the United Kingdom* 82: 579-582.

Thiyagarajan, V., Harder, T. & Qian, P.Y. (2003a). Combined effects of temperature and salinity on larval development and attachment of the subtidal barnacle *Balanus trigonus* Darwin. *Journal of Experimental Marine Biology and Ecology* 287: 223-236.

Thiyagarajan, V., Harder, T., Qiu, J.W. & Qian, P.Y. (2003b). Energy content at metamorphosis and growth rate of the early juvenile barnacle *Balanus amphitrite*. *Marine Biology* 143: 543-554.

Thomason, J.C., Hills, J.M., Clare, A. & Richardson, A.N. (1998). Hydrodynamic consequences of barnacle colonization. *Hydrobiologia* 375/376: 191-201.

Tremblay, R., Oliver, F., Bourget, E. & Rittschof, D. (2007). Physiological condition of *Balanus amphitrite* cyprid larvae determines habitat selection success. *Marine Ecology Progress Series* 340:1-8.

Vial, M.V., López, D.A., Simpfendörfer, R.W. & González, M.L. (1999). Responses to environmental hypoxia of balanomorph barnacles. In M. F. Thompson & R. Nagabhushanam (Eds.), *Barnacles. The Biofoulers*. Regency Publications, New Delhi, India, 217-244.

Vike, S., Nylund S. & Nylund, A. (2009). ISA virus in Chile: Evidence of vertical transmission. *Archives of Virology* 154 (1): 1-8.

Walker, G.(1995). Larval Settlement: Historical and Future Perspective. In F. R. Schram, F. R. & J. T. Hoeg. (Eds), *Crustacean Issues 10, New frontiers in barnacle evolution*, A. A. Balkema, Rotterdam, Netherlands, 69-85.

Yan, Y. & Chan, B.K. (2001). Larval development of *Chthamalus malayensis* (Cirripedia: Thoracica) reared in the laboratory. *Journal of the Marine Biological Association of the United Kingdom* 81 (4):623-632.

Yan, Y. (2003). Larval development of the barnacle *Chinochthamalus scutelliformis* (Cirripedia: Chthamalidae) reared in the laboratory. *Journal of Crustacean Biology* 23 (3): 513-525.

Young, P.S. (1998). Maxillopoda. Thecostraca. *Catalogue of Crustacea of Brazil*. Museo Nacional, Rio de Janeiro, Brazil.

Young, P.S. (2000). Cirripedia thoracica (Crustacea) collected during the "Campagne de La Calypso (1961-1962)" from the Atlantic shelf of South America. *Zoosystema* 22 (1): 58-100.

16

Culture of Harpacticoid Copepods: Understanding the Reproduction and Effect of Environmental Factors

Kassim Zaleha and Ibrahim Busra
Universiti Malaysia Terengganu
Malaysia

1. Introduction

The availability of highly nutritive and inexpensive live food is a subject of major concern in aquaculture industry particularly for the fry production activity. For so long, the aquaculture industry has relied primarily on brine shrimp and rotifers to provide the necessary nutrition for rearing the early life stages of fish and crustaceans. Brine shrimp and rotifers are almost not suitable as first feed for fish larvae due to inappropriate size (too large or too little) (van der Meeren, 1991; Pepin & Penney, 1997), their swimming behavior makes them less susceptible to predation (Buskey et al., 1993; von Herbing & Gallagher, 2000), and insufficiently nutritious (Støttrup & Norsker, 1997). Because of these problems, brine shrimp and rotifers become uneconomical as live food. This initiate the great interest in the identification of alternative live feeds to meet the demands of the aquaculture industry.

Copepods are among the most abundant and important components of aquatic invertebrates in many marine and freshwater ecosystem. They become the major biomass in zooplankton community in the water column, and at many occassions they form significant entity of benthic community structure on bottom sediment.They are a major food source for organisms in higher tropic level such as juvenile fishes and shrimps (Vincx, 1996; Penchenik, 2005) thus they become link between primary producer and other consumers in the natural food-web.

There have been a number of studies providing evidences of the effectiveness of copepods as a food item, and many investigators reported a good growth and survival of fish larvae and crustaceans fed with copepods as test food organisms (van de Meeren, 1991; Holmefjord et al., 1993; Naess et al., 1995; Støttrup & Nosker, 1997). This is due to the good nutritional profile of copepods which meets the nutritional requirements of marine fish and crustaceans larvae as they are rich in essential fatty acids (Støttrup & Norsker, 1997).

Amongs the copepod group, harpacticiod copepods are known to be mostly benthic (Pechenik, 2005), meaning they live on bottom habitat such as sediment or bottom vegetation. They could live as epi-(dwelling on the bottom surface), endo-(burrowing into bottom surface) or mesobenthos (living between grain particles or in pore water in the bottom sediment). As they lead a benthic living mode, they depend on the diets available on bottom habitat including phytobenthos, microbes and detritus.

Harpacticoid copepods has been found to be a good candidate in aquaculture industry since they have a high reproductive potential, short generation time, high population growth, flexible in diet and tolerate a wide range of environmental factors such as temperature and salinity (Sun & Fleeger, 1995; Støttrup & Norsker, 1997). They provide a broad spectrum of prey sizes suitable for fish larvae (Gee, 1989). There are reports that indicate the availability of enzyme in harpacticoid copepods which enable the organisms to convert any type of their organic food into lipids stored in their body (Nanton & Castel 1998; Drillet et al. 2011).

The prominent problem in using copepods as live feed in aquaculture industry is to get high yield due to several reasons which mainly related to the culture technique. Pure culture is difficult to maintain in ponds or in hatchery on a continuous basis. Although continuous cultivation of marine copepods has been achieved, but it is only for a small number of species. Several species of harpacticoid copepods that have been cultivated either for commercial use or laboratory level are *Tisbe holoturidae*, *Tisbe beminiensis*, *Nitokra lacustris* and *Tigriopus japonicus*. As reviewed by Rippingale and Payne (2001), harpacticoid copepods especially *Tisbe* spp (Støttrup & Norsker, 1997) and *Tigriopus* spp (Carli et al., 1995) were the easiest to culture. Nevertheless, the technology to economically culture marine copepods to get highly enough yield is still being developed to date (Marcus & Murray, 2001; Rhodes, 2003; Drillet et al., 2006).

Lacking in information about the basic biology, reproduction and response of copepods in rearing environment could be one of the reason why mass culture technique is yet to be successful for copepod production. Harpacticoid copepods in particular need special attention as they are predominantly benthic in nature. This chapter will discuss on important aspects in rearing harpacticoid copepods as live feeds in aquaculture.

2. Reproductive biology and effect of environmental factors

The first step to understand how to culture harpacticoid copepod is to know the reproductive biology of this animal when kept under culture condition where temperature and salinity would easily change. Copepods such as *Calanoides acutus*, *Rhincalanus gigas*, *Calanus propinquus*, *Paraeuchaeta Antarctica* showed a higher metabolic rate in the summer than in the winter (Kawall et al., 2001). It seemed that increasing in temperature will increase the respiration rate and it has positive correlation with the relative biomass of copepods (Le Borgne, 1982). For harpacticoid copepods, a number of researchers have documented that temperature is the major factor controlling the reproductive activity in the harpacticoid copepods. For example, the optimum temperature for the maximum production of the *Tisbe battagliai* in the laboratory condition was 20°C (Williams & Jones, 1999).

The optimal conditions for the temperate is differ from the tropical species due to its natural environment. Local adaptation in term of reproductive traits could happen in order for the copepod to maximise available energy with changing temperature. This was shown by the population of harpacticoid *Scottolana canadensis* from different lattitude (Lonsdale & Levinton, 1986; 1989). William & Jones (1999) noted that under optimal conditions in the laboratory (high quantity of algal food and temperature 20°C), *Tisbe battagliai* developed rapidly from hatching to the adult stage, have a rapid generation time and produce in quick succession, numerous broods containing large numbers of offspring. Females that were reared at 15°C lived twice as long as compared to 25°C. Increasing in temperature (>25°C) will decrease the number of offspring per day.

Salinity affects the functional and structural responses of invertebrates through many aspect such as changes in total osmo-concentration, relative proportion of solutes, coefficient of absorption and saturation of dissolved gases (Kinne, 1964). Changes in total osmo-concentration could considerably modify rates of metabolism and activity. This is because of the influence of salinity variations on several biochemical and physiological mechanisms which the function is essential for survival in marine organisms (Fava & Martini, 1988). A study on harpacticoid copepod, *Tigriopus californicus* showed that high salinity slowed down their growth rate and delayed the maturation time. A decrease in reproductive rate at increasing salinity caused moderate decrease in total number of clutches (Dybdahl, 1995). Gaudy et al. (1982) reported that the egg sac production and offspring survival of *Tisbe holothuriae* increased at higher salinities (28-30 ppt). However, the net reproductive rate of *T. holothuriae* declined at salinities above and below 38 ppt (Miliou & Moraitou-Apostolopoulou, 1991). The production of fewer offspring per egg sac and an increment in mortality had been observed.

The salinity tolerance was different at each life cycle stage, and in different sex (Damgaard & Davenport, 1994). In most species, the narrowest tolerance rate is during very early stages, which later increased and finally decreased again in the adult stage. The narrowest range of tolerance is often associated to the period of embryonic development. Nauplii are commonly less sensitive, and during early growth could even tolerate wider salinity range and more pronounced to salinity fluctuations than adult individuals (Milione & Zeng, 2008).

2.1 Investigation on reproductive biology of a harpacticoid copepod, *Pararobertsonia* sp.

A harpacticoid copepod, *Pararobertsonia* sp. was first collected during an ecological survey in an estuary in Terengganu, Malaysia (5° 02.260' N, 103° 17.821' E). The estuary is located at the river mouth of Merchang River, meeting the South China Sea. It is inhabited by large coverage of small seagrass, *Halodule pinifolia* and oysters farming activities by local people. During field sampling, a 62μm net was towed horizontally on the surface of exposed seagrass patches to catch the copepods. The trapped samples were placed into container filled with seawater and aeration was provided. The collected samples were transported back to Institute of Oceanography Laboratory, Universiti Malaysia Terengganu (UMT).

In laboratory, isolation process was carried out to separate *Pararobertsonia* sp. from other benthic copepods. Harpacticoids carrying egg sacs were isolated from samples under a Leica ZOOM 2000 dissecting microscope using an Irwin Loop and placed individually into a petri dish 60 X 15 mm containing 5 ml of filtered-autoclaved seawater (Støttrup & Norsker, 1997) and 1 ml of baker's yeast (0.02 g/L) (Nanton & Castell, 1998). The experiment was maintained closest to the normal habitat condition at temperature 26°C to 27°C and salinity 24 to 26 ppt. Each plate was initiated with single gravid copepod to ascertain monospecific culture. The culture was maintained separately until the identification was carried out for species confirmation.

After being isolated and identified, there are several steps need to be completed before culture of harpacticoid can be successfully maintained in laboratory for any investigation. These include activities such as breeding and up-scaling under laboratory condition. Water as culture media need to be treated and conditioned to meet the copepod's environment requirement.

Data recorded were analysed using statistical analysis of variance (ANOVA) at 0.05 level of probability. Kolmogorov–Smirnov tests were used to test for the normality respectively before the comparisons. The differences on the effects of temperature and salinity were compared using Univariate ANOVA. One-way ANOVA was used to test the differences on the effects of pH. Tukey's multiple comparison procedure was used to compare the significant differences between treatment means. All statistical analyses were performed using SPSS program, version 11.5.

2.1.1 Establishing a stock culture

Seawater was filtered through GFC membrane filter after passed through UV treatment before use. A 500 ml and 1000 ml beakers were used as culture vessel. The culture was maintained at temperature between 26°C to 27°C and salinity between 22 to 26 ppt. Culture was fed daily with 1 ml of baker's yeast (0.02 g/L) (Nanton and Castell, 1998) and with mixed algal (*Isochrysis* sp., *Nannochloropsis* sp., *Chaetoceros* sp.) at the density of 1×10^6 cell/ml. No aeration was provided for the 500 ml and 1000 ml stock cultures. The partial replacement of seawater was done every three days and full replacements at every month by passing the cultures through a 45 μm mesh net. The trapped copepods were then transferred into new culture vessel.

2.1.2 Determination of eggs' number

Total of 150 to 200 individual copepodids regardless of sex were isolated from the stock culture and observation was made several times everyday for any gravid female. After few days, 30 gravid females were collected and transferred into a vial and fixed with a few drops of 5% buffered formalin. The number of eggs was counted under Leica DME compound microscope for each individual.

2.1.3 Determination of lifespan, egg production and development

Prior to the experiments, 30 mating pairs were isolated and placed individually in a small petri dish (60 X 15 mm) containing 5 ml new culture medium. The cultures were maintained at temperature 26-27°C and salinity of 22-26 ppt. *Chaetoceros* sp. was offered at the density 1 $\times 10^6$ cell/ml. Each dish was examined daily for the first appearance of the egg sac before removing the males. This was done to ensure the number of sacs per female from its first copulation.

Each dish was examined daily at six hours intervals each day to ascertain the time of egg release and the appearance of the next egg sac. Once a female released all eggs, it will be transferred to a new dish with fresh culture medium for further observation. This was repeated until all of the females died. The daily routine consisted of checking each individual for egg development and duration under a Leica ZOOM 2000 dissecting microscope. Food was added every other day in the same concentration of 1×10^6 cell/ml. The total number of egg sacs per female, maturation time of egg sacs (time between the appearance of the egg sac and its hatching), interval time between egg sacs (time between hatching and the appearance of the next egg sac for one fertilization) and lifespan of female were determined.

2.2 Effect of temperature and salinity on reproductive biology of a harpacticoid copepod, *Pararobertsonia* sp.

Gravid females of *Pararobertsonia* sp. from the first copulation were selected for this study as to ensure that the next generation (nauplii) were exposed to the specified temperature and salinity regime designed in the experiment. Table 1 summarizes the combination of treatments designed for the harpacticoids.

Temperature (oC)	Salinity (ppt)	Treatment Label	Number of replicate
5	5	T_LS_L	5
	25	T_LS_C	5
	45	T_LS_H	5
25	5	T_CS_L	5
	25	T_CS_C	5
	45	T_CS_H	5
45	5	T_HS_L	5
	25	T_CS_C	5
	45	T_HS_H	5

Table 1. Combination of different temperature and salinity treatment for the harpacticoid copepod culture (Key: T=Temperature; S=Salinity; L=Low; H=High; C=Control)

Nine different treatments were prepared ($T_L S_L$, $T_L S_C$, $T_L S_H$, $T_C S_L$, $T_C S_C$, $T_C S_H$, $T_H S_L$, $T_H S_C$ and $T_H S_H$) to determine the effects of different temperature and salinity on the reproduction and development of *Pararobertsonia* sp. The experiment was conducted in three different temperatures; 5°C, 25°C and 45°C. In each temperature setup, copepods were treated with three salinity levels; 5 ppt, 25 ppt, and 45 ppt. All copepods were gradually exposed to the lower (5°C) and higher (45°C) temperature from the control value (25°C) before the experiment started to avoid shock effect which would instantly kill the animal. The same treatment was also applied for salinity exposure.

2.3 Results

2.3.1 Lifespan, egg production and development in *Pararobertsonia* sp.

Lifespan of 30 individual females of *Pararobertsonia* sp. was determined in this experiment by measuring the length of time from first day of hatching as nauplii until they died under laboratory control condition at temperature 25°C and salinity 25 ppt. The average lifespan of a female *Pararobertsonia* sp. was 31.2 ± 3.57 days with a minimum and maximum of 26.0 and 40.0 days respectively.

The interval time between the production of egg sac is measured as time between hatching and the appearance of the next egg sac from one fertilization event. The maturation time of egg sac was determined as the time between the appearance of eggs and its hatching time. A female of *Pararobertsonia* sp. appears to produce several egg sacs from one fertilization event. The total number of egg sac per female had large fluctuation and significantly different among the 30 individual copepods observed. Total number of egg sacs per female was 6.7 ± 2.54 egg sacs in average, where the minimum and maximum was 3.0 and 12.0 egg sacs respectively. Mean total number of eggs per sac was 21.7 ± 4.79 eggs. The total number of eggs per sac ranged between 14.0 and 30.0 eggs per sac.

The time of formation between first egg sacs and the next was recorded within 0 to 11 hours. The egg sac could be seen in the oviduct within minimum hours range from 0 to 11 hours and maximum hours range from 60 to 71 hours after the previous egg release. The time between hatching and the appearance of the next egg sacs was significantly frequent within 0 to 11 hours with 65.88 % and less frequent within 60 to 71 hours with 1.18 % (Figure 1).

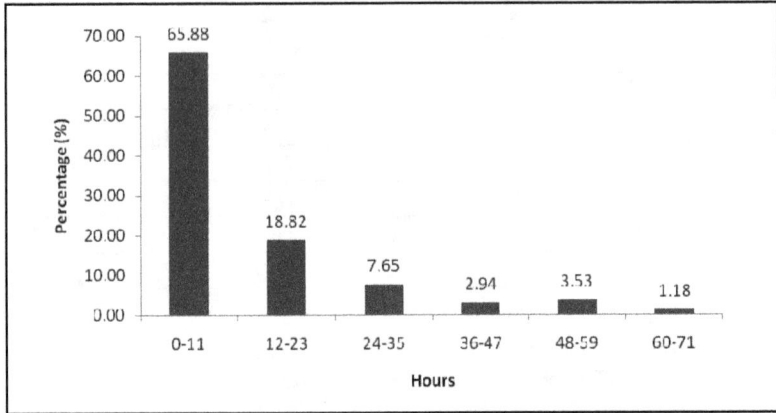

Fig. 1. The percentage (%) of interval time between egg sacs of a female *Pararobertsonia* sp. under laboratory control condition (temperature 25°C; salinity 25 ppt).

A female could have different maturation time for each interval of their production of egg sac. Figure 2 shows the percentage (%) of maturation time for each individual female to carry their eggs until its hatching. Time (hours) recorded for eggs to be released and hatched

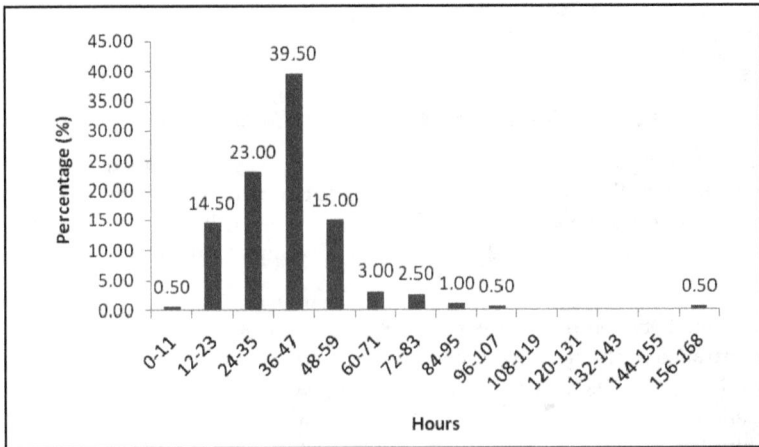

Fig. 2. The percentage (%) of eggs and their maturation time for a female *Pararobertsonia* sp. under laboratory controlled condition (temperature 25°C; salinity 25 ppt).

was within 0 to 11 hours until 156 to 167 hours. However, most of the eggs (39.50 %) were matured and hatched within 36 to 47 hours while only 1 % took 0 to 11 hours, 96 to 107 hours and 156 to 168 hours to hatch. It shows that the maturation time of eggs was shorter at early and longer for the late egg sacs produced.

2.3.2 Effect of temperature and salinity

In general, there is no significant effect of different salinity on the production of eggs/sac of copepod for each temperature treatment (Table 2). Despite the large difference in salinity (5-45ppt) , the number of eggs produced in each egg sac of a female was more or less the same when cultured under the same temperature. Decrement of temperature reduced the number of eggs/sac more than when they were exposed to the increment of temperature to 45°C. Copepods cultured in 25°C produced significantly high number of eggs if compared to 5°C and 45°C.

Temperature (°C)	Salinity (ppt)	Treatment Label	Number of eggs/sac
5	5	T_LS_L	11.6±1.14
	25	T_LS_C	10.8±2.59
	45	T_LS_H	11.6±1.82
25	5	T_CS_L	24.4±4.72
	25	T_CS_C	23.8±2.87
	45	T_CS_H	23.8±8.14
45	5	T_HS_L	22.8±5.76
	25	T_CS_C	19.8±7.85
	45	T_HS_H	15.0±1.41

Table 2. The number of eggs/sac/female copepod *Pararobertsonia* sp. reared under different temperature and salinity

The number of offspring and their survival is shown in Figure 3. No nauplii survived to the next life-stage at 5°C. On the other hand, survival significantly increase (p<0.05) at 25°C. Copepods in 45°C treatment showed the same pattern of development as in low temperature (5°C). Increase of temperature up to 45°C still permit gravid females to produce high number of eggs/sac despite being exposed to different salinity although no survival to adult stages were observed.

Table 3 summarizes the data on maturation and generation time from the present study. Different temperatures were significantly affected (ANOVA, P<0.05) the maturation time of egg sacs while no significant difference (ANOVA, P>0.05) was observed among the salinities in every temperature treatments. The mean maturation time of egg sacs was shortest at high temperature (45°C) and longest at control temperature (25°C).

No data on generation time was recorded for 5°C due to the delay in the development of nauplii into the copepodite stage throughout the study period (20 days). At 45°C, the only data obtained was on generation time from nauplii to copepodite as no copepods survived to the adult stage. The mean generation time from copepodite to adult at 25°C was longest in low salinity (3.2 ± 0.84 days) and shortest in high salinity (2.0 ± 0.71 days). However, the mean generation time from nauplii to gravid female was longest in high salinity with 8.8 ±

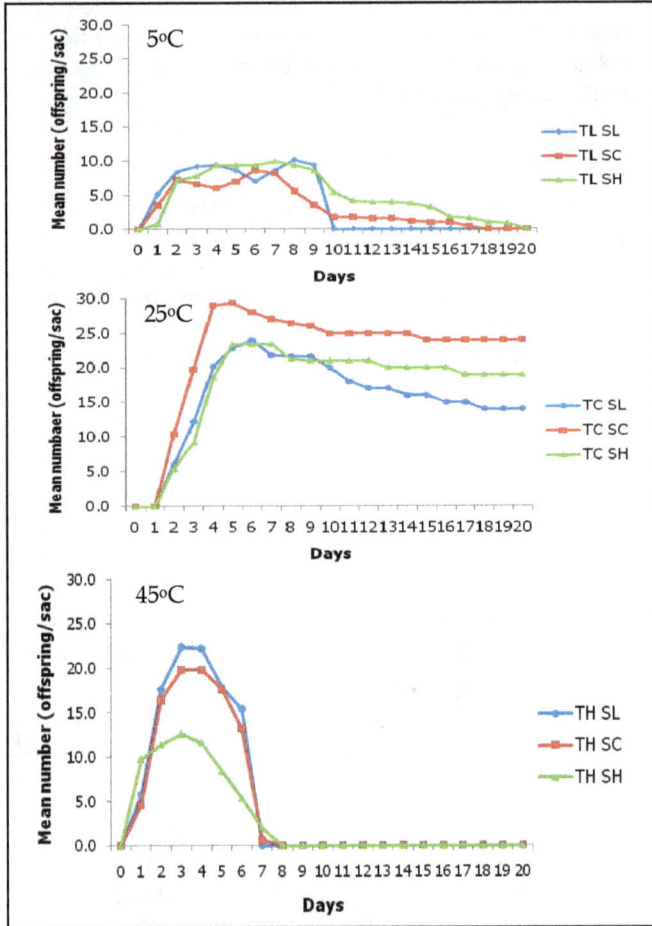

Fig. 3. Mean number of offspring per egg sac per female harpacticoid *Pararobertsonia* sp. at low (5ₒC), control (25ₒC) and high (45ₒC) temperature and low (5ppt), control (25ppt) and high (45ppt) salinity. Key: T=Temperature; S=Salinity; ʟ=Low; ʜ=High; c=Control

1.10 days. The shortest generation time was in control condition 25°C and 25ppt (7.3 ± 0.96 days). The longest generation time from N1 to C1 was obtained from copepod reared in high salinity with 3.8 ± 0.45 days while the shortest of the generation time was observed in low salinity with 2.2 ± 0.45 days.

2.4 Discussion

2.4.1 Reproductive biology and development of *Pararobertsonia* sp.

Detail report on reproduction and development in copepods particularly harpacticoids were mostly from the temperate species (Hicks & Coull, 1983; Sun & Fleeger, 1995; William & Jones, 1999; Rhodes, 2003). *Pararobertsonia* sp. is a tropical harpacticoid that has been

Temperature (°C)	Treatment	Maturation time of egg sacs (days)	Generation time from N1-C1 (days)	Generation time from C1-Adult (days)	Generation time from N1-Gravid (days)
5	$T_L S_L$	1.6[a]±0.55	0.0±0.00	0.0±0.00	0.0±0.00
	$T_L S_C$	2.0[a]±1.22	0.0±0.00	0.0±0.00	0.0±0.00
	$T_L S_H$	2.0[a]±0.71	0.0±0.00	0.0±0.00	0.0±0.00
25	$T_C S_L$	2.6[a]±0.89	2.4[a]±0.55	3.2[a]±0.84	8.6[a]±1.52
	$T_C S_C$	2.0[a]±0.00	2.3[a]±0.50	3.0[a]±0.82	7.3[a]±0.96
	$T_C S_H$	2.2[a]±0.45	3.8[a]±0.45	2.0[a]±0.71	8.8[a]±1.10
45	$T_H S_L$	1.2[b]±0.45	2.2[a]±0.45	0.0±0.00	0.0±0.00
	$T_H S_C$	1.0[b]±0.00	2.4[a]±0.55	0.0±0.00	0.0±0.00
	$T_H S_H$	1.0[b]±0.00	3.3[a]±0.58	0.0±0.00	0.0±0.00

Table 3. Maturation time of egg sacs (time between the appearance of egg and hatching), generation time from Nauplii I to Copepodite I, generation time from Copepodite I to Adult and generation time from Nauplii I to gravid female of *Pararobertsonia* sp. at different temperatures (5°C, 25°C and 45°C) and salinities (5 ppt, 25 ppt and 45 ppt). Means in the column with the same superscript are not significantly different (P>0.05).

successfully cultured in our laboratory environment for many generations since 2007, but the detail of the reproductive biology and development stage has never been reported before. The probability for a female *Pararobertsonia* sp. to produce multiple egg sacs from one fertilization event is very high, the same as for other harpacticoids. The gravid female can be gravid for several times in a short period without remating. Hicks & Coull (1983) reported that the number of sacs produced from a single copulation vary from four to 12 for five species of *Tisbe* and three to 21 for other 21 species of harpacticoids. A female of *Tisbe biminiensis* produced up to nine egg sacs during its life (Pinto et al., 2001), while *Tisbe battagliai* fed on *Isochrysis galbana* Parke produced 5.3 ± 2.2 egg sacs when cultured under temperature 25°C and salinity 25 ppt (William & Jones, 1999).

Many researchers have shown that egg production in copepod is different for each individual. A study by Guérin et al. (2001) showed that the number of eggs per sac produced by *Tisbe holothuriae* has large fluctuation, ranging from the maximum of 133 to a minimum of only three eggs in an egg sac. Sun and Fleeger (1995) found that a harpacticoid *Amphiascoides atopus* (Diosaccidae) typically carries two egg sacs and the average brood size was 24 eggs per ovigerous female. The mean number of eggs per sac for *T. biminiensis* fed on *Nitzchia closterium* was 69.0 ± 24.6, *Tetraselmis gracilis* was 40.6 ± 16.1 and mixed of *N. closterium* and *T. gracilis* was 46.0 ± 17.1 eggs (Pinto et al., 2001). *Pararobertsonia* sp. in the present study produced the lower mean number of eggs per sac (21.7 ± 4.79) than above studies which could be related to the type of the given diet.

Maturation time as well as interval time between egg sacs for every individual of the same species is reported to be different and varies among each other (Tester & Turner, 1990).

These differences may be due to inherent biological variability which is determined partly by genetic differences. In the present study the maturation time of egg sac of individual *Pararobertsonia* sp. showed wide variation (0 – 167 hours). Most egg hatching took placed within 36 to 47 hours, comparable to maturation time of most harpacticoid copepods which is within 48 hours (Dam & Lopes, 2003). Most females took the shortest time (within 11 hours) for the duration between hatching and the appearance of the next egg sac. By comparison, *T. biminiensis* took two days to produce a new egg sac (Pinto et al., 2001).

The lifespan of *Pararobertsonia* sp. cultured in temperature 25°C, salinity 25 ppt and fed on *Chaetoceros* sp. was about 31.2 ± 3.57 days, which is within the normal range. In comparison, lifespan of *T. biminiensis* cultured in temperature 28-30°C, salinity 34 ppt and fed on *N. closterium*, *T. gracilis* and mixed of both diet was about 29.0 ± 7.2, 32.9 ± 4.9 and 32.7 ± 4.6 days respectively (Pinto et al., 2001). William & Jones (1999) reported on lifespan of *T. battagliai* when fed on *Isochrysis galbana* Parke was 23 ± 4.9 at temperature 25°C and salinity 25 ppt. A number of factors including the difference in species, quantity and quality of the food source and environmental condition including temperature and salinity could affect the reproduction and lifespan of the species.

2.4.2 Effects of temperature on reproduction and development

Temperature is often the most important environmental factor affecting the productivity of copepods in natural systems (Christou & Moraitou-Apostolopoulou, 1995; Siokou-Frangou, 1996). In general, the effects of temperature on marine copepods are well studied, but information on the effects of temperature for tropical copepod species are relatively limited. Many previous works have employed the effects of temperature on temperate harpacticoids species sush as *Tisbe* (Hicks & Coull, 1983; Miliou & Moraitou-Apostolopoulaou, 1991; Williams & Jones, 1994). Limited number of tropical species has been used as subject matter to study the effects of temperature in culture condition, which includes *Pararobertsonia* sp. (Zaleha & Farahiyah-Ilyana, 2010) and *Nitocra affinis* (Matias-Peralta et al., 2005).

Results from the present study showed how temperature affecting the offspring production, survival, maturation time of eggs and generation time in *Pararobertsonia* sp.The differences in temperature significantly affects the reproductive and development rate of *Pararobertsonia* sp. Low temperature delayed the development whereas high temperature increased the development rate but extreme temperatures (>45°C) could lead to mortality. Temperature stress may have negative effects on survival and reproduction. These findings were relatively similar to the study of Takahashi & Ohno (1996). The later study found that *Acartia tsuensis* (Copepoda: Calanoida) could develop normally from egg to adult within a temperature range of 17.5 to 30°C while optimum growth and minimum mortality were achieved at around 25°C. However, the development slowed at both lower and higher ends of temperature at 17.5°C and 30°C. Similar trends have also been observed in other copepod species, such as *Diaptomus pallidus* (Geiling & Campbell, 1978) and *Acrocalanus gibber* (McKinnon, 1996).

Increased and decreased temperature also affected egg production of *Pararobertsonia* sp. where the egg production at decrement or increment temperature was lower compared to control temperature. This is because *Pararobertsonia* sp. could not adapt the environmental stress that inhibits normal gonad development and consequently affected the average number of eggs produced per female. This study is consistent with previous study by

Miliou & Moraitou-Apostolopoulou (1991). The later study reported that a reduction in the number of egg sacs and the total number of offspring produced by the Greek strain of *Tisbe holothuriae* was observed when the temperature was lower or higher than the optimum (19°C). In addition, Ambler (1985) and Uriarte et al. (1998) revealed that egg production is normally lower at low temperatures and generally increase with increasing temperature up to a thermal threshold, after which decline begins. Such a trend has been reported for the egg production of *Acartia tonsa* (Ambler, 1985), *A. clausi* (Uye, 1985), *A. bifilosa* (Uriarte et al., 1998), *A. lilljeborgi* (Ara, 2001). A study on *Boeckella hamata* (Copepoda: Calanoida), showed that clutch size decreased with increased temperature (Hall & Burns, 2002).

Survival of *Pararobertsonia* sp. in this present study was solely affected by temperature rather than salinity. Some previous studies showed that temperature has a direct influence on the survival of copepod (Peterson, 2001). The negative effects of higher temperature on the survival of *Pararobertsonia* sp. are similar with other previous studies. Survival of copepods and cladocerans were better at low temperature than at high temperature (Moore et al., 1996). Chinnery & Williams (2004) found that survival, egg production and hatching rate of *A. bifilosa*, *A. clausi*, *A. discaudata*, and *A. tonsa* increased when temperature rise from 5 to 20°C. Although increasing temperature showed the positive effects, the development and survival reduces as temperature rises beyond a certain level (Chinnery & Williams, 2004). For example, temperatures greater than 30°C was unfavourable to survival, percent ovigerous females, and fecundity of *Pseudodiaptomus pelagicus* (Copepoda: Calanoida) (Rhyne et al., 2009).

Slight increment or decrement in temperature affected the maturation time of the egg sacs of *Pararobertsonia* sp. There are only few laboratory studies that have explored the relationship between temperature and development rate in the life history stages of harpacticoid copepods (Palmer & Coull, 1980). The present study revealed similar finding as previously described by Chandler et al., (2003). The later study reported that eggs of harpacticoid copepod, *Amphiasucus tenuiremis* were developed in two to three days at temperature 25°C and salinity 30 ppt. Temperature give significant effects on the maturation time of eggs sac of copepods (McLaren et al., 1969). Evidence from some laboratory studies proved that temperature has strong influence on reproductive and postembryonic development (Hicks & Coull, 1983; William & Jones,1999; Matias-Peralta et al., 2005). Rhyne et al. (2009) found that 26 to 30°C was the best range for nauplii production while 28 to 32°C was the best for fast maturation rate of nauplii. A study by Williams & Jones (1994) also noted that a benthic harpacticoid, *Tisbe battagliai* has their best temperature at 20°C and increasing of temperature towards 25°C decreased the production rate.

Harpacticoid copepods have short generation time as reviewed by Sun & Fleeger (1995), Chandler et al., (2003) and McKinnon et al. (2003). However, data on the duration of the larval stages of marine harpacticoid copepods and the influence of environmental factors (specifically temperature) and their potential interaction on postembryonic development are relatively limited (Williams & Jones, 1994). The generation time of *Pararobertsonia* sp. found from this present study clearly shown that temperature do affect the period between stages of life cycle. Increment in temperature decreased the development time while decrement in temperature increased the development time of *Pararobertsonia* sp. Hicks & Coull, (1983) found the similar finding where the development time of *Tisbe* sp. decreased with increasing temperature. *Tisbe* sp. required high temperature (up to 25°C) for faster development. In addition, mean development time of calanoid copepod, *Pseudocalanus newmani* decreased exponentially with increasing temperature and reached the shortest duration at 32°C (Rhyne et al., 2009). Similarly, Williams & Jones (1994) clarified that fastest development for harpacticoid copepod *T. battagliai* occurred in the warmer months of the year which regarding to the highest production of superior food. However, in the present study, food was given at constant concentration.

The development time of *Pararobertsonia* sp. at increment temperature was two times faster compared to control and decrement temperature. Low temperature can retard activity of organisms and consequently reduces the consumption of oxygen. Whereas, high temperature can increase oxygen consumption to one point where metabolic demands exceed energy reserved. A calanoid copepod, *Pseudocalanus newmani*, was reported to delay the development time from 20.9 to 42.3 days when temperatures decreased from 15 to 6°C (Lee et al., 2003). This trend was also observed in *A. clausi*, where development time delayed from 35.4 to 74.8 days when temperatures decreased from 10 to 5°C (Chinnery & Williams, 2004).

Some previous reports on the respond of tropical copepods to temperature stress presented similar finding with this present study. Milione & Zeng (2008) stressed the effects of temperature on both population growth and hatching rates. The later study suggested that for maximum population growth and egg hatching success of a tropical calanoid copepod, *A. sinjiensis* should be cultured at 30°C with a salinity of 30 psu. Likewise, Matias-Peralta *et al.* (2005) showed that the *N. affinis*, grow well and achieved highest maximum production (124.2 ± 2.6 offspring female[-1]) at temperature 35°C. In comparisons, Zaleha & Farahiyah-Ilyana (2010) reported that temperature of 25 ± 1°C and high salinity (25 ppt - 35 ppt) were the optimum condition for the maximum production (2.3 - 3.7 individual/ml) of a tropical *Pararobertsonia* sp. in the laboratory condition.

In this study, the survival and reproductive parameters of individual *Pararobertsonia* sp. in different temperature treatments showed wide variation. These differences caused by inherent biological variability and physiological response. Thermal stress caused energy to be allocated toward survival processes rather than reproduction. This may also be explained based on the study by Williams & Jones (1999) where they reported that nauplii production of *T. battagliai* ceased after 20 days at 25°C while lower temperature treatments continued to produce nauplii for 36 days. The adaptation of some individuals will be better than others in respond to environmental stress due to the attributes of some individuals to establish their population (Depledge, 1990; 1994). Every metabolic rate of zooplankton such as respiration and feeding rate is dependent on temperature (Heinle, 1969). In this present study, *Pararobertsonia* sp. grew well and achieved high reproductive activity at temperature 25°C, the same temperature for the stock which has been maintained for two years. In contrast, the copepods were exposed to daily change towards the required temperature in the other two experiments. This might be the reason for the different adaptability and productivity found in this study as they need to tolerate and respond to the environmental stress everyday and it could be affecting their physiological response.

2.4.3 Effects of salinity on reproduction and development

Although temperature is recognised as an important factor controlling reproduction in harpacticoid copepods, there are other factors which regulate reproductive activity including food resource availability, environmental stability and their effects on the evolution of particular life-history strategies. Some researchers revealed that salinity is one of the main environmental factors controlling species distribution, the rates of growth, developments in larvae stages and reproduction of harpacticoid copepod, especially on those with restricted capacity of osmoregulation (Miliou & Moraitou-Apostolopoulou, 1991; Miliou, 1996).

Generally, the results of this study apparently showed that *Pararobertsonia* sp. could tolerate in wide variation of salinities ranging from 5 to 45 ppt. Similar results were reported by Matias-Peralta et al. (2005). They clarified that *Nitocra affinis*, a tropical harpacticoid could tolerate in salinities from 10 to 35 ppt. In addition, study done by Sun & Fleeger (1995), they found that a harpacticoid *Amphiascoides atopus* was able to survive in a wide range of salinities (10 – 60 ppt). Different salinities showed to have different effects on offspring production and survival of *Pararobertsonia* sp. Conversely, there are no effects shown on the number of eggs per sac, maturation time and generation time. Devreker et al. (2009) reported that the combination of salinity and temperature have different effects on the physiology of an estuary calanoida copepod, *Eurytemora affinis*. High salinities are most stressful for *E. affinis* at high temperatures (Kimmel & Bradley, 2001). Survival of *E. affinis* could strongly decreased when high salinities are combined with high temperatures (Gonzalez & Bradley, 1994).

High survival (more than 80%) of *Pararobertsonia* sp. was achieved under salinity 25 to 45 ppt. Extreme low salinity (5ppt) could give significant effect on their survival. Nevertheless, they can survive in the laboratory cultures when salinity dropped gradually into 0 ppt (personal observation). Supporting this finding, Gaudy et al. (1982) stated that decreases in salinity resulted to high mortalities of *Tisbe holothuriae*. Staton et al. (2002) found that there is a non-linear survival response of *Microathridion littorale* (estuarine harpacticoid copepod) to short term immersion of 24 hours in 3, 12 and 35 psu. Copepods that were transferred in the 12 psu showed the lowest survival rate.

Although *Pararobertsonia* sp. has been observed to survive better at higher salinity compared to lower salinity, the generation time from nauplii to gravid female was longer under high salinity. However, the generation time from copepodite to adult was shorter under high salinity. This difference could be related to the physiological difference existing between first naupliar and late copepodite stages. Similar result was reported by Hagiwara et al. (1995) for *Tisbe japonicus*. They found that development time of *T. japonicus* was fastest at higher salinities (16 -32 ppt) and growth rate tended to be slower in low salinity.

Pararobertsonia sp. were able to reproduce and survived from the first egg sac hatched as nauplii to gravid female under different salinities ranging from 5 to 45 ppt at temperature 25°C compared to higher and lower temperature. It is clearly shown that *Pararobertsonia* sp. could tolerate to salinity changes rather than temperature. As reported by Devreker *et al.*, (2009), the development of estuary calanoid copepod, *E. affinis* appeared to be more sensitive to temperature than salinity. Lee & Petersen, (2002) revealed that temperature-salinity interaction effect on salinity and temperature tolerance. The tolerance to temperature and salinity stress is controlled by a group of regulatory genes as documented by Kimmel & Bradley (2001). The genotype controls the synthesis of proteins necessary for metabolic activity. Consequently, this difference in genotype modified the metabolic performance as a function of environmental conditions.

Under control temperature (25°C), *Pararobertsonia* sp. showed the ability to survive and even breed in a salinity ranged of 5 to 45 ppt. This might be due to its great adaptation and osmoregulatory ability. Kimmel & Bradley (2001) demonstrated that salinity variations induce synthesis or degradation of amino acids during osmoregulation. This generates an increase in the consumption of protein reserves as well as in energy requirements for enzymatic activity. Without energy renewal, this stress decreases copepod survivorship and causes death of the nauplius in the early stages. The present study suggests that there is no salinity stress at this range of salinity because the nauplii can develop normally to adult stage. However, harpacticoids did not survive when they were transferred directly into the salinity 5 ppt (personal observation). As reviewed by Staton et al. (2002), they noted that exposure of low salinity in more than 24 hours for *Microathridion littorale* (estuarine harpacticoid copepod) could lead to mortality.

Pararobertsonia sp. is regarded as an estuarine harpactiocid copepod which is considered to be exposed to large fluctuations of salinity due to tidal cycle daily. Therefore, this species could already adapted to salinity fluctuation more than the temperature changes, thus become less affected by the salinity changes in the experiment. Goolish & Burton (1989) confirmed that the variability in individual's physiological response to salinity changes was due to the salinity history of organism and species specific hereditary traits.

3. Conclusion

Harpacticoid copepods are potential candidate as live feed in aquaculture. They have most of the required characteristics to replace artemia and rotifers as starter food for newly hatched fish and shrimp larvae. Nevertheless, the mass production of copepods as live feed for aquaculture purposes is still at the experimental stage and success story is limited to only few copepod species. Understanding the basic biology of the species in culture condition will help in planning and handling the copepod culture for mass production. An example given in this chapter is the reproduction and development data of a tropical harpacticoid copepod, *Pararobertsonia* sp. This species could produce multiple egg sacs from a single copulation, with an average of 6.7 ± 2.47 egg sacs (ranging from 3.0 to 12.0 egg sacs) in 31.2 ± 3.57 days (average of lifespan). The production of eggs per sac was 21.7 ± 4.79, varies from 14 to 30 eggs. The maturation time of egg sac is variable with range from 0 to 167 hours (7 days). However, most of the eggs were matured within 36 to 47 hours (2 days). The interval time between egg sacs varies from 0 to 71 hours (3 days), but most of individual took 11 hours to produce the next eggs. In this present study it is clearly shown that reproductive biology of every individual of *Pararobertsonia* sp. are varies among each other. Temperature appears to give significant effect on reproduction and development of *Pararobertsonia* sp. compared to salinity. High temperature increased while low temperature delayed the development of *Pararobertsonia* sp. but extreme temperature could lead to the mortality. This is particularly true if the copepod is drastically exposed to the temperature of beyond the tolerant limit. On the other hand, this species has a wide range of salinity tolerance (5 to 45 ppt). However, direct exposure to that lowest or highest salinity could lead to the mortality as well.

4. Acknowledgment

The study was carried out as part of a research project 'Development of cyst in marine harpacticoid copepods' funded by National Oceanographic Directorate, Ministry of Science, Technology and Innovation, Malaysia.

5. References

Ambler, J.W. (1985). Seasonal factors affecting egg production and viability of eggs of *Acartia tonsa* Dana from East Lagoon, Galveston, Texas. *Estuarine, Coastal and Shelf Science* 20:743–760.

Ara, K. (2001). Daily egg production rate of the planktonic calanoid copepod *Acartia lilljeborgi* Giesbrecht in the Cananéia Lagoon estuarine system, Sao Paulo, Brazil. *Hydrobiologia* 445: 205–215.

Buskey, E.J.; Coulter, C. & Strom, S. (1993). Locomotory patterns of microzooplankton: potential effects on food selectivity of larval fish. *Bulletin of Marine Science* 53:29–43.

Carli, A.; Mariottini, G.L.& Pane, L. (1995). Influence of nutrition on fecundity and survival in *Tigriopus fulvus* Fischer (Copepoda, Harpacticoida). *Aquaculture* 134:113–119.

Chandler, G.T.; Cary, T.M.; Volz, D.C.; Walse, S.S.; Ferry, J.L. & Klosterhaus, S.L. (2003). Fipronil effects on estuarine copepod (*Amphiascus tenuiremis*) development, ferility, and reproduction: A rapid life-cycle assay in 96-well microplate format. *Environmental Toxicology and Chemistry* 23:117–124.

Chinnery, F.E. & Williams, J.A. (2004). The influence of temperature and salinity on *Acartia* (Copepoda: Calanoida) nauplii survival. *Marine Biology* 145:733–738.

Christou, E.D. & Moraitou-Apostolopoulou, M. (1995). Metabolism and feeding of mesozooplankton of the eastern Mediterranean coast (Hellenic waters). *Marine Ecology: Progress Series* 126:39–48.

Dam, H.G. & Lopes, R.M. (2003). Omnivory in the calanoid copepod *Temora longicornis*: feeding, egg production and egg hatching rates. *Journal of Experimental Marine Biology and Ecology* 292:119– 137.

Damgaard, R.M. & Davenport, J. (1994). Salinity tolerance, salinity preference and temperature tolerance in the high shore copepod *Tigriopus brevicornis*. *Marine Biology* 118:443-449.

Depledge, M.H. (1990). Interactions between heavy metals and physiological processes in estuarine invertebrates. In: P.L. Chambers and C.M. Chambers, Editors, Estuarine Ecotoxicology, JAPAGA, Ashford, Co., Wicklow, Ireland. Pp 89–100.

Depledge, M.H. (1994). Genotypic toxicity: implications for individuals and populations. *Environmental Health Perspectives* 102 (Supplement 12):101–104.

Devreker, D.; Souissi, S.; Winkler, G.; Forget-Leray, J. & Leboulenger, F. (2009). Effects of salinity, temperature and individual variability on the reproduction of *Eurytemora affinis* (Copepoda; Calanoida) from the Seine estuary: A laboratory study. *Journal of Experimental Marine Biology and Ecology* 368:113–123.

Drillet, G.; Iversen, M.H.; Sørensen, T.F.; Ramløv, H.; Lund, T. & Hansen, B.W. (2006). Effect of cold storage upon eggs of a calanoid copepod *Acartia tonsa* Dana and their offspring. *Aquaculture* 254:714–729.

Drillet, G.; Frouël, S.; Sichlau,M.H.; Jepsen P.M.; Højgaard,J.K.; Joarder, A.K. & Hansen, B.W. (2011). Status and recommendations on marine copepod cultivation for use as live feed. *Aquaculture* 315, 155-166.

Dybdahl, M.F. (1995). Selection of life history traits across a wave exposure gradient in the tidepool copepod *Tigriopus californicus* (Baker). *Journal of Experimental Marine Biology and Ecology* 192:195-210.

Fava, G. & Martini, E. (1988). Effect of inbreeding and salinity on quantitative characters and asymmetry of *Tisbe holothuriae* (Humes). *Hydrobiologia* 167/168:463-467.

Gaudy, R.; Guerin, J.P. & Moraitou-Apostolopoulou, M. (1982). Effect of temperature and salinity on the population dynamics of *Tisbe holothuriae*, Humes (Copepoda; Harpacticoida) fed two different diets. *Journal of Experimental Marine Biology and Ecology* 57:257–271.

Gee, J.M. (1989). An ecological and economic review of meiofauna as food for fish. *Zoological Journal of the Linnean Society* 96:243-261.

Geiling, W.T. & Campbell, R.S. (1978). The effect of temperature on the development rate of the major life stages of *Diaptomus pallidus* Herrick. *Hydrobiologia* 61:304–307.

Gonzalez, C.R. & Bradley, B.P. (1994). Salinity stress proteins in *Eurytemora affinis*. *Hydrobiologia* 292/293:461–468

Goolish, R. & Burton, R. (1989). Energetics of osmoregulation in an intertidal copepod: effects of anoxia and lipid reserves on the pattern of free amino acid accumulation. *Functional Ecology* 3:81–89.

Guérin, J.P.; Kirchner, M. & Cubizolles, F. (2001). Effects of *Oxyrrhis marina* (Dinoflagelata), bacteria and vitamin D2 on population dynamics of *Tisbe holothuriae* (Copepoda). *Aquaculture* 261:1 –16.

Hagiwara, A.; Jung, M. & Hiramaya, K. (1995). Interspecific relations between marine rotifers *Branchionus rotundiformis* and zooplankton species in the rotifer mass culture tanks. *Fisheries Science* 61:619-623.

Hall, C.J. & Burns, C.W. (2002). Effects of temperature and salinity on the survival and egg production of *Gladioferens pectinatus* Brady (Copepodas: Calanoida). *Estuarine, Coastal and Shelf Science* 55:557–564.

Heinle, D.R. (1969). Temperature and zooplankton. *Chesapeake Science* 10:186-209.

Hicks, G.R.F. & Coull, B.C. (1983). The ecology of marine meiobenthic harpacticoid copepods. *Oceanography and Marine Biology*: Annual Review 21:67–175.

Holmefjord, I.; Gulbrandsen, J.; Lein, I.; Refstie, T.; Léger, Ph. Harboe, T.; Huse, I.; Sorgeloos, P.; Bolla, S.; Olsen, Y.; Reitan, K.I.; Vadstein, O.; Øie G. & Danielsberg, A. (1993). An intensive approach to Atlantic halibut fry production. *Journal of World Aquaculture Society* 24:275-284.

Kawall, H.G.; Torres, J.J. & Geiger, S.P. (2001). Effects of the ice-edge boom and season on the metabolism of copepods in the Weddell Sea, Antarctica. *Hydrobiologia* 453/454:66-77.

Kimmel, D.G. & Bradley, B.P. (2001). Specific protein responses in the calanoid copepod *Eurytemora affinis* (Poppe, 1880) to salinity and temperature variation. *Journal of Experimental Marine Biology and Ecology* 266:135–149.

Kinne, O. (1964). The effects of temperature and salinity on marine and brackish water animals: II. Salinity and temperature salinity combinations. *Oceanography and Marine Biology: Annual Review* 2:281–339.

Le Borgne, R. (1982). Zooplankton production in the eastern tropical Atlantic Ocean: net growth efficiency and P:B in term of carbon, nitrogen and phosphorus. *Limnology and Oceanography* 27:681-698.

Lee, C.E. & Petersen, C.H. (2002). Genotype-by environment interaction for salinity tolerance in the freshwater invading copepod *Eurytemora affinis*. *Physiological and Biochemical Zoology* 75:335-344.

Lee, C.E.; Remfert, J.L. & Gelembuik, G.W. (2003). Evolution of physiological tolerance and performance during freshwater invasions. *Integrative and Comparative Biology* 43:439–449

Lonsdale, D.J. & Levinton, J.S. (1986). Growth rate and reproductive differences in a widespread estuarine harpacticoid copepod (*Scottolana canadensis*). *Marine Biology* 91:231–237.

Lonsdale, D.J. & Levinton, J.S. (1989). Energy budgets of latitudinally separated *Scottolana Canadensis* (Copepoda: Harpacticoida). *Limnology and Oceanography* 34:324–331.

Marcus, N.H. & Murray, M. (2001). Copepod diapause eggs: a potential source of nauplii for aquaculture. *Aquaculture* 210:107-115.

Matias-Peralta, H.; Fatimah, M.Y.; Mohamed, S. & Aziz, A. (2005). Effect of some environmental parameters on the reproduction and development of a tropical marine harpacticoid copepod *Nitocra affinis* f. *californica* Lang. *Marine Pollution Bulletin* 51:722-728.

McKinnon, A.D. (1996). Growth and development in the subtropical copepod *Acrocalanus gibber*. *Limnology and Oceanography* 41:1438–1447.

McKinnon, A.D.; Duggan, S.; Nichols, P.D.; Rimmer, M.A.; Semmens, G. & Robino, B. (2003). The potential of tropical paracalanid copepods as live feeds in aquaculture. *Aquaculture* 223:8-106

McLaren, I. A.; Corkett, C. J. & Zillioux, E. J. (1969).Temperature adaptations of copepod eggs from the arctic to the tropics. *Biology Bulletin* 137:486–493.

Milione, M. & Zeng, C. (2008). The effects of temperature and salinity on population growth and egg hatching success of the tropical calanoid copepod, *Acartia sinjiensis*. *Aquaculture* 275:116–123.

Miliou, H. (1996). The effect of temperature, salinity and diet on final size of female *Tisbe holothuriae* (Copepoda: Harpacticoida). *Crustaceana* 69:742-753.

Miliou, H. & Moraitou-Apostolopoulou, M. (1991). Combined effects of temperature and salinity on the population dynamics of *Tisbe holothuriae* Humes (Copepoda: Harpacticoida). *Archives of Hydrobiology* 121:431–448.

Naess, T.; Germain-Henry, M. & Naas, K.E. (1995). First feeding of Atlantic halibut (*Hippolossus hippoglossus*) using different combination of Artemia and wild zooplankton. *Aquaculture* 130:235-250.

Nanton, D.A. & Castell, J.D. (1998). The effects of dietary fatty acids on the fatty acid composition of harpacticoid copepods, *Tisbe* sp., for use as a live food for marine fish larvae. *Aquaculture* 163:251-261.

Palmer, M.A. & Coull, B.C. (1980). The prediction of development rate and the effect of temperature for the meiobenthic copepod, *Microarthridion littorale* (Pope). *Journal of Experimental Marine Biology and Ecology* 48:73-83.

Penchenik, J.A. (2005). Biology of the invertebrates. Fifth Edition. Tufts University. McGraw-Hill. 608pp.

Pepin, P. & Penney, R.W. (1997). Patterns of prey size and taxonomic composition in larval fish: are there general size dependent models? *Journal of Fish Biology* 51:84–100.

Peterson, W.T. (2001). Patterns in stage duration and development among marine and freshwater calanoid and cyclopoid copepods: a review of rules, physiological constraints, and evolutionary significance. *Hydrobiologia* 453/454:91–105.

Pinto, C.S.C.; Souza-Santos, L.P. & Santos, P.J.P. (2001). Development and population dynamics of *Tisbe biminiensis* (Copepoda: Harpacticoida) reared on different diets. *Aquaculture* 198:253-267.

Rippingale, R.J. & Payne, M.F. (2001). Intensive cultivation of a calanoid copepod, *Gladioferens imparipes*. A guide to procedures. Department of Environmental Biology, Curtin University of Technology, Perth. 62pp.

Rhodes, A.C.E. (2003). Marine harpacticoid copepod culture for the production of long chain highly unsaturated fatty acids and carotenoid pigments. A dissertation submitted to the Graduate Faculty of North Carolina State University in partial fulfillment of the requirements for the Degree of Doctor of Philosophy. 161pp.

Rhyne, A.L.; Ohs, C.L. & Stenn, E. (2009). Effects of temperature on reproduction and survival of the calanoid copepod *Pseudodiaptomus pelagicus*. *Aquaculture* 292:53–59.

Siokou-Frangou, I. 1996. Zooplankton annual cycle in a Mediterranean coastal area. *Journal of Plankton Research* 18:203–223.

Staton, J.L.; Schizas, N.V.; Klosterhaus, S.L.; Griffitt, R.J.; Chandler, G.T. & Coull B.C. (2002). Effect of salinity variation and pesticide exposure on an estuarine harpacticoid copepod, *Microarthridion littorale* (Poppe), in the southeastern US. *Journal of Experimental Marine Biology and Ecology* 278:101-110

Støttrup, J.G. & Norsker, N.H. (1997). Production and use of copepods in marine fish larviculture. *Aquaculture* 155:231-247.

Sun, B. & Fleeger, J.W. (1995). Sustained mass culture of *Amphiascoides atopus*, a marine harpacticoid copepod in a recirculating system. *Aquaculture* 136:313-321.

Takahashi, T. & Ohno, A. (1996). The temperature effect on the development of calanoid copepod, *Acartia tsuensis*, with some comments to morphogenesis. *Journal of Oceanography* 52:125–137.

Tester, P.A. & Turner, J.T. (1990). How long does it take copepods to make eggs?. *Journal of Experimental Marine Biology and Ecology* 141:169–182.

Uye, S.I. (1985). Resting egg production as a life history strategy of marine planktonic copepods. *Bulletin of Marine Research* 37:440–449

Uriarte, I.; Cotano, U. & Villate, F. (1998). Egg production of *Acartia bifilosa* in the small temperate estuary of Mundaka, Spain, in relation to environmental variables and population development. *Marine Ecology*: Progress Series 166:197–205

Van der Meeren, T. (1991). Selective feeding and prediction of food consumption in turbot larvae (*Scophthalmus maximus* L.) reared on the rotifer *Brachionus plicatilis* and natural zooplankton. *Aquaculture* 93:35–55.

Vincx, M. (1996). Meiofauna in marine and freshwater sediment. In: Hall, G. S. (editor) Methods for the Examination of Organism Diversity in Soils and Sediments. CAB International. pp 187-195.

Von Herbing, H. & Gallagher, S. (2000). Foraging behavior in early Atlantic cod larvae (*Gadhus morhua*) feeding on a protozoan (*Balanion* sp.) and a copepod nauplius (*Pseudodiaptomus* sp.). *Marine Biology* 136:591–602

Williams, T.D. & Jones, M.B. (1994). Effects of temperature and food quantity on postembryonic development of *Tisbe battagliai* (Copepoda: Harpacticoida). *Journal of Experimental Marine Biology and Ecology* 183:283–298.

Williams, T.D. & Jones, M.B. (1999). Effect of temperature and food quantity on the reproduction of *Tisbe battagliai* (Copepoda: Harpacticoid). *Journal of Experimental Marine Biology and Ecology* 236:273-290.

Zaleha, K. & Farahiyah-Ilyana J. (2010). Culture and growth of a marine harpacticoid, *Pararobertsonia* sp. in different salinity and temperature. *Sains Malaysiana* 39(1): 137-141.

17

New Developments in Biotechnology and IPR in Aquaculture – Are They Sustainable?

Anne Ingeborg Myhr, G. Kristin Rosendal and Ingrid Olesen
Genøk- Centre of Biosafety, Fridtjof Nansen Institute, Nofima
Norway

1. Introduction

The objective of this chapter is to give an overview and analysis of the current trends and developments in biotechnology in aquaculture research and management. The technological developments along with structural changes in the aquaculture sector may affect access and intellectual property rights (IPR) regimes. These issues will be discussed in a wide perspective involving both short and long-term biological effects, ethical and other social aspects (economic, legal and political issues), including their partly inherent contradictions needing compromising for sustainable development. The chapter will focus on current biological challenges within aquaculture as a growing food production sector, with less emphasis on external effects such as environmental effects. Cases from farmed salmon and cod in Norway in addition to shrimp and tilapia in Asia will be highlighted.

2. Concept of sustainable development in aquaculture

Since the publication of the World Conservation Strategy, the concept of 'sustainable development' has received increasing importance in most policy areas. A widely used definition of the concept is 'development that meets the needs of the present without compromising the ability of future generations to meet their own needs' (WCED, 1987). The Rio declaration of 1992 clarified that governments have a global responsibility for resolving conflicts over the environment in ways that protect the interest of humanity and nature. One good example of international obligations that has included the concept is the Convention of Biodiversity (CBD). With regard to aquaculture, recommendations for employment of sustainability can be found in the Holmenkollen guidelines for sustainable aquaculture (1999), in the Norwegian ministry of fisheries and costal affairs strategy for an environmental sustainable seafood industry (2009), and in the EU communication; A strategy for the sustainable development of European aquaculture (2002).

Critics of the concept of sustainable development have, however, argued that the concept is elusive, and highly varying views persist among both scientists and regulators with regard to what the concept constitutes and implications by implementation. The main contested values and practices of sustainable development are: what values are important within sustainable development and how to set priorities between them, and how to achieve maintenance and preservation of nature and biodiversity versus a just society and economic development (Kamara et al., 2006). For example, the rapid spread of aquaculture has raised concern about land-use change in coastal areas, impacts on wild fish by escapees, environmental pollution, and extensive use of marine resources for fish feed production. A conceptual framework for sustainable aquaculture has been presented from three perspectives: environmental, economic and sociological (Caffey et al., 1998). This implies that introduction of modern biotechnology must be explored both with regard to the adequacy of present approaches and with regard to the problem solving nature of the new technology. Moreover that there needs to be an awareness that application of modern biotechnology in aquaculture also influences socio-economic values as employment, income, and local economic activity as well as ethics, which are all important elements of sustainability as understood by most users. Hence, sustainable development requires a renewed focus on stakeholders and their needs, it demands clearer understanding of stakeholders perspectives and public concerns as well as attention to issues of institutional structure and representation in decision-making processes.

The next section deals with biological/ecological challenges in aquaculture while section four provides a picture of recent technological developments that may have a bearing on these challenges. In section five we present international and domestic regulations relevant to both modern biotechnology and the access issues, thus pertaining to the discussion of sustainability. We then briefly account the present structural developments and management trends within aquaculture. With this broad framework in mind, we turn to examine actor perceptions of how biotechnology and IPR may affect sustainability in aquaculture. This section builds on surveys and interviews with key actors. Then in section eight we highlight some of the major issues for understanding how IPR and biotechnology may affect sustainability in aquaculture. Finally we discuss implications and give some recommendations for how developments in biotechnology and IPR in aquaculture can contribute to sustainability.

3. Current management and biological challenges within aquaculture

Aquaculture industry is currently facing many challenges. These involve animal health and welfare, environmental effects and social effects including economics, global and fair utilization and sharing of resources, rural viability etc. Within the format of a book chapter, only a limited number of challenges can be handled properly. The focus of this section will therefore be limited to biological challenges with emphasis on animal health and welfare, and the management of fish breeding including biotechnological methods.

3.1 Search for improved animal health

The growth in aquaculture has been accompanied with an increase in diseases caused by pathogens that includes a wide range of bacterial, viral, parasitic and fungal infections. At present diseases in aquaculture are causing big economic problems and are affecting animal welfare significantly. The high density of fish together with the effective pathogen transportation in water creates favourable living conditions for these pathogens. Hence, diseases tend to multiply in farm environments, a situation that represent potential ecological threats both to the farmed fish in itself and to the farm environment including wild fish. In salmon aquaculture disease prevention with antibiotics and chemicals was for many years the solution preferred. However, the potential pollution associated with chemicals and the excessive use of antibiotics together with the emergence of multiple resistance to antibiotics created concerns and initiated a search for alternative ways, as selection for increased disease resistance, to deal with the problem.

Selection for increased disease resistance in fish has mainly been based on challenge tests carried out under controlled conditions. Challenge-tested fish cannot be used as parents for the next generation of elite salmon, meaning that selection cannot be applied directly on the breeding candidates. To circumvent this problem, geneticists have been searching for genes controlling the degree of resistance to different diseases. Markers for such genes may be ideal criteria for selection, because they can be applied directly without requiring challenge testing (see also section 4.2). Thus, the accuracy of selection can be increased while the need to sacrifice fish in challenge tests is reduced.

Selection for genetic disease resistance has been emphasized in Norwegian salmon breeding since 1995. In 2007, Moen et al. (2009) identified markers for a gene that explains most (80 %) of the genetic variation in resistance to infectious pancreatic necrosis (IPN) in both fry and post-smolts. Based on these findings, Aqua Gen has developed and applied a tool using these markers for selecting IPN-resistant fish directly. This tool can, with very high accuracy, determine whether individual fish have zero, one or two copies of the gene variant (allele) that give high resistance. This approach may also be useful for pancreas disease (PD), which is an important economic disease of farmed Atlantic salmon that cause significant losses through mortality and reduced production (SalmoBreed, 2011).

Currently, salmon lice (*Lepeophtheirus salmonis*) represent a major health and welfare problem in the salmon industry. Furthermore, it is also an ecological problem, since the lice multiply in fish farms, and then spread to the wild salmon population. Chemical treatment is commonly used to combat the lice, but use of biological measures such as cleaner fish has increased lately due to development of resistant lice to the chemicals. However, moderate genetic variation has been shown for resistance to the salmon louse, and thus it may be possible to reduce problems caused by lice through selective breeding programs (Kolstad et al., 2005). Breeding for disease and parasite resistance in Norwegian salmon and trout is considered to be important for the fish themselves, producers and consumers alike and would increase the sustainability of the industry, and the know-how could be transferred to other aquaculture species.

3.2 Genetic diversity and fish breeding strategies

Substantial long term selection responses of 10-15% higher growth rate per generation have been documented for several species of farmed fish, such as Atlantic salmon in Norway since the 1970'ies and Nile tilapia in Asia since the 1990'ies. Highly favorable benefit cost ratios ranging from 8 to 60 is reported for fish breeding. Due to the high fertility of fish and convenience of handling and distributing seeds, such high benefit cost ratio can be obtained for fish breeding programs. For this it is, however, important to note that an efficient dissemination structure and organization reaching a high number of farmers is crucial. In Norway, seed from the improved farmed salmon was sold to the farmers as eyed eggs and smolts. As the genetic gain became more apparent, the demand for genetically improved salmon increased rapidly during the first decade. Until the market demand and dissemination was appropriately developed, public funding allowed for establishment of the salmon breeding program. During the early 1980'ies, the Norwegian Fish Farmers Association got involved in the program, and in 1992, the breeding program was turned into a private company, Aqua Gen, a sustainable business with long term profitability. The GIFT (Genetic Improvement of Farmed Tilapia) program is another example of a successful breeding program resulting from public funded research and technology development. The GIFT seeds have been disseminated to several countries in Asia and Latin America to support intensive large scale farms and small subsistence farms.

Genetic variation is essential for selection response, and a sufficiently large and genetically diverse breeding population is, therefore, fundamental when establishing and running an animal breeding program. A breeding design with appropriate family structure is critical to maintain a large effective population size and obtain a long-term selection response with low rates of inbreeding in fish breeding. For mass selection, Bentsen & Olesen (2002) concluded from a simulation study that a minimum of 50 families (pairs of parents) are required to prevent inbreeding and obtain a long-term response in a mass selection program for aquaculture. Gjerde et al. (1996) presented optimum designs for fish breeding programs with constrained inbreeding and mass selection. Various breeding designs for between-family, within-family, and combined selection (between- and within-family) are presented and evaluated by Bentsen & Gjerde (1994). Combined family based selection designs may improve the accuracy of selection substantially, particularly for traits with low heritability. Due to the higher probability of selecting large numbers of sibs from a few families the number of broodstock selected per family needs to be restricted to avoid a high rate of inbreeding and reduced genetic variation.

Less than 10% of the fish stocked for aquaculture in 2005 originated from family based selection programs (Gjedrem et al., 2011). This situation has not improved much recently, and for many aquaculture species with huge production quantities, such as carps, only a few efficient selection programs are active. Furthermore, the effective population sizes are often limited and in some cases too low, because the high reproductive capacity allows the use of a low number of broodstock. Such populations may still gain sufficient short-term advantage above non-improved populations and capture much of the market share. In turn, this discourages further genetic introductions into the breeding nucleus. Long-term inbreeding and loss of genetic variability because of genetic drift may then affect performance and further the long-term genetic progress. In such populations, strategies for continuous (re)introduction of genetic variability from outside the breeding nucleus without adverse performance consequences are, therefore, required. Furthermore, initiation of additional breeding programs is expected for different environments for the most important farmed species and this may improve the situation.

3.3 Prospects for genetic improvement of fish welfare

During the initial stages, breeding programmes for farmed fish usually focus on improving productivity traits such as growth. During later stages, disease resistance, survival, and product quality traits are often emphasized during selection to develop a more robust fish. Domesticated fish fit better for a life in captivity and farm environment, and are therefore less stressed and will be more robust and perform better with respect to growth and survival. Hence, maintaining good fish welfare by reducing stress load and making sure the fish is thriving will be a key to promote a more robust fish and profitable farming.

Huntingford et al. (2006) list several factors in aquaculture that represent fish welfare challenges including aggressive interactions, handling and removal from water, diseases and permanent adverse physical states and possibly increased levels of aggressiveness due to selection for fast growth. However, it has been shown in both salmon and cod that after a few generations of selection for growth and domestication in hatcheries and farms, we obtain calmer, less aggressive carnivorous fish. In a review of the effects of domestication on aggressive and schooling behaviour in fish, Ruzzante (1994) conclude that domestication may strongly affect behavioural traits, but it is the intensity of the behaviour rather than the behavioural pattern itself that is affected. Olesen et al. (2011) emphasised possible correlated effects on stress coping for fast growing fish.

Selection for high production efficiency in terrestrial animals is known to give undesirable effects in traits like health and reproduction (Rauw et al., 1998). However, in the Nordic countries broader breeding goals including functional and welfare traits have been selected for. Olesen et al. (2000) discussed definition of breeding goals for sustainable farm animal production, and suggest a procedure including non-market values for appropriate weighing of traits providing public goods (e.g. welfare traits). Since 1995, farmed salmon in Norway have been selected for resistance to diseases. Such selection will obviously reduce stress and suffering connected to diseases. Particularly for farmed fish, there is a lack of information on genetic variances and covariances of many welfare related traits such as behaviour (e.g. aggression) and stress coping. Consequently, we do not know possible unfavourably correlated responses in some fish welfare indicators, e.g. poorer ability to cope with stress, resulting from the current selection for productivity traits. Hence, more knowledge and research is needed on fish welfare traits and their genetic parameters. Regarding survival and maturation, behaviour, dominance, aggressiveness and activity level, it is reported genetic differences between wild and farmed salmon (Fleming & Gross, 1992; McGinnity et al., 1997; Metcalfe et al., 2003; Petersson & Järvi, 2006). Hatchery reared salmonids showed a weaker antipredator response (Johnsson et al., 1996) and less physiological stress due to higher stocking densities (Mazur & Iwama, 1993) when compared to the wild. As farmed fish adapt to the farm environment, such domesticated fish will suffer less in the farm environment. Relevant fish welfare indicators or traits that currently can be taken into account in selective breeding are growth, survival (or mortality), social interactions/behaviour (e.g. cannibalism for carnivorous species) and frequency of injuries (e.g. fin injuries) (Turnbull et al., 1998).

3.4 Animal welfare and animal ethics

Promoting good husbandry practices and ensuring the welfare of farmed fish are well-established parts of the European Union policy for sustainable aquaculture development. However, there are often conflicts and trade-offs between short term profit of the industry and demand for cheap animal products on one hand and animal welfare on the other, that the animals do not gain from. Animal welfare has mostly been discussed in relation to research animals, land based animals for food production, and pets. Some of these issues will be highlighted before we move into implications of fish farming on animal welfare.

An important question with regard to animal husbandry is if it is morally legitimate to use animals merely as a resource or means to meet our needs, or if there are moral considerations that place restrictions on such an approach. Many difficult questions have arisen with regard to animals' intrinsic value. Assuming that animals do have intrinsic value, all encroachments on their lives (by humans) become moral issues and demand carefully considered answers and actions. The Norwegian Animal Welfare Act of 2010, states that animals have an intrinsic value. This term implies that animal welfare must be prioritised irrespective of the value the animal may have for people, which also contributes to clarifying the animal's status.

The word 'welfare' is derived from *well* + *fare*, i.e., how well (or dignified) an animal 'fares' (travels) through life. How well is an animal able to regulate its biological functions in relation to its environment? A function based definition of animal welfare is given (Broom, 1986): 'The welfare of an animal is its state as regards its attempts to cope with its environment'. Other definitions focus on an animal's subjective experience or awareness of its condition (feeling based) and/or on whether it can live a natural life (nature based). Hence, the term 'animal welfare' applies to both the mental/emotional and physical health of the individual animal or the animal's condition while trying to cope with its environment. The term also includes behaviour, as well as physiological and immunological factors. In this context, health is defined more broadly than merely the absence of disease. An important basis for ensuring animal health is the animals' well-being. It also includes positive welfare, with the implication that denying animals positive experiences and stimuli is also an ethical problem with regard to animal protection. 'Animal protection' is here seen as the protection of the emotional and the physical health of individual animals.

Most current animal ethicists use animal ability of sentience for ascribing direct moral considerations. Lund et al. (2007) claimed that fish welfare should be given serious moral considerations depending on their possession of the morally relevant similarities of sentience. The same authors reason further that fish are likely to be sentient and therefore deserve serious consideration. They also concluded from a simple risk analysis that the probability that the fish can feel pain is not negligible, and that if they really experience pain the consequence is great due to the possibly high number of suffering animals. Hence, farmed fish should be given the benefit of doubt. Even from a more egoistic standpoint, we can argue for fair treatment of animals. If we inflict suffering upon animals, we violate human dignity and may contribute to the development of a crueller society, as also indicated by Mahatma Gandhi ('The greatness of a nation and its moral progress can be judged by the way its animals are treated.').

4. Biotechnology in aquaculture and its role in innovation

Modern biotechnology does involve new tools to meet several of the challenges that aquaculture is at present striving with. In this chapter we will therefore limit our presentation to breeding and vaccine development and then to the promising possibilities by chromosome manipulation, DNA marker selection and genetic engineering.

4.1 Reproduction technology involving chromosome manipulation

In most aquaculture species, external fertilization is natural, and opens many powerful methods of genetic engineering, including manipulation of chromosome number such as haploids and polyploids with three (tetraploids) and four sets of chromosomes (tetraploids). Furthermore, animals with chromosomes from only the dam (gynogenesis) or from only the sire (androgenesis) can be produced. A more comprehensive overview of the techniques involved and applications is given by Refstie & Gjedrem (2005).

Production of sterile fish may solve the problem of escaped fish interacting with wild fish and the need for protecting improved genetic material against 'piracy copying'. Early sexual maturation cause problems in commercial farming due to poorer filet quality, higher mortality and reduced growth. Sterile triploid fish will not produce gonads and can continue growing and being slaughtered at any time. For many species, one of the sexes gets earlier sexual mature and hence lower growth and body weights. Chromosome manipulations can be used to produce either all male (as in e.g. tilapia and salmon) or all female fish (as in e.g. halibut, hake and angler fish) depending on the sex preferred. Triploid trout and all male tilapia are the most common applications of chromosome manipulations in aquaculture.

Questions of cost/benefit analysis that need to be addressed before using reproduction technology are: is one gender more highly priced by the market; is one gender of higher production value to producer; is gender determination known for the species or is it mainly environmentally induced, will 'clean green' market perception/sales pitch be jeopardized by the use of this technology (Robinson, 2002).

4.2 Application of molecular genetics in aquaculture breeding

The two areas of modern biotechnology that has been expected to have most significant impact on genetic improvement of aquaculture species are DNA markers and transgenics (Hayes & Andersen, 2005). A DNA marker is an identifiable physical location on a chromosome whose inheritance can be monitored (Hyperdictionary, 2003). A comprehensive overview of DNA markers and linkage mapping together with a discussion of potential applications of DNA markers in aquaculture breeding programmes is given by Hayes & Andersen (2005). Furthermore, whole genome sequencing and application of genomics in aquaculture breeding programs is discussed by Quinn et al. (2011, see chapter in this book). As mentioned, DNA markers have already been applied in aquaculture breeding for direct and highly accurate selection of IPN-resistant fish (Aqua Gen, 2010). So called marker assisted selection (MAS) can double genetic gain for traits that can not be measured on selection candidates (e.g. disease resistance), because it utilizes the between family variance, and may also contribute to reduce inbreeding. For MAS, quantitative trait loci (QTL) must be mapped and their effect determined. This is not the case for genomic selection (GS), where the effects of a large number of loci are first estimated using a test group. Selection can then be carried out on genome wide breeding values of the breeding candidates predicted as the sum of the marker effects estimated, assuming an additive genetic model (Meuwissen et al., 2001). However, the high genotyping costs for GS has so far limited its application in aquaculture breeding. Therefore, a scheme with pre-election of parents for growth combined with selective genotyping of large and pooled family groups has been suggested to obtain high accuracies while reducing number of genotypes and costs many fold (Sonesson et al., 2010).

Furthermore, application of new tools of molecular genetics for gaining understanding about genetic regulation of complex traits such as disease resistance may be important (see section 3.1).

4.3 Genetically modified organisms

The possibilities within modern biotechnology related to the ability to identify genes endowing specific phenotypes together with projects intended to map genomes have opened the possibility for the development of genetically modified organisms (GMOs). Of special relevance for aquaculture are research and development of transgenic fish, GM vaccines (here also included DNA vaccines) as well as present and future GM plants to be used in feed. Genetic engineering can also be a useful tool for increased use of IPR, as it may make it easier to fulfil patent criteria such as the inventive step and the demand for reproducibility.

4.3.1 Introduction of transgenic fish

Most focus on trangenic fish is on the possibilities for enhancement of the quality of cultured stocks by improving growth rate and increasing resistance to disease and stress (Melamed et al., 2002). Improved growth rate has been possible by the introduction of growth hormone (GH) genes, in species such as Atlantic salmon, coho salmon, Nile tilapia and hybrid tilapia. The most known example of transgenic salmon is the *AquAdvantage*, developed by Aqua Bounty, which contains a gene construct composed of the regulatory elements of an ocean pout antifreeze protein gene controlling a chinook salmon GH gene. The antifreeze promotor enables stimulation of the growth hormone gene also during cold periods with the result that the transgenic salmon grows much faster than its non-GM counterpart. The company is seeking approval for commercial use of this transgenic fish, and the application has been under evaluation in more than ten years by the US Food and Drug Administration (Niiler, 2000). Other highly relevant approaches are development of disease and parasite resistant fish. At present there is a lack of understanding of genes responsible for disease resistance in fish and different strategies are discussed. These strategies include antisense technology for production of complementary RNA for foreign RNA, expression of antimicrobial substances and peptides (as lysozyme) and efforts to increase production of the fish cytokines and other genes involved in immune defence. Moreover transgenic approaches that combine interesting characteristics, as enhanced growth and disease resistance, together with approaches for development of sterile fish or fish where reproductive activity can be down-regulated is also highly relevant since this will minimise the risk of transgenic fish breeding with wild populations after accidental release or escape.

The potential of transgenic fish to escape and enter the natural environment is an important concern for regulators (Le Curie-Belfond et al., 2009). Unless transgenic fish is used in contained facilities, transgenic fish will certainly escape into the environment. The environmental impact is difficult to predict and will depend on the number of escaped fish, their phenotypic characteristics (related to ability for reproduction and survival over time), and the aquatic biodiversity present in the receiving ecosystem (Kapuscinski & Brister, 2001).

Another potential problem related to transgenic fish that is disease or parasite resistant may be similar to what has been experienced with insect resistant crops (Le Curie-Belford et al., 2009). It has for example been reported that insects have developed resistance to insect resistant crops. Hence, the benefit achieved may over time develop into a long-term problem. If the same unexpected events develops with disease or parasite resistant fish the consequence will be a need for more or other antibiotics and chemicals to cope with resistant pathogens and parasites or new emerging pathogens.

Development of transgenic fish does also raise ethical concern. One implication by the genetic modification process itself is that it may affect fish welfare, behaviour and reproduction (see also Le Curie-Belford et al., 2009). It has been reported pleiotropic effects as changes in coloration, cranial, opercula and lower jaw deformation in transgenic coho salmon (Devlin et al., 1995). Concerns have also been raised that genetic modification strategies may affect the animal's integrity (Verhoog, 2001). This is controversial, and for example Sandøe & Holtug (1998) argue that only welfare of animals and humans are relevant ethical considerations, and that these considerations imply to:

1. Clarify realistic alternatives for improving animals welfare through breeding or biotechnology
2. Consider positive and negative effects on animal welfare
3. Find what benefits are at stake for humans
4. If animals welfare is reduced, consider if the costs of animals will weigh up for humans benefits
5. Weigh conflicting concerns, including risk of possible long term side effects.

When weighing conflicting ethical concerns we often have to compromise between efficiency and animal welfare, where different ethical theories of animal ethics may affect acceptance of biotechnology on animals (Sandøe & Christensen, 2008):

- Contractarianism: implies no problem with instrumental use of animals as a biotechnological resource or as a tool for biotechnology.
- Utilitarianism: can accept the use of biotechnology to improve human and animal welfare if the benefits are bigger than the costs in terms of harmful effects.
- The animal rights view: makes the use of biotechnology in animals more problematic. One can for example not sacrifice a live animal and make it suffer from e.g. health problems in order to improve animal health of future generations of an animal population. If there is a risk that e.g. transgenic fish have unintended health problems, this will be particularly troublesome. Use of both animal breeding and biotechnological methods on animals can be considered as instrumental use of animals and does therefore interfere with this view although it may not imply important harm or suffering for the animal.
- Respect for nature often implies moral problems with both animal and plant breeding and biotechnology. Those who sympathise with this view will often be critical to tamper with the nature although it may not cause animal or human harm or suffering. Hence, they will only support a very cautious and restrictive policy to animal breeding and biotechnology. For farmed fish with close wild relatives in nature this will likely be even more troublesome than for other farm animals. The risk of escaped farmed transgenic fish mating with wild fish is for example particularly problematic in the view of natural populations almost considered as sacred. Traditionally bred farmed salmon with the same alleles as wild salmon, but with other gene frequencies and other combinations can also be troublesome for the same people. When it comes to salmon, the fact that it is often considered as an iconic species may also be a relevant issue.

In the Norwegian Animal Welfare Act (2010) §25, the following is stated about animal breeding: 'Reproduction, including through methods of gene technology, shall not be carried out in such a way that it:

- changes genes in such a way that they influence the animals' physical or mental functions in a negative way, or passes on such genes,
- reduces the animals' ability to practise natural behaviour, or
- stimulates general ethical reactions.

This may reflect a hybrid view including utilitarian (first two points) and animal rights based (third point) views. The first point (animals' normal functions) may also have an element of respect for nature. One may therefore argue that such a hybrid may be the base for the public ethical view on using transgenic animals in Norway.

4.3.2 IPR and the introduction of GM and DNA vaccines

Modern biotechnology provides tools both for rapid detection and identification of disease and holds promises for new and improved vaccines. Generally spoken, there are two strategies for GM vaccine development: The first is represented by gene-deleted bacteria/ viruses to be used for homologous vaccination, i.e. to achieve protective immunity against the GM vaccine itself. The other strategy involves development of recombinant vaccines by genetic engineering where a gene that is immunologically targeted to a) be expressed after insertion in bacteria or yeast and the proteins produced are then further incorporated into a vaccine preparation, b) be inserted in a virus or bacteria by recombination and the recombinant virus or bacteria is then used as a vaccine, and c) DNA vaccine.

In aquaculture the most effective vaccines at present are multivalent (contains several genes of interest) and target salmon, there is also ongoing research to develop vaccines for other species (as seabass, tilapia, grouper etc.). Present approaches are, however, limited to some bacterial and viral diseases while there are no vaccines against parasites of fish. Especially intracellular pathogens, such as virus and some bacteria, have been found to be difficult to eradicate with traditional vaccines. Hence, DNA vaccines may offer a technological solution to these problems. An example of a DNA vaccine is the plasmid encoding infectious haematopoietic necrosis virus (IHNV) glycoprotein under control of a cytomegalovirus promoter (pCMV), which has been injected in Atlantic salmon with the purpose of achieving resistance to IHNV (Traxler et al., 1999). Following early trials on DNA vaccination in mammalian species, several experiments have been conducted in fish with promising results, such as complete protection against viral diseases (Romøren, 2003 and references therein). A combination DNA vaccine, consisting of multiple plasmids encoding several different antigens of a pathogen, holds prospect for inducing a broad spectrum of antibody responses, and hence be effective for vaccination against viruses that undergo antigenic variation (e.g infectious pancreas necrosis virus (IPNV) and infectious salmon anaemia virus (ISAV)) (Kibenge et al., 2001).

The development of both GM and DNA vaccines against infectious fish diseases has several attractive benefits: low cost, ease of production and improved quality control, heat stability, identical production processes for different vaccines, and the possibility of producing multivalent vaccines (Hew & Fletcher 2002; Kwang 2000). On the other hand, there is at present a limited scientific understanding of the fate of such vaccines after injection into the animal. There is a need for research with focus on the stability of the vectors and of the DNA construct, and if there are any unintended immunological impacts, the biological effect of the vaccine after injection (e.g. persistence, distribution, expression and integration) (Gillund et al., 2008a, 2008b). For GM vaccines it is important to investigate potential recombination with relatives and spread in the environment by vectors (Myhr & Traavik, 2011).

An example of the socioeconomic dilemmas relating to IPR, is the case of Intervet's patent on the Pancreas Disease (PD) virus and whether it is a potential barrier to further development of a PD vaccine or not. As the patent has been given on the virus itself, this gives Intervet full monopoly on developing vaccines against Pancreas Disease (PD). Pharmaq wanted a license to produce a PD vaccine and asked the competition authorities for a compulsory licence without success. They argued that the current Intervet vaccine is inefficient and their production is insufficient, and that their own vaccine is superior in terms of time, costs and animal welfare as it can be given as a component of one injection of multi-vaccination (Haavind Vislie, 2008). The Ministry of Fisheries and Coastal Affairs (FKD) was not directly involved although the competition authorities asked for their opinion. In a trial case, Pharmaq have emphasised that they apply a variant of the PD virus that is different from PD virus patented by Intervet. The case is still open as this book is published.

5. International obligations and examples of domestic regulations: How they address sustainable development in aquaculture

As we have seen in section two to four, the biotechnological developments open for benefits, but do also represent new challenges with regard to how they affect economic, social and environmental conditions. This section describes how these activities are sought regulated at the international and domestic arenas in order to enhance sustainability in the aquaculture sector. The section provides an overview of two different sets of legal provisions; the first bulk is rather voluminous and is aimed at access and benefit sharing legislation and hence, only indirectly dealing with the issue of sustainable development. The second is much less developed at the international and national arenas but these legal acts and instruments are more directly dealing with sustainable development in aquaculture. Both aspects are relevant to the Convention on Biological Diversity (CBD). At its second meeting (November 1995, Jakarta, Indonesia), the Conference of the Parties (COP) to the CBD agreed on a programme of action called the "Jakarta Mandate on Marine and Coastal Biological Diversity," which led to the creation of a work programme in this area.

The Convention on Biological Diversity has three interrelated objectives: Conservation and sustainable use of biodiversity, and access and equitable sharing of benefits from use of genetic resources. The scope of the CBD covers conservation and sustainable use of wild species and improved breeding stocks, as well as equitable sharing of benefits derived from the use of the world's genetic resources. This issue has been subject to controversial negotiations over the years since the establishment of the CBD. Negotiations recently resulted in a Protocol on Access and Benefit Sharing at the 10th Conference of the Parties to the CBD in Nagoya, October 2010.

The World Trade Organization (WTO) establishes global standards for harmonisation of intellectual property rights and the World Intellectual Property Organization has as its mandate to strive towards cooperation and harmonisation of IPR in all member countries. Harmonized IPR regulations target all technological fields similarly, including biotechnology and when biological material forms part of the invention. The tension between the overlapping objectives of the various international treaties is a controversial north-south issue.

Two changes in patent law have made the patent system controversial: First that patents are granted on various forms of biological material; and second that patents are granted to essentially basic research, increasing the commercial aspects of research. These tendencies raise questions about whether patents contribute to innovation or not. Also the link between exclusive rights and access rules is problematic, as very limited amount of benefits arising from utilization of genetic resources have been shared with providers. The CBD seeks to balance expanding patent regimes by establishing a compromise between access to technology and access to the input factors in biotechnology – genetic resources (Koester, 1997; Rosendal, 2000). This interaction between different international objectives has caused North-South conflicts over access to seeds and medicinal plants versus patented technology in the agriculture and medicinal sectors. This is different for aquaculture and animal husbandry where breeding material has usually not moved from south to north.

Access to genetic resources, conservation, equitable sharing of benefits, and IPR systems to boost innovation – are all internationally agreed objectives – but they are not necessarily mutually compatible (Rosendal, 2006). Conservation is basically a prerequisite for all the other objectives, as acknowledged in the three objectives of the CBD. The essence of the CBD is to tie the balance between ABS and IPR to that of conservation: Without access to the genetic resources, there can be little innovation. The CBD attempts to establish a system for innovation based on biodiversity to contribute in a fair manner to the conservation of the diversity. IPR legislation seeks its justification in increasing the incentive for innovation. There are, however, indications that broad patent claims have the potential also to hamper innovation by stifling access to technology and increase the transaction costs for other actors. Without innovation, there may be fewer benefits to share. Without benefit sharing from utilisation of genetic resources, there may be less will and ability to conserve biodiversity in developing countries – although this particular dimension has less immediate relevance in aquaculture compared to the agricultural and pharmaceutical sectors (FAO, 2009).

Turning to domestic norms and regulations and applying Norway as an example, the overall goals for aquaculture are linked to safeguard coastal settlements and increase value, sustainable management and innovation (White Paper, 2005:9, 136). Norway acknowledges responsibility for about one third of the world's remaining populations of wild salmon, as well as environmental responsibilities through Norwegian owned salmon farms including production in other countries (White Paper, 2009:142). This raises interesting questions about the relationship between Norwegian utilization of this resource and Norway's responsibility for managing the wild material according to the CBD.

At the domestic level, Norway has recently developed two relevant legal acts: The Nature Diversity Act of 2009 and the Act on Management of Wild Marine Resources of 6th June 2008. The Wild Marine Resources Act grants discretion for the government to establish a procedure of governmental permission before bioprospecting of wild marine genetic resources and is hence of less immediate relevance for export from breeding programmes. The Nature Diversity Act establishes genetic material as a commons resource that should remain a common property resource in Norway, and it also gives the Ministry the discretion to require permits for accessing genetic resources. Also the Marine Resources Act states that marine resources are a common resource. Both Acts require the respective ministries to supplement the legislation with detailed administrative regulations for access to genetic resources, but several challenges remain in developing these regulations. One is the relationship between access to the resources and the right to use it for a patented invention. Export of fingerlings and breeding material will invariably include the genetic material – and may hence require special protection in order to secure the interests of the exporter to maintain these resources as a commons.

Let us now turn to instruments that are more directly trying to tackle environmental concerns in aquaculture. Most central of these is the emerging Aquaculture Stewardship Council (ASC), which is expected to be in operation by 2011 as the world's leading certification and labelling programme for responsibly farmed seafood. The ASC is the outcome of the Aquaculture Dialogues and it will be responsible for working with independent, third party entities to certify farms that are in compliance with the standards for responsible aquaculture, which is created by the Dialogues. These standards are designed to minimize the key negative environmental and social impacts related to 12 aquaculture species. Similar to those of the Forest Stewardship and Marine Stewardship Councils, these standards prescribe quantitative performance levels that farmers must reach to become certified. More than 2,000 aquaculture producers, conservationists, scientists and others are involved in the process, which is coordinated by World Wildlife Fund (WWF). Along with the Dutch Sustainable Trade Initiative (IDH), the WWF also help fund the development of the ASC, which will be a non-profit organization.

6. Structural developments and management trends in aquaculture

The new promises and threats of the technological developments provide a backdrop also for this section, which is aimed at combining technology and structural trends. The general structural trends in aquaculture are similar to those in agriculture and animal husbandry worldwide, that is to say moving away from public funding and small scale enterprises towards merging, privatization and internationalization. In this section we briefly account for these trends and then in section eight, we point to some of the implications this may have for management and sustainable development in aquaculture.

In Norway, salmon and trout breeding programs were started with public financing in 1971 by a non-profit research institute (Gjøen & Bentsen, 1997). The base populations of these programs were collected from Norwegian rivers (Atlantic salmon) and from Scandinavian farmed populations (rainbow trout) and these breeding populations were transferred in 1985 to a cooperative ownership by salmon farmers' organizations. However, as a result of an economic crisis in the late 1980s, this activity was transferred to a shareholder company in 1992 (at present Aqua Gen AS). At that time it was decided that the value of the breeding material should be secured in a way that took care of the public interest; hence the structure of ownership was divided between private and public shareholders, but that structure was only bound for a five years period. With the next government the largest public shareholder was turned into a private venture company, who in 2007 decided to sell its shares in Aqua Gen AS. The German EW Group, which is also the holding company of the world's leading poultry genetic companies (Aviagen), concluded an agreement to take majority ownership of the shares in Aqua Gen AS. Thus, the Ministry of Fisheries and Coastal Affairs (FKD) gradually lost control over the material from the originally public supported breeding programme for salmon.

Currently, the Norwegian legal system is unclear on regulating genetic material originating in the wild or coming from public breeding programmes and hence a comprehensive management system for aquatic genetic resources is not in place. The sale of Aqua Gen AS to German EW Group is illustrative of the dilemma. This breeding material can now in theory be patented and removed from the public domain. The development has moved from a situation of public control and ownership, via a cooperative situation, to the current situation of increasingly dominating market actors. This raises question about eventual effects of recent Norwegian access regulations in the Nature Diversity Act and the Marine Wild Species Act. The situation has raised questions about the need to regulate access to wild or breeding material in other species, as cod and halibut, because a similar level of international exchange has not taken place for these other species of fish.

Since the early 1990s, public support for Norwegian farm animal breeding has decreased, reflecting a political will to privatize breeding. A counter trend is the initiative from the Ministry of Fisheries and Coastal Affairs to fund the establishment of a breeding program for cod at the research institute, Nofima in Tromsø. In line with the policy goal of safeguarding rural settlements, this may reflect an intact willingness to finance development of breeding programmes in Norway, at least in the districts of Northern Norway. The current official goal is to retain the cod breeding material and associated competence and knowledge bases that are being built up as a Norwegian public good asset. The end goal may be that cod, like salmon, is intended to become profitable and commercialised at some point down the line, but the legal process of how to deal with this has only just started (interview NN1).

7. Actor interests responding to structural and technological changes

In this section we turn to the description of how affected actors perceive the legal, structural and technological developments in the aquaculture sector. The aim is to study their views on whether and how these parallel trends are likely to affect sustainable development in aquaculture.

7.1 Public opinion and attitudes to new biotechnologies

In Europe, surveys have been carried out from 1996 to 2007, to identify Europeans opinion and attitudes to modern biotechnology. Although difficult to draw any conclusion it seems that modern biotechnology used in production for medicines and pharmaceutical products receives highest support while modern biotechnology used in agriculture receives less support (although concern about GM foods have had a small decline since 2001). For example in one recent Eurobarometer poll (GMO compass, 2009), European consumers were asked to identify the environmental themes about which they were most concerned, and on an average, 20% of respondents cited the topic of 'GMOs in farming'. Although these surveys have been on GM crops and medicines, the aspects related to benefits and environmental issues may have similar support and doubt versus the different possibilities for introduction of modern biotechnology in aquaculture (see also 7.1.1 for results from a survey among Norwegian students).

There are several challenges involved in measuring actor perceptions of GMO. A study of the GM debate in the UK, Australia and New Zealand found that access to decision-making and the inability to weigh explicit social value judgements with the broad science consensus were the major obstacles to successful deliberative public debate (Walls et al., 2005). For instance, in the New Zealand experience, non-scientific arguments were implicitly marginalised because the templates (questionnaire) employed for the interest groups made it difficult to use holistic arguments. A 'holistic argument' in this case might imply a consideration of the growing dominance of multinational corporations in the life sciences. These enterprises increasingly decide on options for the development of new medicines and food, they are part and parcel of the GM revolution – but somehow their role seemed to be 'beside the point' in the questionnaire developed to study the public debate (Walls et al., 2005). Most studies on GM plants and products are based on information provided by research laboratories and/or released by industry (Gaskell et al., 2003). It is also important to note that this documentation, along with the GMO applications, is provided by multinational corporations that enjoy little public trust (Gaskell & Bauer, 2001).

7.1.1 Students attitudes to the use of GMO in aquaculture

Knowledge and attitudes regarding the use of GMOs in aquaculture was studied among students at the Norwegian University of Life Sciences (UMB) that can be considered as future stakeholders in food production and management of biological resources (Unpublished data). The survey result showed that most students were well informed about genetic engineering and/or GMOs. Many students were very concerned about environmental impacts from the use of GMOs. For example it was found that most of the respondents disagreed on the statement on whether the use of GMOs results in no negative environmental effects, this was also confirmed by the question on whether the students would buy transgenic salmon if it was more environmentally friendly (showed by an 40 % increase in willingness to buy transgenic salmon if it was more environmentally friendly). However, many students were not willing to buy transgenic salmon if it was more nutritious or more disease resistant. Only 47% of the respondents would buy transgenic salmon if a relevant authority (e.g. Norwegian Food Safety Authority) has approved it as safe.

Many students were totally agreeing that there is a need to reduce the risk by initiating more research on GMOs, that new knowledge about risks must be taken into account and that we need to seek expert advice to get more understanding about the potential risks to health and the environment. The students were also strongly agreeing on the need to reduce the risk of genetic engineering applications by increasing transparency to the public about research and information about the technology. Moreover, the need for improving communication between scientists and the public was recognized.

Spirituality or religion had little or no influence when it came to ethical issues by GMOs in this survey. Students supported use of GMOs for saving human lives (e.g. by producing medicines and vaccines) followed by production of animal feed (e.g. from plants, algae and microorganisms). Surprisingly, it was not found that production of cheaper food could encourage the students to support the use of GMOs. In general, students requested labelling of GM food as well as of salmon fed with GM plant feed.

7.2 Public attitudes to animal welfare and willingness to support and pay for e.g. improved health and welfare of farmed fish

In a recent study by Olesen et al. (2010) it was found that a relatively large percentage of Norwegian consumers agreed (at least partially) that Norwegian animal welfare standards were sufficiently strict (78,1%), and that fish welfare was sufficiently protected in Norwegian fish farming (67.9%). In order to estimate a lower bound for the consumers' willingness to pay for improved welfare for farmed salmon, a real choice experiment with eco-labelled salmon was carried out in Norway (Olesen et al., 2010). It was found that the average respondent preferred eco-labelled salmon to conventional salmon when the colour was the same, and was willing to pay additional 2 euro per kg fillet for eco-labelled salmon. The price premium depends on the conventional and organic salmon being the same colour, and an inferior appearance due to lack of pigmentation significantly reduces consumer interest in organic salmon. This is also consistent with the results in other studies investigating consumer preferences toward organic products with inferior appearance (Thompson & Kidwell, 1998; Yue et al., 2009).

7.3 Actor perceptions of needs and interests in access to aquatic genetic material

Studies of the fish breeding sector for several species and in several countries indicate that they are all prone to rapid structural changes in response to calls for profitability and commercialization (Olesen et al., 2007; Rosendal et al forthcoming). This correspond with a recent survey among fish farmers in Norway where it was found that the most important source of risk for the industry was future salmon prices, institutional risks and diseases (Bergfjord, 2009). In the same study the respondents (farmers) was also asked to identify the most important risk management strategies to the risk issues identified, which was to keep cost low and ensure profitability.

The public breeding programme on cod is currently seen as a public good for Norwegian breeders (interview NN2). The authorities are concerned that the public and private cod breeding programmes can compete on a level playing field, so that the one with public funding is not given unfair competitiveness. At this early stage, it is acknowledged that it is very hard to fund a breeding program, as the economic returns from increased growth may still be a long way off. For salmon, the real growth and economic returns from the breeding program was not apparent until about the fourth generation – and cod is still only in the second or third. There are two major reasons why public funding may be the preferred solution, at least in the early phases: First, during the early phases of breeding, basic mass selection using individual phenotypic information can provide a similar and much cheaper response in growth. This is why more advanced breeding programmes are often less profitable, particularly on a short term, as they are equally costly to start and run the first generations. However, phenotype or mass selection is usually much more limited with respect to selection towards a broader breeding goal with several traits, such as disease resistance. Also, it may be more vulnerable to less control of and rapid increase of inbreeding with resulting genetic erosion. Hence, in the long run, the more advanced family based breeding programmes will become more economically and biologically viable – or sustainable (interview, NN3).

Second, compared to a private breeding programme, aiming at short term profit, a public or cooperative programme usually have a broader range of breeding goals including animal welfare and environmental concerns. Hence, it is expected to obtain apparent gains in growth rate and economic returns later than a private programme – but to give more long term viable fish material and become more sustainable. This gives a competitive edge to the private cod programme in the short run and a competitive edge to the public one in the long run (interview NN1). In combination, they may be in line with the objectives stated in the White Paper for rural/coastal settlements, increased value, sustainable management and innovation (2005). Similar trends have been found for shrimp in India and tilapia in Asia (Ramanna Pathak, forthcoming, Ponzoni et al., 2010).

Considering the problems following Norwegian aquaculture investments and operations in Chile (where the entire salmon farming sector has recently been suffering from widespread outbreaks of infectious salmon anaemia), the enterprises expect to see stricter regulations regarding biological and environmental risks (Marine Harvest, 2011). New regulations will, however, require investment in new technology.

7.4 Researchers opinions and attitudes to new biotechnologies

Very few studies have been carried out with the intention to investigate researchers opinions and attitudes to new biotechnologies. In one study carried out by Kvakkestad et al. (2007) it was found that that different scientists, depending on scientific discipline (ecology, molecular biology, plant breeding), source of funding (public or industry) or whether they worked within industry, government or academia, interpreted data differently in situations characterised with uncertainty, and thus expressed a diversity of opinions about the risks arising from GM crops. In a recent study by Gillund and Myhr (2010) perspectives on alternative feed resources for salmon were identified among stakeholders in Norwegian aquaculture. In this study the sustainability of plant production in industrial agriculture, and particularly the cultivation of GM plants, was contested among the participants. The participants defined a broad range of appraisal criteria concerning health and welfare issues, economical issues, environmental issues, and knowledge and social issues, which illustrates that finding sustainable alternative feed resources is difficult.

8. Discussion

In spite of the tremendous benefit cost ratios and value creation for the society in terms of more efficient fish production and lower fish prices, only a small percentage (ca 10%) of the current world aquaculture production is based on genetically improved material from modern breeding programs (Gjedrem et al., 2011). The reasons that aquaculture is lagging tremendously behind the agriculture sector in this respect may be the lack of tradition and training of applying systematic selective breeding, little interaction with agriculture or aquaculture research groups with the knowledge and technology, low prices of roe and fry and resulting low profit margins of genetic improvement for breeders and hatcheries. The latter has made genetic improvement an insecure investment objective, perhaps also due to the ease of illegally reproducing highly fertile fish and marketing 'pirate copied' material. The last two decades, R&D funds have tended to be prioritized for research in molecular genetics and genomics with less funding for further development and establishment of selective breeding that has proved to give long term genetic gains. However, also for applying genomic information efficiently a selective breeding program is a prerequisite.

With interesting similarities to the ABS debate in the CBD, there seems to be little immediate value in breeding and breeding programmes. Previous studies have shown that the incentives for capitalizing on salmon breeding materials have been virtually non-existent, due to low roe prices and low profit for improved breeding material; a trend that does not seem likely to change in the near future (Olesen et al., 2007; Rosendal et al., 2006). Similarly, there seems to be little profit to be reaped from increased knowledge about and improvements of genetic resources and their traits, as has also been claimed in the ABS debate (Grajal, 1999). At the same time, it is hard to refute the great profits from biotechnology – from traditional breeding to genetic engineering – and there is a growing business interest in access to valuable genetic material (Laird & Wynberg, 2005). This seems paradoxical also with a view to the valuable good of faster growing and hence cheaper salmon, which is resulting from the breeding programs. Here also we see that the willingness to pay is small but the interest in access is paradoxically high. The problem is that bringing forth fast growing, disease free fish is relatively expensive, whereas the result can be copied at very low costs. This has led to a pressure towards profitability and privatisation in the aquaculture sector, including the public breeding programmes. However, the cost of maintaining a good, disease free product is relatively high and the question is whether the market can be expected to deliver this service, when there is such a high degree of uncertainty regarding profits. Due to the high fertility of aquatic species, aquaculture breeding programs have shown to give high benefit/cost ratios and tremendous value creation for the society. This is also the case for family based programs aiming for a broad and sustainable breeding goal with many traits. The paradox illustrates the challenge of securing policy goals of affordable access to genetic improvements in breeding and to stimulate sustainability and innovation in aquaculture. The alternative to the continued funding of public breeding programmes may portend forfeiting the normative ideal of providing improved breeding material on an affordable basis.

In general, aquaculture is experiencing pressure towards higher production efficiency and short term profits. Hence, actors face emerging difficulties pertaining to adequate funding for sustainable breeding programmes and affordable access to improved genetic material. Historically, aquaculture in India and Norway has mainly been based on public investments to increase production, develop and widely disseminate material to as many users as possible, rather than creating proprietary products. The same was true for the original objectives of the GIFT tilapia project. This illustrates the nature of breeding material as a public good. Greater involvement of private sector leads however to stronger need for legal protection of genetic material. As this keeps knowledge out of the public domain, it is perceived to have negative implications for aquaculture. In a study by Rosendal et al. (forthcoming) it was found similarities between Norway, India and the GIFT donors regarding their normative objective to maintain affordable access to improved breeding material. Moreover, a common concern among the actors interviewed was how to avoid the tendency towards monopolisation in a globalised market and how to maintain affordable access to aquatic breeding material. At the same time, the demand for profitability is undermining these goals. This may lead cod (and well as shrimp and carp) breeding programmes on a similar track as that of salmon and GIFT tilapia. A waiver of public control may seem to go against the interests and advice from both private and public actors in the aquaculture sector itself.

Market consolidations and privatisation are among the structural factors that the actors themselves recognize as most important in changing the ground rules within the salmon sector (Olesen et al., 2007). The privatisation and commercialisation can be expected to turn the breeding goals towards developing products for which there are economically viable markets rather than developing new products based on social, ecological and biological criteria (with e.g. disease resistance and fish welfare traits). This development has come a long way in the case of salmon (and tilapia), where those that were previously public collections and publicly funded breeding programmes and breeders' lines have now been privatised; similar trends can be expected in the case of farmed cod. The overall structural traits of the aquaculture sector also go a long way in explaining why the aquaculture sector is much less subject to ABS conflicts between developed and developing countries compared to the plant sector. While not engendering a North-South conflict, the basic interests in access to breeding material remains similar for plant and animal genetic resources. As a result of the structural developments leading to fewer and larger companies, access conflicts may be more likely to evolve between small and large scale actors in the sector rather than between countries.

The biology of breeding suggests that the real value lies in continuous upgrading and improvement, and patents are not useful for this as it freezes innovation. The cost and time of obtaining a patent along with the long protection period in patent law (twenty years) hardly promote rapid innovation in sectors where continuous upgrading in a biological dynamic system is the most viable and sustainable approach. Interestingly, neither of the two salmon breeding companies (AquaGen and Landcatch Natural Selection) that managed to map the QTL marker for IPN resistance have chosen to patent it. Probably, the costs involved with enforcing such patents are considered too high for these companies (even for AquaGen backed by a big international firm). Nevertheless, similar to agriculture and pharmacy, the structural changes within the aquaculture sector seem to be much more influential than biological traits in affecting actors' perceptions of need for access and protection. This also has implication for the broader international debate pertaining to the Nagoya Protocol on ABS and whether it needs to be supplemented by sector based treaties emulated to different types of genetic resources. As the structural traits of monopolization and globalisation are similar across the board (agriculture, pharmacy, aquaculture), that would suggest a reduced rationale for a sector approach to regulating international transactions with genetic resources.

9. Recommendation

Aquaculture is a source for food all around the world (FAO, 2009). Can it with its biotechnological development be sustainable on both short and long-term? As we have presented in this chapter there are several challenges that actors within the aquaculture sector is facing and decisions has to be made.

In the EU communication (2002) 'A strategy for the sustainable development of European aquaculture' it is emphasised that 'The fundamental issue is therefore the maintenance of competiveness, productivity, durability of the aquaculture sector. Further developments of the industry must take an approach where farming technologies, socio-economics, natural resources use and governance are all integrated so that sustainability can be achieved.' This is a very ambitious vision for the future and implies an integrated assessment of environmental, social, economic and legal issues. To do so we recommend that:

- Selective breeding programs with sufficiently low inbreeding rate are initiated to supply domesticated and genetically improved seed fit for a life in farm environment. Moreover, along with these programs there is also a need for breeding programs for farming in different environments for the most important farmed species.
- Breeding strategies towards broad and long terms goals should be included to ensure genetic variability and robust farm animals with good welfare. Initiatives for breeding for e.g. disease and parasite resistance are started up and improved to ensure sustainability of the industry. Application of molecular genetics (MAS and GS) may improve the efficiency of such initiatives. However, efficient application of MAS and GS require well organised selective breeding programs.
- Public support or cooperative ownership and organisation is necessary to promote the recommendations above. Economic support and training to establish lacking selective breeding schemes may for example be needed to apply and exploit powerful molecular techniques, and parallel funding and training are needed for development of local capacity within genetic engineering/biotechnology and selective breeding programs.
- Patent laws and practice should avoid too broad patents (e.g. of virus and bacteria) that hamper further innovation.
- Introduction of GMOs and products such as transgenic fish, DNA vaccines and GM vaccines need to involve application of precautionary approaches that include research on adverse effects by the inserted genes on animal health and welfare and on the environment. Moreover such introduction must also be followed with monitoring strategies for detection of unexpected effects.
- Consumers and public opinions and attitudes to new biotechnologies in aquaculture need to be identified together with studies on willingness to support and pay for e.g. improved health and welfare of farmed fish.
- Potential implications for the industry by introduction of biotechnology is important and need to be elaborated with the intention to find ways that the technology can stimulate sustainable innovations.

10. Acknowlegdements

The study is financed by the Research Council of Norway ("Stimulating sustainable innovation in aquaculture", Project Number: 187970).

11. Interviews

Interview with NN1, Ministry of Fisheries and Coastal Affairs, 26th August 2009.

Interview with NN2, Research Institute, 9th June 2009.

Interview with NN3, Breeding Company, 1st July 2009.

12. References

Animal Welfare Act. In: *GOVERNMENT.NO, Information from the Government and the Ministries.* 10.08.2011, Available from
http://www.regjeringen.no/en/doc/laws/Acts/animal-welfare-act.html?id=571188

Aquagen (2010). QTL-rogn - dokumentert IPN-beskyttelse fra første dag. In: *AquaGen Kunnskapsbrev.* No.1. (2010), p. 1, 19.06.2011, Available from:
http://aquagen.no/filestore/01-2010QTL-rogn-dokumentertIPN-beskyttelsefrafrstedag.pdf

Bentsen, H.B., & Gjerde, B. (1994). Design of fish breeding programs. *Proceedings of the 5th World Congress of Genetics Applied to Livestock Production,* Vol.19, pp. 353- 359, Guelph, Canada, August 1994.

Bentsen, H.B. & Olesen, I. (2002). Designing aquaculture mass selection programs to avoid high inbreeding rates. *Aquaculture,* Vol.204, pp. 349-359 .

Bergfjord, O.J. (2009). Risk perception and risk management in Norwegian aquaculture, *Journal of Risk Research,* Vol.12, No.1, pp.91-104.

Broom, D.M. (1986). Indicators of poor welfare. *British Veterinary Journal,* Vol.142, pp. 524-526.

Caffey, R.H., Kazmierczak, R.F., Romaire, R.P. & Avault, J.W. (1998). Indicators of aquaculture sustainability: a Delphi survey, *Book of Abstracts of The international triennial conference and exposition of the World Aquaculture Society, the National Shellfisheries Association and the Fish Culture Section of the American Fisheries Society,* pp.91, Las Vegas, USA, 1998.

CBD (2000). *Cartagena Protocol on Biosafety to the Convention on Biological Diversity.* (Montreal: Secretariat of the Convention on Biological Diversity). 16.12.2010, Available from
http://bch.cbd.int/protocol/publications/cartagena-protocol-en.pdf

Devlin, R.H., Yesaki, T.Y., Donaldson, E.M., Du, S.J., Shao, J.D. & Hew, C.L. (1995). Production of germline transgenic Pacific salmonids with dramatically increased growth performance. *Canadian Journal of Fisheries and Aquatic Sciences,* Vol. 52, pp. 1376-1384.

European Commission (2002). *A strategy for the sustainable development of European aquaculture.* 29.4.2010, Available from
http://ec.europa.eu/fisheries/cfp/aquaculture_processing/aquaculture/strategy_en.htm.

FAO, (2009). Bartley, D.M., Benzie, J.A.H. et al.: *The Use and Exchange of Aquatic Genetic Resources for Food and Agriculture.* CGRFA Background Study Paper No. 45 , FAO, Rome

Fleming, I.A. & Gross, M.R. (1992). Reproductive-behavior of hatchery and wild coho salmon (oncorhynchus-kisutch) - does it differ. *Aquaculture,* Vol.103, pp. 101-121.

Gaskell, G. & Bauer, M.W. (Eds.) (2001). *Biotechnology, 1996–2000: The years of controversies,* Science Museum Press, London

Gaskell, G., Allum, N. & Stares, S. (2003). *Europeans and Biotechnology in 2002: Eurobarometer 58.0.* A report to the EC Directorate General for Research from the project Life Sciences in European Society. March 21, 2003 (2nd edition).

Gillund, F., Kjølberg, K., Krayer von Krauss, M. & Myhr, A.I. (2008a). Do uncertainty analyses reveal uncertainties? Using the introduction of DNA vaccines to aquaculture as a case. *Science of the Total Environment ,* Vol.407, pp. 185-196.

Gillund, F., Tonheim T., Seternes T., Dalmo R.A. & Myhr A.I. (2008b). DNA vaccination in aquaculture –Expert judgements of impact on environment and fish health, *Aquaculture,* Vol. 284, pp. 25-34.

Gillund, F. & Myhr, A.I. (2010). Perspectives on salmon feed: A deliberative assessment of several alternative feed resources. *Journal of Agricultural and Environmental Ethics,* Vol. 23, pp. 527-550.

Gjedrem, T., Robinson, N., Rye, M. (2011). The importance of selective breeding in aquaculture to meet future demands for animal protein: A review. *Aquaculture.* Submitted

Gjerde, B., Gjøen, H.M. & Villanueva, B. (1996). Optimum designs for fish breeding programmes with constrained inbreeding. Mass selection for a normally distributed trait. *Livestock Production Science,* Vol.47, pp. 57-72.

Gjøen, H.M. & Bentsen, H.B. (1997). Past, present and future of genetic improvement in salmon aquaculture, ICES Journal of Marine Science, Vol.54, pp. 1009-1014

GMO Compass (2009). Opposition decreasing or acceptance increasing?, 28.5.2011, Available from
http://www.gmo-compass.org/eng/news/stories/415.an_overview_european_consumer_polls_attitudes_gmos.html

Grajal, A. (1999). Biodiversity and the Nation State: Regulating Access to Genetic Resources Limits Biodiversity Research in Developing Countries. *Conservation Biology,* Vol.13, No.4, pp. 6-10.

Hayes, B. & Andersen, Ø. (2005). Modern biotechnology in aquaculture. In: *Selection and Breeding Programmes,* T. Gjedrem (Ed.), pp. 301-317, Springer, ISBN -10 1-4020-3341-9, Dordrecht.

Haavind Vislie advokatfirma. (2008). Letter of 19 June to Norwegian competition authorities.

Hew, C.L. & Fletcher, G.L. (2002). The role of aquatic biotechnology in aquaculture. *Aquaculture,* Vol.197, pp. 191-204.

Holmenkollen Guidelines (1999). Holmekollen Guidelines for sustainable aquaculture 1997. In: *Sustainable Aquaculture,* N. Svennevig, H. Reinertsen, M. New (Eds.), pp. 343-347, Proceedings of the Second International Symposium on Sustainable Aquaculture: Food for the future? Oslo, Norway, November, 1997, AA. Balkema, Rotterdam/Brookfield

Hyperdictionary, (2003). *hyperdictionary.* Available from: http://www.hyperdictionary.com.

Huntingford, F.A., Adams, C., Braithwaite, V.A., Kadri, S., Pottinger, T.G., Sandøe, P. & Turnbull, J.F. (2006). Current issues in fish welfare. *Journal of Fish Biology*, Vol.68, pp. 332-372.

Johnsson, J.I., Petersson, E., Joensson, E., Bjoernsson, B.T. & Jaervi, T. (1996). Domestication and growth hormone alter antipredator behavior and growth patterns in juvenile brown trout, Salmo trutta. *Canadian Journal of Fisheries and Aquatic Sciences*, Vol.53, pp. 1546-1554

Kamara, M., Coff, C. & Wynne, B. (2006). *GMOs and Sustainability: Contested visions, routes and drivers*. Report prepared for the Danish Council of Ethics, Copenhagen.

Kapuscinski, A.R. & Brister. D. J. (2001). Genetic impacts of aquaculture. In: *Environmental Impacts of aquaculture*, K.D. Black (Ed.), pp. 128-153, Sheffield, UK, Sheffield Academic press.

Kibenge, F.S.B., Garate, O.N., Johnson, G., Arriagada, R., Kibenge, M.J.T. & Wadowska, D. (2001). Isolation and identification of infectious salmon anaemia virus (ISAV) from Coho salmon in Chile. *Diseases of Aquatic Organisms* , Vol.45, pp. 9–18.

Koester, V. (1997). The Biodiversity Convention Negotiation Process and Some Comments on the Outcome. *Environmental Policy and Law*, Vol. 27, No.3, pp. 175-192.

Kolstad, K., Heuch, P.A., Gjerde, B., Gjedrem, T. & Salte R. (2005). Genetic variation in resistance of Atlantic salmon (Salmo salar) to the salmon louse *Lepeophtheirus salmonis*. *Aquaculture*, Vol. 247, pp. 145-151.

Kvakkestad, V., Gillund, F., Kjølberg, K. & Vatn, A. (2007). Scientists' perspectives on the deliberate release of GM crops. *Environmental Values*, Vol.16, pp. 79-104.

Kwang, J. (2000). Fishing for vaccines. *Nature Biotechnology*, Vol.18, pp. 1145-1146.

Laird, S. & Wynberg, R. (2005). *The Commercial Use of Biodiversity: An Update on Current Trends in Demand for Access to Genetic Resources and Benefit-Sharing, and Industry Perspectives on ABS Policy and Implementation*. UNEP/CBD/WGABS/4/INF/5.

Le Curieux – Belfond, O., Vandelac, L., Caron, J. & Seralini, G.E. (2009). Factors to consider before production and commercialization of aquatic genetically modified organisms: the case of transgenic salmon. *Environmental Science and Policy*, Vol.12, pp. 170-189.

Lund, V., Mejdell, C., Röcklinsberg, H., Anthony, R. & Håstein, T. (2007). Expanding the moral circle: farmed fish as objects of moral concern. *Diseases of Aquatic Organisms*, Vol.75, pp. 109-118.

Marine Harvest. (2011). *Salmon, a small protein source, but a great alternative in the global market place*. Presentation held by A. Aarskog at the North Atlantic Seafood Conference in Oslo 3.3. 2011, Available from http://hugin.info/209/R/1494191/430048.pdf

Mazur, C.F. & Iwama, G.K. (1993). Effect of handling and stocking density on hematocrit, plasma-cortisol, and survival in wild and hatchery-reared chinook salmon (oncorhynchus-tshawytscha). *Aquaculture*, Vol.112, pp. 291-299

McGinnity, P., Stone, C., Taggart, J.B. Cooke, D., Cotter, D., Hynes, R. et al. (1997). Genetic impact of escaped farmed Atlantic salmon (Salmo salar L.) on native populations: use of DNA profiling to assess freshwater performance of wild, farmed, and hybrid progeny in a natural river environment. *ICES Journal of Marine Science*, Vol.54, pp. 998-1008

Melamed, P., Gong, Z., Fletcher, G. & Hew, C.L. (2002). The potential impact of modern biotechnology on fish aquaculture. *Aquaculture*, Vol. 204, pp. 255-269.

Metcalfe, N.B., Valdimarsson, S.K. & Morgan, I.J. (2003). The relative roles of domestication, rearing environment, prior residence and body size in deciding territorial contests between hatchery and wild juvenile salmon. *Journal of Applied Ecology*, Vol.40, pp. 535-544.

Meuwissen, T.H.E., Hayes, B.J. & Goddard M.E. (2001). Prediction of total genetic value using genome-wide dense marker maps. *Genetics*, Vol.157, pp. 1819-1829.

Moen T., Baranski M., Sonesson A.K. & Kjøglum, S. (2009). Confirmation and fine-mapping of a major QTL for resistance to infectious pancreatic necrosis in Atlantic salmon (Salmo salar): population-level associations between markers and trait. *BMC GENOMICS*, Vol.10, pp. 368.

Myhr, A.I. & Traavik, T. (2011). *Genetically Engineered Virus Vaccine Vectors: Environmental Risk Management Challenges*, In: Genetic Engineering, InTech, ISBN 978-953-307-671-3.

Niiler, E. (2000). FOA researchers consider first transgenic fish. *Nature Biotechnology*, Vol. 18, pp. 143.

Norwegian Ministry of Fisheries and Costal Affairs (2009). *Strategi for en miljømessig bærekraftig havbruksnæring*. 10.07.2009, Available from http://www.regjeringen.no/...strategier.../strategi-for-en-miljomessig-barekraftig-.html.

Olesen, I., Groen, A.F. & Gjerde, B. (2000). Definition of animal breeding goals for sustainable production systems. *Journal of Animal Science*, Vol.78, pp. 570-582.

Olesen, I., Rosendal, G.K., Bentsen, H.B., Tvedt, M.W. & Bryde, M. (2007). Access to and protection of aquaculture genetic resources - Strategies and regulations. *Aquaculture*. Vol.272, No.1, pp. S47-S61

Olesen, I., Alfnes, F., Røra, M., Navrud, S. & Kolstad, K. (2010). Eliciting consumers' willingness to pay for organic and welfare labelled salmon by a non-hypothetical choice experiment. *Livestock Science*, Vol.127, pp. 218-226.

Olesen, I., Myhr, A.I. & Rosendal, G.K. (2011). Sustainable aquaculture: are we getting there? Ethical perspectives on salmon farming. *Journal of Agricultural and Environmental Ethics*. Vol.24, No. 4, pp. 381-408.

Petersson, E. & Järvi, T. (2006). Anti-predator response in wild and sea-ranched brown trout and their crosses. *Aquaculture*, Vol.253, pp. 218-228.

Ponzoni, R.W., Khaw. H.L. & Yee, H.Y. (2010). GIFT: the story since leaving ICLARM (now WorldFish Centre) – socioeconomic, access and benefit sharing and dissemination aspects. *FNI Report*, 14/2010, Oslo, Norway

Ramanna Pathak, A. (fortcoming) Balancing biodiversity, access to genetic resources and profits in India's shrimp sector.

Rauw, W.M., Kanis, E., Noordhuizen-Stassen, E.N. & Grommers, F.J. (1998). Undesirable side effects of selection for high production efficiency in farm animals. A review. *Livestock Production Science*, Vol.56, pp. 15-33.

Refstie, T. & Gjedrem, T. (2005). Chromosome Engineering. In *selection and Breeding, Programmes, T.* Gjedrem (Ed.), pp 287-299, Springer, ISBN -10 1-4020-3341-9, Dordrecht

Robinson, N. (2002). Techniques for advanced genetic improvement of fin fish and shellfish in aquaculture. *Report on an NRE scientific exchange visit to aquaculture and marine science institute in Scotland and Norway,* September 2002. Australia, Victorian Institute of Animal Science

Romøren, K. (2003). Liposome- and Chitosan-based formulations used for gene delivery to fish: *-in vivo* and *in vitro* studies of possibilities and limitations. Ph.D. dissertation, ISSN 1501-7710. Faculty of Mathematics and Natural Sciences, Department of Pharmaceutics, School of Pharmacy, University of Oslo, Norway

Rosendal, G.K. (2000). *The Convention on Biological Diversity and Developing Countries.* Dordrecht: Kluwer Academic Publishers

Rosendal, G. K., Olesen, I., Bentsen, H.B., Walløe Tvedt, M. & Bryde, M. (2006). Access to and legal protection of aquaculture genetic resources – Norwegian perspectives. *Journal of World Intellectual Property,* Vol. 9, No.4, pp. 392-412.

Rosendal, G.K. (2006). Regulating the Use of Genetic Resources – Between International Authorities. *European Environment,* Vol.16, No. 5, pp. 265-277.

Ruzzante, D.E. (1994). Domestication effects on aggressive and schooling behavior in fish. *Aquaculture,* Vol.120, pp. 1-24.

SalmoBreed. (2011). SalmoBreed AS har identifisert genetisk markør for sykdommen pancrease disease (PD). In: *Nyhetsbrev fra SalmoBreed.* Vol.7, No.3, (June 2011), p.1, 19.06.2011, Available from:
http://www.salmobreed.no/files/Nyhetsbrev/nyhetsbrev_2011/2011_3_nyhetsbr ev_SalmoBreed_-_Kopi.pdf

Sandøe; P. & Holtug, N. (1998). Ethical aspects of gene and biotechnology. *Acta Agriculturae Scandinavia, Section A: Animal Science,* Suppl. 29, pp. 51-58.

Sandøe, P. & Christensen, S.B. (2008). *Ethics of animal use.* Blackwell Publishing, ISBN 978-1-4051-5120-7, Oxford, UK

Sonesson, A.K., Meuwissen, T.H.E. & Goddard, M.E. (2010). The use of communal rearing of families and DNA pooling in aquaculture genomic selection schemes. *Genetics Selection Evolution,* Vol.42, No.41, pp. 9

Thompson, G.D. & Kidwell, J. (1998). Explaining the choice of organic produce: Cosmetic defects, prices, and consumer preferences. *American Journal of Agriculture Economics,* Vol.80, pp. 277-287.

Traxler, G.S., Anderson, E., LaPatra, S.E., Richard, J., Shewmaker, B. & Kurath, G. (1999). Naked DNA vaccination of Atlantic salmon, *Salmon salar* against IHNV. *Diseases of Aquatic Organisms,* Vol. 38, pp. 183–190.

Turnbull, J.F., Adams, C.E., Richards, R.H. & Robertson, D.A. (1998). Attack site and resultant damage during aggressive encounters in Atlantic salmon (Salmo salar L.) parr. *Aquaculture,* Vol.159, pp. 345-353.

Verhoog, H. (2001). The intrinsic value of animals: its implementation in governmental regulations in the Netherlands and its implications for plants. *Proceedings of Ifgene workshop Intrinsic value and integrity of plants in the context of GE.* pp. 15-18, Dornach, Switzerland, May 9-11, 2001

Walls, J., Tee Rogers-Hayden, A.M. & O'Riordan, T. (2005). 'Seeking citizens' views on GM crops: Experiences from the United Kingdom, Australia and New Zealand', *Environment,* Vol.47, No.7, pp. 23–36

WCED (1987). *Our Common Future.* Oxford: Oxford University Press.

White Paper. (2005). (Stortingsmelding) *Den blå åker (The blue field)* No. 19. 2004-2005.

White Paper. (2009). (Stortingsmelding) *Interesser, ansvar og muligheter. Hovedlinjer i norsk utenrikspolitikk (Interests, responsibilities and options. Main avenues in Norwegian foreign politics)* No.15. 2008-2009.

Yue, C., Alfnes, F. & Jensen, H.H. (2009). Discounting spotted apples: Investigating consumers' willingness to accept cosmetic damage in an organic product. *Journal of Agriculture & Applied Economics* Vol.41, pp. 1-18

18

Omics Methodologies:
New Tools in Aquaculture Studies

María-José Prieto-Álamo, Inmaculada Osuna-Jiménez,
Nieves Abril, José Alhama, Carmen Pueyo and Juan López-Barea
Department of Biochemistry and Molecular Biology, University of Córdoba
Agrifood Campus of International Excellence, ceiA3
Spain

1. Introduction

According to the FAO, a growing percentage of world aquatic production is derived from aquaculture, whose importance is increasing dramatically due to commercial overfishing and a growing demand for seafood (FAO, 2010). In 1980, aquaculture production represented 9% of fishery resources; by 2010, it had increased to 43%. It is thought that such a production will need to double in the next 25 years. The FAO is promoting aquaculture because it is an important source of income and employment and also because of its great contribution to food security and the development of many countries. Currently, there are three main challenges for developing productive, feasible and sustainable aquaculture: 1) diversification of the proteins used for the feeds, 2) resolution of problems derived from stressful conditions, diseases and/or deterioration of environmental conditions, and 3) introduction of new species to make this industry less vulnerable to market demand (COM, 2002). The Senegalese sole (*Solea senegalensis*) is a flatfish species with a high potential for use in marine aquaculture diversification. The cultivation of sole has been successful under several husbandry conditions, but the frequent occurrence of opportunistic diseases and its high sensitivity to different stressors, such as manipulation, pollutants, etc., make sole unable to be produced industrially (Cañavate, 2005; Dinis et al., 1999). Consequently, the identification of biomarkers responsive to pathological situations and pollutants will help to prevent health problems and to improve their farming.

Biomarkers provide evidence of alterations by physiological or environmental conditions (López-Barea, 1995a). The so-called "classic" biomarkers are suggested *a priori* by virtue of their biological roles but are rather biased because they concentrate on a small number of proteins, excluding others that are also altered in the same conditions but whose relationship with the physiological or environmental changes is unknown (López-Barea & Gómez-Ariza, 2006). In 1989, a group at the University of Cordoba (UCO) began to develop a battery of biomarkers sensitive to physiological or environmental changes in several bioindicator species, including bivalves, crustaceans, fish, mammals and mammalian cell lines. A variety of biochemical parameters were included, such as phase I (ethoxyresorufin-O-deethylase, EROD) or phase II biotransforming enzymes (GSH transferase, GST), antioxidative defences (superoxide dismutases, SOD; catalase, CAT; glutathione peroxidases, GSHPx, glucose-6P and 6P-gluconate dehydrogenases, glutathione reductase, GSSGrase), neurotransmission-linked esterase activities, such as acetylcholine (AcChE) and carboxyl esterases (CbE), oxidative damages to biomolecules, including DNA (8-oxo-dG), proteins (protein-SSG mixed disulphides), lipids (malondialdehyde, MDA), and the glutathione content and redox status (total glutathione, GSSG/GSH). The UCO group also developed new biochemical indicators that are altered by physiological or environmental changes, such as the levels of individual GST and SOD isoenzymes, the activation of promutagens to genotoxins by exposure to extracts of reference or exposed animals –a global measure of biotransforming capacity– and the metallothionein (MT) levels using a new and extremely sensitive HPLC-based fluorescent assay.

The utility of these "classic" biochemical biomarkers was later validated by the UCO group in studies carried out preferentially in natural sites in Spain, Slovakia and Tunisia, and contrasted with experimental exposures to model contaminants carried out under controlled conditions. These studies were reported in the following publications, limited in this review to those made in fish, and listed here by their date of publication: Rodriguez-Ariza et al. (1992, 1993, 1994a, 1994b), Martínez-Lara et al. (1992, 1996, 1997), Pedrajas et al. (1993, 1995, 1998), López-Barea (1995b), Lenartova et al. (1997), López-Barea & Pueyo (1998), Cousinou et al. (1999, 2000), Alhama et al. (2006, 2010), Romero-Ruiz et al. (2003, 2008), and Jebali et al. (2008). These "classic" biochemical biomarkers also responded to physiological changes, including oxidative alterations promoted by different feeding schemes, as described in Pascual et al. (1995a, 1995b, 1997, 2003) and Cánovas-Conesa et al. (2007).

While genes typically exert their functions at the protein level, genetic responses to stress are often regulated at the transcriptional level. Therefore, the determination of transcriptional profiles has become an essential approach in understanding the coordinated gene response to various physiological and pathological variables. The construction of cDNA libraries by *suppression subtractive hybridization* (SSH) (Prieto-Álamo et al., 2009; Williams et al., 2003) is a fundamental methodology used in differential expression studies with non-model species because it enables the identification of genes with no previous knowledge of their sequences. SSH is a PCR-based technique for generating cDNAs enriched in differentially expressed genes, useful for large-scale gene identification in non-model organisms (Diatchenko et al., 1996). Unlike SSH, which only provides qualitative results, *DNA microarrays* give semiquantitative (fold-variation) data, and more importantly, permit, in a single experiment, the analysis of the levels of thousands of transcripts, making them a valuable high-throughput methodology in Functional Genomics. Moreover, heterologous hybridization allows the use of microarrays made from transcripts of one species to probe gene expression in other related species. *Real-time qRT-PCR* has become a reference method to detect and quantify transcripts and to validate the results obtained with other techniques such as subtractive libraries or microarrays.

The UCO team gained wide experience in quantifying changes occurring at the mRNA level by RT-PCR. Of relevance is the devise of new approaches for the quantification of the exact number of transcript molecules and their application to a wide variety of organisms and conditions. This team developed, validated and optimised relative quantifications using complex multiplexed RT-PCR (Gallardo-Madueño et al., 1998; Manchado et al., 2000; Michan et al., 1999; Monje-Casas et al., 2001; Prieto-Álamo et al., 2000; Pueyo et al., 2002) and absolute quantification by real-time RT-PCR (Jiménez et al.,

2005; Jurado et al., 2003, 2007; Montes-Nieto et al., 2007; Prieto-Álamo et al., 2003, 2009). They also developed a quantitatively rigorous approach based on a combination of multiplexed and real-time RT-PCR to increase the number of transcripts to be quantified simultaneously without compromising the sensitivity, reliability and repetitiveness of the absolute measurements (Jurado et al., 2003; Michan et al., 2005; Monje-Casas et al., 2004; Ruiz-Laguna et al., 2005, 2006). These studies have demonstrated the potential benefits of absolute transcript quantifications in studies of tissue-specific expression profiles (Jurado et al., 2003, 2007; Prieto-Álamo et al., 2003, 2009; Ruiz-Laguna et al., 2005), of changes associated with growth stages or with the age or sex of an individual and have been particularly useful in studies with free-living animals (Jiménez et al., 2005; Michan et al., 2005; Monje-Casas et al., 2004; Prieto-Álamo et al., 2003; Ruiz-Laguna et al., 2005, 2006). We have demonstrated that the main drawback of relative quantifications is the variability of most popular internal standards. By comparing the differences in the transcript molecules with the conventional fold variations, we have also shown that relative quantifications grossly overestimate changes affecting poorly transcribed genes in comparison with highly abundant mRNAs.

Proteomics addresses the post-genomic challenge of examining the entire complement of proteins (proteome) expressed by a genome in a cell, tissue or organ at a given time under defined conditions (James, 1997). Protein expression is modulated at different levels from transcription to the maturation of the polypeptides produced by the translation of mature mRNAs. Proteins were initially separated by *two-dimensional electrophoresis* (2-DE; Wilkins et al., 1996), and their expression was analysed by 2D software (Melanie, etc.). Proteins were identified by mass spectrometry analysis of their peptide mass fingerprint (MALDI-TOF-PMF) or *de novo* sequencing of some peptides (nESI-MS/MS), comparing the results with public databases (Simpson, 2003). 2-DE, which is labour-intensive and has low reproducibility, requires a large amount of sample, and its narrow dynamic range is problematic with proteins of extreme Mr/pI. Shotgun proteomic methods allow the analysis of complex protein mixtures after full digestion by *multidimensional separation* coupling tandem liquid chromatography (LC/LC) and MS/MS (Washburn et al., 2001). The application of proteomic technology faces the problem of the lack of genomic information on most non-model sentinel organisms. This makes it difficult to identify differentially expressed proteins by high-throughput methods such as MALDI-TOF-PMF (López-Barea & Gómez-Ariza, 2006).

2. Conventional aquaculture studies with *Solea senegalensis*

2.1 Early studies

Studies of sole aquaculture began in Faro (S Portugal) and Cádiz (SW Spain) to produce good quality larvae and juveniles (Dinis et al., 1999). Broodstock spawning studies established optimal feeding regimes by combining squid (*Loligo vulgaris*) and polychaetes (*Hediste diversicolor*) at the final maturation stages. Spawning was studied in terms of temperature (stopped <16 °C), duration (4-6 months), egg fertilisation rate (20-100%) and viable egg rate (72%). Larvae hatch at 2.4 mm and accept *Artemia nauplii* as the first prey two days after hatching (DAH). Metamorphosis spans from 11 to 19 DAH, at which point the fish are fed live *Artemia metanauplii*. They reach 16 mm at 40 DAH and 35 cm/450 g after 1 year, with 8% survival. Pasteurellosis can cause pigmentation abnormalities and malformations associated with eye migration and can progress to death (Dinis et al., 1999).

The potential of sole for aquaculture was reviewed some time later (Cañavate, 2005). Although important progress in reproduction techniques was reached, much basic knowledge remained lacking. Ongrowth was successfully carried out, but progress was limited by opportunistic diseases due to suboptimal rearing conditions resulting in an inability of the sole to achieve an adequate physiological status for resistance. Growth, survival and pigmentation were studied during sole growth in tanks with three bottom types (Rodiles et al., 2005). The final length and weight was similar in the sand, white and dark conditions, but different pigmentation patterns appeared on the sand (clear, dark) and white bottoms (clear, brown, dark). The homogeneous dark pattern, preferred by markets, is only obtained in tanks with a dark bottom. A lower survival rate was found on sand bottoms due to pathologies derived from the difficulties in maintaining the sand bed.

2.2 Organ development and reproductive studies

Digestive tract development was studied in larvae until 30 DAH, which involved the assessment of histology, digestive enzymes, lipids, proteins and carbohydrates in the buccopharyngeal cavity, oesophagus, early stomach, anterior and posterior intestine, pancreas and liver (Ribeiro et al., 1999a). The digestive tract elongates in metamorphosis, increasing absorption. Phosphatases, lipase and aminopeptidase have been detected starting at 2 DAH and the levels increase during development. Proteins abound in the intestinal epithelium and exocrine pancreas, and neutral lipids are found at the yolk sac intestinal epithelium and liver. After 31 DAH larvae ingest, digest and absorb nutrients because they now have a complete digestive tract. A time course of pancreatic and intestinal enzymes was studied in larvae until 31 DAH (Ribeiro et al., 1999b). Digestive enzymes increase until 10 DAH then decrease until 18 DAH, a pattern typical of developing animals. Alkaline phosphatase abounds from 21-27 DAH, during the development of brush border membranes, with a parallel decrease in the cytosolic enzyme, Leu–Ala peptidase.

Thyroid development was studied in sole larvae by histo- and immunohistochemistry to synchronise larval development and improve fish production (Ortiz-Delgado et al., 2006). The first follicle is visible by the first feeding; increases during metamorphosis and has adult characteristics by 30 DAH. Thyroid hormones decrease to undetectable levels at yolk-sac reabsorption. T3 and T4 are detected by 6 DAH and increase during metamorphosis.

Seasonal profiles of sex steroids –17β-estradiol (17β-E), testosterone (T), 11-ketotestosterone (11-KT), and 17,20β-dihydroxy-4-pregnen-3-one (17,20β-P)– were studied in S. senegalensis in an attempt to achieve steroid-induced maturation (García-López et al., 2006a). Females have six maturation stages, as follows: early, intermediate and final ovarian development, then partially, mid and spawned out. By summer´s end, a new gonadal cycle starts, as demonstrated by increased reproductive parameters. By mid-autumn some females reach advanced maturation stages, which coincide with a peak of running males. By the start of spring, ovarian development reaches its peak, and plasma steroid levels are maximal at the start of the spawning period, which occurs from March to June. In parallel with oocyte and sperm release, the proportion of spawned out fish and non-running males increases, and steroid levels decline. The high levels of 17,20β-P during spawning make it a candidate for a maturation-inducing steroid.

Testicular development was also studied (García-López et al., 2006b). The spermatogenetic cycle consists of the following five stages: early (I), mid (II), and late (III) spermatogenesis, maturation (IV), and recovery (V). In the summer, stage I and V testes are found with low values of sex steroids and I_G (gonadosomatic index). Recrudescence begins in autumn, with an initial increase of I_G 11-KT and T and the appearance of stage II and III testes. In the winter, 11-KT and T peak and soon decrease, and I_G slightly declines. In the spring, 11-KT and T decline further, while I_G slightly increases and running males peak with stage IV testes. Sperm production and quality was assessed in wild-captured and F1 broodstock fish (Cabrita et al., 2006). Males produce motile sperm from February to November, with specific peaks of high spermiation and fluent males. Sperm volume and cell density is lower in F1 males than in wild-captured broodstock.

Ovarian development was also studied (García-López et al., 2007). In the autumn/winter, oocytes progress to vitellogenic stages in parallel with high levels of K (condition factor), I_G, and plasma 17β-E and T. In the late winter/early spring, development is maximal, with females at intermediate and final maturation and K, I_G, 17β-E and T peaking. Steroid levels are lower in cultured sole than in naturally spawning females, leading to atresia and lack of oocyte maturation, thus reducing ovary size with declining K, I_G, and 17β-El and T levels and many perinucleolar oocytes. The amount of circulating 17,20β-P, the putative maturation-inducing steroid, remains near constant through the period, suggesting that oocytes are unresponsive to its stimulation.

Skeletal development and malformations are a bottleneck in sole aquaculture. Maturation and abnormalities of the vertebral column and caudal skeleton have been studied in sole (Gavaia et al., 2002). Different defects are found in the caudal complex and the vertebral column, and 44% of fish show at least one defect. While the causes are unknown, their high incidence may reflect rearing and/or feeding problems. The tissue distribution and evolution of bone Gla (Bgp) and matrix Gla proteins (Mgp) and Ca^{2+} deposition were studied in zebrafish during larval development and in adult tissues as well as sole metamorphosis (Gavaia et al., 2006). In zebrafish, Bpg and Mpg accumulate mainly in the matrix of skeletal structures already calcified or under calcification. In sole metamorphosis, Bpg and Mpg increase in parallel to the calcification of the axial skeleton. In both species, Mpg also accumulates in non-mineralised vessel walls.

2.3 Nutrition studies

Studies on the requirements, catabolism and assimilation of amino acids (AAs) were carried out in early larval, metamorphic and post-larval sole. Initial studies on indispensable (IAA) and dispensable (DAA) amino acids (Rønnestad et al., 2001), showed that sole assimilated most (85%) of the dietary IAAs and catabolised most of the DAAs. Such results were confirmed after studying the bioavailability of several AAs in larvae (Conceição et al., 2003). The demand and availability of AAs and proteins in relation to digestive capacity were reviewed, and AAs sources were described, highlighting the regulatory role of cholecystokinin and peristaltic activity (Rønnestad et al., 2003). A balanced AA profile improved amino acid assimilation in post-larval sole (Aragão et al., 2004a). Changes in AA requirements and dietary imbalances were studied in *Sparus aurata* and *S. senegalensis* (Aragão et al., 2004b); the AA profiles of both changed during ontogeny, especially in sole due to its marked metamorphosis. AA imbalances were found during development. In both species, Phe/Tyr addition was studied to assess the effects on metamorphosis after their conversion into thyroid hormones (Pinto et al., 2010). While Phe did not affect sole metamorphosis, dietary Tyr increased the production of thyroid hormones, which was beneficial for sole metamorphosis.

The nutritional physiology of sole development was studied to optimise diets and understand limiting factors in weaning (Conceiçao et al., 2007). Larvae have a high capacity to digest live prey, even at the early stages. Use of inert microdiets in co-feeding with *Artemia* resulted in the development of intestinal activity and enhanced survival, although it was also accompanied by low growth and high size dispersal. Fatty acid absorption increases with their degree of unsaturation, and larvae spare DHE from catabolism. Rotifers and *Artemia* are deficient in one or more AAs, such as His, Lys, Arg, Thr, or those containing sulphur and aromatic rings, depending on the larval stage; balancing the dietary AA profile with dipeptides increases retention and decreases catabolism in *Artemia*-fed larvae.

The effects of non-protein energy levels on growth and oxidative status were studied in sole fed diets with 4 energy levels (Rueda-Jasso et al., 2004). Cellular energy allocation showed differences in liver, but not in muscle. TBARS were higher in fish fed a diet with high lipid content, in parallel to high CAT and SOD activity. Yet, the protein source or energy levels had no major impact on sole growth, nutrient utilisation or fatty acid composition (Valente et al., 2011). Quantitative lipid imbalances and a low protein/neutral lipid ratio increased the accumulation of lipid droplets in the enterocytes and lowered fatty acid absorption in larvae (Morais et al., 2005). The effects of a neutral lipid level and source were studied in marine fish larvae (Morais et al., 2007). A growth-depressing effect of high neutral lipids, as assessed by lower digestive enzyme activity, absorption and/or food intake was reported. In larvae, lipid transport from enterocytes to the body is more critical than lipolytic activities. Phospholipid digestion is more efficient than that of neutral lipids, whose excess leads to the accumulation of large lipid droplets in the enterocytes, reducing fatty acid absorption and growth.

The feed transit, protein and energy digestibility of practical feed were assessed in sole (Dias et al., 2010). Protein digestibility is high for fishmeal and corn gluten, intermediate for soybean meal, and moderate for wheat meal. Energy digestibility varies from 88 to 93% for soybean meal, corn gluten and anchovy fishmeal and is 73% in wheat meal. Thus, flatfish, despite its high dietary protein requirement, digest vegetable ingredients quite well, suggesting that the development of practical feeds with high levels of plant-protein sources would be beneficial.

2.4 Conventional biomarker studies in *S. senegalensis* aquaculture

Nearly 15 years after the studies of the UCO group, the use of biomarkers to follow fish physiology and pollution effects has become popular and is now applied by most groups. Antioxidant enzymes, stress proteins, lipid peroxides and histology were studied in sole larvae (Fernández-Díaz et al., 2006) fed on the following 3 diets: live *Artemia nauplii*, microcapsules, and vitamin A-supplemented microcapsules. Live-fed larvae grow larger and undergo faster metamorphosis than microcapsule-fed larvae, although all groups have near 80% survival. Vitamin A improves the growth and development compared to an inert diet. *Artemia*-fed larvae have organs with normal development, but histological alterations are seen in larvae fed an inert diet. Catalase (CAT), superoxide dismutase (SOD), total GSH-peroxidase (t-GPX), lipid peroxides (MDA) and stress proteins (HSP70, not HSP60) are diet- and age-dependent. Inert diet-fed larvae have similar biomarker responses, but different (p<0.05) from *Artemia*-fed larvae. Higher antioxidant defences are attributed to the start of metamorphosis and the use of inert food.

A similar approach was used by Cañavate et al. (2007) to assess the effect of light on the development of sole larvae with or without adding β-carotene-rich *Dunaliella salina*. SOD, CAT, t-GPX and MDA were used as biomarkers. Growth and survival after metamorphosis were unaffected by light or *D. salina*. Light affects CAT and t-GPX throughout development but does not affect MDA, and SOD is only affected in metamorphosis. *D. salina* does not affect SOD, CAT or t-GPX and no interaction with light intensity was found. MDA lowers significantly only when *D. salina* is added, and its effect was found only in metamorphosing larvae, whose MDA levels are much higher than in earlier stages. These results confirm the antiperoxidative effect of β-carotene from live algae in the larval rearing process.

2.5 The Pleurogene project

After the initial studies on the growth and reproduction of the Senegalese sole, "Genoma España" and "Genome Canada" promoted the "Pleurogene" project, in order to develop new technology to assess gene and protein expression during the reproduction and breeding of two flatfish, Senegalese sole and Atlantic halibut. The project [http://www.gen-es.org] aimed to improve basic knowledge of reproduction, larval development and survival, and had de following objectives: 1) Establishment of an EST database and shotgun proteome analysis; 2) Construction of a microarray for high-throughput analysis of gene expression; 3) Construction of genetic linkage map; 4) Development of methods for gene expression profiling through laser capture microdissection RNA; 5) Identification of changes in gene expression during gamete development and maturation; 6) Genomic analysis of sex determination and differentiation; 7) Determination of the pattern of gene expression during larval metamorphosis, the ontogeny of the gastrointestinal tract and the effects of dietary treatments; and 8) Development of E-mold, an integrative bioinformatics platform for genomic, proteomic and morphological information from flatfish.

The Pleurogene project developed genomic and proteomic tools to help achieve these goals (Douglas et al., 2007; Cerdà et al., 2008), and also had the following major research results: 1) Development of genomics tools for the Senegalese sole, including 10 different cDNA libraries from adult and larvae tissues and 1 normalised multi-tissue library, 10,300 new sole EST sequences and nearly 500 peptides, and a sole oligonucleotide microarray with probes to detect 4550 different RNAs; 2) Development of a hormone treatment to increase sperm motility in sole; 3) Generation of a sole genetic linkage map; 4) Development of a progeny test or paternity kit; 5) Development of indirect approaches for sex control; 6) Production of recombinant gonadotropin hormone; 7) Generation of gene and protein expression maps associated to larval development, metamorphosis, and nutrition; 8) Generation of gene and protein expression maps of testes producing high quality sperm; 9) Generation of gene expression maps for sexual differentiation and maturation; and 10) Development of the "Solea-mold" that allows for new data to be included, *in silico* experiments to be performed, and comparative studies to be undertaken.

3. "Classic" biomarkers versus "*omics*" methodologies

3.1 Assessing pollution effects in fish

After the initial studies of the UCO group, the use of "classic" biomarkers became increasingly popular to assess the effects of pollutants on fish. One example was the "Prestige" oil spill in November 2002 off the Galician coast (NW Spain). Due to the heavy nature of the crude oil and its low solubility in sea water, its dispersion was low and it remained *in situ* in oil patches adhered to rocks and sediments. The biological effects of these oil patches were tested in *S. aurata* (Morales-Caselles et al., 2006), using MT levels, EROD activity and histology as biomarkers, without significant results. Biomarkers including CAT, GSHPx, GSSGrase and DT-diaphorase, AcChE, CbE, GST and MDA were also analysed in *S. senegalensis*, with modest alterations (Solé et al., 2008).

Antioxidative and phase II and III biotransforming enzymes were used to follow the effects of linear alkylbenzene sulphonates in sole via an *in vivo* continuous-flow assay (Alvarez-Muñoz et al., 2007) that was also used in *S. aurata* and *S. senegalensis* to assess the toxicity of sediments from littoral areas in northern and southern Spain (Jiménez-Tenorio et al., 2007), using MTs, EROD and histopathology. Jimenez-Tenorio et al. (2008) assessed the sediment toxicity caused in *S. senegalensis* by acute or chronic spills using histopathology, EROD and GST as biomarkers. Oxidative stress biomarkers and PAH contents were studied in Huelva soles near a petrochemical plant (Oliva et al., 2010). Significant correlations were found among the levels of GST, GPx and CAT, hepatic levels of PAH metabolites and PAH contents in sediments.

The effect of waterborne copper (Cu) was studied in *S. senegalensis* in static conditions (Fonseca et al., 2009) by assessing the following characteristics: biomarkers such as MTs or MDA, mass indices, and biochemical condition indices such as the RNA/DNA ratio and the lipid and protein content. Cu triggers a biomarker response and lowers growth and condition, without changes in morphometric indices. Decreased condition shows that lipid reserves enable fish to respond to toxicity and to maintain growth and protein synthesis, although with lower rates than control fish.

The effects of Mexel®432 and NaClO antifoulings were studied in *S. senegalensis* by assessing osmolality, Na^+/K^+-ATPase, stress, histology, oxidative damage, antioxidant defences and detoxification (López-Galindo et al., 2010a,b). NaClO increases plasma cortisol, glucose and lactate after an acute stress, with a later recovery. Gill GST and AChE are sensitive to NaClO. Hepatic markers initially respond to NaClO but longer exposures are toxic. Mexel®432 initially increases cortisol, which later returns to basal values, but glucose, lactate and triglycerides decrease. Gills have a lowered Na^+,K^+-ATPase activity, causing an imbalance in osmoregulation. Moderate changes are found in KAT, GSSGrase and GPX but not in MDA, GST or CbE. A multi-biomarker approach was used in *Dicentrarchus labrax*, *S. senegalensis* and *Pomatoschistus microps*, from the Aveiro and Tejo (Portugal) estuaries (Fonseca et al., 2011), which were affected by anthropic activities, without highly significant results.

Detection of DNA damage by biomarkers for genotoxicity was used in *S. senegalensis* (Costa et al., 2008a) exposed to sediments from three Sado Estuary (W Portugal) sites. The two blood parameters used were erythrocyte nuclear abnormalities (ENA) and DNA strand-breaks (DNA-SB). The levels of metals, PAHs, PCBs and DDTs were determined in the sediments. Scarcely polluted sediments are weaker inducers of genotoxic damage, whereas those under urban, industrial or agricultural influences significantly increase ENA and DNA-SB. A strong correlation exists between PAH and PCB content and genotoxicity, while metals have a weaker correlation. In a parallel study, *S. senegalensis* were exposed to Sado Estuary sediments (Costa et al., 2008b). Livers had more histological lesions than gills, and sediments contaminated by organics caused more damage to both organs than those contaminated by metals. Two "classic" biomarkers, MT and CYP1A, were also assessed. Lethality and biomarker responses do not linearly depend on the cumulative levels of contaminants but rather of their bioavailability and synergistic effects (Costa et al., 2009). In a parallel study, exposure to contaminated sediments induced DNA fragmentation and clastogenesis (Costa et al., 2011). Still, the most contaminated sediment revealed an antagonistic effect between metals and organics, enhanced by higher bioavailability. The laboratory assay caused a more pronounced increase in ENA, whereas a significant increase in DNA-SB exists in field-tested fish exposed to reference sediment.

3.2 The advent of "omics" methodologies

In contrast to the "classic" biomarker strategy, *"omics"* approaches, plus *in vitro* and *in silico* methods, are becoming a powerful multidisciplinary strategy. Their use is still at an early stage because most popular bioindicators are poorly represented in gene/protein sequence databases (Ruiz-Laguna et al., 2006; González-Fernández et al., 2008). *Omics* includes *genomics* to study DNA variations, *transcriptomics* for genome-wide characterisation of gene expression by measuring mRNAs, *proteomics* to assess the cell and tissue-wide expression of proteins, and *metabolomics* for global assessment of metabolite concentrations. These technologies provide detailed molecular information that helps to identify response pathways and to define mechanisms and modes of action without requiring previous knowledge. In 2003, the UCO group began a search for new biomarkers through the use of "omics" methods, which allowed the study of the Aznalcóllar spill and the status of Doñana National Park and its surroundings (SW Spain). These studies were collected in the following publications, listed here by publication date: Rodriguez-Ortega et al. (2002, 2003), Ruiz-Laguna et al. (2005, 2006), Bonilla-Valverde et al. (2004), López-Barea & Gómez-Ariza (2006), Romero-Ruiz et al. (2006), Montes-Nieto et al. (2007, 2010), Vioque-Fernández et al. (2007a, 2007b, 2009a, 2009b), González-Fernández et al. (2008), Abril et al. (2011), and Pueyo et al. (2011).

All organisms are adapted to certain extracellular salinity ranges. Osmoregulatory mechanisms are central to adaptation. Osmotic stress was studied in tilapia *Oreochromis mossambicus*, spiny dogfish shark *Squalus acanthias*, and an intertidal sponge *Tetilla mutabilis*, by genomics and proteomics methods (Kültz et al., 2007). SSH, RACE-PCR and proteomics allowed the identification of genes and proteins involved in adaptation to salinity or other environmental stresses. Algorithms based on sequence homology searches (MSBLASTP2) are powerful tools for protein identification. Gene ontology and pathway analysis can subsequently use identified genes and proteins for modelling molecular mechanisms of environmental adaptation. The dependence on information about biochemical pathways and gene ontology databases for model species is a severe barrier for work with non-model species. To minimise this dependence, focusing on a single biological process is key when applying "omics" methods to non-model organisms.

Environmental metabolomics allows for the characterisation of the metabolism of organisms from the natural environment and of those reared under laboratory conditions. Viant (2007) used this approach to characterise the responses of organisms to natural and anthropogenic stressors, discussing the challenges of measuring metabolites and highlighting the dynamic nature of the metabolome, whose variability is a challenge in environmental studies. The normal metabolic operating range (NMOR) is defined as the region in metabolic space in which 95% of individuals reside, and stress is a deviation from NMOR. The importance of genotypic and phenotypic anchoring (e.g., knowing species, gender, and age) is emphasised to facilitate the interpretation of multivariate metabolomics data.

An NRC-UK sponsored international consortium from government agencies, academia and industry in Canada, Japan, the UK, and the USA was carried out on fish toxicogenomics (Van Aggelen et al., 2010). The following three topics were addressed: progress in ecotoxicogenomics, perspectives on roadblocks for practical implementation of toxicogenomics into risk assessment, and dealing with variability in data sets. Although examples of successful application of "omic" technologies were identified, it is critical to perform studies that relate molecular changes to ecologically adverse outcomes. Although there are hurdles to pass on the road to regulatory acceptance, "omics" are already useful for elucidating modes of action of toxicants and can contribute to the risk assessment process as part of a weight-of-evidence approach.

A qRT-PCR approach was used to assess how *Lactobacillus rhamnosus* IMC 501 added to *Amphiprion ocellaris* larvae, alters development, and also to study the responses after probiotic exposure (Avella et al., 2010). Larvae and juveniles had 2-fold higher weight after probiotics were supplied. Metamorphosis occurs 3 days early, and factors involved in growth and development (I-l GF I/II, myostatin, PPAR α/β, vitamin D receptor α, and retinoic acid receptor γ) have higher gene expression. Probiotics lessen the severity of the general stress response as demonstrated by lower levels of glucocorticoid receptor and 70-kDa HSP expression. Improved development of the skeletal head was also found, with 10–20% less deformities in probiotic-treated juveniles.

"Omics" have also been used in chemical screening and perturbation studies in zebrafish (Sukardi et al., 2010). Pharmacological efficacy and selectivity have been evaluated by chemical-induced phenotypic effects, although this has limitations in the identification of action mechanisms. "Omics" also facilitates the translatability of zebrafish studies across species by comparing conserved chemically induced responses. Thus, De Wit et al. (2010) characterised the estrogenic and metabolic effects of 17α-ethinylestradiol (EE2) in *D. rerio*, following the concern regarding the effects of endocrine-disrupting compounds. Oligo microarrays, with 3479 zebrafish-specific oligos, were used to generate differential gene expression levels, and proteomic responses were evaluated by DIGE and MALDI-TOF/TOF. Assessment of the differentially expressed transcripts and proteins showed that both individual platforms could profile clear estrogenic interference and multiple metabolism-related effects and stress responses. Cross-comparison of transcriptomics and proteomics datasets have limited concordance, but a revision of the results shows that transcriptional effects project at the protein level as downstream effects of the affected signalling pathways.

Public databases of coexpressed gene sets are valuable resources for many studies, including gene targeting for functional identification and investigations of regulatory mechanisms or protein–protein interactions. While coexpressed gene databases are highly popular in plant biology, those with animal data are limited due to the lower reliability of coexpression data. The COXPRESdb (http://coxpresdb.jp) represents the coexpression relationship in humans and mouse (Obayashi & Kinoshita, 2011). Updates focusing on enhancing the reliability of gene coexpression data in animals have been reported. A new coexpression measure, Mutual Rank, has been implemented, and five other animal species, such as the rat, chicken, zebrafish, fly and nematode, were included to assess the conservation of coexpression. In addition, different layers of "omics" data have been added into the integrated network of genes. Functioning as a gene network representation, COXPRESdb can help researchers to clarify the functional and regulatory networks of genes in a broad array of animal species.

4. Transcriptomic studies in *Solea senegalensis*

Cultivation of the Senegalese sole is hampered by its sensitivity to different stresses and infectious diseases that can cause high mortality. Consequently, there is a need to identify sole genes responsive to stress, infections and pollutants in order to improve productivity, management and fish welfare. Transcriptomic responses of sole stimulated with lipopolysaccharide (LPS), a mimetic of bacterial infections, and copper sulphate, a zoosanitary compound, were studied by different experimental and methodological approaches, such as SSH libraries, DNA microarrays and real-time qRT-PCR (Prieto-Álamo et al., 2009; Osuna-Jiménez et al., 2009).

4.1 SSH libraries

The construction of subtractive libraries in *S. senegalensis* allowed the identification of differentially expressed genes in response to LPS in the head-kidney, a hematopoietic and lymphoid organ involved in immune response and to $CuSO_4$ in the liver, a central metabolic organ in xenobiotic detoxification and in the defence system (Prieto-Álamo et al., 2009). In both cases, forward (F) and reverse (R) libraries were designed to obtain clones of genes that were up- or down-regulated in response to LPS of $CuSO_4$ relative to the PBS control. To offset inter-individual variations and temporal differences in the responses, the libraries were constructed with total RNA from pooled head-kidney or liver (≥ 10 fish/condition) of soles treated with LPS or $CuSO_4$ for 6 and 24 h. Four hundred sixty clones were sequenced and the products of the ESTs were identified by comparison with the open access databases. A total of 222 unique sequences were detected, and 185 were identified as related to major physiological functions (Table 1).

A high percentage of identified ESTs were related to immune response (Figure 1). Their presence in a sole head-kidney library stimulated with LPS agreed with the immune role of this organ and the immunostimulating effects of LPS (Swain et al., 2008). The number of genes classified as being immune-related was even larger in the liver than in the head-kidney. Most of these immune-related ESTs coded for acute phase proteins (e.g., lysozyme, coagulation factors, proteinase inhibitors, complement components, and Fe transport/homeostasis proteins) according to several genomics studies indicating that, in teleost, the liver is an important source of immune transcripts (Ewart et al., 2005), mediating a powerful acute phase response (Bayne & Gerwick, 2001). In addition, these results supported the capacity of Cu (like other metals) to alter immunological competence, in agreement with a report showing Cu up-regulation of the cytokine TGF-β in striped bass (Geist et al., 2007).

	LPS (head-kidney)	CuSO₄ (liver)
Number of clones sequenced	231	229
Number of clones analysed	222	226
Average sequence length (bp)[a]	411 ± 168	414 ± 155
Number of unique ESTs	133	89
Up-regulated ESTs	62	48
Identified ESTs	49	44
Non-identified ESTs	13	4
Down-regulated ESTs	71	41
Identified ESTs	58	34
Non-identified ESTs	13	7

Table 1. Characteristics of the sole SSH libraries and sequences. [a]Mean ± SEM

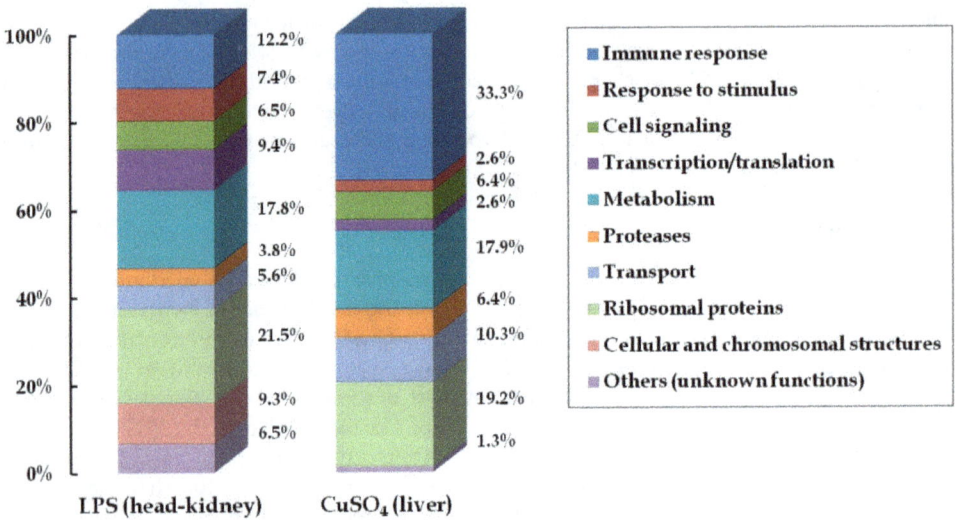

Fig. 1. Functional classification of unique ESTs obtained in SSH libraries from LPS or CuSO₄-treated soles. "Others" indicates genes that had significant identities to databases entries of unknown function.

Genes coding for products involved in osmoregulation and nitrogen excretion (e.g., liver angiotensinogen, sodium potassium ATPase beta subunit, kininogen 1, angiotensin I converting enzyme 1, and alanine-glyoxylate aminotransferase 2-like) were also identified in libraries from the livers of CuSO₄-treated soles, in agreement with the previously reported effects of copper on osmoregulation, acid-base balance and nitrogen excretion (e.g., Blanchard & Grosell, 2006; Evans et al., 2005).

4.2 DNA microarrays

The microarray developed for the European flounder *Platichthys flessus* (GENIPOL platform, Williams et al., 2006) had been used to assess hepatic gene expression of many species of flatfish, confirming that heterologous microarray analyses between closely related species is a suitable approach (Cohen et al., 2007). This platform turned out a very useful tool for analysing the transcriptional expression of *S. senegalensis*.

First, the hepatic response of soles exposed to $CuSO_4$ or LPS was studied using pooled samples. Statistical analyses showed that 405 genes were differentially expressed after Cu treatment at 6 h, 468 with Cu at 24 h, 271 with LPS at 6 h and 664 with LPS at 24 h (Table 2).

	LPS		$CuSO_4$	
	6 h	24 h	6 h	24 h
Up-regulated genes	172	251	101	341
Down-regulated genes	233	217	170	323
Total	405	468	271	664

Table 2. Number of differentially expressed genes in *S. senegalensis* liver in response to LPS or $CuSO_4$ treatments that were identified by heterologous DNA microarrays.

The functional analysis of the results (Blast2GO software) permitted the identification of response-specific genes to $CuSO_4$ (cell junction and cell signalling), to LPS (glutathione transferase and immune response) or to both treatments (immune response, digestive enzymes, unfolded protein binding, intracellular transport and secretion, and proteasome) (Table 3). This way, the functional category cell junction was statistically significantly over-represented amongst the genes induced by copper at 6 h. This term grouped genes related to cellular adhesion, such those coding for claudins (CLDN26) and genes related to cell signalling, such as GIT2 that encodes G-protein-coupled receptor kinase interactor 2, which is in agreement with the ability of copper to alter the tight junction permeability in human intestinal mucosa (Ferruzza et al., 2002). Glutathione-S-transferases were more prevalent amongst the transcripts down-regulated by LPS at 24 h. The capacity of LPS and bacterial infection to down-regulate biotransformation activities such as GSTs has been described in a number of fish species (Reynaud et al., 2008). As shown in Table 3, genes related to the immune response were specifically induced by LPS. These included the antimicrobial peptide hepcidin (HAMP), TNFα-induced protein 9 (TNFAIP9), cytokines (IL8, IL25) and chemotaxins (LECT2).

Other immune-related genes were induced by both LPS and $CuSO_4$ treatments. This is the case for classic piscine acute phase proteins like haptoglobin (HP) or C7 (Bayne & Gerwick, 2001). C7 is a component of the complement system, whose up-regulation by copper is in line with complement proteins being engaged in novel biological functions distinct from their well-established role in innate immunity (Mastellos et al., 2005). Although soles were fasted prior to and during the experiments, digestive enzymes such as trypsin (PRSS2), chymotripsin (CTRB), elastase (ELA4) and carboxypetidase A (CPA1) and B (CPA2) were down-regulated in response to both treatments. This might be due to a general stress caused by the treatment (Auslander et al., 2008). Furthermore, LPS and copper treatments resulted in the up-regulation of genes encoding unfolded protein-binding, which are induced in fish in response to different kinds of stress conditions (e.g., bacterial infection or exposure to heavy metals) (Basu et al., 2002), intracellular transport and secretion, in order to accommodate the rapid onset of cytokine secretion and for membrane traffic associated with the phenotypic changes of immune activation (Pagan et al., 2003), and proteasomal proteins in agreement with previous results in mammalian cells (Qureshi et al., 2003; Fernandes et al., 2006).

	Response to LPS	Response to CuSO$_4$	Selected genes
Cell junctions	---	Up-regulation	CLDN26, GIT2
Glutathione-S-transferases	Down-regulation	---	GST-A, GST1, GST3
Immune response	Up-regulation	---	HAMP, TNFAIP9, IL8, IL25, LECT2
	Up-regulation	Up-regulation	C7, HP
Digestive enzymes	Down-regulation	Down-regulation	PRSS2, CTRB, ELA4, CPA1, CPA2
Unfolded protein binding	Up-regulation	Up-regulation	GP96, HSP70
Intracellular transport/secretion	Up-regulation	Up-regulation	ARF5, TMED7, SEC22
Proteasome	Up-regulation	Up-regulation	PMSD3, MSUG1

Table 3. Selected genes that were significantly differentially expressed in the liver of *S. senegalensis* during LPS or CuSO$_4$ treatments.

Because the GENIPOL cDNA microarray was constructed from ESTs derived from flounder liver and had been used in studies of hepatic expression (Cohen et al., 2007; Williams et al., 2006, 2007, 2008), we investigated whether this platform would be valid for analysing the transcriptional response in the head-kidney. To this end, the GENIPOL microarray was used to compare the basal transcriptional expression in the *Solea senegalensis* head-kidney and liver. We determined that 1004 genes were statistically differentially expressed in both organs: 418 transcripts were more abundant in the liver than in the head-kidney and 586 transcripts were more abundant in the head-kidney than in the liver. Thus, although the microarray was constructed with hepatic ESTs, the number of genes identified as over-expressed is similar in both organs, though slightly higher in the head-kidney than in the liver. The analysis of transcriptional patterns showed that the most represented biological processes amongst the genes up-regulated in the liver in comparison with the head-kidney were those involved in innate immune response, digestion, lipid transport, and monooxygenase activity (Table 4). The category "innate immune response" grouped genes coding for acute phase proteins because, as has been previously discussed, the liver is the main source of these plasmatic proteins. Amongst the genes related to lipid transport that were particularly remarkable were those coding for several apolipoproteins. The term "monooxygenase activity" encompassed genes belonging to the cytochrome P450 family, in agreement with what is known about the detoxifying capacity and the biotransformation activity of the liver in teleosts (Thorgaard et al., 2002).

Functional terms over-represented in the list of transcripts that were more abundant in the head-kidney than in the liver related to cellular division and protein turnover (protein degradation and the proteasome and ribosomal proteins), among others (Table 5). These results agree with the role of the head-kidney as a major haematopoietic organ in teleosts and with its function as a secondary lymphoid organ in the clearance of soluble and particulate antigens from circulation (Whyte, 2007). It is worth noting that when the lists of genes that are up-regulated in both organs were compared, two sequences corresponding to

Innate immune response
alpha-1-antitrypsin, alpha-2-macroglobulin, anticoagulant protein C precursor, coagulation factor VIIc, chemotaxin, complement component C3, complement component C8, complement component C9, complement regulatory plasma protein, fibrinogen alpha, fibrinogen beta chain precursor, fibrinogen gamma chain precursor, haptoglobin, hepcidin precursor, interleukin 8 precursor, kininogen 1, plasma protease C1 inhibitor precursor, prothrombin precursor, putative complement factor, transferrin
Digestive enzymes
chymotrypsinogen 1, chymotrypsinogen 2, trypsinogen 2 precursor
Lipid transport
apolipoprotein A-I, apolipoprotein A-IV, apolipoprotein C-I precursor, apolipoprotein E, apolipoprotein H, 14kDa apolipoprotein, fatty acid-binding protein
Monooxygenase activity
cytochrome P450 2F2, cytochrome P450 2X, cytochrome P450 3A, cytochrome P450 3A45, cytochrome P450 8B1, cytochrome P450 monooxygenase

Table 4. Selected genes up-regulated in the liver in comparison with the head-kidney in *S. senegalensis*.

Cellular division/cytoskeleton
alpha-tubulin, actin-related protein 3 homolog, actin related protein 2/3 complex subunit 4, beta-actin, cofilin 2, coronin 1A, lamin B1, microtubule-based motor protein, mitotic spindle assembly checkpoint protein, myosin regulatory light chain 2, nuclear movement protein PNUDC, thymosin beta-4
Protein degradation/proteasome
polyubiquitin, proteasome alpha 1 subunit isoform 2, proteasome (prosome, macropain) subunit alpha type 7, proteasome beta-subunit C5, proteasome subunit beta type 3, proteasome (prosome, macropain) subunit beta type 5, proteasome 26S ATPase subunit 5, proteasome subunit N3, ubiquitin carboxyl-terminal hydrolase isozyme L, ubiquitin specific protease 9
Ribosomal proteins
40S ribosomal protein S3a, 40S ribosomal protein S4, 60S ribosomal protein L3, 60S ribosomal protein L4, 60S ribosomal protein L13

Table 5. Selected genes up-regulated in the head-kidney in comparison with the liver in *S. senegalensis*.

the GAPDH (glyceraldehyde 3-phosphate dehydrogenase) gene, but with different expression patterns, were detected. The first of these sequences was more abundant in the liver, while the second was prevalent in the head-kidney. *Solea senegalensis* possesses two different GADPH paralogous genes that exhibit different tissue expression patterns, with GAPDH1 being more abundant in the liver and the GAPDH2 isoform more prevalent in the head-kidney (Manchado et al., 2007). A detailed analysis of the sequences detected in the microarray study revealed that these sequences match the two described isoforms. Altogether, these results demonstrated that the GENIPOL microarray platform was valid for analysing the transcriptional response of the sole head-kidney, discriminating between genes coding for transcripts with specific patterns of hepatic or renal expression.

Consequently, the GENIPOL platform was used to assess the response to LPS in the head-kidney. After 24 h of LPS treatment, a total of 224 genes was statistically differentially expressed in the head-kidney (117 up-regulated and 107 down-regulated). The functional analysis of the results revealed that the biological processes altered by LPS treatment in the head-kidney were very similar to those detected in the liver. Most notably amongst the up-regulated genes were the functional groups immune response, unfolded protein-binding, intracellular transport/secretion and proteasome, and digestive enzymes in the list of down-regulated genes. In contrast, the glutathione transferases category, whose transcript levels were down-regulated by LPS in the liver, was not affected in the head-kidney.

4.3 Real-time qRT-PCR

Although the determination of absolute transcript abundance (the exact number of molecules of a transcript in all samples of the study) is the only adequate procedure to accurately assess the expression of a gene (Prieto-Álamo et al., 2003), these kinds of studies are infrequent due to the experimental difficulties that impair this type of analysis. A set of more than 20 sole transcripts was selected for the quantification of their levels by this rigorous approach (Table 6) (Prieto-Álamo et al., 2009; Osuna-Jiménez et al., 2009), based on the physiological relevance of their products and on the magnitude and specificity of their responses.

Immune Response
complement component C3, complement component C7, transferrin, haptoglobin, ferritin M, natural killer enhancing factor, tumor necrosis factor alpha-induced protein 9, hepcidin, nonspecific cytotoxic cell receptor protein-1, angiotensinogen, sequestosome 1, tumor necrosis factor receptor associated factor
Response to stress
CCAAT/enhancer binding protein, cold inducible RNA binding protein, DNA-damage-inducible transcript 4-like, NHP2 non-histone chromosome protein
Energetic Metabolism
glyceraldehyde-3P-DHase, transketolase, NADH-DHase 1 alpha subcomplex 4
Protein Synthesis, Folding and Degradation
asparaginil-tRNA synthetase, heat shock protein GP96, proteasome 26S non-ATPase subunit 3, cathepsin Z
Transport
α-globin

Table 6. Biological functions of proteins encoded by the sole transcripts selected to be validated by real-time qRT-PCR (Prieto-Álamo et al., 2009; Osuna-Jiménez et al., 2009).

The absolute quantification of the selected transcripts was not limited to the samples used for global analysis, but rather extended to samples coming from different individuals, organs, treatments and exposure times. Consequently, this gene-to-gene analysis provided valuable additional information on transcriptional expression patterns of the selected genes. Individual quantification is also mandatory to prevent biased interpretations of specimens with abnormal expression levels. The absolute real-time qRT-PCR on individual samples demonstrated that inter-individual variations of most examined transcripts in treated animals was in the range of that in control fish, indicating similar susceptibility to LPS or $CuSO_4$ challenge among individuals. The qRT-PCR quantifications confirm, in general, the results obtained with the subtractive libraries and the DNA microarrays. As expected, substantial differences in abundance were found depending on the transcript and tissue examined. In general, each transcript displayed a characteristic expression profile, distinguishing between constitutive or up-/down-regulated, early or late responsive, stressor-specific or not, etc., as a function of the organ analysed.

5. Proteomics studies of GBD

Fish reared in earth ponds are eventually affected by Gas Bubble Disease (GBD) outbreaks if ponds are not correctly handled, particularly under high temperature and radiation conditions. GBD is a non-infectious pathology occurring when the partial pressures of atmospheric gases in solution exceed their respective partial pressures in the atmosphere. GBD was initially observed in farmed species, although outbreaks in wild fish, both fresh-water and marine animals, have also been reported. GBD can have serious adverse economic repercussions in fish cultures by reducing productivity and the commercial value of the fish as well as the farm profitability. Thus, the development of biomarkers responsive to hyperoxia stress would be a valuable tool to apply in systems where oxygen supersaturation might be possible, and would also contribute to basic knowledge of oxidative stress. Oxygen supplementation is a common practice in intensive fish farming, in order to allow high density cultivation while reducing the amount of water demanded in aquaculture facilities. It is also required during fish transportation. In intensive aquaculture, the use of oxygen is regulated by sophisticated mechanisms to keep its concentration close to desired values. However, in open ponds, the likelihood of oxygen supersaturation conditions is higher, because primary producers are in a high nutrient environment, occasionally combined with high temperature and radiation. Photosynthetic oxygen overproduction is a factor of concern for pond aquaculture, particularly when species such as sole, which exhibit nocturnal and benthic habits, are considered. These features complicate water management compared to pond operation in pelagic fish farming.

Environmental and physicochemical conditions inducing hyperoxia, such as radiation, temperature and dissolved O_2, were monitored in two independent land-based ponds of an aquaculture research centre (IFAPA Centro El Toruño, Puerto de Santa María, Cádiz, Spain) in which S. senegalensis were reared, after a GBD outbreak was detected in some of these animals (Salas-Leyton et al., 2009). Fig. 2 (upper) shows the appearance of the earth pond used as control (100 m2, water renewed 4-fold/day) and of that in which GBD developed (900 m2, water unrenewed) in which algal blooms can be observed at its border. As shown in Fig. 2 (centre), the dissolved oxygen profile detected in the hyperoxic earth pond was typical of that in environments dominated by macroalgal biomass and high photosynthetic activity, where extreme oxygen levels are reached during a great part of the daily cycle (including night hours), always above saturation without a desaturation phase. As shown in Fig. 2 (lower), the following typical GBD symptoms were detected in fish from this pond: exophthalmia caused by retrobulbar bubbles (A), subcutaneous emphysemas, obstruction of gill lamellas (B), big bubbles located at caudal (C) and dorsal

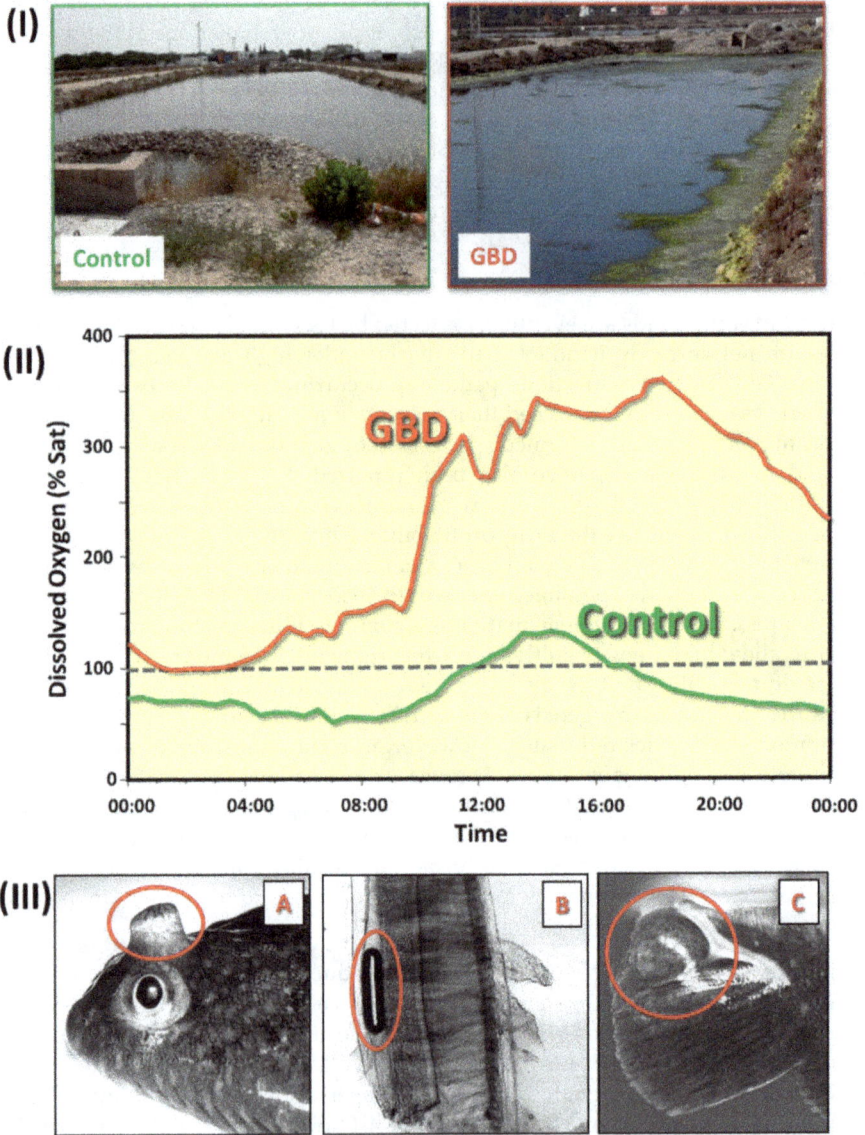

Fig. 2. Earth ponds, dissolved oxygen monitoring and visible symptoms in GBD-affected soles. (I) The experiments were carried out in two independent land-based ponds, one a 900 m² rectangular pond where oxygen supersaturation developed spontaneously, named the GBD pond (right) and the other a 100 m² control pond (left) in which water was renewed four times per day. (II) Dissolved oxygen was simultaneously monitored throughout the daily cycle in both the control (green line) and GBD ponds (red line). (III) Visible symptoms observed in GBD-affected soles included the following: exophthalmia caused by retrobulbar bubbles (A), a bubble obstructing a gill lamella (B), and a large bubble located at caudal fin (C).

fins, haemorrhages, anomalous swimming accompanied by loss of orientation, near-lethargy status and individual isolation were the main effects of O_2 supersaturation (Salas-Leyton et al., 2009).

Under the described aquaculture conditions, a parallel proteomic study was carried out in search of protein alteration patterns that might be used as potential new and unbiased biomarkers of hyperoxic stress (Fig. 3) (Salas-Leyton et al., 2009). The following three health statuses were studied in sole individuals: (1) healthy control fish, (2) GBD-affected but asymptomatic fish, and (3) GBD-affected fish with visible symptoms. Protein expression profiles were studied by 2-DE in cytosolic fractions of gills and livers. A total of 1,525 and 1,632 spots were detected in the four gill and liver gels, respectively, that were run in each of the three situations studied. Fig. 3 (upper) shows the master gels of cytosolic gill (left) and liver (right) proteins. A total of 205 protein spots were differentially expressed in the gills and 498 in the liver in each health status. Fig. 3 (middle) shows the number of spots which are present only, absent, increased or diminished in each of the studied conditions, and the total number of changes found. A significantly higher number of differentially expressed spots were found in GBD-affected soles, mainly in fish with visible symptoms. Of these, 25 spots in the gills and 23 in the liver were selected for identification using tandem mass spectrometry (nESI-IT MS/MS), *de novo* sequencing and a bioinformatics search. Fig. 3 (lower) shows the percentage of the relative intensity of each spot in each health status. Sequence tags were obtained from 9 (gills) and 5 (liver) of the selected spots, resulting in a total of 14 identified spots in the GBD status, which are indicated in Fig. 3 (lower).

Due to the central role of gills in oxygen exchange, the proteins identified in the gills of GBD-affected fish were, unsurprisingly, related to oxidative alteration of the cytoskeleton structure or function (β-tubulin and β-actin), motility (light myosin chain and α-tropomyosin), regulatory pathways (calmodulin, Raf kinase inhibitor protein [RKIP]) and carbohydrate metabolism (glyceraldehyde 3-P dehydrogenase). The hyperoxia-linked effects of these proteins and higher-level responses, related to inflammatory response, apoptosis or cell death, or derived from alterations in the regulatory proteins, have been discussed in depth (Salas-Leyton et al., 2009). In gills, only RKIP was most intense in GBD-affected animals without symptoms, while the other seven identified proteins were most intense in GBD-affected fish with symptoms. Proteins identified in the liver were related to protein oxidative damages (β-globin and fatty acid-binding protein [FABP]), protection from oxidative stress (dicarbonyl/L-xylulose reductase and glycine N-methyltransferase) and inflammatory response (complement component C3), in agreement with the predominant metabolic role of this organ. In the liver, only FABP was most intense in healthy fish, while the other four identified proteins were most intense in GBD-affected fish with visible symptoms. Approximately 50% of the identified proteins corresponded to "unusual" proteins not often found in proteomics screens, while the rest were preferential targets of oxidative stress. Most of these latter proteins were found as truncated forms of oxidised proteins that might trigger cellular defences against hyperoxygenation preceding GBD outbreaks. Some of the identified proteins might be considered to be good hyperoxia stress biomarkers and could be used in the early-warning detection of GBD outbreaks. Massive proteomic approaches have been applied for the first time to the study of this non-infectious pathological condition, gas bubble disease, in fish.

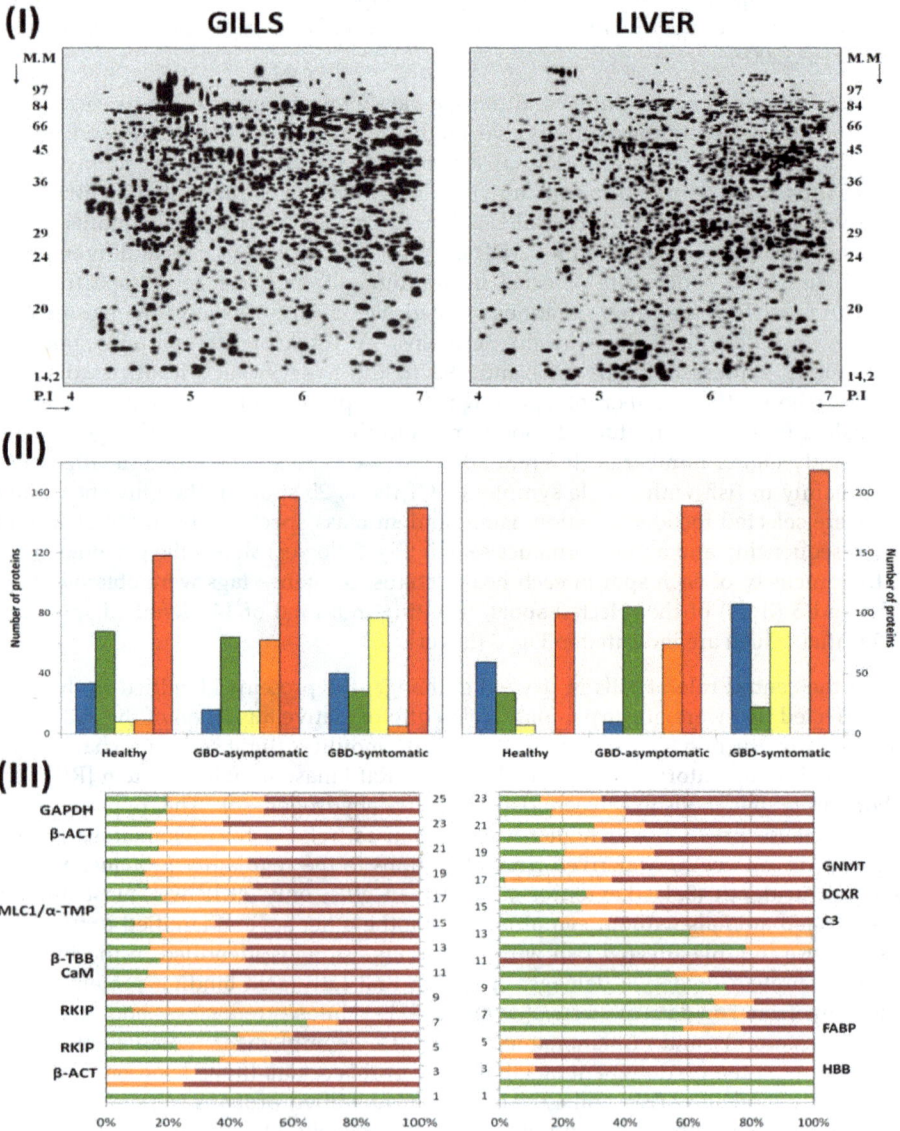

Fig. 3. Proteomic study of GBD-affected juvenile soles after 2-DE of cytosolic gill (left) and liver (right) proteins. (I) Master gels. (II) Number of proteins showing distinct protein expression patterns in healthy, GBD-asymptomatic and GBD-symptomatic fish. Bars represent the number of protein spots present only (■), absent (■), increased (■), diminished (■) and the total number of changes (■) found in each of the conditions. (III) Spots selected for possible identification by mass spectrometry analysis. Bars represent the percentage of the relative intensity of each spot in every status: healthy (■), GBD-asymptomatic (■) and GBD-symptomatic (■) animals. Successfully identified proteins are highlighted.

6. Conclusions

Given the economic importance of *S. senegalensis*, an aquaculture species that remains largely unexplored at the genomic level, the application of postgenomic methodologies provides results that may be highly relevant at a genetic, immunological and toxicological level, contributing to the improvement of management and welfare of this organism in aquaculture. The utility of high-throughput proteomic methods for unveiling the molecular basis of a cumbersome disease in aquaculture has been also demonstrated.

7. Acknowledgments

This work was funded by grants CTM2009-12858-C02 (Spanish Ministry of Science and Education), CYCE-00516 and P08-CVI-03829 (Innovation and Science Agency, Andalusian Government).

8. References

Abril, N.; Ruiz-Laguna, J.; Osuna-Jiménez, I.; Vioque-Fernández, A.; Fernández-Cisnal, R.; Chicano-Galvez, E.; Alhama, J.; López-Barea, J. & Pueyo, C. (2011). Omic approaches in environmental issues. *Journal of Toxicology and Environmental Health, Part A*, 74: 1-19.

Alhama, J.; Romero-Ruiz, A. & López-Barea, J. (2006). Metallothionein quantification in clams (*Chamaelea gallina*) by RP-HPLC and fluorescence detection after monobromobimane derivatization. *Journal of Chromatography A*, 1107, 52-58.

Alhama, J.; Romero-Ruiz, A.; Jebali, J. & López-Barea, J. (2010). Total metallothionein quantification by reversed-phase high-performance liquid chromatography coupled to fluorescence detection after monobromobimane derivatization. In: *Cadmium in the Environment*, Parvau, R.G. (Ed.), 389-405, Nova Science Publishers, Inc., ISBN 978-1-60741-934-1, Hauppauge, NY, USA.

Álvarez-Muñoz, D.; Lara-Martin, P.A.; Blasco, J.; Gómez-Parra, A. & González-Mazo, E. (2007). Presence, biotransformation and effects of sulfophenylcarboxilic acids in the benthic fish *Solea senegalensis*. *Environment International*, 33, 565-570.

Aragão, C.; Conceiçao, L.E.C.; Martins, D.; Rønnestad, I.; Gomes, E. & Dinis, M.T. (2004a). A balanced dietary amino acid profile improves amino acid retention in post-larval Senegalese sole (*Solea senegalensis*). *Aquaculture*, 233, 293-304.

Aragão, C.; Conceiçao, L.E.C.; Fyhn, H.J. & Dinis, M.T. (2004b). Estimated amino acid requirements during early ontogeny in fish with different life styles: gilthead seabream (*Sparus aurata*) and Senegalese sole (*Solea senegalensis*). *Aquaculture*, 242, 589-605.

Auslander, M.; Yudkovski, Y.; Chalifa-Caspi, V.; Herut, B.; Ophir, R.; Reinhardt, R.; Neumann, P.M. & Tom, M. (2008). Pollution-affected fish hepatic transcriptome and its expression patterns on exposure to cadmium. *Marine Biotechnology, (NY)*, 10(3), 250-261.

Avella, M.A.; Olivotto, I.; Silvi, S.; Place, A.R. & Carnevalli, O. (2010). Effect of dietary probiotics on clownfish : a molecular approach to define how lactic bacteria modulate development in a marine fish. *American Journal of Physiology. Regulatory, Integrative and Comparative Physiology*, 298, R359-R371.

Basu, N.; Todgham, A.E.; Ackerman, P.A.; Bibeau, M.R.; Nakano, K.; Schulte, P.M. & Iwama G.K. (2002). Heat shock protein genes and their functional significance in fish. *Gene*, 295(2), 173-183.

Bayne, C.J. & Gerwick, L. (2001). The acute phase response and innate immunity of fish. *Developmental & Comparative Immunology*, 25(8-9), 725-743.

Blanchard, J. & Grosell, M. (2006). Copper toxicity across salinities from freshwater to seawater in the euryhaline fish *Fundulus heteroclitus*: is copper an ionoregulatory toxicant in high salinities? *Aquatic Toxicology*, 80(2), 131-139.

Bonilla-Valverde, D.; Ruíz-Laguna, J.; Muñoz, A.; Ballesteros, J.; Lorenzo, F.; Gómez-Ariza, J.L. & López-Barea, J. (2004). Evolution of biological effects of Aznalcollar mining spill in the Algerian mouse (*Mus spretus*) using biochemical biomarkers. *Toxicology*, 197: 123-138.

Cabrita, E.; Soares, F. & Dinis, M.T. (2006). Characterization of Senegalese sole, *Solea senegalensis*, male broodstock in terms of sperm production and quality. *Aquaculture*, 261, 967-975.

Cánovas-Conesa, B.; Salas, E.; Alhama, J.; López-Barea, J. & Cañavate J.P. (2007). Efecto de la densidad de cultivo sobre los biomarcadores bioquímicos y la carga energética en el lenguado senegalés (*Solea senegalensis*). In: *Libro de Actas del XI Congreso Nacional Acuicultura*, Cerviño, A.; Guerra, A. & Pérez, C. (Eds.), 1463-1466, Universidad de Vigo, ISBN 978-84-611-9085-0, Vigo, España.

Cañavate, J.P. (2005). Potential of the Senegal sole *Solea senegalensis* Kaup, 1858 for marine aquaculture diversification. *Boletín del Instituto Español de Oceanografía*, 21, 147-154.

Cañavate, J.P.; Prieto, A.; Zerolo, R.; Solé, M.; Sarasquete, C. & Fernández-Díaz, C. (2007). Effects of light intensity and addition of carotene rich *Dunaliella salina* live cells on growth and antioxidant activity of *Solea senegalensis* Kaup (1858) larval and metamorphic stages. *Journal of Fish Biology*, 71, 781-794.

Cerdà, J.; Mercadé, J.; Lozano, J.J.; Manchado, M.; Tingaud-Sequeira, A.; Astola, A.; Infante, C.; Halm, S.; Viñas, J.; Castellana, B.; Asensio, E.; Cañavate, P.; Martínez-Rodríguez, G.; Piferrer, F.; Planas, J.V.; Prat, F.; Yúfera, M.; Durany, O.; Subirada, F.; Rosell, E. & Maes, T. (2008). Genomic resources for a commercial flatfish, the Senegalese sole (*Solea senegalensis*): ETS sequencing, oligo microarray design, and development of the Soleamold bioinformatic platform. *BMC Genomics*, 9, 508.

Cohen, R.; Chalifa-Caspi, V.; Williams, T.D.; Auslander, M.; George, S.G.; Chipman, J.K. & Tom, M. (2007) Estimating the efficiency of fish cross-species cDNA microarray hybridization. *Marine Biotechnology*, 9, 491-499

COM/0511 (2002). Communication from the Commission on *A Strategy for the Sustainable Development of European Aquaculture*.

Conceiçao, L.E.C.; Grasdalen, H. & Rønnestad, I. (2003). Amino acid requirements of fish larvae and post-larvae: new tools and recent findings. *Aquaculture*, 227, 221-232.

Conceiçao, L.E.C.; Ribeiro, L.; Engrola, S.; Aragão, C.; Morais, S.; Lacuisse, M.; Soares, F. & Dinis, M.T. (2007). Nutritional physiology during development of Senegalese sole (*Solea senegalensis*). *Aquaculture*, 268, 64-81.

Costa, P.M.; Lobo, J.; Caeiro, S.; Martins, M.; Ferreira, A.M.; Caetano, M.; Vale, C.; DelValls, T.A. & Costa, M.H. (2008a). Genotoxic damage in *Solea senegalensis* exposed to sediments form the Sado Estuary (Portugal): Effects of metallic and organic contaminants. *Mutation Research*, 654, 29-37.

Costa, P.M.; Diniz, M.S.; Caeiro, S.; Lobo, J.; Martins, M.; Ferreira, A.M.; Caetano, M.; Vale, C.; DelValls, T.A. & Costa, M.H. (2008b). Histological markers in liver and gills of juvenile *Solea senegalensis* exposed to contaminated estuarine sediments: A weighted indices approach. *Aquatic Toxicology*, 92, 202-212.

Costa, P.M.; Caeiro, S.; Diniz, M.S.; Lobo, J.; Martins, M.; Ferreira, A.M.; Caetano, M.; Vale, C.; DelValls, T.A. & Costa, M.H. (2009). Biochemical endpoints on juvenile *Solea senegalensis* exposed to estuarine sediments: the effects of contaminant mixtures on metallothionein and CYP1A induction. *Ecotoxicology*, 18, 988-1000.

Costa, P.M.; Neuparth, T.S.; Caeiro, S.; Lobo, J.; Martins, M.; Ferreira, A.M.; Caetano, M.; Vale, C.; DelValls, T.A. & Costa, M.H. (2011). Assessment of the genotoxic potential of contaminated estuarine sediments in fish peripheral blood: Laboratory versus *in situ* studies. *Environmental Research*, 111, 25-36.

Cousinou, M.; Dorado, G. & López-Barea, J. (1999). Amplification and cloning of cDNAs of cytochrome P4501A1 and metallothionein genes from *Sparus aurata* Linnaeus, 1758 and *Liza aurata* (Risso, 1810) by RACE-PCR. *Boletín del Instituto Español de Oceanografía*, 15, 473-484.

Cousinou. M.; Nilsen, B.; López-Barea, J. & Dorado, G. (2000). New methods to use of fish cytochrome P4501A to assess marine organic pollutants. *Science of Total Environment*, 247, 213-225.

De Wit, M.; Keil, D.; van der Ven, K.; Vandamme, S.; Witters, E. & De Coen, W. (2010). An integrated transcriptomic and proteomic approach characterizing estrogenic and metabolic effects of 17 α-ethinylestradiol in zebrafish (*Danio rerio*). *General & Comparative Endocrinology*, 167, 190-201.

Dias, J.; Yúfera, M.; Valente, L.M.P. & Rema, P. (2010). Feed transit and apperent protein, phosphorous and energy digestibility of practical feed ingredients by Senegalese sole (*Solea senegalensis*). *Aquaculture*, 302, 94-99.

Diatchenko, L.; Lay, Y.F.; Campbell, A.P.; Chenchik, A.; Moqadam, F.; Huang, B.; Lukyanov, K.; Lukyanov, K.; Gurskaya, N.; Sverdlov, E.D. & Siebert, P.D. (1996). Suppresion subtractive hybridization: A method for generating differentially regulated or tissue-specific cDNA probes and librairies. *Proceedings of the National Academy of Sciences of the USA*, 93, 6025-6030.

Dinis, M.T.; Ribero. L.; Soares, F. & Sarasquete, C. (1999). A review on the cultivation potential of *Solea senegalensis* in Spain and Portugal. *Aquaculture*, 176, 27-38

Douglas, S.E.; Knickle, L.C.; Kimball, J. & Reith, M.E. (2007). Comprehensive EST analysis of Atlantic halibut (*Hippoglossus hippoglossus*), a commercially relevant aquaculture species. *BMC Genomics*, 8, 144

Evans, D.H.; Piermarini, P.M. & Choe, K.P. (2005). The multifunctional fish gill: dominant site of gas exchange, osmoregulation, acid-base regulation, and excretion of nitrogenous waste. *Physiological Reviews*, 85(1), 97-177.

Ewart, K.V.; Belanger, J.C.; Williams, J., Karakach, T.; Penny, S.; Tsoi, S.C.; Richards, R.C. & Douglas, S.E. (2005). Identification of genes differentially expressed in Atlantic salmon (*Salmo salar*) in response to infection by *Aeromonas salmonicida* using cDNA microarray technology. *Developmental & Comparative Immunology*, 29(4), 333-347.

FAO: Food and Agriculture Organization of the United Nations (2010). Aquaculture planning. Policy formulation and implementation of sustainable development. FAO Fisheries and Aquaculture Technical Paper 542. Rome FAO.

Fernandes, R.; Ramalho, J. & Pereira, P. (2006). Oxidative stress upregulates ubiquitin proteasome pathway in retinal endothelial cells. *Molecular Vision*, 12, 1526-1535.

Fernández-Díaz, C.; Kopecka, J.; Cañavate, J.P.; Sarasquete, C. & Solé, M. (2006). Variations on development and stress defences in *Solea senegalensis* larvae fed on live and microencapsulated diets. *Aquaculture*, 251, 573-584.

Ferruzza, S.V; Scacchi, M.; Scarino, M.L. & Sambuy, Y. (2002). Iron and copper alter tight junction permeability in human intestinal Caco-2 cells by distinct mechanisms. *Toxicology In Vitro*, 16(4), 399-404.

Fonseca, V.; Serafim, A.; Company, R.; Bebianno, M.J.; & Cabral, H.N. (2009) Effect of copper on growth, condition indices and biomarker response in juvenile sole *Solea senegalensis*. *Scientia Marina*, 73, 51-58.

Fonseca, V.; Franca, S.; Serafim, A.; Company, R.; Lopes, B.; Bebianno, M.J.; & Cabral, H.N. (2011) Multi-biomarker response to estuarine habitat contamination in three fish species: *Dicentrarchus labrax*, *Solea senegalensis* and *Pomatochistus microps*. *Aquatic Toxicology*, 102, 216-277.

Gallardo-Madueño, R.; Leal, J.F.; Dorado, G.; Holmgren, A.; López-Barea, J. & Pueyo, C. (1998). *In vivo* transcription of *nrdAB* operon and of *grxA* and *fpg* genes is triggered in *Escherichia coli* lacking both thioredoxin and glutaredoxin 1 or thioredoxin and glutathione, respectively. *The Journal of Biological Chemistry*, 273 (29), 18382-18388.

García-López, A.; Fernández-Pasquier, V.; Couto, E.; Canario, A.V.M.; Sarasquete, C. & Martínez-Rodriguez, G. (2006a). Non-invasive assessment of reproductive status and cycle of sex steroid levels in a captive broodstock of Senegalese sole *Solea senegalensis* (Kaup). *Aquaculture*, 254, 583-593.

García-López, A.; Anguis, V.; Couto, E.; Canario, A.V.M.; Cañavate, J.P.; Sarasquete, C. & Martínez-Rodriguez, G. (2006b). Testicular development and plasma sex steroid levels in cultured male Senegalese sole *Solea senegalensis* Kaup. *General and Comparative Endocrinology*, 147, 343-351.

García-López, A.; Couto, E.; Canario, A.V.M.; Sarasquete, C. & Martínez-Rodriguez, G. (2007). Ovarian development and plasma sex steroid levels in cultured female Senegalese sole *Solea senegalensis*. *Comparative Biochemistry and Physiology - Part A*, 146, 342-354.

Gavaia, P.J.; Dinis, M.T. & Cancela, M.L. (2002). Osteological development and abnormalities of the vertebral column and caudal skeleton in larval and juvenile stages of hatchery-reared Senegal sole (*Solea senegalensis*). *Aquaculture*, 211, 305-323.

Gavaia, P.J.; Simes, D.C.; Ortiz-Delgado, J.B.; Viegas, C.S.B.; Pinto, J.P.; Kelsh, R.N.; Sarasquete, C. & Cancela, M.L. (2006). Osteocalcin and matrix Gla protein in zebrafish (*Danio rerio*) and Senegal sole (*Solea senegalensis*): Comparative gene and protein expression during larval development through adulthood. *Gene Expression Patterns*, 6, 637-652.

Geist, J.; Werner, I.; Eder, K.J. & Leutenegger, C.M. (2007). Comparisons of tissue-specific transcription of stress response genes with whole animal endpoints of adverse effect in striped bass (*Morone saxatilis*) following treatment with copper and esfenvalerate. *Aquatic Toxicology*, 85(1), 28-39.

Gonzalez-Fernández, M.; García-Barrera, T.; Jurado, J.; Prieto-Álamo, M.J.; Pueyo, C.; López-Barea, J. & Gómez-Ariza, J.L. (2008). Integrated application of transcriptomics, proteomics, and metallomics in environmental studies. *Pure and Applied Chemistry*, 80, 2609-2626.

James, P. (1997). Protein identification in the post-genome era: the rapid rise of proteomics. *Quarterly Review of Biophysics*, 30, 279–331.

Jebali, J.; Banni, M.; Gerbej, H.; Boussetta, H.; López-Barea, J. & Alhama, J. (2008). Metallothionein induction bu Cu, Cd and Hg in *Dicentrarchus labrax* liver: assessment by RP-HPLC with fluorescent detection and spectrophotometry. *Marine Environmental Research*, 65, 358-363.

Jiménez, A.; Prieto-Álamo, M.J.; Fuentes-Almagro, C.A.; Jurado, J.; Gustafsson, J.A.; Pueyo, C. & Miranda-Vizuete, A. (2005) Absolute mRNA levels and transcriptional regulation of the mouse testis-specific thioredoxins. *Biochemical and Biophysical Research Communications*, 330, 56-74.

Jiménez-Tenorio, N.; Morales-Caselles, C.; Kalman, J.; Salamanca, M.J.; González de Canales, M.L.; Sarasquete, C. & DelValls, T.A. (2007). Determining sediment quality for regulatory proposes using fish chronic assays. *Environment International*, 33, 474-480.

Jiménez-Tenorio, N.; Salamanca, M.J.; García-Luque, E.; González de Canales, M.L. & DelValls, T.A. (2008). Chronic bioassay in benthic fish for the assessment of the quality of sediments in different areas of the coast of Spain impacted by acute and chronic oilspills. *Environmental Toxicology*, 23, 634-642.

Jurado, J.; Prieto-Álamo, M.J.; Madrid-Risques, J. & Pueyo, C. (2003) Absolute gene expression patterns of thioredoxn and glutaredoxin redox systems in mouse. *The Journal of Biological Chemistry*, 278 (46), 45546-45554.

Jurado, J.; Fuentes-Almagro, C.A.; Prieto-Álamo, M.J. & Pueyo, C. (2007) Alternative splicing of *c-fos* pre-mRNA : contribution of the rates of synthesis and degradation to the copy number of each transcript isoform and detection of a truncated c-Fos immunoreactive species. *BMC Molecular Biology*, 8, 83.

Kültz, D.; Fiol, D.; Valkova, N.; Gomez-Jimenez, S.; Chan, S.Y.; & Lee, J. (2007) Functional genomics and proteomics of the cellular stress response in «non-model» organisms. *The Journal of Experimental Biology*, 210, 1593-1601.

Lenartova, V.; Holovska, K.; Pedrajas, J.R.; Martínez-Lara, E.; Peinado, J.; López-Barea, J.; Rossival, I. & Kosuth, P. (1997). Antioxidant and detoxifying fish enzymes as biomarkers of river pollution. *Biomarkers*, 2, 247-252.

López-Barea, J. (1995a). Biomarkers in Ecotoxicology: an Overview, In: *Toxicology in Transition*, Degen, G.H.; Seiler, J.P. & Bentely, P. (Eds.), 57-79, Springer, ISBN 978-3540587811, Berlin, Germany.

López-Barea, J. (1995b). Biomarcadores que detectan contaminantes ambientales. In: *Actas del V Congreso Nacional de Acuicultura*, Castelló, F. & Calderer, A. (Eds.), 953-958, Publicaciones de la Universidad de Barcelona, Barcelona, Spain.

López-Barea, J. & Pueyo, C. (1998). Mutagen content and metabolic activation of promutagens by molluscs as biomarkers of marine pollution. *Mutation Research*, 399, 3-15.

López-Barea, J. & Gómez-Ariza, J.L. (2006). Environmental proteomics and metallomics. *Proteomics*, 6 (Suppl 1), S51-S62.

López-Galindo, C.; Vargas-Chacoff, L.; Nebot, E.; Casanueva, J.F.; Rubio, D.; Solé, M. & Mancera, J.M. (2010a). Biomarker responses in *Solea senegalensis* exposed to sodium hypochlorite used as antifouling. *Chemosphere*, 78, 885-893.

López-Galindo, C.; Vargas-Chacoff, L.; Nebot, E.; Casanueva, J.F.; Rubio, D.; Solé, M. & Mancera, J.M. (2010b). Sublethal effects or the organic antifoulant Mexel®432 on ormoregulation and xenobiotic detoxification in the flatfish *Solea senegalensis*. *Chemosphere*, 79, 78-85.

Manchado, M.; Michán, C. & Pueyo, C. (2000). Hydrogen peroxide activates the *SoxRS* regulon *in vivo*. *The Journal of Bacteriology*, 182(23), 6842-6844.

Manchado, M.; Infante, C.; Asensio, E. & Cañavate, J.P. (2007). Differential gene expression and dependence on thyroid hormones of two glyceraldehyde-3-phosphate dehydrogenases in the flatfish Senegalese sole (*Solea senegalensis* Kaup). *Gene*, 400(1-2), 1-8.

Martínez-Lara, E.; Pascual, P.; Toribio, F.; López-Barea, J. & Bárcena, J.A. (1992). Rapid method for the quantification of glutathione transferase isoenzymes in crude extracts. *Journal of Chromatography A*, 609, 141-146.

Martínez-Lara, E.; Toribio, F.; López-Barea, J. & Bárcena, J.A. (1996). Glutathione-S-transferase isoenzyme patterns in the gilthead seabream (*Sparus aurata*) exposed to environmental contaminants. *Comparative Biochemistry and Physiology - Part C: Toxicology & Pharmacology*, 113, 215-220.

Martínez-Lara, E.; George, S.G.; López-Barea, J. & Bárcena, J. (1997). Purification and characterization of multiple glutathione transferase isoenzymes from grey mullet liver. *Cellular and Molecular Life Sciences*, 53, 759-768.

Mastellos, D.; Andronis, C.; Persidis, A. & Lambris, J.D. (2005). Novel biological networks modulated by complement. *Clinical Immunology*, 115(3), 225-235.

Michan, C.; Manchado, M.; Dorado, G. & Pueyo, C. (1999). *In vivo* transcription of the *Escherichia coli oxyR* regulon as a function of growth phase and in response to oxidative stress. *The Journal of Bacteriology.*, 181(9), 2759-2764.

Michan, C.; Monje-Casas, F. & Pueyo, C. (2005). Transcript copy number of genes for DNA repair and translesion synthesis in yeast: contribution of transcription rate and mRNA stability to the steady-state level of each mRNA along with growth in glucose-fermentative medium. *DNA Repair*, 4(4), 469-478.

Monje-Casas, F.; Jurado, J.; Prieto-Álamo, M.J.; Holmgren, A. & Pueyo, C. (2001) Expression analysis of the *nrdHIEF* operon from *Escherichia coli*. Conditions that trigger the transcript level *in vivo*. *Journal of Biological Chemistry*, 276 (21), 18031-18037.

Monje-Casas, F.; Michan, C. & Pueyo, C. (2004). Absolute transcript levels of thioredoxin- and glutathione-dependent redox systems in *Saccharomyces cerevisiae*: response to stress and modulation with growth. *Biochemical Journal*, 383(Pt 1), 139-147.

Montes-Nieto, R.; Fuentes-Almagro, C.A.; Bonilla-Valverde, D.; Prieto-Álamo, M.J.; Jurado, J.; Carrascal, M.; Gómez-Ariza, J.L.; López-Barea, J. & Pueyo, C. (2007). Proteomics in free-living *Mus spretus* to monitor terrestrial ecosystems. *Proteomics*, 7, 4376-4387.

Montes-Nieto, R.; García-Barrera, T.; Gómez-Ariza, J.L. & López-Barea, J. (2010). Environmental monitoring of Domingo Rubio stream (Huelva Estuary, SW Spain) by combining conventional biomarkers and Proteomic analysis in *Carcinus maenas*. *Environmental Pollution*, 158, 401-408.

Morales-Caselles, C.; Jiménez-Tenorio, N.; González de Canales, M.L.; Sarasquete, C. & DelValls, T.A. (2006). Ecotoxicity of sediments contaminated by the oil spill associated with the tanker "Prestige" using juveniles of the fish *Sparus aurata*. *Archives of Environmental Contamination and Toxicology*, 51, 652-660.

Morais, S.; Koven, W.; Rønnestad, I.; Dinis, M.T. & Conceiçao, L.E.C (2005). Dietary protein/lipid ratio affects growth and amino acid and fatty acid absorption and metabolism in Senegalese sole (*Solea senegalensis* Kaup 1858) larvae. *Aquaculture*, 246, 347-357.

Morais, S.; Conceiçao, L.E.C.; Rønnestad, I.; Koven, W.; Cahu, C.; Zambonino Infante, J.L. & Dinis, M.T. (2007). Dietary neutral lipid level and source in marine fish larvae: Effects on digestive physiology and food intake. *Aquaculture*, 268, 106-122.

Obayashi, T. & Kinoshita, K. (2011). COEXPRESdb: a database to compare gene coexpression in seven model animals. *Nucleic Acids Research*, 39 (Suppl 1), D1016-D1022.

Oliva, M.; González de Canales, M.L.; Gravato, C.; Guilhermino, L. & Perales, J.A. (2010). Biochemical effects and polycyclic aromatic hydrocarbons (PAHs) in senegal sole (*Solea senegalensis*) from a Huelva estuary (SW Spain). *Ecotoxicology and Environmental Safety*, 73, 1842-1851.

Ortiz-Delgado, J.B.; Ruane, N.M.; Pousao-Ferreira, P.; Dinis, M.T. & Sarasquete, C. (2006). Thyroid gland develoment in Senegalese sole (*Solea senegalensis* Kaup 1858) during early life stages: A histochemical and immunohistochemical approach. *Aquaculture*, 260, 346-356.

Osuna-Jiménez, I.; Williams, T.D.; Prieto-Álamo, M.-J.; Abril, N.; Chipman, J.K. & Pueyo, C. (2009). Immune- and stress-related transcriptomic responses of *Solea senegalensis* stimulated with lipopolysaccharide and copper sulphate using heterologous cDNA microarrays. *Fish & Shellfish Immunology*, 26, 699-706.

Pagan, J.K.; Wylie, F.G.; Joseph, S.; Widberg, C.; Bryant, N.J.; James, D.E. & Stow, J.L. (2003). The t-SNARE syntaxin 4 is regulated during macrophage activation to function in membrane traffic and cytokine secretion. *Current Biology*, 13(2), 156-160.

Pascual, P.; Rodríguez-Ariza, A.; Navas, J.I.; Toribio, F. & López-Barea, J. (1995a). Cambios bioquímicos estacionales durante el cultivo de doradas (*Sparus aurata*). In: *Actas del V Congreso Nacional de Acuicultura*, Castelló, F. & Calderer, A. (Eds.), 582-587, Publicaciones de la Universidad de Barcelona, Barcelona, Spain.

Pascual, P.; Navas, J.I.; Toribio, F. & López-Barea, J. (1995b). Indicadores bioquímicos de esters en doradas (*Sparus aurata*). In: *Actas del V Congreso Nacional de Acuicultura*, Castelló, F. & Calderer, A. (Eds.), 754-759, Publicaciones de la Universidad de Barcelona, Barcelona, Spain.

Pascual, P.; Navas, J.I.; Toribio, F. & López-Barea, J. (1997). Cambios clínicos estacionales en la carga energética adenílica y guanílica de doradas (*Sparus aurata*). In: *Actas del VI Congreso Nacional de Acuicultura*, de Costa, J.; Abellán, E.; García, B.; Ortega, A. & Zamora, S. (Eds.), 832-837, Publicaciones de la Universidad de Murcia, Murcia, Spain.

Pascual, P.; Pedrajas, J.R.; Toribio, F.; López-Barea, J. & Peinado, J. (2003). Effect of food deprivation on oxidative stress biomarkers in fish (*Sparus aurata*). *Chemico-Biological Interactions*, 145, 191-199.

Pedrajas, J.R.; Peinado, J. & López-Barea J. (1993). Purification of Cu,Zn-superoxide dismutase isoenzymes from fish liver: appearance of new isoforms as consequence of pollution. *Free Radical Research Communications*, 19, 29-41.

Pedrajas, J.R.; Peinado, J. & López-Barea J. (1995). Oxidative stress in fish exposed to model xenobiotics. Oxidatively modified forms of Cu,Zn-superoxide dismutase as potential biomarkers. *Chemico-Biological Interactions*, 98, 267-282.

Pedrajas, J.R.; Gavilanes, F.; López-Barea, J. & Peinado, J. (1998). Incubation of superoxide dismutase with malondialdehyde and 4-hydroxy-2-nonenal forms new active isoforms and adducts. An evaluation of xenobiotics in fish. *Chemico-Biological Interactions*, 116, 1-17.

Pinto, W.; Rodrigues, V.; Dinis, M.T. & Aragão, C. (2010). Can dietary amino acid supplementation be beneficial during fish metamorphosis?. *Aquaculture*, 310, 200-205.

Prieto-Álamo, M.J.; Jurado, J.; Gallardo-Madueño, R.; Monje-Casas, F.; Holmgren, A. & Pueyo, C. (2000). Transcriptional regulation of glutaredoxin and thioredoxin pathways and related enzymes in response to oxidative stress. The *Journal of Biological Chemistry*, 275(18), 13398-13405.

Prieto-Álamo, M.J.; Cabrera-Luque, J.M. & Pueyo C. (2003). Absolute quantitation of normal and ROS-induced patterns of gene expression: an *in vivo* real-time PCR study in mice. *Gene Expression*, 11, 23-34.

Prieto-Álamo, M.J.; Abril, N.; Osuna-Jimenez, I. & Pueyo, C. (2009). *Solea senegalensis* genes responding to lipopolysaccharide and copper sulphate challenges: Large-scale identification by suppression subtractive hybridization and absolute quantification of transcriptional profiles by real-time RT-PCR. *Aquatic Toxicology*, 91, 312-319.

Pueyo, C.; Jurado, J.; Prieto-Álamo, M.J.; Monje-Casas, F. & López-Barea, J. (2002). Multiplex reverse transcription-polymerase chain reaction for determining transcriptional regulation of thioredoxin and glutaredoxin pathways. *Methods in Enzymology*, 247, 441-451.

Pueyo, C.; Gómez-Ariza, J.L.; Bello-López, M.A.; Fernández-Torres, R.; Abril, N.; Alhama, J.; García-Barrera, T. & López-Barea, J. (2011). New methodologies for assessing the presence and ecological effects of pesticides in Doñana National Park (SW Spain). In: *Pesticides in the Modern World / Book 1*, Bakic, S. (Ed.), 165-196, InTech Open Access Publisher, ISBN 978-953-307-437-5, Rijeka, Croatia.

Qureshi, N.; Perera, P.Y.; Shen, J.; Zhang, G.; Lenschat, A.; Splitter, G.; Morrison, D.C. & Vogel, S.N. (2003). The proteasome as a lipopolysaccharide-binding protein in macrophages: differential effects of proteasome inhibition on lipopolysaccharide-induced signaling events. *The Journal of Immunology*, 171(3), 1515-1525.

Reynaud, S.; Raveton, M. & Ravanel, P. (2008). Interactions between immune and biotransformation systems in fish: a review. *Aquatic Toxicology*, 87(3), 139-145.

Ribeiro, L.; Sarasquete, C. & Dinis, M.T. (1999a). Histological and histochemical development of the digestive tract system of *Solea senegalensis* (Kaup, 1858) larvae. *Aquaculture*, 171, 293-308.

Ribeiro, L.; Zambonino-Infante, J.L.; Cahu, C. & Dinis, M.T. (1999b). Development of digestive enzymes in larvae of *Solea senegalensis*, Kaup 1858. *Aquaculture*, 179, 293-308.

Rodiles, A.; Herrera, M.; Hachero, I.; Rosano, M.; Ferrer, J.R.; Márquez, J.M. & Navas, J.I. (2005). Influence of different bottom types on the Senegal sole *Solea senegalensis* Kaup, 1858 ongrowing. *Boletín del Instituto Español de Oceanografía*, 21, 195-199.

Rodríguez-Ariza, A.; Abril, N.; Navas, J.I.; Dorado, G.; López-Barea, J. & Pueyo, C. (1992). Metal, mutagenicity and biochemical studies on bivalve molluscs from Spanish coasts. *Environmental and Molecular Mutagenesis*, 19, 112-124.

Rodríguez-Ariza, A.; Peinado, J.; Pueyo, C. & López-Barea, J. (1993). Biochemical indicators of oxidative stress in fish from polluted littoral areas. *Canadian Journal of Fisheries and Aquatic Sciences*, 50, 2568-2573.

Rodríguez-Ariza, A.; Dorado, G.; Navas, J.I.; Pueyo, C. & López-Barea, J. (1994a). Promutagen activation by fish liver as a biomarker of littoral pollution. *Environmental and Molecular Mutagenesis*, 24, 116-123.

Rodríguez-Ariza, A.; Toribio, F. & López-Barea, J. (1994b). Rapid determination of glutathione status in fish liver using high-performance liquid chromatography and electrochemical detection. *The Journal of Chromatography B*, 656, 311-318.

Rodríguez-Ortega, M.J.; Alhama, J.; Funes, V.; Romero-Ruíz, A.; Rodríguez-Ariza, A. & López-Barea, J. (2002). Biochemical biomarkers of pollution in the clam *Chamaelea gallina* from South-Spanish littoral. *Environmental Toxicology & Chemistry*, 21, 542-549.

Rodríguez-Ortega, M.J.; Grosvik, B.E.; Rodríguez-Ariza, A.; Goksoyr, A. & López-Barea, J. (2003). Changes in protein expression profiles in bivalve molluscs (*Chamaelea gallina*) exposed to four environmental pollutants. *Proteomics*, 3, 1535-1543.

Romero-Ruíz, A.; Amezcua, O.; Rodríguez-Ortega, M.J.; Muñoz, J.L.; Alhama, J.; Rodríguez-Ariza, A.; Gómez-Ariza, J.L. & López-Barea, J. (2003). Oxidative stress biomarkers in the bivalve *Scrobicularia plana* transplanted to the Guadalquivir Estuary (S Spain) after Aznalcóllar mining spill. *Environmental Toxicology & Chemistry*, 22, 92-100.

Romero-Ruiz, A.; Carrascal, M.; Alhama, J.; Gómez-Ariza, J.L.; Abian, J. & López-Barea, J. (2006). Environmental proteomic studies in clams from the Doñana bank of Guadalquivir Estuary (SW Spain). *Proteomics*, 6 (Suppl 1), S245-S255.

Romero-Ruiz, A.; Alhama, J.; Blasco, J.; Gómez-Ariza, J.L. & López-Barea, J. (2008). New metallothionein assay in *Scrobicularia plana*: heating effect and correlation with other biomarkers. *Environmental Pollution*, 156, 1340-1347.

Rønnestad, I.; Conceiçao, L.E.C.; Aragão, C. & Dinis, M.T. (2001). Assimilation and catabolism of dispensable and indispensable free amino acids in post-larval Senegal sole (*Solea senegalensis*). *Comparative Biochemistry and Physiology - Part C: Toxicology & Pharmacology*, 130, 461-466.

Rønnestad, I.; Tonheim, S.K.; Fyhn, H.J.; Rojas-García, C.R.; Kamisaka, Y.; Koven, W.; Finn, R.N.; Terjesen, B.F.; Barr, Y. & Conceiçao, L.E.C. (2003). The supply of amino acids during early feeding stages of marine fish larvae: a review of recent findings. *Aquaculture*, 227, 147-164.

Rueda-Jasso, R.; Conceiçao, L.E.C.; Dias, J.; De Coen, W.; Gomes, E.; Rees, J.F.; Soares, F.; Dinis, M.T. & Sorgeloos, P. (2004). Effect of dietary non-protein energy levels on condition and oxidative status of Senegalese sole (*Solea senegalensis*) juveniles. *Aquaculture*, 231, 417-433.

Ruiz-Laguna, J.; Abril, N.; Prieto-Álamo, M.J.; López-Barea, J. & Pueyo, C. (2005). Tissue, species and enviromental differences in absolute quantities of murine mRNAs coding for α, μ, ω, π and θ glutathione S-transferases. *Gene Expression*, 12, 165-176.

Ruiz-Laguna, J.; Abril, N.; García-Barrera, T.; Gómez-Ariza, J.L.; López-Barea, J. & Pueyo, C. (2006). Absolute transcript expression signatures of *Cyp* and *Gst* genes in *Mus spretus* to detect environmental contamination. *Environmental Science & Technology*, 40, 3646-3652.

Salas-Leiton, E.; Cánovas-Conesa, B.; Zerolo, R.; López-Barea, J.; Cañavate, J.P. & Alhama, J. (2009). Proteomics of juvenile Senagal sole (*Solea senegalensis*) affected by gas bubble disease in hyperoxygenated ponds. *Marine Biotechnology*, 11, 473-487.

Simpson, R.J. (2003). *Proteins and Proteomics. A Laboratory Manual.* Cold Spring Harbor, USA: Cold Spring Harbor Laboratory Press.

Solé, M.; Lima, D.; Reis-Henriques, M.A. & Santos, M.M. (2008). Stress biomarkers in juvenile Sole, *Solea senegalensis*, exposed to water-accommodated fraction of the «Prestige» fuel oil. *Bulletin of Environmental Contamination and Toxicology*, 80, 19-23.

Sukardi, H.; Ung, C.Y.; Gong, Z. & Lam, S.H. (2010). Incorporating zebrafish Omics into Chemical Biology and Toxicology. *Zebrafish*, 7, 41-52.

Swain, P.; Nayak S.K.; Nanda P.K. & Dash S. (2008). Biological effects of bacterial lipopolysaccharide (endotoxin) in fish: A review. *Fish & Shellfish Immunology*, 25(3), 191-201.

Thorgaard, G.H.; Bailey, G.S.; Williams, D.; Buhler, D.R.; Kaattari, S.L.; Ristow, S.S.; Hansen, J.D.; Winton, J.R.; Bartholomew, J.L.; Nagler, J.J.; Walsh, P.J.; Vijayan, M.M.; Devlin, R.H.; Hardy, R.W.; Overturf, K.E.; Young, W.P.; Robison, B.D.; Rexroad, C. & Palti, Y. (2002). Status and opportunities for genomics research with rainbow trout. *Comparative Biochemistry and Physiology*, 133(4), 609-646.

Valente, L.M.P.; Linares, F.; Villanueva, J.L.R.; Silva, J.M.G.; Espe, M.; Escórcio, C.; Pires, M.A.; Saavedra, M.J.; Borges, P.; Medale, F.; Alvarez-Blázquez, B. & Peleteiro, J.B. (2011). Dietary protein source or energy levels have no major impact on growth performance, nutrient utilisation or flesh fatty acids composition of market-sized Senegalese sole. *Aquaculture*, 318, 128-137.

Van Aggelen, G.V.; Ankley, G.T.; Baldwin, W.S.; Bearden, D.W.; Benso, W.H.; Chipman, J.K.; Collette, T.W.; Craft, J.A.; Denslow, N.D.; Embry, M.R.; Falciani, F.; George, S.G.; Helbing, C.C.; Hoekstra, P.F.; Iguchi, T.; Kagami, Y.; Katsiadaki, I.; Kille, P.; Liu, L.; Lord, P.G.; McIntyre, T.; O'Neill, A.; Osachoff, H.; Perkins, E.J.; Santos, E.M.; Skirrow, R.C.; Snape, J.R.; Tyler, C.R.; Versteeg, D.; Viant, M.R.; Volz, D.C.; Williams, T.D. & Yu, L. (2010). Integrating Omic technologies into aquatic Ecological Risk Assessment and Environmental Monitoring: hurdles, achievements, and future outlook. *Environmental Health Perspectives*, 118, 1-5.

Viant, M.R. (2007). Metabolomics of aquatic organisms: the new «omics» on the block. *Marine Ecology Progress Series*, 332, 301-306.

Vioque-Fernández, A.; Alves-de-Almeida, E.; Ballesteros, J.; García-Barrera, T.; Gómez-Ariza, J.L. & López-Barea, J. (2007a). Doñana National Park survey using *Procambarus clarkii* as bioindicator: esterase inhibition and pollutant levels. *Toxicology Letters*, 168, 260-268.

Vioque-Fernández, A.; Alves-de-Almeida, E. & López-Barea, J. (2007b). Esterases as pesticide biomarkers in crayfish (*Procambarus clarkii*, Crustacea): Tissue distribution, sensitivity to model compounds and recovery from inactivation. *Comparative Biochemistry and Physiology - Part C: Toxicology & Pharmacology*, 145, 404-412.

Vioque-Fernández, A.; Alves-de-Almeida, E. & López-Barea, J. (2009a). Assessment of Doñana National Park contamination in *Procambarus clarkii*: integration of conventional biomarkers and proteomic approaches. *Science of the Total Environment*, 407, 1784-1797.

Vioque-Fernández, A.; Alves-de-Almeida, E. & López-Barea, J. (2009b). Biomarker responses and protein expression profile changes in *Procambarus clarkii* after chlorpyrifos or carbaryl exposure. *Biomarkers*, 14, 299-310.

Washburn, M.P.; Wolters, D. & Yates, J.R. (2001). Large-scale analysis of the yeast proteome by multidimensional protein identification technology. *Nature Biotechnology*, 19 (3), 242-247.

Whyte, S.K. (2007). The innate immune response of finfish--a review of current knowledge. *Fish & Shellfish Immunology*, 23(6), 1127-1151.

Wilkins, M.R.; Pasquali, C.; Appel, R.D.; Ou, K.; Golaz, O.; Sanchez, J.-C.; Yan, J.X.; Gooley, A.A.; Hughes, G.; Humphery-Smith, I.; Williams, K.L. & Hochstrasser, D.F. (1996). From proteins to proteomes: large scale protein identification by two-dimensional electrophoresis and amino acid analysis. *Nature Biotechnology*, 14, 61–65.

Williams, T.D.; Gensberg, K.; Minchin, S.D. & Chipman, J.K. (2003). A DNA expression array to detect toxic stress response in European flounder (*Platichthys flesus*). *Aquatic Toxicology*, 65, 141-157.

Williams, T.D.; Diab, A.M.; George, S.G.; Godfrey, R.E.; Sabine, V.; Conesa, A.; Minchin, S.D.; Watts, P.C. & Chipman, J.K. (2006). Development of the GENIPOL European Flounder (*Platichthys flesus*) microarray and determination of temporal transcriptional responses to cadmium at low dose. *Environmental Science & Technology*, 40, 6479-6488.

Williams, T.D.; Diab, A.M.; George, S.G.; Sabine, V. & Chipman, J.K. (2007). Gene expression responses of European flounder (*Platichthys flesus*) to 17-beta estradiol. *Toxicology Letters*, 168(3), 236-248.

Williams, T.D.; Diab, A.; Ortega, F.; Sabine, V.S.; Godfrey, R.E.; Falciani., F.; Chipman, J.K. & George, S.G. (2008). Transcriptomic responses of European flounder (*Platichthys flesus*) to model toxicants. *Aquatic Toxicology*, 90(2), 83-91.

Permissions

The contributors of this book come from diverse backgrounds, making this book a truly international effort. This book will bring forth new frontiers with its revolutionizing research information and detailed analysis of the nascent developments around the world.

We would like to thank Z. A. Muchlisin, Ph.D, for lending his expertise to make the book truly unique. He has played a crucial role in the development of this book. Without his invaluable contribution this book wouldn't have been possible. He has made vital efforts to compile up to date information on the varied aspects of this subject to make this book a valuable addition to the collection of many professionals and students.

This book was conceptualized with the vision of imparting up-to-date information and advanced data in this field. To ensure the same, a matchless editorial board was set up. Every individual on the board went through rigorous rounds of assessment to prove their worth. After which they invested a large part of their time researching and compiling the most relevant data for our readers. Conferences and sessions were held from time to time between the editorial board and the contributing authors to present the data in the most comprehensible form. The editorial team has worked tirelessly to provide valuable and valid information to help people across the globe.

Every chapter published in this book has been scrutinized by our experts. Their significance has been extensively debated. The topics covered herein carry significant findings which will fuel the growth of the discipline. They may even be implemented as practical applications or may be referred to as a beginning point for another development. Chapters in this book were first published by InTech; hereby published with permission under the Creative Commons Attribution License or equivalent.

The editorial board has been involved in producing this book since its inception. They have spent rigorous hours researching and exploring the diverse topics which have resulted in the successful publishing of this book. They have passed on their knowledge of decades through this book. To expedite this challenging task, the publisher supported the team at every step. A small team of assistant editors was also appointed to further simplify the editing procedure and attain best results for the readers.

Our editorial team has been hand-picked from every corner of the world. Their multi-ethnicity adds dynamic inputs to the discussions which result in innovative outcomes. These outcomes are then further discussed with the researchers and contributors who give their valuable feedback and opinion regarding the same. The feedback is then collaborated with the researches and they are edited in a comprehensive manner to aid the understanding of the subject.

Apart from the editorial board, the designing team has also invested a significant amount of their time in understanding the subject and creating the most relevant covers. They scrutinized every image to scout for the most suitable representation of the subject and create an appropriate cover for the book.

The publishing team has been involved in this book since its early stages. They were actively engaged in every process, be it collecting the data, connecting with the contributors or procuring relevant information. The team has been an ardent support to the editorial, designing and production team. Their endless efforts to recruit the best for this project, has resulted in the accomplishment of this book. They are a veteran in the field of academics and their pool of knowledge is as vast as their experience in printing. Their expertise and guidance has proved useful at every step. Their uncompromising quality standards have made this book an exceptional effort. Their encouragement from time to time has been an inspiration for everyone.

The publisher and the editorial board hope that this book will prove to be a valuable piece of knowledge for researchers, students, practitioners and scholars across the globe.

List of Contributors

Michael P. Craig, Mitul B. Desai, Kate E. Olukalns, Scott E. Afton, Joseph A. Caruso and Jay R. Hove
University of Cincinnati, USA

Marina Paolucci, Ettore Varricchio and Elena Coccia
Department of Biological, Geological and Environmental Sciences, University of Sannio, Italy

Adele Fabbrocini
C.N.R. National Research Council, I.S.MAR, Italy

Maria Grazia Volpe
C.N.R. National Research Council, I.S.A., Italy

Marit Espe
National Institute of Nutrition and Seafood Research (NIFES), Bergen Norway, Norway

Adel El-Mowafi and Kari Ruohonen
Ewos Innovation AS, N-4335 Dirdal, Norway

A. Catarina Guedes and F. Xavier Malcata
CIMAR/CIIMAR – Centro Interdisciplinar, de Investigação Marinha e Ambiental, Porto, Portugal

F. Xavier Malcata
ISMAI – Instituto Superior da Maia, Castelo da Maia, Avioso S. Pedro, Portugal

F. Xavier Malcata
Instituto de Tecnologia Química e Biológica, Universidade Nova de Lisboa, Avenida da República, Oeiras, Portugal

Didier Montet, Doan Duy Le Nguyen and Amenan Clementine Kouakou
CIRAD, UMR 95 Qualisud, TA B-95/16, Montpellier, France

Doan Duy Le Nguyen
Can tho University, Faculty of Agriculture, Viet Nam

Amenan Clementine Kouakou
University of Abobo-Adjame, Laboratory of Microbiology and Molecular Biology, Ivory Coast

Kazue Nagasawa, Carlo Lazado and Jorge M. O. Fernandes
Faculty of Biosciences and Aquaculture, University of Nordland, Bodø, Norway

Anastasia Imsiridou, Elena Aggelidou, Vassilios Katsares and Sofia Galinou-Mitsoudi
Department of Fisheries and Aquaculture Technology, Alexander Technological Educational Institute of Thessaloniki, Nea Moudania, Halkidiki, Greece

Nikoleta Karaiskou
Department of Genetics, Development and Molecular Biology, School of Biology, Aristotle University of Thessaloniki, Thessaloniki, Macedonia, Greece

Nicole L. Quinn, Alejandro P. Gutierrez and William S. Davidson
Department of Molecular Biology and Biochemistry, Simon Fraser University, Burnaby, British Columbia, Canada

Ben F. Koop
Department of Biology, University of Victoria, Victoria, British Columbia, Canada

Angel Balisi Encarnacion, Huynh Nguyen Duy Bao, Reiko Nagasaka and Toshiaki Ohshima
Tokyo University of Marine Science and Technology, Japan

Carlos Yanes Roca
Stirling University, UK

Kevan L. Main
Mote Marine Laboratory, USA

Lucía E. Ocampo Q., Mónica Botero A. and Luis Fernando Restrepo
Antioquia University, Colombia

Martha P. Hernández-Vergara
Native Crustacean Aquaculture Laboratory, Graduate Study and Research Division, Instituto Tecnológico de Boca del Río, Boca del Río, Veracruz, México

Carlos I. Pérez-Rostro
Genetic Improvement and Production Laboratory, Graduate Study and Research Division, Instituto Tecnológico de Boca del Río, Boca del Río, Veracruz, México

Alfonso Mardones
Escuela de Acuicultura, Facultad de Recursos Naturales, Universidad Católica de Temuco, Casilla 15-D, Temuco, Chile

Patricio De los Ríos-Escalante
Escuela de Ciencias Ambientales, Facultad de Recursos Naturales, Universidad Católica de Temuco, Casilla 15-D, Temuco, Chile

Satit Kovitvadhi
Department of Agriculture, Faculty of Science and Technology, Bansomdejchaopraya, Rajabhat University, Bangkok, Thailand

Uthaiwan Kovitvadhi
Department of Zoology, Faculty of Science, Kasetsart University, Bangkok, Thailand

Daniel A. López and Boris A. López
Department of Aquaculture and Aquatic Resources, Universidad de Los Lagos, Osorno, Chile

Daniel A. López
Advanced Research Center, Universidad de, Playa Ancha, Valparaíso, Chile

Christopher K. Pham and Eduardo J. Isidro
Department of Oceanography and Fisheries, Universidade dos Açores, Horta, Portugal

Kassim Zaleha and Ibrahim Busra
Universiti Malaysia Terengganu, Malaysia

Anne Ingeborg Myhr, G. Kristin Rosendal and Ingrid Olesen
Genøk- Centre of Biosafety, Fridtjof Nansen Institute, Nofima, Norway

María-José Prieto-Álamo, Inmaculada Osuna-Jiménez, Nieves Abril, José Alhama, Carmen Pueyo and Juan López-Barea
Department of Biochemistry and Molecular Biology, University of Córdoba, Agrifood Campus of International Excellence, ceiA3m, Spain